Algorithms and Computation in Mathematics · Volume 15

Petteri Kaski
Patric R.J. Östergård

Classification Algorithms for Codes and Designs

With 61 Figures and 30 Tables

Springer

Authors

Petteri Kaski

Department of Computer Science and Engineering
Helsinki University of Technology
P. O. Box 5400
2015 HUT, Helsinki
Finland
e-mail: petteri.kaski@hut.fi

Patric R.J. Östergård

Department of Electrical and Communications, Engineering
Helsinki University of Technology
P.O. Box 3000
2015 HUT, Helsinki
Finland
e-mail: patric.ostergard@hut.fi

Mathematics Subject Classification (2000): 05-02, 05Bxx, 05Cxx, 05E20, 51Exx, 68-02, 68Rxx, 94-02, 94Bxx

ISSN 1431-1550

Additional material to this book can be downloaded from http://extras.springer.com.

ISBN-10 3-642-43250-6 Springer Berlin Heidelberg New York
ISBN-13 978-3-642-43250-7 Springer Berlin Heidelberg New York

Springer is a part of Springer Science+Business Media
springeronline.com

Softcover re-print of the Hardcover 1st edition 2006

Typesetting: by the authors and TechBooks using a Springer LaTeX macro package
Cover design: *design & production* GmbH, Heidelberg

Printed on acid-free paper SPIN: 11427841 46/TechBooks 5 4 3 2 1 0

Contents

Preface

> A new starting-point and a new method are requisite, to insure a complete [classification of the Steiner triple systems of order 15]. This method was furnished, and its tedious and difficult execution undertaken, by Mr. Cole.
>
> F. N. Cole, L. D. Cummings, and H. S. White (1917) [129]

The history of classifying combinatorial objects is as old as the history of the objects themselves. In the mid-19th century, Kirkman, Steiner, and others became the fathers of modern combinatorics, and their work – on various objects, including (what became later known as) Steiner triple systems – led to several classification results. Almost a century earlier, in 1782, Euler [180] published some results on classifying small Latin squares, but for the first few steps in this direction one should actually go at least as far back as ancient Greece and the proof that there are exactly five Platonic solids.

One of the most remarkable achievements in the early, pre-computer era is the classification of the Steiner triple systems of order 15, quoted above. An onerous task that, today, no sensible person would attempt by hand calculation. Because, with the exception of occasional parameters for which combinatorial arguments are effective (often to prove nonexistence or uniqueness), classification in general is about algorithms and computation.

The approach of using computers to obtain mathematical results used to be controversial – and still is, in some circles – but has grown into an invaluable tool in many areas of contemporary mathematics. Notably, in the recent decades we have, for example, seen computer-aided solutions of two challenging problems: the four-color theorem and the nonexistence of a projective plane of order 10. As a matter of fact, the latter result is surveyed in Chap. 12.

Looking back at the history again, the reader may contemplate whether the tedious calculations by Mr. Cole should be trusted more than a computer search. The results in [129] *are* correct; a verification of this result published in 1955 was among the first scientific computations when computers became available to researchers [247]. In the past there are many examples of manual calculations that have later been corrected in a computer search. Note, however, that the results on Steiner triple systems of order 15 in a paper published by Fisher [189] in 1940, occasionally claimed to be erroneous (incomplete, with one design missing), were never meant to form a complete classification, a fact that a careful reading of the paper reveals. Certainly, correctness of computational results is of central importance and will be discussed separately in the main text.

Mr. Cole and his co-authors found (all) 80 Steiner triple systems of order 15. But when are two objects different? This question is here answered through a proper definition (and motivation) of the concepts of equivalence and isomorphism for the objects considered.

Among the vast number of combinatorial objects, there is an idea behind the choice of classifying codes and designs and objects closely related to these. Namely, they can all be viewed as some sort of incidence structures, whereas, on the other hand, graphs, which are not considered from a classification perspective, are adjacency structures. The same applies to algebraic objects such as groups.

In studying this book, Chaps. 2, 3, and 4 are essential. Chapter 2 treats the foundations of the combinatorial objects considered; Chap. 3 the concepts of isomorphism, representations, and symmetry; and Chap. 4 presents the generic algorithms for classifying combinatorial objects. Chapter 5 contains several algorithms for solving subproblems in the classification of the objects discussed later. This chapter may be omitted unless one wants to implement these particular algorithms (but it may also be studied separately by anyone who wants to get an insight into contemporary algorithms for several important hard problems). Chapters 6 to 8 contain specific results for, respectively, designs, codes, and related structures. There is some dependency between these chapters, but hardly any overlapping. Constructions of objects with prescribed automorphism groups are studied in Chap. 9, validity of computational results is addressed in Chap. 10, and complexity issues are considered in Chap. 11. Finally, the celebrated nonexistence proof for projective planes of order 10 is surveyed in Chap. 12.

This book serves several different audiences. We have attempted to completely cover all important classification results in the areas of the book, and hope that researchers will find it an invaluable reference showing the state of the art. In fact, some of the results presented were obtained during the very process of writing this text. Most notably, these include a classification of the Steiner triple systems of order 19, the next open case after order 15, and a classification of the Steiner quadruple systems of order 16.

Due to its multidisciplinary nature, the book can be used as course material for graduate courses in computer science and discrete mathematics, as well as in coding theory (and selectively even for undergraduate courses). Elementary background knowledge in group theory will make the book more accessible, but it has been our intention to make the book as self-contained as possible.

Anyone wanting to implement classification algorithms (for any conceivable objects) will here find a basis for such work. Further research in the area is encouraged by a number of open research problems. Many of these problems are descriptions of new research ideas rather than specific problems that no one has managed to solve so far (although there are problems of this type, too). Whenever classification results are tabulated in the text, there is a smallest open case. Such instances are not explicitly stated as open problems, with a few exceptions where the instances are of greater general interest.

We also hope that the supplementary DVD with its comprehensive lists of combinatorial objects will prove a useful source for those whose main interest is in studying and using the classified objects for various purposes.

Last but not least, the following colleagues contributed to this project via valuable discussions and remarks: Gunnar Brinkmann, Harri Haanpää, Tommi Junttila, Gerzson Kéri, Clement Lam, Ilkka Niemelä, Pekka Orponen, Alexander Rosa, and Gordon Royle. Needless to say, we are more than grateful for this assistance. The suggestions of the anonymous reviewers were also of great help in preparing the final version of this book. The work was supported in part by the Helsinki Graduate School in Computer Science and Engineering (HeCSE), the Foundation of Technology (Tekniikan Edistämissäätiö), the Nokia Foundation, and by the Academy of Finland under Grants No. 44517, No. 100500, No. 107493, and No. 202315.

Espoo,
September 2005

Petteri Kaski
Patric Östergård

1

Introduction

Combinatorial design theory (or just design theory) is a branch of discrete mathematics which originated in the design of statistical experiments for agriculture and through generalization of various recreational problems.

A combinatorial design can be described as an arrangement of a finite set of points into a finite collection of blocks with some prescribed properties. As an example [504, 561], the following famous recreational problem was posed by Thomas P. Kirkman in the *Lady's and Gentleman's Diary* of 1850.

> Fifteen young ladies in a school walk out three abreast for seven days in succession: it is required to arrange them daily, so that no two walk twice abreast.

A solution is given below.

Day 1	Day 2	Day 3	Day 4	Day 5	Day 6	Day 7
A B C	A H I	A J K	A D E	A F G	A L M	A N O
D J N	B E G	B M O	B L N	B H J	B I K	B D F
E H M	C M N	C E F	C I J	C L O	C D G	C H K
F I O	D K O	D H L	F K M	D I M	E J O	E I L
G K L	F J L	G I N	G H O	E K N	F H N	G J M

A generalization of this problem to an arbitrary number of girls and days was unsolved for well over a hundred years until it was settled in the 1960s independently by Lu [380] and Ray-Chaudhuri and Wilson [499].

Design theory has strong connections [13, 89] to coding theory, a more recent branch of discrete mathematics, with origins in the engineering problem of finding good error-correcting and error-detecting codes for noisy communication channels. It was the seminal work of Claude E. Shannon [530] in the middle of the 20th century that laid the foundations for the new fields of coding theory and information theory, whose importance in the modern world shows no signs of declination.

A code consists of a set of codewords, which are tuples with element taken from a prescribed set. Golay [209] and Hamming [251] found the first few codes and code families of importance, which have been named after their originators. To give one example, we list a code with the parameters of the ternary Hamming code of length 4:

$$\begin{array}{ccc} 0000 & 1012 & 2021 \\ 0111 & 1120 & 2102 \\ 0222 & 1201 & 2210 \end{array} \tag{1.1}$$

Note that if you take any two of these nine codewords, they differ in at least (actually, exactly) three positions. This means that if you send one of these codewords over a noisy channel and it is corrupted in one bit, then only the codeword that was transmitted differs in one bit from the received word, and the error can be corrected. Similarly, two errors can be detected.

In addition to the traditional, prevailing applications in, respectively, communication systems [373, 606] and design of statistical experiments [496, 564], modern coding theory and design theory are fields of active research with connections to, for example, graph theory [89], finite group theory [163, 223], and computer science and cryptography [118, 122, 603].

The fifteen schoolgirls problem and its generalizations are examples of an existence problem in discrete mathematics. We here define this and some central related types of problems. The problems are listed in order of increasing difficulty: the solution to an instance of a problem in the list implies the solution to the corresponding instances of the problems of the earlier types.

existence problem Given a collection of properties, decide whether there exists an object realizing these properties.

counting problem Given a collection of properties, count, up to some criterion of isomorphism, the number of distinct objects meeting the properties.

classification problem Given a collection of properties, describe, up to some criterion of isomorphism, all the objects that have the desired properties.

In the literature, the words *cataloguing*, *census*, *generation*, and *listing* are used as synonyms for *classification*. Out of these, *generation* and *listing* are generally used for easy problems and the others for hard problems: we generate permutations but classify designs. The term *enumeration* is somewhat ambiguous since it has been used in the literature both for counting and classification.

Solving classification problems is of interest for both practical and theoretical reasons. A complete catalogue of objects of a particular type can be searched for objects meeting some additional requirements, or for counterexamples to a conjecture. Additionally, an inspection of classified objects often provides new insights into the structure of these, and could, ideally, lead to a thorough understanding of the entire class of objects (in group theory the word *classification* is often used in this stronger sense).

As for general mathematical interest in studying classification problems, already Felix Klein in his *Erlanger Programm* (1872) [319] described mathematics as the study of properties of sets that remain invariant under certain specified groups of transformations. We quote Rotman [509, p. 17]:

> Two basic problems occurring in mathematics are: classification of all systems of a given type (e.g., groups, semigroups, vector spaces, topological spaces); classification of all the "maps" or transformations from one such system into another. By a classification of systems, we mean a way to distinguish different systems or, what is the same thing, a way to tell when two systems are essentially the same (isomorphic).

Algorithmic methods have been widely used in design theory for solving existence and classification problems [205], whereas, until recently, computer-aided classifications of codes have received less attention. The success of algorithmic methods is based on the fact that the configurations studied are finite and thus amenable to local and exhaustive search methods.

Local search methods [342, Chap. 5] are typically incomplete in the sense that they are not guaranteed to find a solution to a problem even if one exists. Thus, local search methods are primarily applicable to solving existence problems by explicit construction, where they often outperform exhaustive methods. Exhaustive search [342, Chap. 4], on the other hand, considers all candidate solutions in turn and is guaranteed to find a solution if one exists. Exhaustive search can thus be applied to generate all combinatorial structures with particular properties, or to settle the nonexistence of such a structure. Indeed, often an exhaustive computer search has been the only instrument for demonstrating the nonexistence of a particular object. The nonexistence of projective planes of order 10 is perhaps the most notable such result to date [351]; that result is thoroughly discussed in Chap. 12.

A central issue in the design of practical exhaustive search algorithms for the generation of combinatorial configurations is the detection and elimination of *equivalent*, or *isomorphic*, copies of (sub)configurations. This is because failure to do so results in redundant work proportional to the number of such copies – which is typically exponential in the size of the configuration – making the search very inefficient or altogether infeasible to conduct with available computational resources. Furthermore, from a mathematical point of view, isomorphic configurations are by definition identical in the structure of interest, so it is desirable to eliminate all but one of such generated structures. As an example, consider a permutation of the values in the first position of the code in (1.1):

$$0 \mapsto 1, \quad 1 \mapsto 2, \quad 2 \mapsto 0. \tag{1.2}$$

With respect to the error-correcting and error-detecting properties, the new, distinct code cannot be distinguished from the original one.

In this book, we study the problem of isomorph-free exhaustive generation in an algorithmic and group-theoretic framework. The main structures considered are codes and designs, but we also devote one chapter, Chap. 8, to other combinatorial objects that are so closely related to codes and designs that the developed algorithms can be applied more or less directly. These objects include orthogonal arrays, Latin squares, and Hadamard matrices. Objects outside the scope of the book include graphs (which have been extensively studied and deserve a book on their own) and algebraic objects (such as groups).

The organization of the book is as follows. Chapter 2 focuses on introducing the terminology and the necessary theoretical background on designs, codes, and certain related objects. The central concepts of group and isomorphism are discussed in Chap. 3, and Chap. 4 describes three generic algorithmic frameworks for the problem of generating an exhaustive collection of nonisomorphic objects: orderly generation, generation by canonical augmentation, and utilization of homomorphisms and group-theoretic localization. Some auxiliary algorithms needed as parts of the classification algorithms are considered in Chap. 5. In Chaps. 6 to 8, specific algorithms for, respectively, designs, codes, and related structures are presented. Exhaustive tables of classification results for the most important structures have also been included. Classification of objects with prescribed automorphism groups is the theme of Chap. 9, and complexity results (for various problems occurring in the earlier chapters and related to classification) and the efficiency of algorithms that generate all solutions to a given problem are discussed in Chap. 11. The book ends in Chap. 12 with a survey of the celebrated result that no projective plane of order 10 exists.

Before proceeding with the main text, it is necessary to mention some conventions for mathematical notations and the pseudocode algorithms.

The set $\{0, 1, \ldots, q-1\}$ is denoted by Z_q. For a given set Ω, $\binom{\Omega}{k}$ denotes the set of k-element subsets – briefly, k-subsets – of Ω, and Ω^k the set of ordered k-tuples with elements from Ω. The ring of order q over Z_q with addition and multiplication carried out modulo q is denoted by $\mathbb{Z}_q$, and the finite field of order q by $\mathbb{F}_q$. Moreover, the multiplicative group of $\mathbb{F}_q$ is $\mathbb{F}_q^* = \mathbb{F}_q \setminus \{0\}$. A basic knowledge of finite fields is assumed in the discussion of linear codes; more information about finite fields can be found in any textbook on abstract algebra or [372, 409].

It is our aim to make the algorithms as self-explanatory as possible. The syntax is close to that of the main contemporary programming languages (such as C) and needs no further explanation. The variables are not declared, but the types of the input parameters to procedures are indicated. In addition to the ordinary types, we use variables that are sets. The **return** statement is omitted when it occurs at the end of a procedure.

Catalogues of objects considered in this book are provided in digital form on a supplementary DVD. Specifically, for object types where the tabulated classification results exhaust a certain range of parameters, exhaustive lists of nonisomorphic objects are made available (exceptions are Table 8.4, which is directly related to Table 8.3, and instances with more than 100 million objects). The tables in question are Tables 6.2–6.10 (BIBDs), Tables 6.12–6.16 (RBIBDs), Table 7.2 (error-correcting codes), Table 7.4 (covering codes), Table 7.7 (one-error-correcting linear codes), Table 8.1 (1-factorizations of complete graphs), Table 8.2 (Latin squares), and Table 8.3 (Hadamard matrices).

2

Graphs, Designs, and Codes

This chapter gives a brief introduction to graphs, designs, codes, and concepts related to these. Although classification algorithms for graphs are not considered in this book, graphs provide a convenient framework for several of the concepts and algorithms to be discussed.

For a general introduction to discrete mathematics, see [87, 376]. For an introduction to design theory, the reader is referred to the monographs [7, 36, 274, 562]; an encyclopedia of central results for combinatorial designs is [116]. For the theory of codes, see the classical reference [388] or [487]; introductory texts include [264, 273, 374, 484]. Introductory graph theory texts include [54, 159, 326, 604].

2.1 Graphs

Probably the most important – at least, the most studied – combinatorial object is a *graph*. For our purposes, the definition of a graph as a set system is appropriate.

Definition 2.1. *A* set system *is an ordered pair* $(X, \mathcal{S})$*, where* X *is a finite set and* $\mathcal{S}$ *is a collection of subsets of* X*.*

Definition 2.2. *A* graph G *is an ordered pair* (V, E)*, where* V *is a finite set of* vertices *and* E *is a set of two-element subsets of vertices, called* edges*.*

For a graph G we write $V(G)$ and $E(G)$ for the vertex set and the edge set, respectively. If the graph is clear from the context we may write simply V and E.

The number of vertices in a graph is called its *order* and the number of edges its *size*. An edge $\{u, v\}$ *joins* the vertices u and v. Two vertices are *adjacent* if they are joined by an edge. An edge e and a vertex u are *incident* if $u \in e$. The *neighborhood* $\Gamma(u)$ of a vertex u is the set of all vertices adjacent to u. The *degree* $d(u)$ of a vertex u is the number of vertices adjacent to u, that is, $d(u) = |\Gamma(u)|$. A graph in which all vertices have the same degree k is *regular* (k-regular). A k-regular graph of order v where every adjacent pair of vertices has λ common neighbors and every nonadjacent pair has μ common neighbors is said to be a (v, k, λ, μ) *strongly regular graph*. The *complement* of a graph $G = (V, E)$ is the graph $\bar{G} = (V, \binom{V}{2} \setminus E)$.

There are a number of ways to represent a graph.

Example 2.3. By definition, to describe a graph it suffices to specify the vertex set and the edge set, for example,

$$\begin{aligned} V &= \{1, 2, 3, 4, 5, 6\}, \\ E &= \{\{1, 2\}, \{2, 3\}, \{2, 6\}, \{3, 4\}, \{3, 6\}, \{4, 5\}\}. \end{aligned}$$

Example 2.4. Perhaps the visually most pleasing way to represent a graph is to draw it. Figure 2.1 contains a drawing of the graph in Example 2.3.

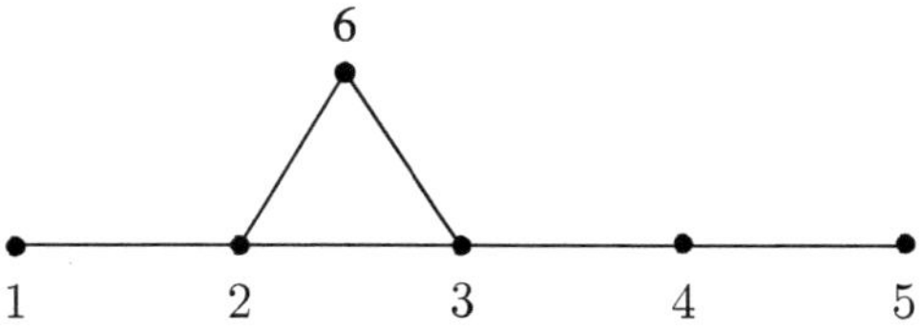

Fig. 2.1. A graph

A third common way to represent a graph is via an adjacency matrix.

Definition 2.5. *Let $G = (V, E)$ be a graph of order n with $|V| = n$ vertices labeled as $\{v_1, v_2, \ldots, v_n\}$. Subject to this labeling, the* adjacency matrix *of G is the $n \times n$ matrix $\mathbf{A} = (a_{ij})$ defined for all $1 \le i, j \le n$ by*

$$a_{ij} = \begin{cases} 0 & \text{if } \{v_i, v_j\} \notin E, \\ 1 & \text{if } \{v_i, v_j\} \in E. \end{cases}$$

Note that an adjacency matrix is always symmetric ($a_{ij} = a_{ji}$) with the diagonal entries a_{ii} equal to zero. If the vertex set has a natural order – in particular, when $V = \{1, 2, \ldots, n\}$ – then we assume that the adjacency matrix is defined subject to the labeling $v_1 < v_2 < \cdots < v_n$ induced by the order.

Example 2.6. The adjacency matrix of the graph in Example 2.3 is

$$\mathbf{A} = \begin{bmatrix} 0\,1\,0\,0\,0\,0 \\ 1\,0\,1\,0\,0\,1 \\ 0\,1\,0\,1\,0\,1 \\ 0\,0\,1\,0\,1\,0 \\ 0\,0\,0\,1\,0\,0 \\ 0\,1\,1\,0\,0\,0 \end{bmatrix}.$$

To describe a graph it is necessary to distinguish the vertices from each other by labeling them. However, this labeling is irrelevant as long as the properties of interest are independent of the labeling used.

Definition 2.7. *A graph G is* isomorphic *to a graph H if there exists a bijection $f : V(G) \to V(H)$ such that, for all $u, v \in V(G)$, we have $\{u, v\} \in E(G)$ if and only if $\{f(u), f(v)\} \in E(H)$. Such a bijection f is called an* isomorphism *of G onto H. An isomorphism of G onto itself is an* automorphism.

If G is isomorphic to H, then also H is isomorphic to G. It is therefore common practice to say that two such graphs are isomorphic, and write $G \cong H$.

Example 2.8. Two isomorphic graphs are shown in Fig. 2.2. Also an isomorphism is given.

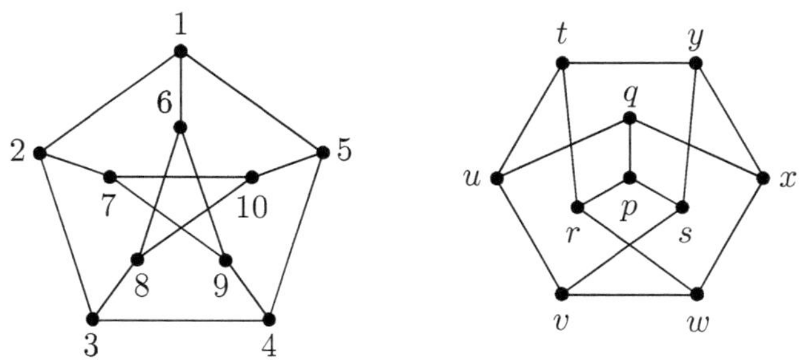

$f(1) = p, \quad f(2) = q, \quad f(3) = x, \quad f(4) = w, \quad f(5) = r,$
$f(6) = s, \quad f(7) = u, \quad f(8) = y, \quad f(9) = v, \quad f(10) = t$

Fig. 2.2. Two isomorphic graphs and an associated isomorphism

Example 2.9. Up to isomorphism there are 11 graphs of order 4. These are depicted in Fig. 2.3.

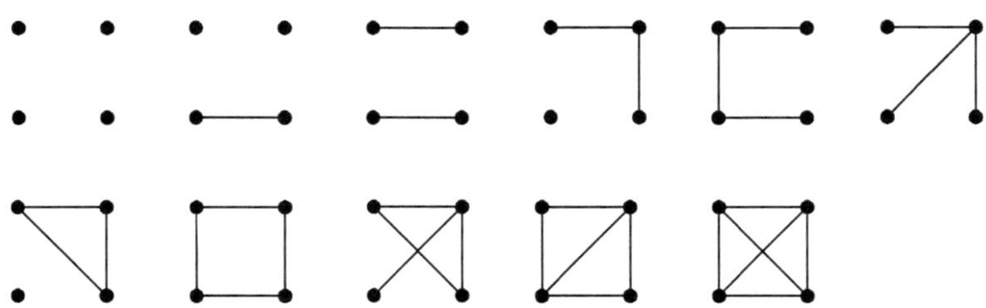

Fig. 2.3. The 11 graphs of order 4

Special notations and names are used for certain important graphs and families of graphs. We list a few such families that occur later in this book. Example graphs from some of these families are depicted in Fig. 2.4.

complete bipartite graph A graph isomorphic to the graph $K_{m,n}$ with vertex set $V_1 \cup V_2$, such that $|V_1| = m$, $|V_2| = n$, $|V_1 \cap V_2| = 0$, and all vertex pairs with one vertex from V_1 and the other from V_2 are joined by an edge.

complete graph A graph isomorphic to the graph K_n, which is of order n and where all pairs of distinct vertices are joined by an edge.

cube or **n-cube** A graph isomorphic to the graph Q_n whose vertices are the 2^n binary words of length n and whose edges are the pairs of words that differ in exactly one position. By a generalization to words over an alphabet of size q, we get the graph K_q^n ($K_2^n \cong Q_n$).

cycle A graph isomorphic to the graph C_n with vertex set $\mathbb{Z}_n$ and edge set $\{i, i+1\}$, $i \in \mathbb{Z}_n$.

empty graph A graph isomorphic to the graph $\bar{K}_n$, which is of order n and has no edges.

path A graph isomorphic to the graph P_n with vertex set $\{0, 1, \ldots, n\}$ and edge set $\{i-1, i\}$, $1 \leq i \leq n$, where n is the *length* of the path. The vertices 0 and n are the *endvertices* of the path. A path *connects* its endvertices.

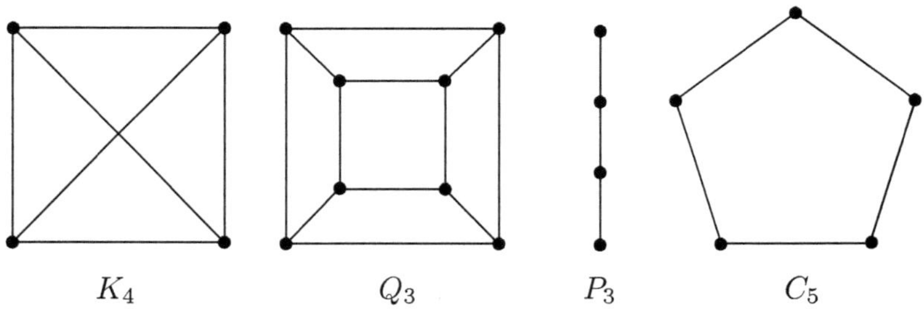

Fig. 2.4. Some common types of graphs

Weights may be associated to a graph, either to the vertices (this is more common) or to the edges. An unweighted graph is a special case of a weighted graph with weights 1.

The vertices of a graph may be colored by associating a color – an element from a given set, the "palette" – to each vertex. A coloring is *proper* if adjacent vertices have different colors. Edge colorings are defined analogously; an edge coloring is *proper* if edges incident to a common vertex have different colors.

Definition 2.10. *A graph H is a* subgraph *of a graph G if $V(H) \subseteq V(G)$ and $E(H) \subseteq E(G)$ such that for all $\{u, v\} \in E(H)$ both $u \in V(H)$ and $v \in V(H)$. A subgraph is* spanning *if $V(G) = V(H)$.*

A spanning path or cycle in a graph is called a *Hamiltonian* path or cycle, respectively.

Definition 2.11. *For a graph G, the subgraph* induced *by $W \subseteq V(G)$ has W as its vertex set and its edge set consists of all edges $\{u, v\} \in E(G)$ for which $u, v \in W$ holds.*

Definition 2.12. *A* clique *in a graph G is a set of vertices whose induced subgraph is a complete graph. An* independent set *in a graph G is a set of vertices whose induced subgraph is an empty graph.*

A clique whose vertices are not a proper subset of the vertices of a (larger) clique is called *maximal*. The cliques in a graph with largest size among all cliques in the graph are called *maximum*. Note that there may be several maximum cliques. In a graph with weights associated to the vertices, a *maximum-weight clique* is a clique that has the largest sum of weights among all cliques in the graph. These concepts apply to independent sets in an analogous way.

Two vertices in a graph G are *connected* if there exists a path in the graph connecting the two vertices. Connectedness of vertices is an equivalence relation on $V(G)$; the subgraphs induced by the associated equivalence classes in $V(G)$ are called the *connected components* of G. A graph with only one connected component is said to be *connected*.

Definition 2.13. *A* tree *is a connected graph where the path connecting any two vertices is unique. A* forest *is a graph where all the connected components are trees.*

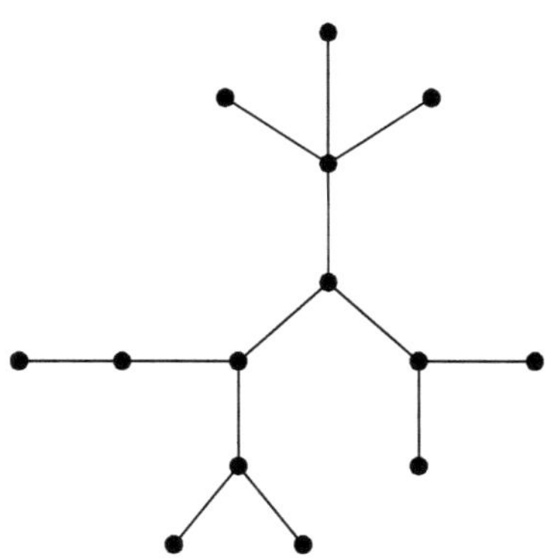

Fig. 2.5. A tree

Definition 2.14. *A* rooted tree *is a pair (T, r) where $T = (V, E)$ is a tree and $r \in V$ is the* root *of T.*

If x, y are vertices of a rooted tree, then we say that x is a *descendant* of y if y occurs on the path connecting x to the root. We say that x is an *ancestor* of y if y is a descendant of x. If $x \neq y$, then we speak of a *proper* descendant (ancestor). The *parent* of a nonroot vertex y is the vertex that is adjacent to y in the path connecting y to the root. Conversely, the vertices y with x as parent are the *children* of x. Two vertices having the same parent are called *siblings*. A vertex with no children is a *leaf*. The *depth* or *level* of a vertex x is the length of the path connecting x to the root. The *height* of a rooted tree is the maximum depth of a vertex. The *subtree rooted* at a vertex x is the rooted tree with root x induced by all descendants of x.

Definition 2.15. *A* factor *of a graph G is a spanning subgraph of G. A* factorization *of G is a set of factors of G such that each edge of G occurs in exactly one factor. Two factorizations are isomorphic if there exists an isomorphism of the underlying graphs that maps the factors in one factorization onto factors in the other factorization.*

Of particular interest are factorizations in which every factor is regular.

Definition 2.16. *A factor that is k-regular is called a* k-factor. *A factorization into k-factors is called a* k-factorization.

Example 2.17. A 1-factorization of K_6 is shown in Fig. 2.6. Note that this factorization can be used to schedule a round robin tournament – that is, a tournament where each teams meets each other exactly once – with six teams and five rounds.

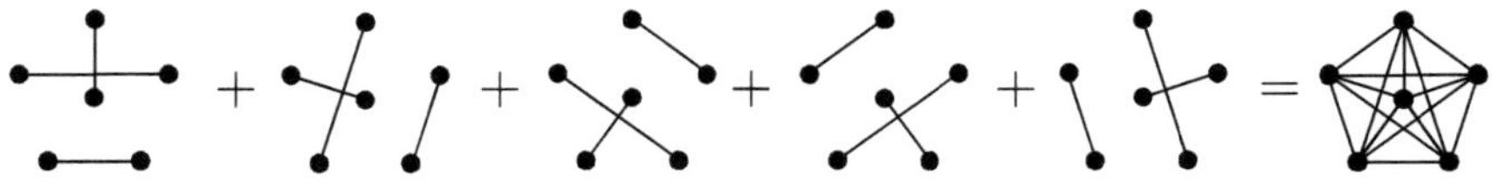

Fig. 2.6. One-factorization of K_6

Example 2.18. The complete graph K_{2n} admits a 1-factorization for all $n \geq 1$. Let $V = \mathbb{Z}_{2n-1} \cup \{\infty\}$ be the vertex set, and let the edge set of the ith 1-factor be $F_i = \{\{i, \infty\}\} \cup \{\{i - j, i + j\} : 1 \leq j \leq n - 1\}$ for all $0 \leq i \leq 2n - 2$. The resulting 1-factorization of K_{2n} is known as GK_{2n}; the 1-factorization in Fig. 2.6 is GK_6.

Example 2.19. The six nonisomorphic 1-factorizations of K_8 are given in Fig. 2.7. The 1-factorization GK_8 is isomorphic to the 1-factorization in the last row.

For more on 1-factorizations, see [597].

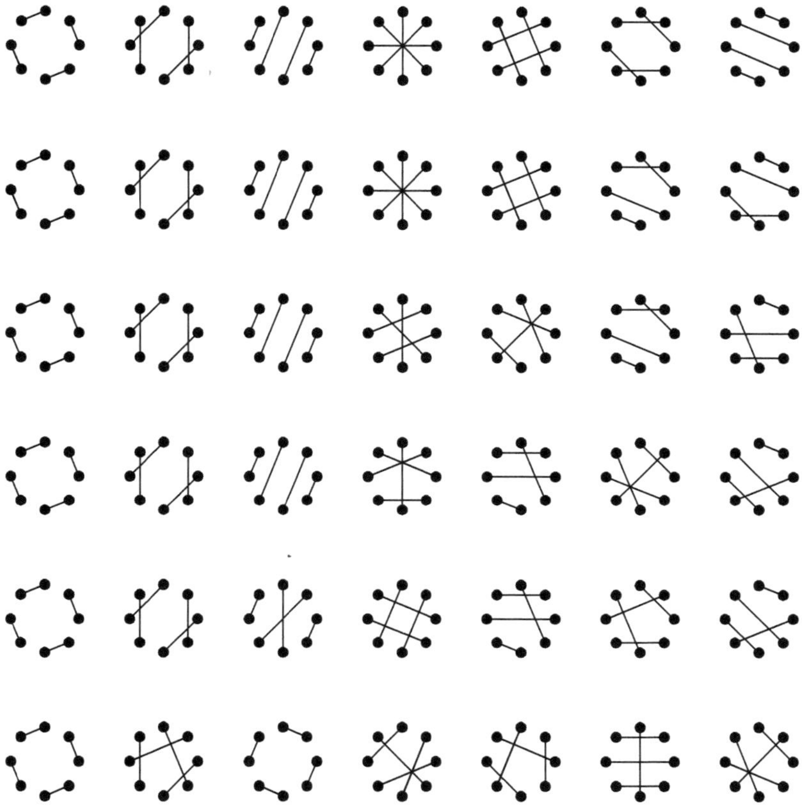

Fig. 2.7. The six nonisomorphic 1-factorizations of K_8

2.2 Designs

In general terms, recalling the origin of design theory, a design is simply a formal description of a testbed for statistical experiments. It is when one adds constraints to these experiments that one starts facing structures amenable to combinatorial construction and classification questions. In this section, the main definitions and some important results needed later are presented. Our consideration of designs proceeds from the most general one to more and more specific ones, from incidence structures, through t-designs and block designs, and finally ending at Steiner triple systems. A few other types of designs are also briefly mentioned. Even though an exhaustive treatment of all possible types of designs (cf. [116]) is not possible here, the theory and algorithms to be presented can be used with some (often very small) modifications for many other classes of objects.

2.2.1 Incidence Structures

The most basic object studied in design theory is an incidence structure.

Definition 2.20. *An* incidence structure $\mathcal{X}$ *is a triple* $(P, \mathcal{B}, I)$, *where* P *and* $\mathcal{B}$ *are finite sets and* $I \subseteq P \times \mathcal{B}$. *The set* P *is a set of* points *and the set* $\mathcal{B}$ *is a set of* blocks. *The relation* I *is an* incidence relation.

If $(p, B) \in I$, then we say that the point p and the block B are *incident*. The pair (p, B) is a *flag*. In the literature, the elements of $\mathcal{B}$ are sometimes called *lines* – especially when incidence structures are discussed in a geometry framework.

For an incidence structure $\mathcal{X}$ we write $P(\mathcal{X})$, $\mathcal{B}(\mathcal{X})$, and $I(\mathcal{X})$ for the point set, the block set, and the incidence relation of $\mathcal{X}$, respectively. If the incidence structure is clear from the context, we write simply P, $\mathcal{B}$, and I.

For a point p, we write $[p]$ for the set of blocks incident with p and $|p|$ for the number of blocks incident with p. Similarly, for a block B we write $[B]$ for the set of points incident with B and $|B|$ for the cardinality of this set.

An incidence structure is *simple* if no two different blocks are incident with the same set of points. An incidence structure that is not simple is said to have *repeated blocks*.

Example 2.21. A structure often encountered in this book is the *Pasch configuration*, which consists of six points $P = \{u, v, w, x, y, z\}$, four blocks $\mathcal{B} = \{P, Q, R, S\}$, and the incidence relation

$$I = \{(u, P), (v, P), (w, P), (u, Q), (x, Q), (y, Q), \\ (v, R), (x, R), (z, R), (w, S), (y, S), (z, S)\}.$$

The Pasch configuration is depicted in Figure 2.8.

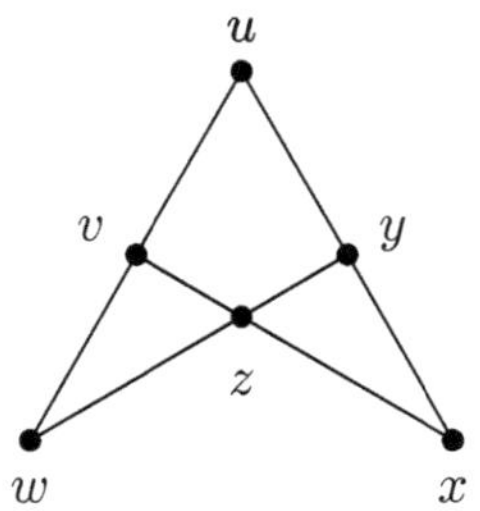

Fig. 2.8. Pasch configuration

We observed for graphs that the labels of the vertices are necessary only to describe the graph, but the properties studied are in general independent of the labels. A similar situation occurs with incidence structures, which motivates the following definition of isomorphism.

Definition 2.22. *Two incidence structures, $\mathcal{X}$ and $\mathcal{Y}$, are* isomorphic *if there exists a pair of bijections, $f_P : P(\mathcal{X}) \to P(\mathcal{Y})$ and $f_{\mathcal{B}} : \mathcal{B}(\mathcal{X}) \to \mathcal{B}(\mathcal{Y})$, such that for all $p \in P(\mathcal{X})$ and $B \in \mathcal{B}(\mathcal{X})$, we have $(p, B) \in I(\mathcal{X})$ if and only if $(f_P(p), f_{\mathcal{B}}(B)) \in I(\mathcal{Y})$. Such a pair $f = (f_P, f_{\mathcal{B}})$ is an* isomorphism. *An isomorphism of $\mathcal{X}$ onto itself is an* automorphism.

Up to isomorphism we may (and often will) treat a simple incidence structure as a set system, where each block B is identified with its set of incident points $[B]$. For example, the blocks of the Pasch configuration can be described as the set

$$\{\{u, v, w\}, \{u, x, y\}, \{v, x, z\}, \{w, y, z\}\},$$

which clearly carries all the relevant incidence information except for the labels of the blocks. Analogously, an incidence structure with repeated blocks may be treated as a set system if it is not necessary to distinguish between blocks with the same set of incident points.

A convenient algebraic representation for an incidence structure is the incidence matrix.

Definition 2.23. *Let $\mathcal{X} = (P, \mathcal{B}, I)$ be an incidence structure with v points and b blocks, labeled as $\{p_1, p_2, \ldots, p_v\}$ and $\{B_1, B_2, \ldots, B_b\}$, respectively. Subject to this labeling, the* incidence matrix *of $\mathcal{X}$ is the $v \times b$ matrix $\mathbf{N} = (n_{ij})$ defined for all $1 \le i \le v$ and $1 \le j \le b$ by*

$$n_{ij} = \begin{cases} 0 & \textit{if } (p_i, B_j) \notin I, \\ 1 & \textit{if } (p_i, B_j) \in I. \end{cases}$$

If there is a natural order relation defined on the points and blocks, then we assume that the incidence matrix is constructed subject to the order-induced labelings $p_1 < p_2 < \cdots < p_v$ and $B_1 < B_2 < \cdots < B_b$.

Example 2.24. An incidence matrix of the Pasch configuration in Example 2.21 is as follows (the rows are labeled u, v, w, x, y, z from top to bottom and the columns P, Q, R, S from left to right):

$$\mathbf{N} = \begin{bmatrix} 1\,1\,0\,0 \\ 1\,0\,1\,0 \\ 1\,0\,0\,1 \\ 0\,1\,1\,0 \\ 0\,1\,0\,1 \\ 0\,0\,1\,1 \end{bmatrix}.$$

Definition 2.25. *An incidence structure $\mathcal{Y}$ is a* substructure *of an incidence structure $\mathcal{X}$ if $P(\mathcal{Y}) \subseteq P(\mathcal{X})$, $\mathcal{B}(\mathcal{Y}) \subseteq \mathcal{B}(\mathcal{X})$, and $I(\mathcal{Y}) \subseteq I(\mathcal{X})$.*

Example 2.26. The incidence structure in Fig. 2.9 contains seven substructures isomorphic to the Pasch configuration in Fig. 2.8. (Remove any point and the blocks incident with the point to obtain a Pasch configuration.)

We will also use the term *configuration* for a substructure, implying that the substructure itself has certain regularity properties, or that it occurs as a substructure of a regular structure such as a design (cf. [36, 123]).

Definition 2.27. *For an incidence structure $\mathcal{X}$, the substructure* induced *by a set of points $Q \subseteq P(\mathcal{X})$ has the point set Q, the block set consisting of all blocks incident with at least one point in Q, and the incidence relation obtained by restricting the incidence relation of $\mathcal{X}$ to these points and blocks.*

The substructure induced by a set of blocks is defined by exchanging the roles of points and blocks in Definition 2.27.

Two structures related to an incidence structure are its dual and complement.

Definition 2.28. *The* dual *of an incidence structure $\mathcal{X}$ is the incidence structure $\mathcal{X}^*$ defined by*

$$P(\mathcal{X}^*) = \mathcal{B}(\mathcal{X}), \quad \mathcal{B}(\mathcal{X}^*) = P(\mathcal{X}), \quad I(\mathcal{X}^*) = \{(B, p) : (p, B) \in I(\mathcal{X})\}.$$

Example 2.29. The dual of the Pasch configuration consists of four points and six blocks formed by all distinct pairs of the four points.

Definition 2.30. *The* complement *of an incidence structure $\mathcal{X}$ is the incidence structure $\bar{\mathcal{X}}$ defined by*

$$P(\bar{\mathcal{X}}) = P(\mathcal{X}), \quad \mathcal{B}(\bar{\mathcal{X}}) = \mathcal{B}(\mathcal{X}), \quad I(\bar{\mathcal{X}}) = (P(\mathcal{X}) \times \mathcal{B}(\mathcal{X})) \setminus I(\mathcal{X}).$$

Example 2.31. The Pasch configuration is isomorphic to its complement.

Incidence structures that satisfy $\bar{\mathcal{X}} \cong \mathcal{X}$ are called *self-complementary*. Analogously, incidence structures that satisfy $\mathcal{X}^* \cong \mathcal{X}$ are *self-dual*.

2.2.2 t-Designs

Definition 2.32. *Let t, v, k, λ be integers such that $v \geq k \geq t$ and $\lambda \geq 1$. A t-(v, k, λ)* design *is an incidence structure with the following properties:*

1. *$|P| = v$,*
2. *$|B| = k$ for all $B \in \mathcal{B}$,*
3. *for any $T \subseteq P$ with $|T| = t$, there are exactly λ blocks incident with all points in T.*

Example 2.33. A single block incident with all the v points is clearly a t-design for any $t \leq v$. Similarly, if we take as blocks all the k-subsets of P, we obtain a t-(v, k, λ) design for any $t \leq k$. These types of designs are usually omitted from consideration as *trivial*.

Example 2.34. A 2-$(7,3,1)$ design can be formed by using the points $\mathbb{Z}_7 = \{0,1,\ldots,6\}$ and taking as blocks all 3-subsets of the form $\{0+x, 1+x, 3+x\}$, $x \in \mathbb{Z}_7$. The resulting design is known as the *Fano plane*, and is depicted in Fig. 2.9.

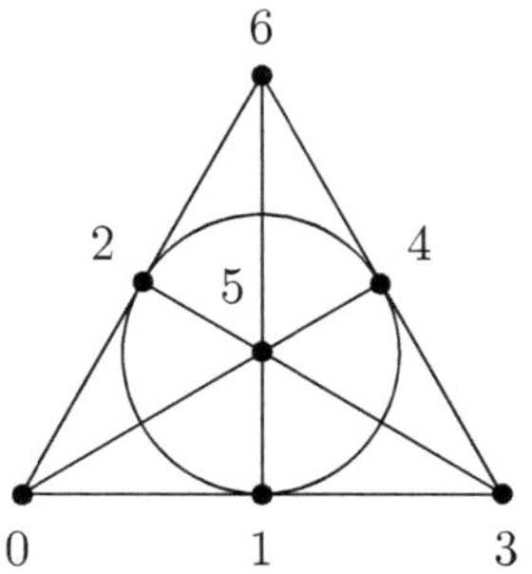

Fig. 2.9. Fano plane

Example 2.35. Given a 3×3 matrix of nine distinct elements, we obtain a 2-$(9,3,1)$ design by listing the triples formed by the rows, the columns, and the forward and back diagonals of the matrix.

The designs in Examples 2.34 and 2.35 are the smallest nontrivial designs in two important infinite families of 2-designs, the projective and affine geometries.

Example 2.36. Let q be a prime power and consider the 3-dimensional vector space $\mathbb{F}_q^3$. Take as points the 1-dimensional subspaces of $\mathbb{F}_q^3$ and as blocks the 2-dimensional subspaces. A point is incident with a block if and only if the 1-dimensional subspace is contained in the 2-dimensional subspace. The resulting 2-$(q^2+q+1, q+1, 1)$ design is known as a *projective geometry* of order q and dimension 2 or a *projective plane* of order q.

Example 2.37. Let q be a prime power and consider the 2-dimensional vector space $\mathbb{F}_q^2$. A *line* in $\mathbb{F}_q^2$ consists of all points $(x,y) \in \mathbb{F}_q^2$ that satisfy an equation of the form $ax + by = c$, where $a, b, c \in \mathbb{F}_q$ and at least one of a, b is nonzero. Take as points the elements of $\mathbb{F}_q^2$ and as blocks all lines in $\mathbb{F}_q^2$. A point is incident with a line if and only if the line contains the point. The resulting 2-$(q^2, q, 1)$ design is known as an *affine geometry* of order q and dimension 2 or an *affine plane* of order q.

We proceed to derive some necessary existence conditions for t-designs.

Theorem 2.38. *Consider an arbitrary* t-(v,k,λ) *design* $(P, \mathcal{B}, I)$ *and let* $L \subseteq P$ *with* $0 \le |L| = \ell \le t$. *The number of blocks incident with all the points of* L *is*

$$b_\ell = \lambda \binom{v-\ell}{t-\ell} \Big/ \binom{k-\ell}{t-\ell}. \tag{2.1}$$

Proof. We count in two ways the number of pairs (B, T), where $B \in \mathcal{B}$, $|T| = t$, and $L \subseteq T \subseteq [B] \subseteq P$. First, each of the b_ℓ blocks B incident with all points in L leads to $\binom{k-\ell}{t-\ell}$ such pairs. Second, the number of subsets $T \subseteq P$ with $T \supseteq L$ and $|T| = t$ is $\binom{v-\ell}{t-\ell}$, each contributing λ pairs by Definition 2.32. □

Corollary 2.39. *A t-(v,k,λ) design is an ℓ-(v,k,b_ℓ) design for arbitrary $0 \leq \ell \leq t$ and with b_ℓ defined as in* (2.1).

Corollary 2.40. *The number of blocks in a t-(v,k,λ) design is*

$$b = \lambda \binom{v}{t} \Big/ \binom{k}{t}.$$

The number of blocks incident with any point is

$$r = \lambda \binom{v-1}{t-1} \Big/ \binom{k-1}{t-1}.$$

In what follows we always write b for the number of blocks in a design, v for the number of points, and r for the number of blocks incident with a point.

Since each b_ℓ must be an integer in Theorem 2.38, we obtain the following necessary existence condition.

Corollary 2.41. *A t-(v,k,λ) design exists only if, for all $0 \leq \ell \leq t$,*

$$\lambda \binom{v-\ell}{t-\ell} \text{ is divisible by } \binom{k-\ell}{t-\ell}. \tag{2.2}$$

Perhaps the most fundamental problems in design theory are to determine for which parameters t, v, k, λ there exists a design and to give explicit constructions for such designs. Although much is known, these problems are far from being completely settled, especially in the interesting case when λ is small. As an example, for $\lambda = 1$ not a single example of a t-design with $t \geq 6$ is known, and only finitely many designs with $t \geq 4$ and $\lambda = 1$ are known (see [36] and [116]).

2.2.3 Balanced Incomplete Block Designs

The main topic of interest in our later discussion of classification algorithms for designs is 2-designs, which are often called *balanced incomplete block designs* (BIBDs) or simply *block designs*. For BIBDs, the parameters v, k, λ, b, r are connected by the following equations:

$$vr = bk, \quad \lambda(v-1) = r(k-1). \tag{2.3}$$

Since r and b must be integers by (2.2) we obtain the necessary existence conditions

$$\lambda(v-1) \equiv 0 \pmod{k-1}, \quad \lambda v(v-1) \equiv 0 \pmod{k(k-1)}. \tag{2.4}$$

We develop some further necessary existence conditions for BIBDs. Let $\mathbf{N}$ be an incidence matrix of a BIBD. Then

$$\mathbf{N}\mathbf{N}^T = (r-\lambda)\mathbf{I} + \lambda\mathbf{J}, \tag{2.5}$$

where $\mathbf{N}^T$ denotes the transpose of $\mathbf{N}$, $\mathbf{I}$ denotes an identity matrix, and $\mathbf{J}$ denotes a matrix with all entries equal to 1. If it is necessary to stress the size of the latter two matrices, we write $\mathbf{I}_v$ and $\mathbf{J}_v$, respectively.

Theorem 2.42 (Fisher's inequality). *A 2-(v,k,λ) design with $v > k$ has $b \geq v$.*

Proof. Let $\mathbf{N}$ be an incidence matrix of a 2-(v,k,λ) design with $v > k$. By (2.5) and the fact that

$$\det(x\mathbf{I} + y\mathbf{J}) = (x + ym)x^{m-1} \tag{2.6}$$

for an $m \times m$ matrix – see [89, Lemma 1.12] for a proof of (2.6) – we get

$$\det(\mathbf{N}\mathbf{N}^T) = rk(r-\lambda)^{v-1} \neq 0 \tag{2.7}$$

because $v > k$ implies $r > \lambda$ by (2.3). Since $\mathbf{N}$ and $\mathbf{N}\mathbf{N}^T$ have the same rank and $\mathbf{N}\mathbf{N}^T$ has full rank by (2.7), the rank of $\mathbf{N}$ is v. Since the rank cannot exceed the number of columns, $b \geq v$. □

An important family of BIBDs consists of those designs whose incidence matrices are a square matrices.

Definition 2.43. *A BIBD is called* square *if $v = b$.*

We remark that many authors use by tradition the term *symmetric* design for a square design, which is somewhat misleading because it is neither required that the matrix $\mathbf{N}$ be symmetric nor that the design have symmetry in the form of nontrivial automorphisms.

The equality $v = b$ implies $r = k$ by (2.3). The following fundamental result was first proved by Ryser [513]; our proof is from [36, Proposition II.3.2].

Theorem 2.44. *Let $\mathcal{D}$ be a square 2-(v,k,λ) design. Then, the dual $\mathcal{D}^*$ is also a square 2-(v,k,λ) design.*

Proof. The claim is trivial if $v = k$ so suppose $v > k$. By the proof of Theorem 2.42 the incidence matrix $\mathbf{N}$ has (full) rank v and is therefore invertible. We observe that $r\mathbf{J} = \mathbf{N}\mathbf{J} = \mathbf{J}\mathbf{N} = k\mathbf{J}$ and hence $\mathbf{N}^{-1}\mathbf{J} = r^{-1}\mathbf{J}$. Thus,

$$\mathbf{N}^T\mathbf{N} = \mathbf{N}^{-1}\mathbf{N}\mathbf{N}^T\mathbf{N} = \mathbf{N}^{-1}\big((r-\lambda)\mathbf{I} + \lambda\mathbf{J}\big)\mathbf{N} = (r-\lambda)\mathbf{I} + \lambda\mathbf{J},$$

which shows that $\mathbf{N}^T$ is an incidence matrix of a 2-(v,k,λ) design. □

Corollary 2.45. *Every nontrivial square* 2-(v,k,λ) *design is simple.*

The projective planes from Example 2.36 constitute an infinite family of self-dual square designs.

Square designs allow the construction of two types of designs with smaller parameters.

Definition 2.46. *Let $\mathcal{D}$ be a square* 2-(v,k,λ) *design and let B be a block of $\mathcal{D}$. The* derived design *of $\mathcal{D}$ with respect to B is the incidence structure induced by the points $[B]$, with the block B itself deleted. The* residual *of $\mathcal{D}$ with respect to B is the incidence structure induced by the points $P \setminus [B]$.*

It follows from Theorem 2.44 that a derived design of a square 2-(v,k,λ) design $\mathcal{D}$ is a 2-$(k,\lambda,\lambda-1)$ design, and a residual of $\mathcal{D}$ is a 2-$(v-k,k-\lambda,\lambda)$ design.

A fundamental necessary existence condition for square BIBDs is given in the following theorem due to Bruck, Ryser, and Chowla [74, 107].

Theorem 2.47 (Bruck–Ryser–Chowla). *Let v,k,λ be integers satisfying $\lambda(v-1) = k(k-1)$ for which a square* 2-(v,k,λ) *design exists. Then*

1. *if v is even, then $k-\lambda$ is a square;*
2. *if v is odd, then the equation $z^2 = (k-\lambda)x^2 + (-1)^{(v-1)/2}\lambda y^2$ has a solution in integers x,y,z, not all zero.*

Proof. It follows from (2.7) and $k = r$ that $\det(\mathbf{N}) = \sqrt{k^2(k-\lambda)^{v-1}}$. Since $\det(\mathbf{N})$ is an integer, $k-\lambda$ must be a square if $v-1$ is odd, that is, if v is even. For the case v odd, see, for example, [36, p. 93]. □

2.2.4 Steiner Triple Systems

For a t-design, the smallest nontrivial parameters t,k,λ are $t = 2$, $k = 3$, $\lambda = 1$.

Definition 2.48. *A* 2-$(v,3,1)$ *design is called a* Steiner triple system *of* order v, *or briefly an* STS(v).

The family of Steiner triple systems is among the most studied – if not the most studied – family of designs. For a comprehensive treatment of Steiner triple systems, see [123].

Example 2.34 contains an STS(7) and Example 2.35 contains an STS(9). From (2.2) we obtain the necessary condition $v \equiv 1 \pmod 6$ or $v \equiv 3 \pmod 6$ for the parameter v. These conditions are also sufficient as the following two constructions from [376, Chap. 19] demonstrate.

Example 2.49. Construction for an STS$(6j+3)$. Let $n = 2j+1$ and use $\mathbb{Z}_n \times \mathbb{Z}_3$ as the point set. The blocks consist of all triples of the form $\{(x,0),(x,1),(x,2)\}$ with $x \in \mathbb{Z}_n$ and all triples $\{(x,i),(y,i),((x+y)(j+1),i+1)\}$ with $x,y \in \mathbb{Z}_n$, $x \neq y$, and $i \in \mathbb{Z}_3$. (Addition and multiplication are carried out coordinatewise modulo n and 3, respectively.)

Example 2.50. Construction for an STS$(6j+1)$. We take as point set $\mathbb{Z}_{2j} \times \mathbb{Z}_3 \cup \{\infty\}$. The $6j+1$ *base blocks* are

$$\{(0,0),(0,1),(0,2)\}; \tag{2.8}$$
$$\{\infty,(0,0),(j,1)\},\ \{\infty,(0,1),(j,2)\},\ \{\infty,(0,2),(j,0)\}; \tag{2.9}$$
$$\{(0,0),(i,1),(-i,1)\},\ \{(0,1),(i,2),(-i,2)\},\ \{(0,2),(i,0),(-i,0)\}; \tag{2.10}$$
$$\{(j,0),(i,1),(1-i,1)\},\ \{(j,1),(i,2),(1-i,2)\},\ \{(j,2),(i,0),(1-i,0)\}, \tag{2.11}$$

where the blocks (2.10) are included once for every $i = 1,2,\ldots,j-1$ and the blocks (2.11) once for every $i = 1,2,\ldots,j$. The block set is now constructed by adding the element $(\ell,0)$ to each of the $6j+1$ base blocks for all $\ell = 0,1,\ldots,j-1$. (Addition is carried out coordinatewise modulo $2j$ and 3 with $\infty + x = x + \infty = \infty$ for all $x \in \mathbb{Z}_{2j} \times \mathbb{Z}_3$.)

These two constructions solve the existence problem for Steiner triple systems. The next step is of course to classify all the nonisomorphic STS(v). This, however, is not an easy problem because the number of nonisomorphic designs increases very rapidly with increasing v by a result of Wilson [610].

Theorem 2.51. *The number of nonisomorphic* STS(v) *is at least* $(e^{-5}v)^{v^2/6}$ *for all* $v \equiv 1,3 \pmod 6$.

2.2.5 Some Other Families of Designs

This section lists several other families of designs that we will require later in this book. Two generalizations of block designs are pairwise balanced designs and group divisible designs. Both of these families are of fundamental importance in the construction of many other types of designs including BIBDs; see [36, 116].

Definition 2.52. *Let* v, λ *be positive integers and let* K *be a nonempty set of positive integers. A* pairwise balanced design *with parameters* v, K, λ *– a* PBD$(v, K; \lambda)$ *– is an incidence structure with the following properties:*

1. $|P| = v$,
2. $|B| \in K$ *for all* $B \in \mathcal{B}$,
3. *for any pair of distinct points, there are exactly* λ *blocks incident with both points.*

To simplify the notation we write PBD(v, K) if $\lambda = 1$.

Definition 2.53. *Let* v, λ_1, λ_2 *be positive integers and let* K *be a nonempty set of positive integers. A* group divisible design *with parameters* $v, K, \lambda_1, \lambda_2$ *– a* GDD$(v, K; \lambda_1, \lambda_2)$ *– is an incidence structure with the following properties:*

1. $|P| = v$,
2. $|B| \in K$ *for all* $B \in \mathcal{B}$,

3. *there exists a partition of the point set P into disjoint* groups *such that, for any pair of distinct points, the number of blocks incident with both points is λ_1 if the points belong to the same group, and λ_2 if the points belong to different groups.*

We remark that the term "group" in this context is in no way related to the more familiar algebraic notion of a group.

The *type* of a group divisible design is the multiset consisting of the sizes of all the groups. It is customary to use exponential notation for the group type. We write $g_1^{a_1} g_2^{a_2} \cdots g_m^{a_m}$ to indicate that there are a_i groups of size g_i for $i = 1, 2, \ldots, m$. If $\lambda_1 = 0$ we write $\mathrm{GDD}(v, K; \lambda)$, where $\lambda = \lambda_2$. Similarly, if $K = \{k\}$ for some positive integer k, then we write $\mathrm{GDD}(v, k; \lambda_1, \lambda_2)$. If $\lambda_1 = 0$ and $\lambda_2 = 1$, we write $\mathrm{GDD}(v, K)$.

The following list contains more families of designs.

Steiner system A t-$(v, k, 1)$ design. The notation $S(t, k; v)$ is also used in the literature.

projective plane A square 2-$(n^2 + n + 1, n + 1, 1)$ design. The parameter n is the *order* of the plane; cf. Example 2.36.

affine plane A 2-$(n^2, n, 1)$ design. The parameter n is the *order* of the plane; cf. Example 2.37.

quasi-residual design A 2-(v, k, λ) design with $r = k + \lambda$. A quasi-residual design that is a residual design of a square 2-$(v + k + \lambda, k + \lambda, \lambda)$ design is called *embeddable*.

quasi-derived design A 2-(v, k, λ) design with $k = \lambda + 1$. A quasi-derived design that is a derived design of a square 2-$(v(v - 1)/k + 1, v, k)$ design is called *embeddable*.

Hadamard design A square 2-$(4n - 1, 2n - 1, n - 1)$ design. Hadamard designs will be encountered in connection with Hadamard matrices in Sect. 2.4.3.

Hadamard 3-design A 3-$(4n, 2n, n - 1)$ design. To be encountered in Sect. 2.4.3.

transversal design A $\mathrm{GDD}(kn, k)$ of type n^k. The notation $\mathrm{TD}(k, n)$ is commonly used for transversal designs.

The following basic embeddability result will be required later. Proving the case $\lambda = 1$ is straightforward. A proof of the case $\lambda = 2$, called the *Hall–Connor theorem* after the originators, is presented in [245].

Theorem 2.54. *Quasi-residual designs with $\lambda = 1$ or $\lambda = 2$ are embeddable.*

Consequently, any square designs with $\lambda = 1$ or $\lambda = 2$ may be constructed via their residuals.

2.2.6 Resolutions of Designs

Resolvability is a desirable property of a design for many practical applications such as the design and analysis of experiments and the design of tournaments

[7, 200]. Resolutions of BIBDs are also relevant because of their connection to coding theory (see Sect. 2.3.2 and Theorem 3.82). A comprehensive monograph on construction techniques for resolvable designs is [200].

Definition 2.55. *A* parallel class *in an incidence structure is a set of blocks P such that every point is incident with exactly one block in P. A* resolution *of an incidence structure is a partition of the blocks into parallel classes. An incidence structure is* resolvable *if it has a resolution.*

In practice we will view resolutions as ordered pairs $\mathcal{R} = (\mathcal{X}, R)$, where R is the partition of the blocks into parallel classes and $\mathcal{X}$ is the underlying incidence structure. A resolvable BIBD is called an RBIBD.

Example 2.56. The solution to the fifteen schoolgirls problem that was presented in Chap. 1 is a resolution of a STS(15).

A resolution of a Steiner triple system is a *Kirkman triple system* of *order* v, or a KTS(v). There is some ambiguity with this term in the literature, since in some places a resolvable Steiner triple system, rather than a resolution, is called a Kirkman triple system [36].

Example 2.57. A resolution of the unique 2-$(2n, 2, 1)$ design corresponds to a 1-factorization of the complete graph K_{2n}.

Resolvability of a design is a nontrivial property in the sense that it cannot be decided based on the parameters t, v, k, λ alone.

Example 2.58. Up to isomorphism there are four 2-$(8, 4, 3)$ designs, exactly one of which is resolvable. We encourage the reader to find the resolvable design from the four isomorphism class representatives:

$\mathcal{D}_1$: `abcd abce abfg acfh adef adgh aegh`
`bcgh bdeg bdfh befh cdeh cdfg cefg`

$\mathcal{D}_2$: `abcd abce abfg acfh adef adgh aegh`
`bcgh bdeh bdfg befh cdeg cdfh cefg`

$\mathcal{D}_3$: `abcd abce abfg acfh adeh adfg aegh`
`bcgh bdef bdgh befh cdeg cdfh cefg`

$\mathcal{D}_4$: `abcd abef abgh aceg acfh adeh adfg`
`bceh bcfg bdeg bdfh cdef cdgh efgh`

For $v = 2k$, the following theorem of Alltop [5] connects families of resolvable t-designs with consecutive values of t.

Theorem 2.59. *A resolvable $2t$-$(2k, k, \lambda)$ design is simultaneously a resolvable $(2t+1)$-$(2k, k, \lambda')$ design with $\lambda' = \lambda(k-2t)/(2k-2t)$ and vice versa.*

Example 2.60. By Theorem 2.59, the resolvable 2-$(8,4,3)$ design in Example 2.58 is a resolvable 3-$(8,4,1)$ design.

In general, a given incidence structure may have several resolutions. An important special case in which a resolution is always unique is formed by the affine incidence structures.

Definition 2.61. *An incidence structure is* affine *if it admits a resolution and a positive integer μ such that any two blocks from different parallel classes are incident to precisely μ common points. The integer μ is called the* intersection parameter.

It follows directly from Definition 2.61 that an affine design has a unique resolution. Affine planes, that is, 2-$(n^2,n,1)$ designs, are indeed affine in the sense of Definition 2.61.

Theorem 2.62. *A* 2-$(n^2,n,1)$ *design is affine with intersection parameter* $\mu = 1$.

Proof. Consider a 2-$(n^2,n,1)$ design and define a binary relation on the blocks by setting $B_1 \sim B_2$ if the blocks B_1, B_2 are either equal or disjoint. We claim that $\sim$ is an equivalence relation whose equivalence classes form the parallel classes of a resolution.

It is obvious that $\sim$ is reflexive and symmetric. We argue by contradiction to establish transitivity. Suppose that $B_1 \sim B_2$, $B_2 \sim B_3$, $B_1 \neq B_2 \neq B_3 \neq B_1$, and $B_1 \not\sim B_3$. Thus, there exists a point x incident with B_1 and B_3. Since $\lambda = 1$, each point pair $\{x,y\}$ with $y \in B_2$ occurs in a unique block distinct from B_1, B_3. Thus, x is incident with $k+2 = n+2$ blocks, which is a contradiction since $r = n+1$.

Let B be a block in an equivalence class with j blocks. Clearly, $j \le n$. Since $\sim$ is an equivalence relation, the $b - j = n(n+1) - j$ blocks not in the equivalence class of B are incident with at least one point incident with B. There are exactly $(r-1)k = n^2$ such incidences, thus $j = n$. Since $\lambda = 1$, any two blocks are incident with at most one common point, implying $\mu = 1$. □

For a t-(v,k,λ) design to be resolvable an obvious necessary condition is that k divides v. Another necessary condition that involves the parameters v, b, r, k, λ is given in the following theorem [58].

Theorem 2.63 (Bose's condition). *Let $\mathcal{D}$ be a resolvable* 2-(v,k,λ) *design with $v > k$. Then, $b \ge v + r - 1$, or equivalently $r \ge k + \lambda$. Equality holds if and only if $\mathcal{D}$ is affine. In this case,*

$$v = s^2\mu, \quad k = s\mu, \quad \lambda = \frac{s\mu - 1}{s-1}, \quad r = \frac{s^2\mu - 1}{s-1}, \quad b = sr,$$

where μ is the intersection parameter of $\mathcal{D}$ and s is an integer.

Proof. See [36, Theorem II.8.7]. □

Example 2.64. A design with the parameters of an affine design need not be affine if the design is not resolvable. For example, the parameters 2-$(8, 4, 3)$ are the parameters of an affine design with $\mu = 2$, but three of the designs in Example 2.58 are not resolvable and hence not affine.

Example 2.65. Bose's condition and the requirement that k divide v are not sufficient to guarantee that a resolvable 2-(v, k, λ) design exists. Theorems 6.39 and 6.40 give further conditions under which a design is not resolvable. Further examples are provided by the nonexistence of certain projective planes – by the Bruck–Ryser–Chowla theorem or by computer search (Chap. 12) – which implies the nonexistence of affine planes by Theorem 2.120. A sporadic example is the nonexistence of resolvable 2-$(15, 5, 4)$ designs (see Table 6.13).

Only two families of affine resolvable 2-(v, k, λ) design are known; see [36, Examples II.8.9] for more details on the following theorem (Hadamard matrices will be considered in Sect. 2.4.3).

Theorem 2.66. *Affine resolvable* 2-$(s^2\mu, s\mu, \frac{s\mu-1}{s-1})$ *designs exist for* $s = q$ *and* $\mu = q^n$, *where* q *is a prime power and* n *is a nonnegative integer; and for* $s = 2$ *and* $\mu = n$ *whenever an Hadamard matrix of order* $4n$ *exists.*

Shrikhande [532] conjectured that these are the only values of s and μ for which such designs exist.

Definition 2.67. *Two resolutions,* $(\mathcal{X}, R)$ *and* $(\mathcal{Y}, S)$, *are* isomorphic *if there exists an isomorphism* $f = (f_P, f_B)$ *of* $\mathcal{X}$ *onto* $\mathcal{Y}$ *such that* $f_{\mathcal{B}}$ *maps the parallel classes in* R *onto parallel classes in* S. *Such an* f *is an* isomorphism *of* $(\mathcal{X}, R)$ *onto* $(\mathcal{Y}, S)$. *An isomorphism of a resolution onto itself is an* automorphism.

Example 2.68. There are 80 nonisomorphic STS(15), only four of which are resolvable. Three of the four resolvable STS(15) have two nonisomorphic resolutions each, and one has a unique resolution up to isomorphism [408]. Two nonisomorphic resolutions of an STS(15) over the point set $\{\mathtt{a}, \mathtt{b}, \ldots, \mathtt{o}\}$ are given below.

```
abc dhl ekn fio gjm      abc dhl ekn fio gjm
ade bhj clo fkm gin      ade bhj clo fkm gin
afg bmo chk djn eil      afg bln cij dko ehm
ahi bdf cmn ejo gkl      ahi bdf cmn ejo gkl
ajk bln cef dim gho      ajk bmo cdg eil fhn
alm beg cij dko fhn      alm bik cef djn gho
ano bik cdg ehm fjl      ano beg chk dim fjl
```

When $v = 2k$ a resolvable design has a unique resolution, since each block can then be completed to a parallel class in only one way.

2.3 Codes

The origin of coding theory is in engineering applications, but it soon turned out that this topic is also of purely mathematical interest. In particular, many of the basic mathematical problems of coding theory are related to construction of certain combinatorial objects.

A class of combinatorial objects often has alternative representations [116]; we will later see several ways in which codes are linked to designs. The reason why such connections should be investigated is that they provide the possibility of choosing a proper description for a particular study of the corresponding objects.

We start with a brief introduction to the theory of codes.

2.3.1 Preliminaries

The concept of *distance* is central in the study of codes. There are some obvious – in the sense that Euclidean distance, which is the distance concept of everyday language, fulfills these – requirements that a distance function should fulfill.

Definition 2.69. *A nonempty set* Ω *together with a mapping* $d : \Omega \times \Omega \to \{0, 1, \ldots\}$ *is a* (discrete) metric space *if the mapping* d*, called the* metric *or* distance function*, has the following properties for all* $x, y, z \in \Omega$*:*

1. $d(x, y) = 0$ *if and only if* $x = y$,
2. $d(x, y) = d(y, x)$,
3. $d(x, z) \leq d(x, y) + d(y, z)$.

In the sequel, we consider tuples over the set $Z_q = \{0, 1, \ldots, q-1\}$, called the *alphabet* – any alphabet with q symbols could actually have been chosen – and denote the set of all such tuples of length n by Z_q^n. The proof of the following lemma is straightforward and is omitted.

Lemma 2.70. *The mapping* $d_H : Z_q^n \times Z_q^n \to \{0, 1, \ldots\}$ *defined by*

$$d_H(\mathbf{x}, \mathbf{y}) = |\{i \in \{1, 2, \ldots n\} : x_i \neq y_i\}|$$

is a metric for Z_q^n.

Definition 2.71. *The metric space* (Z_q^n, d_H) *is called the* q*-ary* Hamming space *of dimension* n*. The corresponding metric* d_H *is called the* Hamming metric *or the* Hamming distance.

The tuples $\mathbf{x} \in Z_q^n$ can be interpreted as vectors in that space and are in the context of coding theory called *words*. The components of a word $(x_1, x_2, \ldots, x_n)$ are called *coordinates* and the values $x_i \in Z_q$ are called *coordinate values*. If there is no risk for confusion, we may simply write a word as $x_1 x_2 \cdots x_n$, cf. (1.1). The *Hamming weight*, or just *weight*, $\mathrm{wt}(\mathbf{x})$ is the number of nonzero coordinates in the word $\mathbf{x}$. The weight and distance are closely related, as $\mathrm{wt}(\mathbf{x}) = d_H(\mathbf{x}, \mathbf{0})$, where $\mathbf{0}$ is the all-zero word.

Definition 2.72. *A* q*-ary* (block) code C *of* length n *is a nonempty subset of* Z_q^n. *The words of* C *are called* codewords, *and the* cardinality *of this code is* $|C|$. *If all codewords have the same weight, then the code is said to be a* constant weight code.

Note that we define codes to be sets and not multisets. For some special types of codes it is necessary to consider multisets over Z_q^n, but such codes are not treated here. One may consider a space where different coordinates have different alphabets; codes in such a space are said to be *mixed*.

There are several common ways of manipulating a code.

puncturing Delete a coordinate of the code.
extending Add a new coordinate to the code. A common way of extending a binary code is to add a parity check bit to each codeword.
shortening Delete a coordinate and retain the codewords with a given value – often 0 – in the deleted coordinate.
lengthening The inverse of shortening. Add a coordinate and codewords. The new coordinate has a given value in the old codewords – often 0 – and other values in the new codewords.

If q is a prime power and we particularize the elements of the alphabet to be elements of the finite field [372, 409] of order q, thereby considering $(\mathbb{F}_q^n, d_H)$, we get a finite vector space. In this case we can use tools from linear algebra to manipulate linear codes.

Definition 2.73. *A code* $C \subseteq \mathbb{F}_q^n$ *is* linear *if it forms a subspace of the vector space* $\mathbb{F}_q^n$. *Otherwise* C *is* nonlinear.

Unless mentioned otherwise, the codes we study in this work are *unrestricted*, that is, either nonlinear or linear. Linear codes are also considered in their own right; an introduction to these is given in Sect. 2.3.3. Whenever unrestricted codes are considered, we assume without loss of generality that $C \subseteq Z_q^n$. There are a few central parameters related to codes.

Definition 2.74. *The* minimum (Hamming) distance *of a code* C *with at least two codewords is* $d(C) = \min\{d_H(\mathbf{x}, \mathbf{y}) : \mathbf{x}, \mathbf{y} \in C,\ \mathbf{x} \neq \mathbf{y}\}$.

Definition 2.75. *The* covering radius *of a code* C *is* $R(C) = \max\{d_H(\mathbf{x}, C) : \mathbf{x}$ *arbitrary*$\}$, *where* $d_H(\mathbf{x}, C) = \min\{d_H(\mathbf{x}, \mathbf{c}) : \mathbf{c} \in C\}$.

The minimum distance $d(C)$ of a code C is the most important parameter in the study of error-correcting and error-detecting codes because it measures the ability of the code to sustain transmission errors. Namely, if only words of C are used in the transmission of information, then up to $d(C) - 1$ bit errors in a word can be detected and up to $\lfloor (d(C) - 1)/2 \rfloor$ bit errors can be corrected (by correcting a received word to the closest codeword). For the practical significance of the covering radius, see [109]. When the minimum distance or the covering radius is studied, this is pointed out by talking about *error-correcting codes* and *covering codes*, respectively.

Codes can now be specified according to the parameters of the space, the minimum distance, and the covering radius.

Definition 2.76. *An $(n, M, d)_q R$ code is a q-ary code of length n, cardinality M, minimum distance at least d, and covering radius at most R. Either of the parameters d and R, or both, may be omitted; if $q = 2$, that parameter may also be omitted.*

Note that we in this definition give bounds for the minimum distance and covering radius, instead of exact values. Both of these forms occur in the literature; our choice simplifies the subsequent discussion of codes.

Example 2.77. The ternary code in (1.1) is a $(4, 9, 3)_3 1$ code.

An $(n, M, d)_2$ binary code with constant weight w is also called a *packing design* since the codewords can be viewed as w-subsets (blocks) of an n-set such that every t-subset of the n-set occurs in *at most* one block, where $t = w + 1 - \lfloor d/2 \rfloor$. We may in fact define even more general packing designs with every t-subset occurring in at most λ blocks, where λ is called the *index* of the packing design.

For covering codes of length n and constant weight w we have to specify the weight w_C of the words to be covered with covering radius R. Binary constant weight covering codes, or *covering designs*, can then be viewed as w-subsets (blocks) of an n-set of points such that every w_C-subset of the n-set intersects at least one block – or, more generally, at least λ blocks – in at least $\lceil (w + w_C - R)/2 \rceil$ points. Covering designs of this type are coined *B-coverings* in [552], and the term *covering design* usually refers to the case $R = w - w_C$; in this case every w_C-set of points occurs in at least one block.

Fundamental problems in coding theory are to determine, for fixed parameter values, the minimum cardinality of an error-correcting code and the minimum cardinality of a covering code:

$$\begin{aligned} A_q(n, d) &= \max\{M : \text{there exists an } (n, M, d)_q \text{ code}\}, \\ K_q(n, R) &= \min\{M : \text{there exists an } (n, M)_q R \text{ code}\}. \end{aligned}$$

One may also look at similar functions when both the values of d and R are prescribed [464], but such exceptional codes are not discussed here. Codes that attain the bound $A_q(n, d)$ are called called *optimal* error-correcting codes, and codes attaining $K_q(n, R)$ are called *optimal* covering codes. The *Hamming sphere* or *ball* of radius r around $\mathbf{x} \in Z_q^n$ is defined by

$$B_r(\mathbf{x}) = \{\mathbf{y} \in Z_q^n : d_H(\mathbf{x}, \mathbf{y}) \le r\}. \tag{2.12}$$

The following elementary bound is called the *sphere-covering bound* or the *Hamming bound*.

Theorem 2.78.

$$A_q(n, 2R+1) \le \frac{q^n}{\sum_{i=0}^{R}(q-1)^i\binom{n}{i}} \le K_q(n, R).$$

Proof. The Hamming spheres of radius R must contain all q^n words of the space if the code has covering radius (at least) R, and they must be non-overlapping if the code has minimum distance (at most) $2R+1$. The theorem now follows as

$$|B_R(\mathbf{x})| = \sum_{i=0}^{R}(q-1)^i\binom{n}{i}.$$

□

Through Hamming spheres it is convenient to introduce the concept of *index* for error-correcting and covering codes: such a code has index λ if every word in Z_q^n occurs in, respectively, at most or at least λ spheres (of prescribed radius). Codes with index $\lambda > 1$ are called *multiple packing codes* and *multiple covering codes*.

Codes that attain both bounds in Theorem 2.78 – then $A_q(n, 2R+1) = K_q(n, R)$ – are called *perfect*. Before leaving the general discussion of codes, we give an example that illustrates some of these basic properties and functions.

Example 2.79. Consider codes in Z_2^3. With $R = 1$, it follows from the sphere-covering bound that $A_2(3,3) \le 2 \le K_2(3,1)$. The $(3,2,3)_2 1$ code

$$C = \{000, 111\}$$

proves that $A_2(3,3) = K_2(3,1) = 2$. The code C is perfect.

Example 2.80. Consider codes in Z_2^4. With $R = 1$, the sphere-covering bound gives that $A_2(4,3) \le 16/5 \le K_2(4,1)$. Since there exist $(4,4)_2 1$ codes, one example being

$$\{0000, 0001, 1110, 1111\},$$

we have that $K_2(4,1) = 4$. It is not difficult to see that a $(4,3,3)_2$ code cannot be found. Since $(4,3,2)_2$ codes exists, such as

$$\{0000, 1111\},$$

we get that $A_2(4,1) = 2$.

2.3.2 Equidistant Codes

A code is *equidistant* if $d_H(\mathbf{x}, \mathbf{y}) = d(C)$ for all distinct $\mathbf{x}, \mathbf{y} \in C$. It turns out that certain equidistant optimal codes are related to resolutions of designs, this will be shown in Theorem 3.82. The *Plotkin bound* is of central importance in the study of these codes. This bound was proved by Plotkin [489] for the binary case. The q-ary generalization to be presented here is from [53]. As a preliminary step we require the following discrete analog of the Cauchy–Schwarz inequality:

Lemma 2.81. *Let* $\sum_{i=0}^{n-1} a_i = A$, *where the variables* a_i *are nonnegative integers. Then* $\sum_{i=0}^{n-1} a_i^2$ *attains its smallest value when* $a_i = \lfloor (A+i)/n \rfloor$.

Proof. We first show that the smallest value cannot be attained if there exist i and j such that $a_i - a_j \geq 2$. Namely, then we can substitute the values a_i and a_j by $a_i - 1$ and $a_j + 1$, respectively, to get an even smaller value, since $(a_i-1)^2 + (a_j+1)^2 = a_i^2 + a_j^2 + 2(a_j - a_i + 1) < a_i^2 + a_j^2$. The unique solution (up to permutation of the indices) to $\sum_{i=0}^{n-1} a_i = A$ with $|a_i - a_j| \leq 1$ for all i, j is $a_i = \lfloor (A+i)/n \rfloor$. □

Theorem 2.82 (Generalized Plotkin bound). *If there exists an* $(n, M, d)_q$ *code, then*

$$\binom{M}{2} d \leq n \sum_{i=0}^{q-2} \sum_{j=i+1}^{q-1} M_i M_j, \tag{2.13}$$

where $M_i = \lfloor (M+i)/q \rfloor$. *If equality holds, then the code is equidistant, and the distribution of the values in a coordinate is up to permutation uniquely given by the values of* M_i.

Proof. As in the proof of the binary case [489], we sum the distances between all ordered pairs of distinct codewords in two different ways. An arbitrary column contributes with $\sum_{i=0}^{q-1} m_i(M - m_i) = M \sum_{i=0}^{q-1} m_i - \sum_{i=0}^{q-1} m_i^2 = M^2 - \sum_{i=0}^{q-1} m_i^2$ to this sum, where m_i is the number of codewords with value i in this coordinate. Using Lemma 2.81, this expression attains its largest value when $m_i = M_i = \lfloor (M+i)/q \rfloor$, and the maximum is $2 \sum_{i=0}^{q-2} \sum_{j=i+1}^{q-1} M_i M_j$. Summing over all coordinates gives $2n \sum_{i=0}^{q-2} \sum_{j=i+1}^{q-1} M_i M_j$.

On the other hand, all pairwise distances between codewords are greater than or equal to d, and summing the distances in this way gives a total sum of at least $M(M-1)d$. The theorem is now proved by combining this lower bound and the previously obtained upper bound. □

Corollary 2.83 (Binary Plotkin bound). *If there exists an* $(n, M, d)_2$ *code with* $2d > n$, *then*

$$M \leq 2 \left\lfloor \frac{d}{2d-n} \right\rfloor.$$

Corollary 2.84. *If* q *divides* M *and the parameters of an* $(n, M, d)_q$ *code* C *satisfy*

$$\binom{M}{2} d = n \binom{q}{2} (M/q)^2, \tag{2.14}$$

then C *is equidistant and every coordinate value occurs exactly* $k = M/q$ *times in every coordinate of the code.*

A code with the property that every coordinate value occurs equally often in every coordinate is called *equireplicate*. The codes characterized by

Corollary 2.84 were first studied by Semakov and Zinov'ev [526], who called these ED_m-codes (equidistant with maximal distance). We follow [580] and call them *optimal equidistant* (OE) codes.

Example 2.85. The code in (1.1) is an $(4,9,3)_3$ OE code.

In the next theorem we will see that the word *optimal* in the name of the class of OE codes is justified, they are indeed optimal.

Theorem 2.86. *An OE code is optimal.*

Proof. Consider an $(n, M, d)_q$ OE code. Because the minimum distance does not decrease after deleting codewords, it suffices to show that an $(n, M+1, d)_q$ code cannot exist. Suppose, for the sake of contradiction, that such a code C exists. Then, we can delete any codeword $\mathbf{x} \in C$ and obtain a code C', which is an OE code by Corollary 2.84.

We will now carry out double counting for the sum of distances between x and the words in C'. Since C' is equireplicate, every coordinate contributes with $(q-1)M/q$ to the sum, so the total sum is $n(q-1)M/q$, which equals $(M-1)d$ by Corollary 2.84. On the other hand, all pairwise distances between $\mathbf{x}$ and the M words in C' must be at least d, so the total sum must be at least Md, and we have a contradiction. □

In general, equality in (2.13) is not sufficient to establish optimality. For example, the parameters of $(4,8,3)_3$ and $(4,9,3)_3$ codes (which exist, see Example 2.85) lead to equality in (2.13), whereas only codes with the latter parameters, which are OE codes, are optimal. One should be careful about the word order and not confuse OE (optimal equidistant) codes with equidistant optimal codes, which are equidistant and attain $A_q(n, d)$ (it is not difficult to see that the former class is a proper subclass of the latter class).

Given parameters that lead to equality in the Plotkin bound, it is in the general case a highly nontrivial task to find a corresponding equidistant code or to prove nonexistence. However, we are able to prove that nonexistence of a certain OE code implies nonexistence of a (not necessarily equidistant) code with the same parameters and one codeword less.

Theorem 2.87. *If q divides M and a putative $(n, M, d)_q$ OE code does not exist, then there is no $(n, M-1, d)_q$ code either.*

Proof. It is straightforward to verify that if q divides M and the parameters of a putative $(n, M, d)_q$ code give equality in the Plotkin bound, then the parameters of a putative $(n, M-1, d)_q$ code give equality in the Plotkin bound as well (carry out double counting of distances between a removed word and the rest of the codewords). Assume now that such an $(n, M-1, d)_q$ code C exists.

Consider a new word $\mathbf{x}$ that in each coordinate takes the (unique) value that occurs $M/q-1$ times in C. For an arbitrary $\mathbf{y} \in C$, consider the code

$C' = C \cup \{\mathbf{x}\} \setminus \{\mathbf{y}\}$. The codes C and C' have the same value distributions in all coordinates, so a double counting of distances between codewords will give the same sum for C' and C. But $d_H(\mathbf{y}, \mathbf{z}) = d$ for all $\mathbf{z} \in C \setminus \{\mathbf{y}\}$, so the total sum depends solely on $d_H(\mathbf{x}, \mathbf{y})$. Since the sum is the same for all choices of $\mathbf{y}$, $d_H(\mathbf{x}, \mathbf{y})$ is the same for all choices of $\mathbf{y}$. The code $C \cup \{\mathbf{x}\}$ is equireplicate so the sum of distances between codewords equals that of an OE code. But the distances $d_H(\mathbf{x}, \mathbf{y})$ for $\mathbf{y} \in C'$ must then be exactly d, otherwise the total distance sum would be smaller than or greater than $\binom{M}{2}d$. Hence, an $(n, M, d)_q$ code exists, and we have arrived at a contradiction. □

For appropriate parameters, the trick of studying value distributions in the proof of Theorem 2.87 can be used to get an $(n, M, d)_q$ OE code from an $(n, M-1, d)_q$ code attaining the Plotkin bound (and this can be done in a unique way).

In Theorem 3.82 a correspondence between OE codes and resolutions of BIBDs is presented.

The Plotkin bound for unrestricted codes has an analogous version for constant weight codes, presented by Johnson in [284] and strengthened in [388, p. 526].

Theorem 2.88 (Second Johnson bound). *If there exists an $(n, M, 2\delta)_2$ code with constant weight w, then*

$$(n-b)a(a-1) + ba(a+1) \leq (w-\delta)M(M-1), \tag{2.15}$$

where $a = \lfloor wM/n \rfloor$ and $b = wM - na$.

If equality holds in (2.15), then C is equidistant. Furthermore, the number of 1s is $a+1$ in b of the coordinates and a in the remaining $n-b$ coordinates. Johnson's original bound is slightly weaker.

Corollary 2.89. *If there exists an $(n, M, 2\delta)_2$ code with constant weight w, then*

$$M \leq \frac{\delta n}{w^2 - wn + \delta n} \tag{2.16}$$

provided that the denominator is positive.

Semakov and Zinov'ev [527] observed that if equality holds in (2.16), then the codewords of any such code define the rows of an incidence matrix of a 2-(v, k, λ) design with $v = M$, $k = wM/n$, $\lambda = w - \delta$, $b = n$, and $r = w$; and vice versa.

In a manner analogous to that of the proof of Theorem 2.87 (see also the subsequent remark), the second Johnson bound can be used to prove that in constructing a BIBD point by point along the lines of Sect. 6.1.1, one missing row can always be completed (in a unique way). A similar result – proved with a different technique – holds in constructing Steiner systems block by block [1, Theorem 13] (such methods are considered in Sects. 6.1.3 and 6.2.2).

2.3.3 Linear Codes

According to Definition 2.73, a linear code is a subspace of $\mathbb{F}_q^n$. We know from algebra that the size of such subspaces is a power of q, say q^k, where k is called the *dimension* of the code, and we may now simplify the notation of codes by expressing the dimension instead of the cardinality.

Definition 2.90. *An $[n,k,d]_qR$ code is a q-ary linear code of length n, dimension k, minimum distance at least d, and covering radius at least R, where either of the parameters d and R, or both, may be omitted. If $q = 2$, that parameter may also be omitted.*

The dimension of a given code C is expressed by $\dim(C)$. The value $r = n - k$ is called the *co-dimension* of the code, and the value k/n is the *rate* of the code. The *weight enumerator* of a binary $[n,k]$ code (the binary case is sufficient for our needs) is the polynomial

$$W(x,y) = \sum_{i=0}^{n} A_i x^{n-i} y^i,$$

where A_i is the number of codewords of weight i. If $A_i = 0$ whenever i is not divisible by 2, then the code is *even*, and if $A_i = 0$ whenever i is not divisible by 4, then the code is *doubly-even*.

As in the unrestricted case, we may also define functions related to possible parameters of linear codes. Among several reasonable functions and definitions of optimality, we here study

$$d_{\max}(n,k) = \max\{d : \text{there exists an } [n,k,d]_2 \text{ code}\}$$

and call an $[n,k,d_{\max}(n,k)]_2$ code *optimal*. Similar functions may be introduced for $q > 2$.

We will next discuss some basic properties of linear codes; see [388] or any introductory textbook for proofs of these results. Two types of matrices are closely related to linear codes.

Definition 2.91. *A $k \times n$ matrix $\mathbf{G}$ is a* generator matrix *for an $[n,k]_q$ linear code C if $\mathbf{uG}$ generates the codewords of C for $\mathbf{u} \in \mathbb{F}_q^k$.*

Definition 2.92. *An $(n-k) \times n$ matrix $\mathbf{H}$ is a* parity check matrix *for an $[n,k]_q$ linear code C if $\mathbf{Hx}^T = 0$ exactly when $\mathbf{x} \in C$.*

The vector $\mathbf{Hx}^T$ is called the *syndrome* of $\mathbf{x}$. Vectors with the same syndrome form *cosets* of the linear code.

Observe that row operations (row addition, scalar multiplication) on the matrices $\mathbf{G}$ and $\mathbf{H}$ do not affect the code to which they are related. If we permute columns, the code will be affected, but we then get an equivalent code; equivalence of linear codes is discussed in Sect. 7.3.1. These operations can now be used to transform a parity check matrix or a generator matrix into, for example, the forms $[\mathbf{A}\ \mathbf{I}]$ or $[\mathbf{I}\ \mathbf{B}]$. The following theorem – see [388, p. 5] for a proof – shows how a generator matrix with the latter form can be transformed into a parity check matrix for the same code (and conversely).

Theorem 2.93. *A linear code with generator matrix* $\mathbf{G} = [\mathbf{I}_k\ \mathbf{A}]$ *has parity check matrix* $\mathbf{H} = [-\mathbf{A}^T\ \mathbf{I}_{n-k}]$ *and vice versa.*

Example 2.94. The code in (1.1) is linear over $\mathbb{F}_3$, with generator matrix

$$\mathbf{G} = \begin{bmatrix} 1\,0\,1\,2 \\ 0\,1\,1\,1 \end{bmatrix}.$$

Out of several possible generator matrices, this one was chosen to be able to apply Theorem 2.93 to get the following parity check matrix:

$$\mathbf{H} = \begin{bmatrix} 2\,2\,1\,0 \\ 1\,2\,0\,1 \end{bmatrix}.$$

The parity check matrix is useful if one wants to check the minimum distance or the covering radius of a linear code. Proofs of the following two results can be found in [388, Chap. 1, Theorem 10] and [109, Theorem 2.1.9], respectively.

Theorem 2.95. *The minimum distance of a code with parity check matrix* $\mathbf{H}$ *is* d *if every* $d-1$ *columns of* $\mathbf{H}$ *are linearly independent and some* d *columns of* $\mathbf{H}$ *are linearly dependent.*

Theorem 2.96. *The covering radius of a code with parity check matrix* $\mathbf{H}$ *is the smallest integer* R *such that every vector in* $\mathbb{F}_q^r$ *can be obtained as a linear combination of at most* R *columns of* $\mathbf{H}$.

It is easy to verify that the code in Example 2.94 has minimum distance 3 and covering radius 1. Therefore, since $A_3(4,3) = K_3(4,1) = 9$, it is a perfect code (see the discussion after Theorem 2.78). In fact, it belongs to the class of perfect *Hamming codes* with parameters

$$[(q^m - 1)/(q-1), (q^m-1)/(q-1) - m, 3]_q 1,$$

where $m > 0$ and q is a prime power. Parity check matrices of Hamming codes can be obtained by taking as the column vectors a maximal set of non-zero vectors such that no vector is a multiple of another vector.

Example 2.97. A parity check matrix for the binary Hamming code of length 15 ($q = 2$, $m = 4$) is

$$\mathbf{H} = \begin{bmatrix} 0\,0\,0\,0\,0\,0\,0\,1\,1\,1\,1\,1\,1\,1\,1 \\ 0\,0\,0\,1\,1\,1\,1\,0\,0\,0\,0\,1\,1\,1\,1 \\ 0\,1\,1\,0\,0\,1\,1\,0\,0\,1\,1\,0\,0\,1\,1 \\ 1\,0\,1\,0\,1\,0\,1\,0\,1\,0\,1\,0\,1\,0\,1 \end{bmatrix}.$$

The (Euclidean) *inner product* of two codewords in $\mathbb{F}_q^n$, $\mathbf{c} = (c_1, c_2, \ldots, c_n)$ and $\mathbf{c}' = (c'_1, c'_2, \ldots, c'_n)$, is

$$\mathbf{c} \cdot \mathbf{c}' = c_1 c'_1 + c_2 c'_2 + \cdots + c_n c'_n,$$

with addition and multiplication carried out in $\mathbb{F}_q$. There are also other possible inner products, such as the Hermitian inner product, which is often preferred for $\mathbb{F}_4$ [497].

Definition 2.98. *Given a linear code C, its* dual *code is defined as $C^\perp = \{\mathbf{c}' : \mathbf{c} \cdot \mathbf{c}' = 0$ for all $\mathbf{c} \in C\}$. A linear code C is called* self-orthogonal *if $C \subseteq C^\perp$ and* self-dual *if $C = C^\perp$.*

Self-dual and self-orthogonal codes are surveyed in [497]. Using basic linear algebra, one may prove that the dual code is necessarily linear and that for codes of length n,

$$\dim(C^\perp) = n - \dim(C).$$

2.3.4 Equivalence of Codes

In the study of codes, there are several possible ways of defining which codes should be considered indistinguishable. For some background and motivation, we look at the basic engineering problem of sending information across various communication channels. In this informal consideration, we assume that the codes are unrestricted. One channel, the *binary memoryless channel*, is depicted in Fig. 2.10.

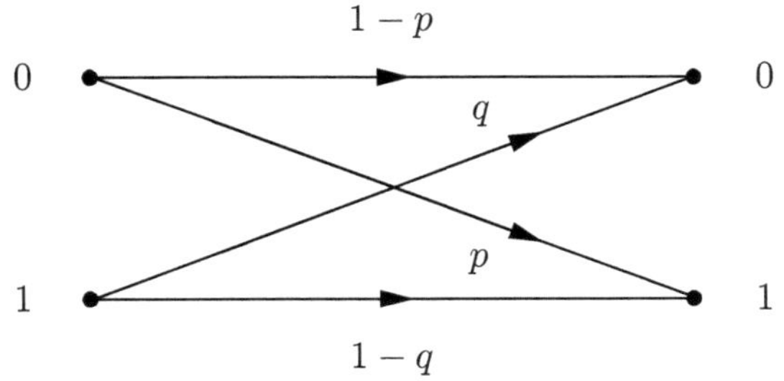

Fig. 2.10. The binary memoryless channel

In the binary channel, a transmitted 0 is corrupted with probability p and a transmitted 1 is corrupted with probability q. The channel is said to be memoryless as the probability of error is conditionally independent of previous transmissions and errors. With such a channel, it is obviously irrelevant, with respect to error probabilities, in what order the bits are transmitted, and this leads us to the following definition (for codes over any alphabets).

Definition 2.99. *Two codes are said to be* isomorphic *if one can be transformed into the other by a permutation of the coordinates in the codewords.*

In many practical situations, the error probabilities, p and q, take the same value. The binary memoryless channel with $p = q$ is called the *binary symmetric channel* (BSC). Then also the coordinate values may be permuted in a code without affecting the performance of the code.

Definition 2.100. *Two codes are said to be* equivalent *if one can be transformed into the other by a permutation of the coordinates in the codewords followed by permutations of the coordinate values, independently for each coordinate.*

Example 2.101. The $(4, 9, 3)_3$ codes

$$\begin{aligned} C &= \{0000, 0111, 0222, 1021, 1102, 1210, 2012, 2120, 2201\}, \\ C' &= \{0012, 0101, 0220, 1021, 1110, 1202, 2000, 2122, 2211\} \end{aligned}$$

are equivalent. To obtain C' from C, first permute the coordinates:

$$1 \mapsto 2, \quad 2 \mapsto 3, \quad 3 \mapsto 4, \quad 4 \mapsto 1.$$

Then exchange the coordinate values 0 and 1 in every coordinate but the last.

We will later see that this definition of equivalence coincides with a definition we arrive at after a more formal treatment. Actually, the main question is whether there are other transformations of the codes than those in Definition 2.100 that should be allowed. In a formal treatment, one possible way of approaching this issue is to regard as indistinguishable all codes of equal cardinality that have the same collection of pairwise codeword distances and thereby the same error-detecting capabilities (but they may differ with respect to the possibility of correcting errors, as we will see in Example 2.103).

Definition 2.102. *Two codes $C, C' \subseteq Z_q^n$ are said to be* isometric *if there exists a bijection $f : C \to C'$ satisfying $d_H(\mathbf{x}, \mathbf{y}) = d_H(f(\mathbf{x}), f(\mathbf{y}))$ for all $\mathbf{x}, \mathbf{y} \in C$.*

Only sporadic results on code isometry have been published in the literature [131, 543]. In particular, we observe that all equidistant codes of fixed length, cardinality, and minimum distance are isometric. The following example is in part from [131].

Example 2.103. The ternary codes

$$\begin{aligned} C_1 &= \{000, 011, 022\}, \\ C_2 &= \{000, 110, 011\} \end{aligned}$$

are equidistant with minimum distance 2 and are therefore isometric. They are, however, not equivalent: the code C_1 has a coordinate with only one value and any equivalence transformation maintains this property. The codes are distinguishable with respect to error correction. For each codeword in C_1, there are two words with one bit-error that are correctable (for example, if 000 is sent: 100 and 200), but for each codeword in C_2 three bit-errors are correctable (if 000 is sent: 200, 020, and 002).

2.4 More Combinatorial Objects

We conclude the introduction to combinatorial objects by presenting three families of objects that are closely related to designs and codes, and, moreover, link together several types of such objects.

2.4.1 Orthogonal Arrays

Orthogonal arrays are primarily used in designing experiments in statistics. They are treated in-depth in [257], which has been the source for some of the proofs in this section; several chapters of [116] are also devoted to these objects.

Definition 2.104. *An* orthogonal array *of* size N, *with* n constraints, q levels, strength t, *and* index λ, *denoted by* $\mathrm{OA}_\lambda(t,n,q)$, *is an* $n \times N$ *array with entries from* Z_q, *such that in every* $t \times N$ *subarray, every tuple in* Z_q^t *appears exactly* $\lambda = N/q^t$ *times.*

The parameters t and λ of $\mathrm{OA}_\lambda(t,n,q)$ may be omitted and are then understood to be 2 and 1, respectively. It is important to observe that many other notations, in particular $\mathrm{OA}(N,n,q,t)$, have been used instead of $\mathrm{OA}_\lambda(t,n,q)$ in the literature; cf. [257, p. 2].

Example 2.105. One example of an $\mathrm{OA}(2,4,3)$ is

$$\begin{bmatrix} 0&0&0&1&1&1&2&2&2 \\ 0&1&2&0&1&2&0&1&2 \\ 0&1&2&1&2&0&2&0&1 \\ 0&1&2&2&0&1&1&2&0 \end{bmatrix}.$$

You have actually seen this structure earlier, namely the columns of this array are the codewords of the ternary Hamming code in (1.1).

As Example 2.105 indicates, one may as well view an orthogonal array as a code $C \subseteq Z_q^n$, where the columns of the array are the codewords. Codes corresponding in this way to orthogonal arrays with $\lambda = 1$ are called *maximum distance separable* (MDS) and have been extensively studied in coding theory [388, Chap. 11].

Theorem 2.106. *An* $(n, q^t, d)_q$ *code is MDS if and only if it has minimum distance* $d = n - t + 1$.

Proof. Consider an $(n, q^t, d)_q$ MDS code. Since this is an $\mathrm{OA}_1(t, n, q)$, which is also an $\mathrm{OA}_q(t-1, n, q)$, there are codewords that coincide in $t-1$ coordinates, so the minimum distance cannot exceed $n-(t-1) = n-t+1$. If the minimum distance is less than $n - t + 1$, then there are codewords that coincide in (at least) t coordinates, implying that there are t-tuples in these coordinates that occur in more than one codeword. But this contradicts the fact that we have an $\mathrm{OA}_1(t, n, q)$.

In the other direction, any two words of an $(n, q^t, n-t+1)_q$ code coincide in at most $t - 1$ coordinates, so all codewords differ in any given set of t coordinates. Since the code has q^t codewords, it is MDS. □

The connection between orthogonal arrays and codes also motivates the definition of isomorphic orthogonal arrays, which follows Definition 2.100.

Definition 2.107. *Two orthogonal arrays are* isomorphic *if one can be transformed into the other by row and column permutations followed by independent value permutations in the rows.*

Orthogonal arrays of strength 2 and index 1 correspond to transversal designs.

Theorem 2.108. *An orthogonal array* $\mathrm{OA}(k, n)$ *can be transformed into a transversal design* $\mathrm{TD}(k, n)$ *and vice versa.*

Proof. Given an $\mathrm{OA}(k, n)$, let the points of the transversal design to be constructed be the ordered pairs (i, v), where $v \in Z_n$ ranges over the elements of a group and $1 \le i \le k$ ranges over the groups. Moreover, let each column j of the orthogonal array define a block B_j of the transversal design: B_j is incident with (i, v) if and only if the entry at row i, column j of the orthogonal array contains $v \in Z_n$.

In the converse transformation, each group of the transversal design defines one row of the orthogonal array, and each block defines one column. □

Orthogonal arrays have subarrays with smaller strength.

Theorem 2.109. *If there exists an* $\mathrm{OA}_\lambda(t+1, k+1, q)$, *then there exists an* $\mathrm{OA}_\lambda(t, k, q)$.

Proof. Given an $\mathrm{OA}_\lambda(t+1, k+1, q)$, choose a row i and a value $v \in Z_q$. Delete the row i and all columns that do not contain v in row i to get an $\mathrm{OA}_\lambda(t, k, q)$. □

For orthogonal arrays with $q = 2$ levels, the following theorem connects arrays with even and odd strength; the proof follows that of [257, Theorem 2.24].

Theorem 2.110. *There exists an* $\mathrm{OA}_\lambda(2t,k,2)$ *if and only if there exists an* $\mathrm{OA}_\lambda(2t+1,k+1,2)$.

Proof. Given an $\mathrm{OA}_\lambda(2t+1,k+1,2)$, there is an $\mathrm{OA}_\lambda(2t,k,2)$ by Theorem 2.109.

Given an $\mathrm{OA}_\lambda(2t,k,2)$, we form an $\mathrm{OA}_\lambda(2t+1,k+1,2)$ by adding a new row of 0s, and juxtaposing (that is, putting side by side) this array and its complement (obtained by replacing the 0s by 1s and the 1s by 0s). To check the parameters of the new array, we have to look at sets of $2t+1$ rows. Obviously, if the new row is among the $2t+1$ rows, the result is clear, so it suffices to consider sets of $2t+1$ rows that do not contain the new row. We denote by $n(\mathbf{v})$ the number of columns of the original $\mathrm{OA}_\lambda(2t,k,2)$ whose $(2t+1)$-tuple in these rows is $\mathbf{v}$. The number of columns in the constructed $\mathrm{OA}_\lambda(2t+1,k+1,2)$ that have a $(2t+1)$-tuple $\mathbf{v}$ in the chosen rows then equals $n(\mathbf{v})+n(\bar{\mathbf{v}})$.

For arbitrary $(2t+1)$-tuples $\mathbf{v}$ and $\mathbf{v}'$ with $d_H(\mathbf{v},\mathbf{v}')=1$, we get $n(\mathbf{v})+n(\mathbf{v}')=\lambda$ since the original array has strength $2t$ and index λ. If $d_H(\mathbf{v},\mathbf{v}')=2$, then there there is a $\mathbf{v}''$ such that $d_H(\mathbf{v},\mathbf{v}'')=d_H(\mathbf{v}',\mathbf{v}'')=1$, and

$$n(\mathbf{v})-n(\mathbf{v}')=n(\mathbf{v})+n(\mathbf{v}'')-(n(\mathbf{v}')+n(\mathbf{v}''))=0.$$

By induction, $n(\mathbf{v})=n(\mathbf{v}')$ if $d_H(\mathbf{v},\mathbf{v}')$ is even, and $n(\mathbf{v})+n(\mathbf{v}')=\lambda$ if $d_H(\mathbf{v},\mathbf{v}')$ is odd. Since $d_H(\mathbf{v},\bar{\mathbf{v}})=2t+1$ is odd, we have $n(\mathbf{v})+n(\bar{\mathbf{v}})=\lambda$. □

2.4.2 Latin Squares

Certain families of orthogonal arrays are better known under other names; Latin squares form one such family [149].

Definition 2.111. *A* Latin square *of* side *– or* order *–* n *is an* $n\times n$ *array with elements from* Z_n *such that each value occurs exactly once in each row and column. For* $k\le n$*, a* $k\times n$ Latin rectangle *is an array with elements from* Z_n *such that each value occurs exactly once in each row and at most once in each column.*

The element of a Latin square L in row i and column j is denoted by $L(i,j)$. Throughout the text we index the rows and columns with elements from $\{0,1,\ldots,n-1\}$, viewed as elements of Z_n or $\mathbb{Z}_n$, and call a Latin square *reduced* if $L(i,0)=L(0,i)=i$ for all such i. It is easy to construct Latin squares of arbitrary side n; for example, take $L(i,j)=i+j$ for $i,j\in\mathbb{Z}_n$. Latin squares have connections to a variety of other combinatorial objects, cf. [116, Part II]. The following connection to orthogonal arrays suffices for our present purposes.

Theorem 2.112. *A Latin square of side* n *can be transformed into an orthogonal array* $\mathrm{OA}(3,n)$ *and vice versa.*

Proof. From a Latin square of side n, construct on OA$(3,n)$ with columns $(i,j,L(i,j))^T$ for $i,j \in Z_n$. To construct a Latin square from an orthogonal array, proceed in the opposite direction. □

Example 2.113. The first three rows of the orthogonal array in Example 2.105 form an OA$(3,3)$. By Theorem 2.112, this can be transformed into a Latin square of side 3.

0	1	2
1	2	0
2	0	1

There are several definitions concerning indistinguishable Latin squares. We follow [415] for the terminology – see also [117] – and utilize the connection to orthogonal arrays given by (the proof of) Theorem 2.112 for conveniently introducing these concepts.

Definition 2.114. *Two Latin squares are said to be in the same*

1. main class *if the corresponding orthogonal arrays are isomorphic,*
2. type *if the corresponding orthogonal arrays can be mapped onto each other using a permutation of the first two rows, value permutations in the rows, and a permutation of the columns,*
3. isotopy class *if the corresponding orthogonal arrays can be mapped onto each other using value permutations in the rows and a permutation of the columns,*
4. isomorphism class *if the corresponding orthogonal arrays can be mapped onto each other using a value permutation that is the same for all rows and a permutation of the columns.*

Also OA(k,n) with $k > 3$ are related to Latin squares. We require a preliminary definition.

Definition 2.115. *Two Latin squares, L_1 and L_2, are said to be* orthogonal *if for every pair $(s_1,s_2) \in Z_n^2$ the system*

$$L_1(x,y) = s_1, \quad L_2(x,y) = s_2$$

has exactly one solution (x,y).

A set of pairwise orthogonal Latin squares is said to be *mutually orthogonal*, and we then say that we have a set of MOLS (mutually orthogonal Latin squares). The following theorem generalizes Theorem 2.112.

Theorem 2.116. *A set of k MOLS of side n can be transformed into an* OA$(k+2,n)$ *and vice versa.*

Proof. Denote the k MOLS by $L_1, L_2, \ldots, L_k$. Then construct an orthogonal array with columns $(i, j, L_1(i,j), L_2(i,j), \ldots, L_k(i,j))^T$ for $i, j \in Z_n$. In the opposite direction, k MOLS are obtained from an $\mathrm{OA}(k+2, n)$. □

The correspondence given in the proof of Theorem 2.116 is needed in the following definition.

Definition 2.117. *Two sets of MOLS are* isomorphic *if the corresponding orthogonal arrays are isomorphic.*

Example 2.118. By Theorem 2.116, the OA(4, 3) in Example 2.105 can be transformed into two MOLS of side 3.

0	1	2
1	2	0
2	0	1

0	1	2
2	0	1
1	2	0

How many MOLS of side n can there be (more precisely, what is the maximum k, such that an $\mathrm{OA}(k, n)$ exists)? An upper bound on this number is given by the following theorem.

Theorem 2.119. *There are at most $n - 1$ MOLS of side n.*

Proof. Given a set of MOLS, these Latin squares still form a MOLS after a permutation of the values in Z_n in any one Latin square. Therefore, we may transform these into a set of MOLS such that the first row of every square is $(0, 1, \ldots, n-1)$. For any two distinct squares in this set, the value pair $(1, 1)$ occurs among the pairs in the first row, so the 1 in the first column must occur in different rows (distinct from the first row, which contains a 0 in that position). By the pigeonhole principle, the total number of MOLS is at most $n - 1$. □

A set of $n - 1$ MOLS of side n is called a *complete* set of MOLS. We finish the discussion of Latin squares by connecting them to certain affine and projective designs, defined in Sect. 2.2.5.

Theorem 2.120. *Given one of the following objects, it can be transformed into the others:*

1. *a complete set of MOLS of side n,*
2. *an* $\mathrm{OA}(n+1, n)$*,*
3. *an* $(n+1, n^2, n)_n$ *OE code,*
4. *an affine plane of order n,*
5. *a projective plane of order n.*

Proof. (1) $\leftrightarrow$ (2): This follows immediately from Theorem 2.116.

(2) $\rightarrow$ (3): By the definition of an $\mathrm{OA}(n+1, n)$, two columns can agree in at most one row; otherwise there would exist a $2 \times n^2$ subarray that would contain

a pair of values more than once. Hence, the columns of such an array can be viewed as codewords of a code with minimum distance n, an $(n+1, n^2, n)_n$ code. By the Plotkin bound (Corollary 2.84) such a code is equidistant, which implies that any two distinct columns indeed agree in exactly one row.

(3) $\rightarrow$ (2): Since an $(n+1, n^2, n)_n$ OE code has minimum distance n, two codewords cannot agree in two coordinates. Therefore, viewing the codewords as columns of an $(n+1) \times n^2$ matrix, for any pair of distinct rows, each of the n^2 possible pairs of values must occur exactly once.

(3) $\rightarrow$ (4): Associate a unique point with each codeword of an $(n+1, n^2, n)_n$ OE code. Now each coordinate of the code defines a parallel class: two points are incident with the same block if and only if the values that occur in these codewords agree. From the definition of an OE code it follows that every value occurs n times in every coordinate, so every block in every parallel class has size n.

To establish that these $n+1$ parallel classes together define an affine plane, it suffices to prove that every pair of distinct points occurs together in a unique block. In other words, we must prove that every two distinct codewords agree in exactly one coordinate. But this is also a property of an $(n+1, n^2, n)_n$ OE code.

(4) $\rightarrow$ (2): An affine plane of order n always admits a resolution with $n+1$ parallel classes by Theorem 2.62. For every parallel class, label each block in the parallel class with an element of Z_n so that no two blocks in the same parallel class get the same label. Define an $(n+1) \times n^2$ array so that each parallel class in the affine plane corresponds to a row and each point in the plane corresponds to a column. The array element at row i, column j is the label of the block in which j occurs in parallel class i. This results in an $\mathrm{OA}(n+1, n)$ because in an affine plane any two blocks from different parallel classes have exactly one point in common (Theorem 2.62).

(4) $\rightarrow$ (5): An affine plane is quasi-residual and is therefore embeddable by Theorem 2.54.

(5) $\rightarrow$ (4): A residual of a projective plane is an affine plane (Definition 2.46). □

Theorem 2.121. *An affine resolvable* 2-(v, k, λ) *design with the auxiliary parameters* s *and* μ *introduced in Theorem* 2.63 *can be transformed into an* $\mathrm{OA}_\mu(\frac{s^2\mu-1}{s-1}, s)$ *and vice versa.*

Proof. We construct an orthogonal array by taking one row for each parallel class of the 2-(v, k, λ) design and one column for each point. Within a parallel class, the blocks are labeled with elements from Z_s so that no two blocks get the same label. The array element at row i, column j is the label of the block in which j occurs in parallel class i. It is straightforward to verify the parameters of the orthogonal array obtained. In particular, its index is given by the intersection parameter μ. Reversion of the construction proves implication in the opposite direction. □

2.4.3 Hadamard Matrices

Hadamard matrices originate from the problem of maximizing the determinant among matrices with entries −1 and 1; see [525] for a survey.

Definition 2.122. *An*[1] Hadamard matrix *of order* n *is an* $n \times n$ *matrix with entries* −1 *and* 1 *whose rows and columns are pairwise orthogonal, that is, have inner product* 0.

Theorem 2.123. *Hadamard matrices of order* n *exist only if* $n = 1$, $n = 2$, *or* $n \equiv 0 \pmod 4$.

Proof. Let $n \geq 3$. We may clearly multiply any column of the matrix by −1 and the result is still an Hadamard matrix. Thus, without loss of generality we may assume that the first row of the matrix consists only of 1s. Divide the columns of the matrix into four types depending on whether the values in the second and third row are $(1,1)$, $(1,-1)$, $(-1,1)$, or $(-1,-1)$. Denote by a, b, c, d the number of columns of respective type. From the inner product requirement among the first three rows we obtain

$$a + b - c - d = 0, \quad a - b + c - d = 0, \quad a - b - c + d = 0,$$

which implies $a = b = c = d$. The claim follows because $a + b + c + d = n$. □

Sufficiency of the condition in Theorem 2.123 is a longstanding open problem, which has defied many serious attempts. The smallest unresolved instance is $n = 668$ at the time of writing [312].

Conjecture 2.124. Hadamard matrices of order n exist if and only if $n = 1$, $n = 2$, or $n \equiv 0 \pmod 4$.

If we replace the entries that are −1 with 0, we may view an Hadamard matrix of order n as a binary equidistant $(n, n, n/2)_2$ code (in fact, we will later see that it may also be viewed as a binary equidistant $(n-1, n, n/2)_2$ code). This code obviously has the equivalence transformations of any unrestricted code, Definition 2.100. Moreover, one may observe that by transforming a codeword of such a code by transposing the coordinate values in all its coordinates, it remains a binary equidistant $(n, n, n/2)_2$ code. Although this transformation is not related to the isometry of the space Z_2^n, it is included in defining Hadamard equivalence. With the matrix formulation, this gives the following definition.

Definition 2.125. *Two Hadamard matrices are equivalent if one can be transformed into the other by row and column permutations followed by negations of rows and columns.*

[1] In French pronunciation the letter H in Hadamard is silent.

Using the transformations of Definition 2.125, any Hadamard matrix may be transformed into a matrix whose first row and column consist entirely of 1s. Such a matrix is said to be *normalized*.

Example 2.126. The Hadamard matrices of orders 1, 2, 4, and 8 are unique up to equivalence (we use the standard convention of writing -1 as $-$):

$$[1], \quad \begin{bmatrix} 1 & 1 \\ 1 & - \end{bmatrix}, \quad \begin{bmatrix} 1 & 1 & 1 & 1 \\ 1 & 1 & - & - \\ 1 & - & 1 & - \\ 1 & - & - & 1 \end{bmatrix}, \quad \begin{bmatrix} 1 & 1 & 1 & 1 & 1 & 1 & 1 & 1 \\ 1 & 1 & - & - & 1 & 1 & - & - \\ 1 & - & 1 & - & 1 & - & 1 & - \\ 1 & - & - & 1 & 1 & - & - & 1 \\ 1 & 1 & 1 & 1 & - & - & - & - \\ 1 & 1 & - & - & - & - & 1 & 1 \\ 1 & - & 1 & - & - & 1 & - & 1 \\ 1 & - & - & 1 & - & 1 & 1 & - \end{bmatrix}.$$

Hadamard matrices are directly related to several types of codes and designs.

Theorem 2.127. *Given one of the following objects, it can be transformed into the others:*

1. *an Hadamard matrix of order* $4n$,
2. *a* 2-$(4n-1, 2n-1, n-1)$ *square design,*
3. *a* $(4n-1, 4n, 2n)_2$ *OE code,*
4. *a resolvable* 2-$(4n, 2n, 2n-1)$ *design,*
5. *a resolvable* 3-$(4n, 2n, n-1)$ *design,*
6. *an* $\mathrm{OA}_n(4n-1, 2)$.

Proof. It is straightforward to verify that the following transformations give the desired objects:

(1) → (2): Normalize an Hadamard matrix and delete the first row and column. Then replace all entries -1 by 0 to get an incidence matrix of the design.

(2) → (1): Take an incidence matrix of a design, replace all entries 0 by -1, and add one column of 1s and one row of 1s to obtain an Hadamard matrix.

(1) → (3): Negate rows in an Hadamard matrix if necessary so that the first column consists only of 1s. Delete the first column and replace all entries -1 by 0. Every row of the matrix now defines a codeword.

(3) → (1): Place the codewords as rows of a matrix, replace all entries 0 by -1, and add a column of 1s to obtain an Hadamard matrix.

(3) ↔ (4): Follows from Theorem 3.82.

(4) ↔ (5): No transformation is required. A 3-$(4n, 2n, n-1)$ design is by Corollary 2.39 a 2-$(4n, 2n, 2n-1)$ design, and a 2-$(4n, 2n, 2n-1)$ design is by Theorem 2.59 a 3-$(4n, 2n, n-1)$ design.

(1) → (6): Negate columns in an Hadamard matrix if necessary so that the first row consists only of 1s. Delete the first row and replace all entries -1 by 0.

(6) → (1): Add a row of 1s and replace all entries 0 by -1 in the orthogonal array to obtain an Hadamard matrix. □

As we saw in Sect. 2.2.5, the designs in items 2 and 5 of Theorem 2.127 are called Hadamard designs and Hadamard 3-designs, respectively. Actually, Norman [445] showed that item 5 in the list can be somewhat relaxed.

Theorem 2.128. *Any* 3-$(4n, 2n, n-1)$ *design is resolvable.*

Note that the number of nonisomorphic objects of various types in Theorem 2.127 may differ; cf. [616] and Sect. 3.2.

3

Representations and Isomorphism

A quick browse through the *Handbook of Combinatorics* [217] or *The CRC Handbook of Combinatorial Designs* [116] indicates that, perhaps more than any other discipline in contemporary mathematics, combinatorics is characterized by the fact that the objects of interest can be represented in a large number of different but nevertheless equivalent ways. Typically there is no single best representation for a particular object; each representation has its advantages and drawbacks. This is especially true when we are designing a classification algorithm for objects of a given type.

For reasons of efficiency and practicality it is often the case that a classification algorithm employs multiple representations for the objects under consideration. For example, we construct the objects of interest using one representation – say, we construct resolutions of BIBDs by representing them as equidistant codes – but carry out isomorphism computations using a different representation – for example, we represent an equidistant code as a particular type of colored graph and perform the isomorphism computations on the graph. To employ multiple representations we must understand how the isomorphism classes of different types of objects are related. A look at Theorems 2.120 and 2.127 in Chap. 2 should convince the reader that this is not always straightforward.

To study representations and isomorphism, we require an appropriate framework on which to base the study. At a very general level, isomorphism is simply an equivalence relation on the set of objects being studied. In practice, however, the isomorphism relation between objects usually has more structure than an arbitrary equivalence relation. As demonstrated by the families of objects encountered in Chap. 2, in most cases isomorphism is defined through the existence of an appropriate bijection (an isomorphism) substantiating that two given objects have the same structure of interest.

There are at least two possible theoretical frameworks that enable an abstract study of representations and isomorphism in this setting; namely, group actions and category theory, to be discussed in Sects. 3.1 and 3.2, respectively. Group actions provide a finite framework that is more accessible and better suited for algorithms and computation. Also certain notions of isomorphism – such as those associated with codes – are arguably best defined through group actions to enable a more detailed study. Category theory generalizes the framework given by group actions and is more suited for the mathematical study of representations and the relationships between objects of different types, such as those occurring in Theorems 2.120 and 2.127.

Once we have the necessary general concepts available, we proceed to discuss isomorphism computations. Our discussion in Sect. 3.3 occurs on two levels. On one hand, we adopt an abstract viewpoint to isomorphism through categories and group actions with the aim of emphasizing the common underlying principles and computational issues. On the other hand, we want to illustrate how to perform isomorphism computations in practice. Here the main technique that we employ is to transform an object into a colored graph and then to solve the relevant problem using a graph-based algorithm. Actual isomorphism algorithms are not discussed in detail until Sect. 5.6, where we have available a sufficient algorithmic and conceptual background, namely backtrack search and search trees from Chap. 4 together with certain permutation group algorithms from Sect. 5.5.

3.1 Finite Groups and Group Actions

Any serious discussion of equivalence and isomorphism relies on concepts of group theory, the basics of which we review here. For a thorough introduction to group theory we recommend [275, 509]. Permutation groups are discussed in detail in [88, 163, 607] and group actions in [308].

Throughout this book all groups are finite.

Definition 3.1. *A* group *consists of a finite set G and a map $\cdot : G \times G \to G$ that has the following three properties:*

1. *for all $g_1, g_2, g_3 \in G$, we have $(g_1 \cdot g_2) \cdot g_3 = g_1 \cdot (g_2 \cdot g_3)$,*
2. *there exists an element $1_G \in G$ such that $g \cdot 1_G = 1_G \cdot g = g$ for all $g \in G$,*
3. *for all $g \in G$, there exists an element $g^{-1} \in G$ such that $g \cdot g^{-1} = g^{-1} \cdot g = 1_G$.*

Example 3.2. Let $G = \{1, a, b, c, d, e\}$ and define the group operation $\cdot$ by the following table, where the entry at row x, column y gives $x \cdot y$.

	1	a	b	c	d	e
1	1	a	b	c	d	e
a	a	b	1	d	e	c
b	b	1	a	e	c	d
c	c	e	d	1	b	a
d	d	c	e	a	1	b
e	e	d	c	b	a	1

It follows from the properties required from a group that the *identity element* 1_G is unique. Similarly, for every $g \in G$, the *inverse* g^{-1} is unique. We write simply 1 for the identity of G if the group is clear from the context and there is no danger of confusion with the integer 1. The *order* of a group is $|G|$.

Example 3.3. In Example 3.2 the element 1 is the identity element; the inverses are $1^{-1} = 1$, $a^{-1} = b$, $b^{-1} = a$, $c^{-1} = c$, $d^{-1} = d$, and $e^{-1} = e$.

Let G be a group. A subset $H \subseteq G$ is a *subgroup* of G if the group operation of G restricted to H makes H a group. We write $H \leq G$ to indicate that H is a subgroup of G. A subgroup $H \leq G$ is a *normal* subgroup of G if for all $g \in G$ and $h \in H$ it holds that $ghg^{-1} \in H$.

Example 3.4. The subgroups of the group $\{1, a, b, c, d, e\}$ in Example 3.2 are $\{1, a, b, c, d, e\}$, $\{1, a, b\}$, $\{1, c\}$, $\{1, d\}$, $\{1, e\}$, and $\{1\}$. The first, second, and last subgroup are normal subgroups.

For a subset $S \subseteq G$, the set of the intersection of all subgroups $H \leq G$ satisfying $S \subseteq H$ is a subgroup of G and is called the group *generated* by S. We write $\langle S \rangle$ for the subgroup of G generated by S. Equivalently, $\langle S \rangle$ is the subgroup of G that consists of all $g \in G$ that can be obtained as a product $g = g_1 \cdot g_2 \cdot \cdots \cdot g_n$ with $g_i \in S \cup \{1\}$, $n \geq 1$.

Example 3.5. In Example 3.2 we have

$$\langle \{a, c\} \rangle = \{1, a, b, c, d, e\}, \quad \langle \{a\} \rangle = \{1, a, b\}, \quad \langle \{c\} \rangle = \{1, c\}.$$

Let G be a group and let $H \leq G$. A *left coset* of H in G is a set of the form $gH = \{gh : h \in H\}$ for some $g \in G$. Similarly, a *right coset* of H in G is a set of the form $Hg = \{hg : h \in H\}$ for some $g \in G$. We write G/H (respectively, $H\backslash G$) for the set of all left (right) cosets of H in G.

Example 3.6. In Example 3.2 we have

$$\begin{aligned}
\{1, a, b, c, d, e\}/\{1, a, b\} &= \{\{1, a, b\}, \{c, d, e\}\}, \\
\{1, a, b\}\backslash\{1, a, b, c, d, e\} &= \{\{1, a, b\}, \{c, d, e\}\}, \\
\{1, a, b, c, d, e\}/\{1, c\} &= \{\{1, c\}, \{a, d\}, \{b, e\}\}, \\
\{1, c\}\backslash\{1, a, b, c, d, e\} &= \{\{1, c\}, \{a, e\}, \{b, d\}\}.
\end{aligned}$$

Theorem 3.7 (Lagrange). *Let G be a group and let $H \leq G$. Then, the set G/H (respectively, $H\backslash G$) is a partition of G. Moreover, $|G/H| = |H\backslash G| = |G|/|H|$.*

Proof. We work with left cosets only – the proof for right cosets is analogous. To establish the first claim, it suffices to prove that two left cosets, g_1H and g_2H, are either equal or disjoint. Suppose $g_1H \cap g_2H \neq \emptyset$, that is, there exist $h_1, h_2 \in H$ such that $g_1 h_1 = g_2 h_2$. For an arbitrary $g_1 h \in g_1 H$ we thus have $g_1 h = g_2 h_2 h_1^{-1} h \in g_2 H$. Conversely, for $g_2 h \in g_2 H$ we have $g_2 h = g_1 h_1 h_2^{-1} h \in g_1 H$. Thus, $g_1 H = g_2 H$. To establish the latter claim, observe that for an arbitrary $gH \in G/H$ the map $h \mapsto gh$ defines a bijection from H onto gH. Thus, $|H| = |gH|$ because G is finite. The claim now follows because G/H is a partition of G. □

An immediate corollary is that two elements $g_1, g_2 \in G$ are in the same left (right) coset of H in G if and only if $g_1^{-1} g_2 \in H$ (respectively, $g_1 g_2^{-1} \in H$).

A *left (right) transversal* for H in G is a subset of G that contains exactly one element from every left (right) coset of H in G.

Let G and H be groups. A function $f : G \to H$ is a *(group) homomorphism* if $f(g_1 \cdot g_2) = f(g_1) \cdot f(g_2)$ for all $g_1, g_2 \in G$. The *kernel* of a homomorphism is the normal subgroup $\ker f = \{g \in G : f(g) = 1\}$.

Example 3.8. In Example 3.2 let

$$f(1) = 1, \quad f(a) = 1, \quad f(b) = 1, \quad f(c) = c, \quad f(d) = c, \quad f(e) = c.$$

This defines a group homomorphism from $\{1, a, b, c, d, e\}$ onto the subgroup $\{1, c\}$ with $\ker f = \{1, a, b\}$.

A bijective homomorphism is an *isomorphism*. An isomorphism of G onto itself is an *automorphism*. The groups G and H are *isomorphic* if there exists an isomorphism from G to H.

3.1.1 Permutation Groups

Permutation groups constitute the most important family of groups encountered in this book.

Let Ω be a finite nonempty set. A *permutation* of Ω is a bijection of Ω onto itself. For a permutation g of Ω, we write $g(x)$ for the image of $x \in \Omega$ under g. The *product* or *composition* $g \circ h$ of two permutations g, h of Ω is the permutation of Ω defined for all $x \in \Omega$ by

$$(g \circ h)(x) = g(h(x)). \tag{3.1}$$

In practice we often omit the composition operator $\circ$ to lighten the notation and write simply gh.

Throughout this book we use the "right to left" order of composition given by (3.1), but the reader should be aware that many authors follow the alternative "left to right" order of composition, in which case it is customary to use either "left to right" notation "$x(gh) = (xg)h$" or exponential notation "$x^{gh} = (x^g)^h$" to denote the image of x.

Definition 3.9. *The group formed by the set of all permutations of Ω with composition as the group operation is called the* symmetric group *on Ω and is denoted by* Sym(Ω). *A* permutation group *on Ω is a subgroup of* Sym(Ω).

The *degree* of a permutation group $G \leq \mathrm{Sym}(\Omega)$ is $|\Omega|$. Whenever the set Ω is clear from the context we abbreviate $\mathrm{Sym}(\Omega)$ to S_n, where n is the degree of the group.

A permutation $g \in \mathrm{Sym}(\Omega)$ *moves* an element $x \in \Omega$ if $g(x) \neq x$; otherwise x is *fixed* by g. The *identity* permutation is the permutation that fixes all $x \in \Omega$. We write ϵ_Ω – or simply ϵ – for the identity permutation on Ω. Two permutations are *disjoint* if every point moved by one is fixed by the other. A permutation $g \in \mathrm{Sym}(\Omega)$ is a *cycle* if there exist distinct $x_1, x_2, \ldots, x_k \in \Omega$, $k \geq 2$, such that

$$g(x_1) = x_2, \quad g(x_2) = x_3, \quad \ldots, \quad g(x_{k-1}) = x_k, \quad g(x_k) = x_1,$$

and all other $x \in \Omega$ are fixed. We denote such a cycle by $(x_1\ x_2\ \cdots\ x_k)$. The integer k is the *length* of the cycle. A cycle of length k is also called a k-*cycle*. A 2-cycle is called a *transposition*.

Every nonidentity permutation can be expressed as a product of disjoint cycles. This product is unique up to ordering of the cycles.

Example 3.10. Consider the permutation $g \in \mathrm{Sym}(\{1,2,3,4,5,6,7,8\})$ with

$$\begin{aligned} &g(1) = 5, \quad g(2) = 2, \quad g(3) = 8, \quad g(4) = 1, \\ &g(5) = 4, \quad g(6) = 3, \quad g(7) = 7, \quad g(8) = 6. \end{aligned}$$

The decomposition of g into cycles is depicted in Fig. 3.1. In notation, $g = (1\ 5\ 4)(3\ 8\ 6)$.

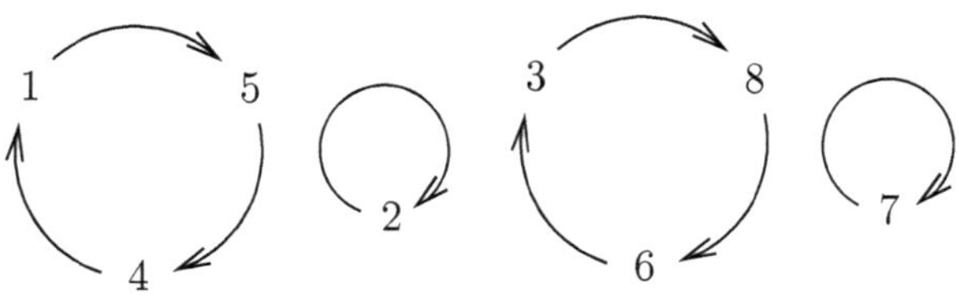

Fig. 3.1. Cycle decomposition

The automorphism groups of combinatorial and algebraic objects are a rich source of permutation groups.

Example 3.11. The cube graph Q_3 in Fig. 3.2 has 48 automorphisms, which we list here. In Theorems 3.53 and 3.54 we will prove that these are all the automorphisms that the cube graph $Q_3 \cong K_2^3$ admits. With composition as the group operation, the automorphisms form a permutation group on $V(Q_3) = \{1,2,3,4,5,6,7,8\}$; this group is called the *automorphism group* of Q_3, or $\mathrm{Aut}(Q_3)$.

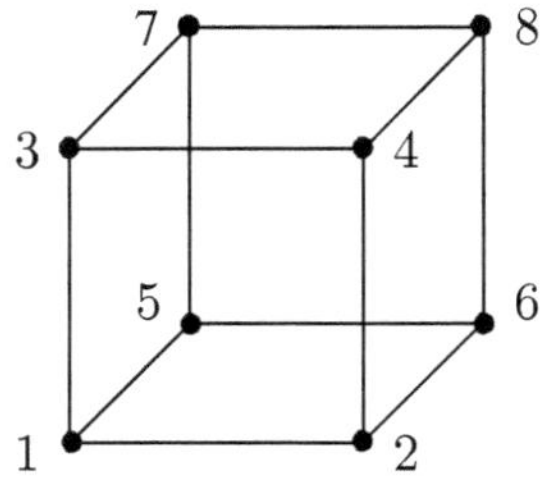

Fig. 3.2. The cube Q_3

ϵ,	(3 5)(4 6),	(2 3)(6 7),
(2 3 5)(4 7 6),	(2 5 3)(4 6 7),	(2 5)(4 7),
(1 2)(3 4)(5 6)(7 8),	(1 2)(3 6)(4 5)(7 8),	(1 2 4 3)(5 6 8 7),
(1 2 4 8 7 5)(3 6),	(1 2 6 8 7 3)(4 5),	(1 2 6 5)(3 4 8 7),
(1 3 4 2)(5 7 8 6),	(1 3 7 8 6 2)(4 5),	(1 3)(2 4)(5 7)(6 8),
(1 3 7 5)(2 4 8 6),	(1 3)(2 7)(4 5)(6 8),	(1 3 4 8 6 5)(2 7),
(1 4)(5 8),	(1 4 6)(3 8 5),	(1 4)(2 3)(5 8)(6 7),
(1 4 7 6)(2 3 8 5),	(1 4 6 7)(2 8 5 3),	(1 4 7)(2 8 5),
(1 5 7 8 4 2)(3 6),	(1 5 6 2)(3 7 8 4),	(1 5 7 3)(2 6 8 4),
(1 5)(2 6)(3 7)(4 8),	(1 5 6 8 4 3)(2 7),	(1 5)(2 7)(3 6)(4 8),
(1 6 4)(3 5 8),	(1 6)(3 8),	(1 6 7 4)(2 5 8 3),
(1 6)(2 5)(3 8)(4 7),	(1 6 7)(2 8 3),	(1 6 4 7)(2 8 3 5),
(1 7 6 4)(2 3 5 8),	(1 7 6)(2 3 8),	(1 7 4)(2 5 8),
(1 7 4 6)(2 5 3 8),	(1 7)(2 8),	(1 7)(2 8)(3 5)(4 6),
(1 8)(2 4)(3 6)(5 7),	(1 8)(2 4 3 7 5 6),	(1 8)(2 6 5 7 3 4),
(1 8)(2 6)(3 7)(4 5),	(1 8)(2 7)(3 4)(5 6),	(1 8)(2 7)(3 6)(4 5).

To motivate the connection between symmetry and automorphisms, we encourage the reader to find a geometric interpretation in terms of Fig. 3.2 for each of the automorphisms listed. For example, (3 5)(4 6) mirrors the cube with respect to the plane containing the vertices $1, 2, 7, 8$.

Example 3.12. Let G be a group. The set $\mathrm{Aut}(G)$ consisting of all the automorphisms of G is a permutation group on G. For example, consider the group $G = \langle(1\ 2\ 3\ 4\ 5)\rangle = \{g_1, g_2, g_3, g_4, g_5\}$, where

$$g_1 = \epsilon, \quad g_2 = (1\ 2\ 3\ 4\ 5), \quad g_3 = (1\ 3\ 5\ 2\ 4),$$
$$g_4 = (1\ 4\ 2\ 5\ 3), \quad g_5 = (1\ 5\ 4\ 3\ 2).$$

To determine $\mathrm{Aut}(G)$, observe that any automorphism a is uniquely determined by the image $a(g_2)$ since all elements of G can be expressed as compositions of g_2 with itself (powers of g_2). Thus, $\mathrm{Aut}(G)$ has order 4 and consists of the permutations

$$a_1 = \epsilon_G, \quad a_2 = (g_2\ g_3\ g_5\ g_4), \quad a_3 = (g_2\ g_4\ g_5\ g_3), \quad a_4 = (g_2\ g_5)(g_3\ g_4).$$

3.1.2 Group Actions

Group actions are the main tool that we use in the sequel in studying isomorphism from a computational perspective. Before proceeding with the abstract definition, let us start with a brief illustrative example.

Example 3.13. Consider the cube Q_3 in Fig. 3.2. In terms of vertices and edges, the graph can be expressed as

$$\begin{aligned} V(X) &= \{1,2,3,4,5,6,7,8\}, \\ E(X) &= \{\{1,2\},\{1,3\},\{1,5\},\{2,4\},\{2,6\},\{3,4\}, \\ &\qquad \{3,7\},\{4,8\},\{5,6\},\{5,7\},\{6,8\},\{7,8\}\}. \end{aligned}$$

An isomorphic graph on the same vertex set is

$$\begin{aligned} V(Y) &= \{1,2,3,4,5,6,7,8\}, \\ E(Y) &= \{\{1,2\},\{1,7\},\{1,8\},\{2,3\},\{2,4\},\{3,5\}, \\ &\qquad \{3,7\},\{4,5\},\{4,8\},\{5,6\},\{6,7\},\{6,8\}\}. \end{aligned}$$

Because both graphs have the same vertex set $V = V(X) = V(Y)$, any isomorphism of X onto Y is a permutation of V. For example, $g = (1\ 2\ 3\ 4\ 5)(6\ 7\ 8) \in \mathrm{Sym}(V)$ is an isomorphism of X onto Y.

There are three fundamental observations to be made. First, we can view the graph Y as being obtained from X via the permutation g *acting* on X by relabeling its vertices and edges, in notation, $g * X = Y$, where

$$\begin{aligned} V(g * X) &= \{g(x) : x \in V(X)\} = V, \\ E(g * X) &= \{\{g(x), g(y)\} : \{x, y\} \in E(X)\}. \end{aligned} \tag{3.2}$$

Second, this notion of acting on a graph makes sense for any permutation in $\mathrm{Sym}(V)$ and any graph with the vertex set V. For example, $h = (1\ 2\ 3)$ acts on Y to yield $Z = h * Y$, where

$$\begin{aligned} V(Z) &= \{1,2,3,4,5,6,7,8\}, \\ E(Z) &= \{\{1,3\},\{1,5\},\{1,7\},\{2,3\},\{2,7\},\{2,8\}, \\ &\qquad \{3,4\},\{4,5\},\{4,8\},\{5,6\},\{6,7\},\{6,8\}\}. \end{aligned}$$

Third, it is easy to check that (3.2) is compatible with the group operation on $\mathrm{Sym}(V)$. That is, if we act on X first by g and then by h, the result is the same as if we initially acted on X by the composition $h \circ g = (1\ 3\ 4\ 5\ 2)(6\ 7\ 8)$; in notation, $h * (g * X) = Z = (h \circ g) * X$. Figure 3.3 illustrates the situation.

The general combinatorial setting is now that we have a finite set Ω of combinatorial objects and a group G whose elements consist of all the possible isomorphisms between the objects in Ω. An action of G on Ω is a mapping that associates with every group element $g \in G$ and every object $X \in \Omega$ the object $g * X$ obtained when we "relabel" X using g.

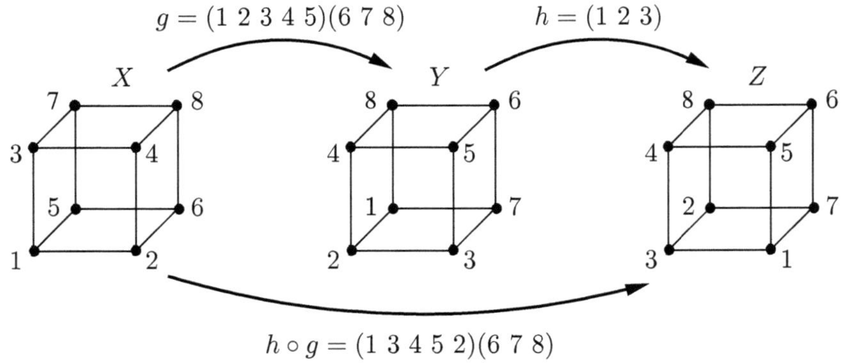

Fig. 3.3. Illustration of Example 3.13

Definition 3.14. *Let G be a group and let Ω be a finite nonempty set. An* action *of G on Ω is a mapping $* : G \times \Omega \to \Omega$ that satisfies the following two properties:*

1. *for all $X \in \Omega$, we have $1_G * X = X$,*
2. *for all $g_1, g_2 \in G$ and $X \in \Omega$, we have $(g_2 \cdot g_1) * X = g_2 * (g_1 * X)$.*

We adopt the convention of using upper-case letters $X, Y, Z, \ldots$ for objects with more internal structure (say, graphs) and lower-case letters $x, y, z, \ldots$ for objects with less structure (say, vertices).

We say that G *acts* on Ω to indicate the presence of an action of G on Ω. Except when formally defining a group action, we often omit the $*$ from the notation and write either $g(X)$ or gX instead of $g * X$.

As with composition of permutations, the reader should be aware that many authors follow an alternative convention for group actions where a group acts "from the right", in which case it is customary to use exponential notation "$(X^{g_1})^{g_2} = X^{(g_1 g_2)}$". In this book a group always acts "from the left" as given by Definition 3.14.

An equivalent description for an action of G on Ω is a group homomorphism of G into $\mathrm{Sym}(\Omega)$. Namely, an action of G on Ω defines a group homomorphism $\gamma : G \to \mathrm{Sym}(\Omega)$ by the rule $\gamma_g(X) = gX$ for all $g \in G$ and $X \in \Omega$, where $\gamma_g \in \mathrm{Sym}(\Omega)$ is the image of g under γ. Conversely, it is easily checked that a group homomorphism $\gamma : G \to \mathrm{Sym}(\Omega)$ defines a group action by $g * X = \gamma_g(X)$.

A group action is *faithful* if the kernel of the associated homomorphism is trivial, that is, 1_G is the only $g \in G$ for which $gX = X$ holds for all $X \in \Omega$.

We proceed to define some elementary group actions frequently occurring in a combinatorial context. A permutation group $G \leq \mathrm{Sym}(\Omega)$ defines an action on Ω by $g * x = g(x)$ for all $g \in G$ and $x \in \Omega$. This is called the *natural* action of the permutation group.

An action of G on Ω defines an elementwise action on objects consisting of elements of Ω. For example, the elementwise action of G on the set of all subsets of Ω is defined for all $g \in G$ and $X \subseteq \Omega$ by $g * X = \{g(x) : x \in X\}$. In an analogous manner G acts on the set of all k-subsets, k-tuples, multisets, set partitions, and set systems consisting of elements of Ω. Such an action is said to be *induced* by the action of G on Ω.

Example 3.15. Consider Example 3.11. The natural action of $\mathrm{Aut}(Q_3)$ on $V(Q_3) = \{1,2,3,4,5,6,7,8\}$ induces an action on the 4-subsets of $V(Q_3)$. For example,

$$\begin{aligned}
\epsilon * \{1,2,3,4\} &= \{1,2,3,4\},\\
(1\ 5\ 7\ 3)(2\ 6\ 8\ 4) * \{1,2,3,4\} &= \{1,2,5,6\},\\
(1\ 5\ 6\ 2)(3\ 7\ 8\ 4) * \{1,2,3,4\} &= \{1,3,5,7\},\\
(1\ 2\ 6\ 5)(3\ 4\ 8\ 7) * \{1,2,3,4\} &= \{2,4,6,8\},\\
(1\ 3\ 7\ 5)(2\ 4\ 8\ 6) * \{1,2,3,4\} &= \{3,4,7,8\},\\
(1\ 7)(2\ 8)(3\ 5)(4\ 6) * \{1,2,3,4\} &= \{5,6,7,8\}.
\end{aligned}$$

A geometric interpretation for these transformations can be found in terms of rotations and faces of the cube in Fig. 3.2.

A group G acts on itself by left multiplication via $g*x = gx$ for all $g, x \in G$. This is the *left regular* action of G on itself. The *right regular* action is given by $g * x = xg^{-1}$.

Analogously, if $H \le G$ then the group G acts on the set G/H of all left cosets by $g * (g_1 H) = (gg_1)H$ for all $g \in G$ and $g_1 H \in G/H$. An action on the set $H\backslash G$ of all right cosets is obtained by setting $g * (Hg_1) = H(g_1 g^{-1})$ for all $g \in G$ and $Hg_1 \in H\backslash G$.

Another basic group-theoretic action is the *conjugation* action of a group G on the set of all of its subgroups; for all $g \in G$ and $H \le G$, the *g-conjugate* of H is the subgroup $g * H = gHg^{-1} = \{ghg^{-1} : h \in H\}$.

3.1.3 Isomorphism and the Orbit-Stabilizer Theorem

Now that we have the basic terminology available, let us proceed to study notions of isomorphism for combinatorial objects using group actions. The following definition is motivated by Example 3.13.

Definition 3.16. *Let G be a group that acts on a finite set Ω. Associate with every two* objects $X, Y \in \Omega$ *the set*

$$\mathrm{Iso}(X,Y) = \{g \in G : gX = Y\}.$$

Each element of $\mathrm{Iso}(X,Y)$ *is an* isomorphism *of X onto Y. The objects X and Y are* isomorphic *if* $\mathrm{Iso}(X,Y)$ *is nonempty.*

We write $X \cong Y$ to indicate that X and Y are isomorphic; if it is necessary to emphasize the acting group, we write $X \cong_G Y$.

The isomorphism relation in Definition 3.16 allows us to capture exactly the isomorphism relation on graphs (Definition 2.7) if we restrict to a set of graphs such that each graph has the same vertex set V. Namely, two graphs X and Y with $V(X) = V(Y) = V$ are isomorphic (in the sense of Definition 2.7) if and only if the graphs are isomorphic (in the sense of Definition 3.16) with respect to the action of $\mathrm{Sym}(V)$ given by (3.2). Furthermore, the isomorphisms of X onto Y (in the sense of Definition 2.7) are exactly the permutations $g \in \mathrm{Sym}(V)$ satisfying $gX = Y$ with respect to (3.2).

For most combinatorial objects encountered in practice – including all objects studied in this book – it is possible to define a group action that characterizes the isomorphism relation and the isomorphisms of objects in an analogous manner. Thus, a group action provides a convenient abstraction from the individual notions of isomorphism that not only allows us to study isomorphism in a unified framework, but also enables the use of group-theoretic concepts and tools in studying isomorphism.

Two basic concepts associated with a group action are the orbits of the action and the stabilizer subgroups of the objects in Ω.

Definition 3.17. *Let G be a group that acts on a finite set Ω, and let $X \in \Omega$. The* orbit *of X under the action of G is the set*

$$GX = \{gX : g \in G\}.$$

The stabilizer *of X in G is the subgroup*

$$N_G(X) = \{g \in G : gX = X\}.$$

We write $G \backslash \Omega$ for the set of all orbits of the action of G on Ω. The set $G \backslash \Omega$ is a partition of Ω. Indeed, for $X, Y \in \Omega$ it is straightforward to check that GX and GY are either equal or disjoint.

An action is *transitive* if it has only one orbit. An action is *regular* if it is transitive and $N_G(X) = \{1\}$ for all $X \in \Omega$.

In terms of Definition 3.16 and combinatorial objects, the orbit of an object $X \in \Omega$ is the set of all objects in Ω isomorphic to X. Furthermore, the stabilizer of X is the automorphism group $\mathrm{Aut}(X) = \mathrm{Iso}(X, X)$ of X. However, to emphasize the role of the acting group G in the present setting, we use the term stabilizer and the notation $N_G(X)$.

Example 3.18. Consider Example 3.11. The natural action of the automorphism group $\mathrm{Aut}(Q_3)$ on $V(Q_3) = \{1, 2, 3, 4, 5, 6, 7, 8\}$ has only one orbit, namely the entire vertex set. Thus, the action is transitive. The induced action of $\mathrm{Aut}(Q_3)$ partitions the set of all 2-subsets of $V(Q_3)$ into three orbits:

$\{\{1,2\},\{1,3\},\{1,5\},\{2,4\},\{2,6\},\{3,4\},\{3,7\},\{4,8\},$
$\{5,6\},\{5,7\},\{6,8\},\{7,8\}\};$
$\{\{1,4\},\{1,6\},\{1,7\},\{2,3\},\{2,5\},\{2,8\},\{3,5\},\{3,8\},$
$\{4,6\},\{4,7\},\{5,8\},\{6,7\}\};$
$\{\{1,8\},\{2,7\},\{3,6\},\{4,5\}\}.$

The reader is encouraged to find a geometric interpretation in terms of Fig. 3.2 for each orbit.

The stabilizer $N_{\mathrm{Aut}(Q_3)}(\{1,2\})$ is the subgroup consisting of the four permutations

$$\epsilon, \quad (3\ 5)(4\ 6), \quad (1\ 2)(3\ 4)(5\ 6)(7\ 8), \quad (1\ 2)(3\ 6)(4\ 5)(7\ 8).$$

Example 3.19. The induced action of the group $\langle(1\ 2\ 3\ 4\ 5\ 6)\rangle$ partitions the edge set of the complete graph K_6 into the three orbits depicted in Fig. 3.4.

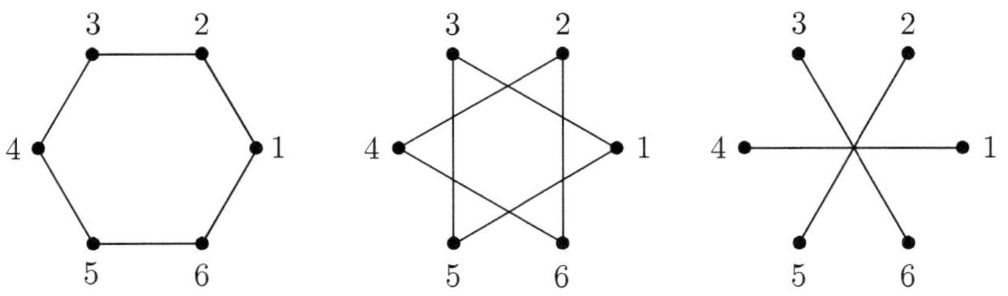

Fig. 3.4. Orbits of $\langle(1\ 2\ 3\ 4\ 5\ 6)\rangle$ on the edge set of K_6

The following theorem gives a fundamental correspondence between the objects in an orbit GX and the left cosets of the stabilizer $N_G(X)$ in G.

Theorem 3.20 (Orbit-stabilizer theorem). *Let G be a group that acts on a finite set Ω and let $X \in \Omega$. Then, the mapping from $G/N_G(X)$ into GX defined by $gN_G(X) \mapsto gX$ is a well-defined bijection. In particular, $|G| = |GX| \cdot |N_G(X)|$.*

Proof. To check that the mapping is well-defined, let $g_1N_G(X) = g_2N_G(X)$, that is, $g_2^{-1}g_1 \in N_G(X)$. It follows that

$$g_1X = (g_2g_2^{-1})g_1X = g_2(g_2^{-1}g_1)X = g_2X,$$

so the mapping is independent of the coset representative g selected. Surjectivity follows because for every $gX \in GX$ the coset $gN_G(X)$ maps to gX. The mapping is injective because $g_1X = g_2X$ implies $g_2^{-1}g_1X = X$, that is, $g_2^{-1}g_1 \in N_G(X)$ and hence $g_1N_G(X) = g_2N_G(X)$. Because G is finite, $|GX| = |G/N_G(X)| = |G|/|N_G(X)|$, where the last equality follows from Lagrange's theorem (Theorem 3.7). □

Example 3.21. We know from Example 3.11 that $|\mathrm{Aut}(Q_3)| = 48$. On the other hand, we know from Example 3.18 that with respect to the induced action of $\mathrm{Aut}(Q_3)$ on 2-subsets of $\{1,2,3,4,5,6,7,8\}$, we have $|N_{\mathrm{Aut}(Q_3)}(\{1,2\})| = 4$. By the orbit-stabilizer theorem, we obtain that the $\mathrm{Aut}(Q_3)$-orbit of $\{1,2\}$ on 2-subsets of $\{1,2,3,4,5,6,7,8\}$ has size $48/4 = 12$. This is in agreement with Example 3.18.

Example 3.22. In Example 3.19 and Fig. 3.4 we have three orbits of sizes 6, 6, and 3, respectively. Since $|\langle(1\ 2\ 3\ 4\ 5\ 6)\rangle| = 6$, the stabilizers of $\{1,2\}$, $\{1,3\}$, and $\{1,4\}$ in $\langle(1\ 2\ 3\ 4\ 5\ 6)\rangle$ thus have orders 1, 1, and 2, respectively.

The orbit-stabilizer theorem gives a way to list the elements in an orbit given a representative from the orbit and the corresponding stabilizer subgroup. Namely, if a left transversal of the stabilizer $N_G(X)$ in the acting group G is available, we can construct the orbit GX by acting on the orbit representative X with each transversal element in turn.

Example 3.23. From Example 3.11 we can find that $|N_{\mathrm{Aut}(Q_3)}(\{1,2,3,4\})| = 8$. By the orbit-stabilizer theorem this implies that the $\mathrm{Aut}(Q_3)$-orbit of $\{1,2,3,4\}$ has size $48/8 = 6$. Thus, the entire orbit of $\{1,2,3,4\}$ is visited in Example 3.15. The acting permutations in Example 3.15 form a left transversal of $N_{\mathrm{Aut}(Q_3)}(\{1,2,3,4\})$ in $\mathrm{Aut}(Q_3)$.

A corollary that frequently finds applications in practice is as follows.

Corollary 3.24. *For all $X, Y \in \Omega$ with $X \cong Y$, the set $\mathrm{Iso}(X,Y)$ is a left coset of the stabilizer $N_G(X)$ in G. Furthermore, $gN_G(X)g^{-1} = N_G(Y)$ for all $g \in \mathrm{Iso}(X,Y)$.*

Proof. The first claim is an immediate corollary of Theorems 3.7 and 3.20. For the second claim, let $g \in \mathrm{Iso}(X,Y)$ and $g_1 \in G$. We have $g_1 \in N_G(Y)$ if and only if $g_1gX = gX$; that is, if and only if $g^{-1}g_1gX = X$; that is, if and only if $g^{-1}g_1g \in N_G(X)$; that is, if and only if $g_1 \in gN_G(X)g^{-1}$. □

Example 3.25. Consider the action (3.2) and the graphs X and Y given in Example 3.13. The stabilizer $N_{\mathrm{Sym}(\{1,2,3,4,5,6,7,8\})}(X) = \mathrm{Aut}(Q_3)$ is listed in Example 3.11. By Corollary 3.24 every isomorphism from X to Y be expressed uniquely as a product of any fixed isomorphism $g \in \mathrm{Iso}(X,Y)$ and an automorphism $a \in \mathrm{Aut}(Q_3)$.

The orbit-stabilizer theorem can often be used to arrive at a double counting consistency check for a classification, provided that we know the stabilizer subgroup orders for the orbit representatives.

Example 3.26. The number of distinct graphs that can be constructed over the vertex set $\{1,2,3,4\}$ is $2^{\binom{4}{2}} = 64$. On the other hand, the classification in Example 2.9 contains 11 graphs, whose stabilizer subgroups with respect to the action (3.2) of $\mathrm{Sym}(\{1,2,3,4\})$ have orders 24, 4, 8, 2, 2, 6, 6, 8, 2, 4, and

24, respectively. Observing that $|\operatorname{Sym}(\{1,2,3,4\})| = 4! = 24$ and using the orbit-stabilizer theorem on each orbit in the classification, we obtain that the total number of distinct graphs over $\{1,2,3,4\}$ is

$$24 \times \left(\frac{1}{24}+\frac{1}{4}+\frac{1}{8}+\frac{1}{2}+\frac{1}{2}+\frac{1}{6}+\frac{1}{6}+\frac{1}{8}+\frac{1}{2}+\frac{1}{4}+\frac{1}{24}\right) = 64,$$

which is in agreement with the earlier count.

More advanced consistency checks of this type are discussed in Chap. 10.

The following lemma due to Cauchy and Frobenius (but often incorrectly attributed to Burnside; see [441]) is useful in counting the number of orbits of a group action.

Theorem 3.27 (Orbit-counting lemma). *Let G be a group that acts on a finite set Ω. Then,*

$$|G| \cdot |G\backslash\Omega| = \sum_{g \in G} \operatorname{fix}_\Omega(g), \tag{3.3}$$

where $\operatorname{fix}_\Omega(g) = |\{X \in \Omega : gX = X\}|$.

Proof. Define a characteristic function $\chi : G \times \Omega \to \{0,1\}$ for all $g \in G$ and $X \in \Omega$ by

$$\chi(g,X) = \begin{cases} 1 & \text{if } gX = X, \\ 0 & \text{if } gX \neq X. \end{cases}$$

On one hand, we have

$$\sum_{g \in G} \sum_{X \in \Omega} \chi(g,X) = \sum_{g \in G} \operatorname{fix}_\Omega(g)$$

On the other hand, applying the orbit-stabilizer theorem and observing that $|GX| = |GY|$ whenever $X \cong_G Y$, we obtain

$$\sum_{X \in \Omega} \sum_{g \in G} \chi(g,X) = \sum_{X \in \Omega} |N_G(X)| = \sum_{X \in \Omega} \frac{|G|}{|GX|} = |G| \sum_{X \in \Omega} \frac{1}{|GX|} = |G| \cdot |G\backslash\Omega|.$$

The two finite sums differ only in the order of summation. □

3.1.4 Semidirect and Wreath Products

Semidirect and wreath products of groups are important tools in studying groups and group actions related to code equivalence and Hadamard matrix equivalence. We begin with an abstract development and then proceed to applications.

Definition 3.28. *Let K and H be groups and let $\theta : H \to \mathrm{Aut}(K)$ be a group homomorphism. The* semidirect product $K \rtimes_\theta H$ *of K by H (subject to θ) is the group consisting of all ordered pairs (k, h), where $k \in K$ and $h \in H$, with the group operation defined for all $k_1, k_2 \in K$ and $h_1, h_2 \in H$ by*

$$(k_1, h_1) \cdot (k_2, h_2) = (k_1 \theta_{h_1}(k_2), h_1 h_2).$$

The direct product is a special case of the semidirect product.

Definition 3.29. *The* direct product $K \times H$ *is the semidirect product of K by H where $\theta_h(k) = k$ for all $h \in H$ and $k \in K$.*

Also the wreath product is a special case of the semidirect product. Let Ω and Δ be nonempty finite sets. Let $\mathrm{Fun}(\Omega, \Delta)$ be the set of all functions from Ω into Δ. If K is a group, then it is straightforward to check that $\mathrm{Fun}(\Omega, K)$ becomes a group if the group operation on $\mathrm{Fun}(\Omega, K)$ is defined for all $f_1, f_2 \in \mathrm{Fun}(\Omega, K)$ and $x \in \Omega$ by

$$(f_1 \cdot f_2)(x) = f_1(x) f_2(x). \tag{3.4}$$

Let H be a group that acts on Ω and let $\gamma : H \to \mathrm{Sym}(\Omega)$ be the group homomorphism that corresponds to this action. We extend the action from Ω to $\mathrm{Fun}(\Omega, \Delta)$ by defining

$$(\gamma_h(f))(x) = f(\gamma_{h^{-1}}(x)) \tag{3.5}$$

for all $h \in H$, $f \in \mathrm{Fun}(\Omega, \Delta)$, and $x \in \Omega$.

Definition 3.30. *The* wreath product $K \wr_\gamma H$ *of K by H (subject to γ) is the group consisting of all ordered pairs (f, h), where $f \in \mathrm{Fun}(\Omega, K)$ and $h \in H$. The group operation is defined for all $f_1, f_2 \in \mathrm{Fun}(\Omega, K)$ and $h_1, h_2 \in H$ by*

$$(f_1, h_1) \cdot (f_2, h_2) = (f_1 \gamma_{h_1}(f_2), h_1 h_2). \tag{3.6}$$

The action (3.5) defines a homomorphism of H into $\mathrm{Aut}(\mathrm{Fun}(\Omega, K))$, where the group operation on $\mathrm{Fun}(\Omega, K)$ is given by (3.4). Thus, the wreath product is a semidirect product of $\mathrm{Fun}(\Omega, K)$ by H.

When H is a permutation group on Ω, and Ω is clear from the context, we always assume that the action of H on Ω in Definition 3.30 is the natural action and simply write $K \wr H$. We remark that this notation is also used by some authors in the case where the action γ is a regular action of H on itself.

A special case that we will encounter often due to its importance in coding theory (see Sect. 2.3.4) is the wreath product $S_q \wr S_n$ of two symmetric groups S_q and S_n. This group has order $(q!)^n \cdot n!$. The wreath product $S_2 \wr S_n$ occurs under many different names, including *hyperoctahedral group* of degree n, the *group of signed permutations*, and the *Coxeter group* of type B_n.

We adopt the convention of viewing the elements of $S_q \wr S_n$ as pairs (k, h), where $h \in S_n$ permutes the integers $1, 2, \ldots, n$ and $k = (k_1, k_2, \ldots, k_n)$ is an

n-tuple of permutations $k_i \in S_q$. In this case the group operation becomes $(k,h) = (a,c) \cdot (b,d)$, where

$$h = cd, \quad k_i = a_i b_{c^{-1}(i)} \quad \text{for all } 1 \le i \le n. \tag{3.7}$$

Example 3.31. Let $q = 3$ and $n = 4$. For

$$a_1 = \epsilon, \quad a_2 = (0\ 1\ 2), \quad a_3 = (1\ 2), \quad a_4 = (0\ 1), \quad c = (1\ 2\ 3\ 4)$$

and

$$b_1 = (0\ 2), \quad b_2 = (1\ 2\ 0), \quad b_3 = \epsilon, \quad b_4 = (0\ 1), \quad d = (2\ 3)$$

we obtain

$$\begin{aligned} k_1 &= a_1 b_4 = (0\ 1), \\ k_2 &= a_2 b_1 = (1\ 2), \\ k_3 &= a_3 b_2 = (0\ 2), \\ k_4 &= a_4 b_3 = (0\ 1), \\ h &= cd = (1\ 2\ 4). \end{aligned}$$

Also other conventions exist (cf. [34, 385, 467]), in particular for the case $q = 2$; later in this section we will discuss one possible representation for $S_2 \wr S_n$ that is convenient in the context of Hadamard matrix equivalence.

The relevance of the group $S_q \wr S_n$ in coding theory is revealed in the following definition and the subsequent discussion.

Definition 3.32. *Suppose that K acts on Δ. The* product action *of $K \wr_\gamma H$ on* $\mathrm{Fun}(\Omega, \Delta)$ *(subject to the action of K) is defined for all $(f,h) \in K \wr_\gamma H$, $g \in \mathrm{Fun}(\Omega, \Delta)$ and $x \in \Omega$ by*

$$((f,h) * g)(x) = \gamma_{f(x)}(g(\gamma_{h^{-1}}(x))). \tag{3.8}$$

We may view an n-tuple $\mathbf{x} = (x_1, x_2, \ldots, x_n) \in Z_q^n$ as a function that associates to every coordinate $1 \le i \le n$ the corresponding coordinate value x_i. If we assume that S_q acts naturally on Z_q, then we obtain a product action of $S_q \wr S_n$ on Z_q^n defined for all $(k,h) \in S_q \wr S_n$, $\mathbf{x} \in Z_q^n$, and $1 \le i \le n$ by

$$\big((k,h) * \mathbf{x}\big)_i = k_i(x_{h^{-1}(i)}). \tag{3.9}$$

In other words, a group element (k,h) first permutes the coordinates in $\mathbf{x}$ so that coordinate i becomes coordinate $h(i)$. Then, the values in the new ith coordinate are permuted according to k_i for every $1 \le i \le n$. A related action is the coordinate-permuting action

$$(h * \mathbf{x})_i = x_{h^{-1}(i)}, \tag{3.10}$$

where no permutation of values occurs.

Definition 2.100 states that two codes are equivalent if one can be obtained from the other by first permuting the coordinates and then the coordinate values separately for each coordinate. In the language of group actions this is the same as saying that two codes are equivalent if they are on the same orbit of the action (3.9) of $S_q \wr S_n$ induced on subsets of Z_q^n. Similarly, Definition 2.99 for code isomorphism is captured by the coordinate-permuting action (3.10).

Example 3.33. In (1.2) and the related text, a transformation of the code in (1.1) is suggested. In terms of the product action (3.9), the group element $(k, h) \in S_3 \wr S_4$ that performs this transformation is

$$k_1 = (0\ 1\ 2), \quad k_2 = k_3 = k_4 = \epsilon, \quad h = \epsilon.$$

Example 3.34. The transformation described in Example 2.101 corresponds to the group element $(k, h) \in S_3 \wr S_4$, where

$$k_1 = k_2 = k_3 = (0\ 1), \quad k_4 = \epsilon, \quad h = (1\ 2\ 3\ 4).$$

For computational purposes it is sometimes necessary to represent a wreath product group as a permutation group. Let K act on Δ with β being the corresponding group homomorphism. We obtain an action of $K \wr_\gamma H$ on $\Delta \times \Omega$ by setting

$$(f, h) * (y, x) = (\beta_{f(\gamma_h(x))}(y), \gamma_h(x))$$

for all $(f, h) \in K \wr_\gamma H$ and $(y, x) \in \Delta \times \Omega$. The action is faithful if and only if the actions given by both β and γ are faithful.

Example 3.35. Consider the group $S_3 \wr S_4$. Let S_3 act naturally on $\{0, 1, 2\}$ and S_4 on $\{1, 2, 3, 4\}$. If we write y_x for the pair (y, x), then a permutation representation for $S_3 \wr S_4$ is generated by

$$\begin{aligned}
&(0_1\ 1_1\ 2_1),\\
&(0_1\ 1_1),\\
&(0_1\ 0_2\ 0_3\ 0_4)(1_1\ 1_2\ 1_3\ 1_4)(2_1\ 2_2\ 2_3\ 2_4),\\
&(0_1\ 0_2)(1_1\ 1_2)(2_1\ 2_2).
\end{aligned}$$

The group elements in Examples 3.33 and 3.34 are represented by

$$(0_1\ 1_1\ 2_1), \quad (0_1\ 1_2\ 0_3\ 0_4\ 1_1\ 0_2\ 1_3\ 1_4)(2_1\ 2_2\ 2_3\ 2_4),$$

respectively.

We conclude this section by discussing the equivalence of Hadamard matrices, which gives us yet another way of representing $S_2 \wr S_n$.

Let $h \in \mathrm{Sym}(\{1,2,\ldots,n\})$ be a permutation. The permutation matrix associated with h is the $n \times n$ integer matrix $\mathbf{P}_h = (p_{ij})$ defined for all $1 \le i, j \le n$ by

$$p_{ij} = \begin{cases} 1 & \text{if } i = h(j), \\ 0 & \text{if } i \ne h(j). \end{cases} \tag{3.11}$$

Any matrix of this form is a *permutation matrix.*

It is easy to check that the map $h \mapsto \mathbf{P}_h$ defines an isomorphism from S_n onto the group of all permutation matrices, with the usual matrix product as the group operation; however, note that this claim depends on the order of composition for permutations (3.1) and on (3.11).

Example 3.36. Let $n = 4$ and $g = (1\ 2\ 3)$, $h = (3\ 4)$. The associated permutation matrices are

$$\mathbf{P}_g = \begin{bmatrix} 0\,0\,1\,0 \\ 1\,0\,0\,0 \\ 0\,1\,0\,0 \\ 0\,0\,0\,1 \end{bmatrix}, \quad \mathbf{P}_h = \begin{bmatrix} 1\,0\,0\,0 \\ 0\,1\,0\,0 \\ 0\,0\,0\,1 \\ 0\,0\,1\,0 \end{bmatrix}, \quad \mathbf{P}_{gh} = \mathbf{P}_g\mathbf{P}_h = \begin{bmatrix} 0\,0\,0\,1 \\ 1\,0\,0\,0 \\ 0\,1\,0\,0 \\ 0\,0\,1\,0 \end{bmatrix}.$$

A *signed* permutation matrix is any matrix of the form $\mathbf{SP}$, where $\mathbf{S}$ is a diagonal matrix with diagonal entries in $\{-1, 1\}$, and $\mathbf{P}$ is a permutation matrix. The product of two signed permutation matrices, $\mathbf{SP}$ and $\mathbf{TQ}$, is a signed permutation matrix

$$(\mathbf{SP}) \cdot (\mathbf{TQ}) = (\mathbf{SPTP}^{-1})(\mathbf{PQ}). \tag{3.12}$$

We leave it as an exercise to prove that the group of all signed $n \times n$ permutation matrices is isomorphic to $S_2 \wr S_n$, cf. (3.7).

Two Hadamard matrices are considered equivalent if one can be transformed into the other by row and column permutations followed by negations of rows and columns (Definition 2.125). This notion of equivalence can be conveniently expressed using signed permutation matrices. Namely, if $\mathbf{H}$ is an Hadamard matrix of order n and $\mathbf{P}_h$ is a permutation matrix, then $\mathbf{P}_h\mathbf{H}$ is the matrix $\mathbf{H}$ with rows permuted so that row i becomes row $h(i)$ for all $1 \le i \le n$; also, $\mathbf{HP}_h^{-1}$ is the matrix $\mathbf{H}$ with columns permuted so that column j becomes column $h(j)$ for all $1 \le j \le n$. Analogously, premultiplication (respectively, postmultiplication) by a diagonal matrix corresponds to multiplying every row (column) by the corresponding diagonal element. It follows that two Hadamard matrices $\mathbf{H}_1$ and $\mathbf{H}_2$ are equivalent if and only if there exist signed permutation matrices $\mathbf{SP}$ and $\mathbf{TQ}$ such that $(\mathbf{SP})\mathbf{H}_1(\mathbf{TQ})^{-1} = \mathbf{H}_2$. Thus, the equivalence classes of Hadamard matrices of order n are the orbits of the action

$$(\mathbf{SP}, \mathbf{TQ}) * \mathbf{H} = \mathbf{SPHQ}^{-1}\mathbf{T}^{-1}, \tag{3.13}$$

where the acting group is the direct product of the group of signed $n \times n$ permutation matrices with itself. The reader is encouraged to check these claims in detail.

3.2 Categories and Equivalence

As we saw in the previous section, group actions provide one possible framework for studying notions of isomorphism associated with combinatorial objects. Another such framework, of which group actions are a special case, is provided by category theory. Category theory provides us with a convenient language for studying different representations of combinatorial objects and their equivalence on the level of isomorphism classes; while such a study could in principle be conducted using group actions, the more restricted finite setting would render a study somewhat cumbersome in many cases. Indeed, our choice of including an elementary discussion of category theory can be motivated by a similar choice made in many algebra textbooks (for example, [355]). Textbooks on category theory include [263, 360, 387].

Up to now we have encountered a variety of different combinatorial objects, where each type of object has been associated with one or more notions of isomorphism. In each case the notion of isomorphism can be defined through certain mappings and/or group elements (isomorphisms) substantiating that two given objects are identical in the structure of interest. For some objects, such as codes and Hadamard matrices, finding the appropriate isomorphisms required some work, cf. Sect. 3.1.4. With the isomorphisms uncovered, in each case the isomorphisms admit an associative composition operation, which leads us to consider each type of object together with one associated notion of isomorphism as an abstract "category". Category theory then offers through functors and reconstructibility – to be discussed in Sects. 3.2.2 and 3.2.3, respectively – the concepts that in many cases enable a rigorous study of representations and relationships between different types of objects. Another ubiquitous concept perhaps best captured in its full generality through category theory is the automorphism group, to be discussed in Sect. 3.2.1.

We begin with some abstract definitions and then proceed to concrete examples of combinatorial origin.

Definition 3.37. *A* category $\mathscr{A}$ *consists of a set* $\mathrm{Ob}(\mathscr{A})$ *of* objects*; associated with every two objects* $X, Y \in \mathrm{Ob}(\mathscr{A})$ *is a finite set* $\mathrm{Mor}(X,Y)$*, the set of* morphisms *of* X *into* Y*. Furthermore, for every three objects* $X, Y, Z \in \mathrm{Ob}(\mathscr{A})$ *there is a map*

$$\circ : \mathrm{Mor}(Y,Z) \times \mathrm{Mor}(X,Y) \to \mathrm{Mor}(X,Z)$$

with the following two properties:

1. *for every* $X \in \mathrm{Ob}(\mathscr{A})$ *there exists a morphism* $1_X \in \mathrm{Mor}(X,X)$ *such that* $f \circ 1_X = f$ *and* $1_X \circ g = g$ *for all* $f \in \mathrm{Mor}(X,Y)$*,* $g \in \mathrm{Mor}(Y,X)$*,* $Y \in \mathrm{Ob}(\mathscr{A})$*,*
2. *for every* $X, Y, Z, W \in \mathrm{Ob}(\mathscr{A})$ *and* $f \in \mathrm{Mor}(X,Y)$*,* $g \in \mathrm{Mor}(Y,Z)$*,* $h \in \mathrm{Mor}(Z,W)$ *it holds that*

$$(h \circ g) \circ f = h \circ (g \circ f). \tag{3.14}$$

Note that the set of morphisms $\mathrm{Mor}(X, Y)$ can be empty if $X \neq Y$. The map $\circ$ defines an associative operation for composition of morphisms; in practice, this map is usually the standard composition operation for mappings or the group operation of a group. Associated with each object X there is a unique morphism 1_X that acts as the identity in composition; in practice, this is often just the identity mapping from some set onto itself or the identity element of a group. Figure 3.5 illustrates the situation in (3.14).

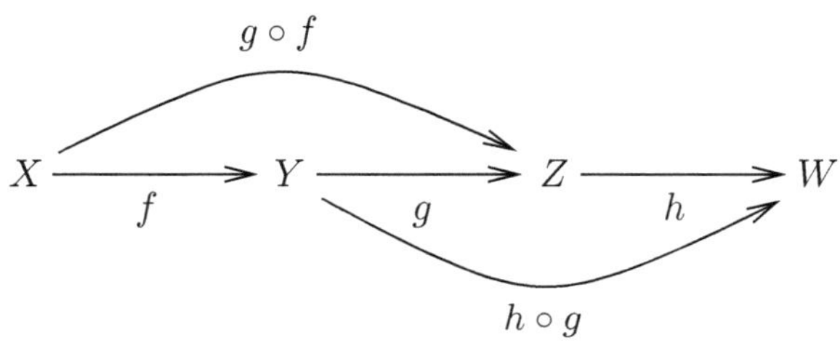

Fig. 3.5. Illustration of Definition 3.37

The following definition of isomorphism generalizes the earlier more specific definitions.

Definition 3.38. *Let* $X, Y \in \mathrm{Ob}(\mathscr{A})$. *A morphism* $f \in \mathrm{Mor}(X, Y)$ *is an* isomorphism *of* X *onto* Y *if there exists a morphism* $f^{-1} \in \mathrm{Mor}(Y, X)$ *such that* $f^{-1} \circ f = 1_X$ *and* $f \circ f^{-1} = 1_Y$.

We write $\mathrm{Iso}(X, Y)$ for the set of all isomorphisms of X onto Y. Two objects X, Y are *isomorphic* if there exists an isomorphism of X onto Y. We write $X \cong Y$ to indicate that X and Y are isomorphic.

In our applications all morphisms are isomorphisms in the sense of Definition 3.38 even though for many categories a wider notion of a morphism (for example, a group homomorphism) can be motivated. A category in which all morphisms are isomorphisms is often also called a *(Brandt) groupoid*.

Each type of object encountered in Chap. 2 together with an associated notion of isomorphism defines a category. Here we make the technical assumption that the objects of a given type have been sufficiently restricted (if necessary) so that they form a *set* in the sense of an appropriate set theory, see [387] for a detailed discussion.

Example 3.39. Graphs form a category with morphisms as in Definition 2.7. To verify the properties in Definition 3.37, let G, H, and K be graphs. Let $f : V(G) \to V(H)$ be an isomorphism of G onto H in the sense of Definition 2.7. Similarly, let $g : V(H) \to V(K)$ be an isomorphism of H onto K. Since f and g are bijections, the composition $g \circ f$ is a bijection of $V(G)$ onto $V(K)$. Furthermore, because f and g are isomorphisms, for all $u, v \in V(G)$ it holds that $\{u, v\} \in E(G)$ if and only if $\{g(f(u)), g(f(v))\} \in E(K)$. Thus, $g \circ f$ is an isomorphism of G onto K in the sense of Definition 2.7. It follows that the standard composition operation $\circ$ for mappings is a composition map for morphisms of graphs in the sense of Definition 3.37. Property (1) in Definition 3.37 is satisfied by the identity map on the vertex set of a graph. Property (2) is a natural property of the composition operation $\circ$ for mappings. Also note that viewing a graph isomorphism f as an isomorphism in the sense of Definition 3.38 is justified because the inverse function f^{-1} meets the required properties.

Example 3.40. Incidence structures form a category with isomorphisms as in Definition 2.22, and where composition is the standard composition operation for mappings.

Example 3.41. Codes come with multiple notions of isomorphism, as witnessed by Definitions 2.99, 2.100, and 2.102. Accordingly, each notion of isomorphism defines a corresponding category of codes. Unless explicitly mentioned otherwise, we assume that Definition 2.100 is used.

1. Isomorphisms associated with Definition 2.99 are permutations $h \in S_n$ that act on a code by (3.10). Composition is given by the group operation (3.1) in S_n.
2. Isomorphisms associated with Definition 2.100 are elements $(k, h) \in S_q \wr S_n$ that act on a code by (3.9). Composition is given by the group operation (3.7) in $S_q \wr S_n$.
3. Isomorphisms associated with Definition 2.102 are distance-respecting bijections $f : C \to C'$. Composition is the standard composition operation for mappings.

In general, an action of a group G on a finite set Ω defines a category if we let Ω be the set of objects, and define morphisms for all $X, Y \in \Omega$ by $\mathrm{Iso}(X, Y) = \{g \in G : gX = Y\}$. The composition operation is the group operation on G. Note that this is in agreement with Definition 3.16.

Example 3.42. Hadamard matrices define a category where isomorphisms are defined by the group action (3.13).

Objects with a particular property in a category define a subcategory whenever the relevant property is invariant under isomorphism.

Example 3.43. BIBDs form a subcategory of the category of incidence structures because the properties that define a BIBD are invariant under isomorphism. Similarly, the resolvable BIBDs form a subcategory of the category of BIBDs.

In this connection one must be careful with the properties invariant under a particular notion of isomorphism. For example, Definition 2.99 preserves linearity in a code, but Definition 2.100 in general does not, cf. Sect. 7.3.1.

We can often individualize "subobjects" of the objects in a category – such as a vertex of a graph or a block of an incidence structure – and obtain a category that inherits its morphisms by appropriate restriction from the parent category. Such categories are important in the context of isomorphism computations as we will see in Sect. 3.3.

Example 3.44. Let the objects of a category consist of the ordered pairs (G, x), where G is a graph and $x \in V(G)$. For two such objects, (G, x) and (H, y), isomorphisms (if any) are defined by

$$\mathrm{Iso}((G, x), (H, y)) = \{f \in \mathrm{Iso}(G, H) : f(x) = y\}.$$

We call this category the category of graphs with one vertex individualized.

An analogous situation is encountered if an object is "built on top" of an existing object, and the new object inherits, again with appropriate restrictions caused by the new structure in the object, its morphisms from the underlying category.

Example 3.45. A resolution is built on top of an incidence structure, cf. Definition 2.67.

3.2.1 Automorphisms and the Automorphism Group

Of fundamental importance are the isomorphisms that transform an object onto itself because such isomorphisms can in many cases be interpreted to record the symmetry present in the object. Example 3.11 provides a geometric illustration of this connection between automorphisms and symmetry.

The following definition generalizes the earlier more specific notions of an automorphism and the automorphism group.

Definition 3.46. *Let $X \in \mathrm{Ob}(\mathscr{A})$. An isomorphism of X onto itself is called an* automorphism. *The set $\mathrm{Aut}(X) = \mathrm{Iso}(X, X)$ of all automorphisms of X with composition as the group operation forms a group, the* automorphism group *of X. A subgroup of $\mathrm{Aut}(X)$ is called a* group of automorphisms *of X.*

An alternative convention used by many authors is to call $\mathrm{Aut}(X)$ the *full* automorphism group; a subgroup of $\mathrm{Aut}(X)$ is in this case called an automorphism group.

The identity morphism 1_X is always an automorphism of $X \in \mathrm{Ob}(\mathscr{A})$. An object is *rigid* if its only automorphism is the identity morphism.

Example 3.47. The graph in Fig. 2.1 is rigid.

Often it is the case that the automorphisms of an object X induce a category on a set of "subobjects" of X.

Example 3.48. The vertices of a graph form a category where the isomorphisms are automorphisms of the graph; cf. Example 3.44. We may also consider other "subobjects" such as the edges or cliques in place of the vertices.

Example 3.49. The resolutions of a BIBD form a category where the isomorphisms are the automorphisms of the BIBD, cf. Definition 2.67.

In the context of this book perhaps the most interesting situation of this kind occurs with the Hamming space (Z_q^n, d_H), whose automorphisms are *isometries*, or distance preserving maps from Z_q^n onto itself. We proceed to characterize the isometries of (Z_q^n, d_H).

Definition 3.50. *Let (Ω, d) be a metric space. A mapping $\psi : \Omega \to \Omega$ is an* isometry *of (Ω, d) if $d(\psi(x), \psi(y)) = d(x, y)$ for all $x, y \in \Omega$. We denote by $\mathrm{Aut}(\Omega, d)$ the set of all isometries of (Ω, d).*

For a finite metric space, $\mathrm{Aut}(\Omega, d)$ is a permutation group on Ω.

Theorem 3.51. *For a finite metric space (Ω, d), $\mathrm{Aut}(\Omega, d) \leq \mathrm{Sym}(\Omega)$.*

Proof. Let $\psi \in \mathrm{Aut}(\Omega, d)$. Suppose $\psi(x) = \psi(y)$ for some $x, y \in \Omega$. Then, $d(\psi(x), \psi(y)) = d(x, y) = 0$. Thus, $x = y$. It follows that ψ is injective since x, y are arbitrary. Since Ω is finite, ψ is bijective. Thus, $\psi \in \mathrm{Sym}(\Omega)$. Two isometries clearly compose to form an isometry, which is sufficient to establish that $\mathrm{Aut}(\Omega, d) \leq \mathrm{Sym}(\Omega)$ because $\mathrm{Sym}(\Omega)$ is finite. □

The following lemma shows that $S_q \wr S_n$ defines a group of isometries of (Z_q^n, d_H).

Lemma 3.52. *Let $S_q \wr S_n$ act on Z_q^n by (3.9) and let $(k, h) \in S_q \wr S_n$. Then, the permutation $\psi \in \mathrm{Sym}(Z_q^n)$ defined for all $\mathbf{x} \in Z_q^n$ by $\psi(\mathbf{x}) = (k, h)\mathbf{x}$ is an isometry of the Hamming space (Z_q^n, d_H).*

Proof. Neither permutation of the coordinate values nor permutation of the coordinates affects the Hamming distance between two words $\mathbf{x}, \mathbf{y} \in Z_q^n$. □

A useful characterization of $\mathrm{Aut}(Z_q^n, d_H)$ is that it is isomorphic to the automorphism group of the graph K_q^n.

Theorem 3.53. $\mathrm{Aut}(Z_q^n, d_H) \cong \mathrm{Aut}(K_q^n)$.

Proof. Construct a graph with one vertex for each word in Z_q^n and with a vertex between edges $\mathbf{x}$ and $\mathbf{y}$ whenever $d_H(\mathbf{x}, \mathbf{y}) = 1$. In this way we get a graph isomorphic to K_q^n. Since, by Definition 3.50, for an isometry ψ we have $d_H(\mathbf{x}, \mathbf{y}) = d_H(\psi(\mathbf{x}), \psi(\mathbf{y}))$ for all $\mathbf{x}, \mathbf{y} \in Z_q^n$, this equality must in particular hold whenever $d_H(\mathbf{x}, \mathbf{y}) = 1$, so the corresponding mapping on K_q^n must be an automorphism. Since all automorphisms of a graph maintain the length of a shortest path between two connected vertices, we get that $\mathrm{Aut}(Z_q^n, d_H) \cong \mathrm{Aut}(K_q^n)$. □

The automorphism group of K_q^n follows as a special case of a more general result in [611] on graph embeddings in K_q^n. More recent proofs and references to some other related results can be found in [131, 255, 287, 543]. The proof here is inspired by [131].

Theorem 3.54. $\mathrm{Aut}(Z_q^n, d_H) \cong S_q \wr S_n$.

Proof. The kernel of the homomorphism of $S_q \wr S_n$ into $\mathrm{Aut}(Z_q^n, d_H)$ defined by Lemma 3.52 is easily seen to be trivial. It remains to show that every isometry in $\mathrm{Aut}(Z_q^n, d_H)$ is given by Lemma 3.52. Let $\psi \in \mathrm{Aut}(Z_q^n, d_H)$ be arbitrary. By composing with an isometry given by Lemma 3.52 if necessary, we may assume that $\psi(\mathbf{0}) = \mathbf{0}$. Thus, ψ must permute the words of any given weight w among themselves:

$$\mathrm{wt}(\psi(\mathbf{x})) = d_H(\psi(\mathbf{x}), \mathbf{0}) = d_H(\psi(\mathbf{x}), \psi(\mathbf{0})) = d_H(\mathbf{x}, \mathbf{0}) = \mathrm{wt}(\mathbf{x}). \qquad (3.15)$$

Denote by $\mathbf{e}_{i,v}$ the word of weight 1 with $v \in Z_q \setminus \{0\}$ in the ith coordinate. Define $(k, h) \in S_q \wr S_n$ by requiring $\psi(\mathbf{e}_{i,v}) = \mathbf{e}_{h(i),k_{h(i)}(v)}$ for all $1 \le i \le n$ and $v \in Z_q \setminus \{0\}$. It follows immediately that $\psi(\mathbf{x}) = (k, h)\mathbf{x}$ for all $\mathbf{x} \in Z_q^n$ of weight at most 1. For any word $\mathbf{y} \in Z_q^n$ of weight $w \ge 2$ there are exactly w words of weight 1 that have distance $w - 1$ to $\mathbf{y}$. Namely, these are precisely those words of weight 1 that agree with $\mathbf{y}$ in exactly one nonzero coordinate. These w words of weight 1 uniquely determine the word $\mathbf{y}$. Consequently, $\psi(\mathbf{x}) = (k, h)\mathbf{x}$ for all $\mathbf{x} \in Z_q^n$. □

When considering classes of codes with additional properties, we must be careful that these properties are not destroyed by an isomorphism. The tradition here is that an isomorphism must respect *every* code that has the special property. For example, an isometry $\psi \in \mathrm{Aut}(Z_q^n, d_H)$ for constant weight codes is required to map all constant weight codes in Z_q^n onto constant weight codes. Typically a restriction of this type amounts to restricting the group action to a subgroup of $\mathrm{Aut}(Z_q^n, d_H)$.

We discuss here constant weight codes as an example; linear codes are discussed in this setting in Sect. 7.3.1.

Theorem 3.55. *An isometry $\psi \in \mathrm{Aut}(Z_q^n, d_H)$ is weight-preserving if and only if ψ fixes the all-zero word.*

Proof. Since ψ is an isometry, $\psi(\mathbf{0}) = \mathbf{0}$ implies by (3.15) that weight is preserved. Conversely, if $\psi(\mathbf{0}) \ne \mathbf{0}$ then $\psi(\mathbf{x}) = \mathbf{0}$ for some $\mathbf{x} \ne \mathbf{0}$, and ψ is not weight-preserving because $\mathrm{wt}(\mathbf{x}) \ne \mathrm{wt}(\psi(\mathbf{x}))$. □

Thus, for a binary code the weight-preserving subgroup of $\mathrm{Aut}(Z_2^n, d_H)$ is given by the coordinate-permuting action (3.10) of S_n. The result in Theorem 3.55 is a general result that holds for all constant weight codes. For a particular weight w, however, the class-preserving subgroup of $\mathrm{Aut}(Z_q^n, d_H)$ may be larger.

Example 3.56. For binary codes of length n and constant weight $w = n/2$, we may add the all-one word to all words of a code to get another code with the same weight. For example, for

$$C = \{1100, 1010, 1001\},$$
$$D = \{0011, 0101, 0110\},$$

this transformation maps C into D. However, by Theorem 3.55 and convention, the codes C and D are not equivalent as constant weight codes.

3.2.2 Functors

A functor transforms the objects and morphisms of one category into objects and morphisms of another category. Functors have two main applications in the context of this book. First, functors can be used to study whether two different families of combinatorial objects (two categories) are really "the same" up to isomorphism of objects, and if not, to point out the difference in subtle cases. Second, functors mediate changes in representation for purposes of isomorphism computations; this will be discussed in Sects. 3.2.3 and 3.3.2.

Definition 3.57. *Let $\mathscr{A}$ and $\mathscr{B}$ be categories. A* (covariant) functor *F of $\mathscr{A}$ into $\mathscr{B}$ associates to each object $X \in \mathrm{Ob}(\mathscr{A})$ an object $F(X) \in \mathrm{Ob}(\mathscr{B})$, and to each morphism $f \in \mathrm{Mor}(X, Y)$ a morphism $F(f) \in \mathrm{Mor}(F(X), F(Y))$ such that the following two properties hold:*

1. *for all $X \in \mathrm{Ob}(\mathscr{A})$ it holds that*
$$F(1_X) = 1_{F(X)}, \tag{3.16}$$
2. *for all $X, Y, Z \in \mathrm{Ob}(\mathscr{A})$ and all $f \in \mathrm{Mor}(X, Y)$, $g \in \mathrm{Mor}(Y, Z)$ it holds that*
$$F(g \circ f) = F(g) \circ F(f). \tag{3.17}$$

Figure 3.6 illustrates how a functor transforms the objects and morphisms of $\mathscr{A}$ into objects and morphisms of $\mathscr{B}$.

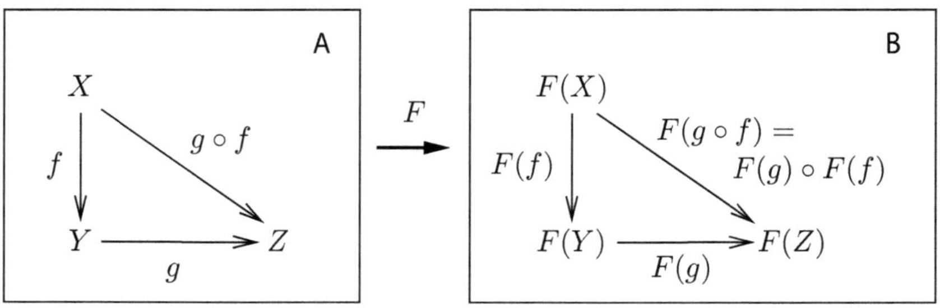

Fig. 3.6. Illustration of Definition 3.57

Here are some simple examples of functors. In each case it is immediate that (3.16) and (3.17) hold, so we will not check these explicitly.

Example 3.58. Consider the category of incidence structures. Associate to each incidence structure $\mathcal{X}$ its dual incidence structure $F(\mathcal{X}) = \mathcal{X}^*$ (Definition 2.28), and to each isomorphism $f = (f_P, f_\mathcal{B}) \in \mathrm{Iso}(\mathcal{X}, \mathcal{Y})$ the isomorphism $F(f) = (f_\mathcal{B}, f_P) \in \mathrm{Iso}(\mathcal{X}^*, \mathcal{Y}^*)$. This defines a functor from the category of incidence structures onto itself. Similarly, a functor is obtained by associating to $\mathcal{X}$ the complement incidence structure $\bar{\mathcal{X}}$ (Definition 2.30) and by associating each isomorphism with itself.

Example 3.59. Let $\mathcal{X}$ be an incidence structure. The *line graph* $\mathrm{LG}(\mathcal{X})$ has the block set $\mathcal{B}(\mathcal{X})$ as vertex set, where any two vertices (blocks) are adjacent if and only if the blocks are incident with at least one common point. Every isomorphism $f = (f_P, f_\mathcal{B}) \in \mathrm{Iso}(\mathcal{X}, \mathcal{Y})$ defines an isomorphism $\mathrm{LG}(f) = f_\mathcal{B} \in \mathrm{Iso}(\mathrm{LG}(\mathcal{X}), \mathrm{LG}(\mathcal{Y}))$.

Alternative names for a line graph in the literature are *block graph* and *block intersection graph*; the latter is also used in a more general context.

Let us continue with two functors related to Theorems 2.120 and 2.127. Here we will only define the functors – their implications to the isomorphism classes of the respective objects will be explored in the context of reconstructibility in Sect. 3.2.3.

Example 3.60. Let us look at the relationship between isomorphism classes of projective planes and affine planes in Theorem 2.120. Consider the category of projective planes with one block individualized, that is, the objects are pairs $(\mathcal{X}, B)$, where $\mathcal{X}$ is a projective plane and $B \in \mathcal{B}(\mathcal{X})$. Let $R(\mathcal{X}, B)$ be the affine plane obtained by taking the residual of $\mathcal{X}$ with respect to B.

Every isomorphism $f = (f_P, f_\mathcal{B}) \in \mathrm{Iso}((\mathcal{X}, B), (\mathcal{Y}, C))$ defines an isomorphism $R(f) \in \mathrm{Iso}(R(\mathcal{X}, B), R(\mathcal{Y}, C))$ obtained by removing B (respectively, C) and the points incident with B (C) from the domain (range) of f. Properties (3.16) and (3.17) are immediate. Thus, R is a functor from the category of projective planes with one block individualized into the category of affine planes. In Example 3.80 we show that these categories are equivalent.

Example 3.61. Here is a concrete instance to illustrate Example 3.60. The lines of a projective plane of order 4 over the point set $\{\mathtt{a}, \mathtt{b}, \ldots, \mathtt{u}\}$ are

```
abcde afghi ajklm anopq arstu bfjnr bgkos
bhlpt bimqu cfkpu cgjqt chmns cilor dflqs
dgmpr dhjou diknt efmot eglnu ehkqr eijps.
```

Taking the residual with respect to the line `abcde`, we obtain the affine plane of order 4 with lines

```
fghi jklm nopq rstu fjnr gkos hlpt imqu fkpu gjqt
hmns ilor flqs gmpr hjou iknt fmot glnu hkqr ijps.
```

Example 3.62. Let us look at the relationship between equivalence classes of Hadamard matrices and isomorphism classes of Hadamard designs in Theorem 2.127. Let $\mathbf{H} = (h_{ij})$ be an Hadamard matrix of order $4n$ with one row $1 \le a \le 4n$ and one column $1 \le b \le 4n$ individualized. Consider the normalization operation from Sect. 2.4.3 and the proof of Theorem 2.127. Performed relative to row a and column b of $\mathbf{H}$, normalization results in a matrix $\mathbf{D}_{ab} = (d_{ij})$ defined for all $1 \le i, j \le 4n$ by

$$d_{ij} = h_{ij}h_{ib}h_{aj}h_{ab}. \tag{3.18}$$

In particular, row a and column b of $\mathbf{D}_{ab}$ are filled with 1s, the rest of the matrix forms an incidence matrix of an Hadamard 2-$(4n-1, 2n-1, n-1)$ design if we replace all -1s by 0s.

Let us look at the normalization operation (3.18) in the context of equivalence and isomorphism. First, it is convenient to represent $\mathbf{H}$ with a, b individualized as a triple $(\mathbf{H}, \mathbf{e}_a, \mathbf{e}_b)$, where $\mathbf{H}$ is an Hadamard matrix of order $4n$, and $\mathbf{e}_k$ denotes the $4n \times 1$ column vector with a 1 in the kth row and 0s elsewhere. Define isomorphism for such triples by extending the action (3.13): a pair $(\mathbf{SP}, \mathbf{TQ})$ of signed permutation matrices acts by

$$(\mathbf{SP}, \mathbf{TQ}) * (\mathbf{H}, \mathbf{e}_a, \mathbf{e}_b) = (\mathbf{SPHQ}^{-1}\mathbf{T}^{-1}, \mathbf{Pe}_a, \mathbf{Qe}_b). \tag{3.19}$$

Call the resulting category $\mathscr{H}_1$. Let $\mathscr{H}_2$ be the category of $\{-1, 1\}$-matrices of size $4n \times 4n$ obtained from objects of $\mathscr{H}_1$ using normalization. Isomorphism in $\mathscr{H}_2$ is given by the action of pre- and postmultiplication by a pair of permutation matrices:

$$(\mathbf{P}, \mathbf{Q}) * \mathbf{D} = \mathbf{PDQ}^{-1}. \tag{3.20}$$

Define a functor HD from $\mathscr{H}_1$ into $\mathscr{H}_2$ by setting

$$\begin{aligned} \mathrm{HD}((\mathbf{H}, \mathbf{e}_a, \mathbf{e}_b)) &= \mathbf{D}_{ab}, \\ \mathrm{HD}((\mathbf{SP}, \mathbf{TQ})) &= (\mathbf{P}, \mathbf{Q}). \end{aligned}$$

To check that isomorphisms transform into isomorphisms, it suffices to observe that

$$\mathrm{HD}((\mathbf{SP}, \mathbf{TQ})(\mathbf{H}, \mathbf{e}_a, \mathbf{e}_b)) = (\mathbf{P}, \mathbf{Q})\, \mathrm{HD}((\mathbf{H}, \mathbf{e}_a, \mathbf{e}_b)).$$

This can be established from (3.18), (3.19), and (3.20) by a direct but slightly lengthy calculation, which we omit. Properties (3.16) and (3.17) follow immediately from (3.12). Thus, HD defines a functor from $\mathscr{H}_1$ into $\mathscr{H}_2$. In Example 3.81 we show that HD establishes a one-to-one correspondence between the isomorphism classes in $\mathscr{H}_1$ and the isomorphism classes in $\mathscr{H}_2$, where the latter category is equivalent to the category of Hadamard 2-$(4n-1, 2n-1, n-1)$ designs.

Example 3.63. Here is a concrete example to illustrate Example 3.62. Let

$$\mathbf{H} = \begin{bmatrix} 1&1&1&1&1&1&1&1 \\ 1&1&-&-&1&1&-&- \\ 1&-&1&-&1&-&1&- \\ 1&-&-&1&1&-&-&1 \\ 1&1&1&1&-&-&-&- \\ 1&1&-&-&-&-&1&1 \\ 1&-&1&-&-&1&-&1 \\ 1&-&-&1&-&1&1&- \end{bmatrix}, \quad a = 6, \quad b = 7.$$

We obtain the normalized matrix

$$\mathbf{D}_{67} = \begin{bmatrix} 1&1&-&-&-&-&1&1 \\ -&-&-&-&1&1&1&1 \\ 1&-&-&1&-&1&1&- \\ -&1&-&1&1&-&1&- \\ -&-&1&1&-&-&1&1 \\ 1&1&1&1&1&1&1&1 \\ -&1&1&-&-&1&1&- \\ 1&-&1&-&1&-&1&- \end{bmatrix}.$$

Observe that if we remove the row and column of 1s from $\mathbf{D}_{67}$, what remains is an incidence matrix of a 2-$(7,3,1)$ design if we replace the -1s with 0s.

The previous examples can all be considered as "canonical" functors, where the transformed objects and morphisms can be defined directly in terms of the original objects and morphisms without introducing any additional structure on top of the original objects. In many cases of combinatorial interest however, a functor can only be defined subject to an appropriate auxiliary "labeling" ℓ_X introduced to every object $X \in \mathrm{Ob}(\mathscr{A})$.

Example 3.64. To give an elementary example of this situation, consider the formal construction of an adjacency matrix of a graph in Definition 2.5. Suppose the vertices of the graph are "○", "□", "▽", and "△". There is no natural ordering for these vertices that would suggest how to label them as v_1, v_2, v_3, v_4 in order to construct an adjacency matrix. Thus, an auxiliary labeling must be introduced to appropriately define the adjacency matrix.

Of course, we would like to abstract away any dependence on a labeling ℓ because it is typically the case that any labeling will do, we just have to select one to appropriately "anchor" the transformations for a rigorous study. The following notion of natural equivalence for functors often provides a convenient abstraction in this respect.

Definition 3.65. *Let F and G be functors from $\mathscr{A}$ to $\mathscr{B}$. A* natural equivalence *t from F to G associates to each $X \in \mathrm{Ob}(\mathscr{A})$ an isomorphism $t_X \in \mathrm{Iso}(F(X), G(X))$ such that*

$$G(f) \circ t_X = t_Y \circ F(f) \tag{3.21}$$

for all $X, Y \in \mathrm{Ob}(\mathscr{A})$ and $f \in \mathrm{Mor}(X, Y)$.

If t is a natural equivalence from F to G, then t^{-1} defined for all $X \in \mathrm{Ob}(\mathscr{A})$ by $t_X^{-1} = (t_X)^{-1}$ is a natural equivalence from G to F. We say that F and G are *equivalent* and write $F \simeq G$ if there exists a natural equivalence from F to G.

As an example of a technically more tedious functor, we study a transformation from resolutions of BIBDs to OE codes that was discovered by Semakov and Zinov'ev [526]. Later we will use this functor to show that the categories of resolutions of BIBDs and OE codes are equivalent (Theorem 3.82).

Example 3.66. Let $\mathcal{R} = (\mathcal{X}, R)$ be a resolution of a 2-(qk, k, λ) design $\mathcal{X}$, $q > 1$. To define the transformation from a resolution to a code, we require a labeling ℓ for the blocks and parallel classes in $\mathcal{R}$ as follows. Let $\ell_{\mathcal{R}} : \{1, 2, \ldots, r\} \times Z_q \to \mathcal{B}(\mathcal{X})$ be a bijection such that $\{\ell_{\mathcal{R}}(i, j) : j \in Z_q\}$ is a parallel class in R for all $1 \le i \le r$, where $r = \lambda(qk-1)/(k-1)$ is the number of parallel classes.

To define the code $\mathrm{SZ}_\ell(\mathcal{R})$, associate with every point $p \in P(\mathcal{X})$ the codeword $\mathbf{x}(p) = (x(p)_1, x(p)_2, \ldots, x(p)_r) \in Z_q^r$ defined for all $1 \le i \le r$ by $(p, \ell_R(i, x(p)_i)) \in I(\mathcal{X})$. Note that each coordinate value $x(p)_i \in Z_q$ is well-defined because p is incident with a unique block in the parallel class $\{\ell_{\mathcal{R}}(i, j) : j \in Z_q\}$. The resulting code

$$\mathrm{SZ}_\ell(\mathcal{R}) = \{\mathbf{x}(p) : p \in P(\mathcal{X})\} \tag{3.22}$$

is equidistant with parameters $(r, qk, r-\lambda)_q$, which satisfy (2.14).

To calculate the minimum distance, observe that any two distinct points $p_1, p_2 \in P(\mathcal{X})$ occur together in exactly λ blocks, which belong to λ different parallel classes. Hence, p_1, p_2 occur in different blocks in the remaining $r - \lambda$ parallel classes, which gives $d_H(\mathbf{x}(p_1), \mathbf{x}(p_2)) = r - \lambda$.

Example 3.67. Here is a concrete instance to illustrate (3.22). Label the parallel classes and the blocks of a resolution $\mathcal{R}$ of an STS(9) over the point set $\{\mathtt{a}, \mathtt{b}, \mathtt{c}, \mathtt{d}, \mathtt{e}, \mathtt{f}, \mathtt{g}, \mathtt{h}, \mathtt{i}\}$ as follows:

$$\begin{aligned}
&\ell_{\mathcal{R}}(1,0) = \mathtt{abc}, \quad \ell_{\mathcal{R}}(1,1) = \mathtt{def}, \quad \ell_{\mathcal{R}}(1,2) = \mathtt{ghi},\\
&\ell_{\mathcal{R}}(2,0) = \mathtt{adg}, \quad \ell_{\mathcal{R}}(2,1) = \mathtt{beh}, \quad \ell_{\mathcal{R}}(2,2) = \mathtt{cfi},\\
&\ell_{\mathcal{R}}(3,0) = \mathtt{aei}, \quad \ell_{\mathcal{R}}(3,1) = \mathtt{bfg}, \quad \ell_{\mathcal{R}}(3,2) = \mathtt{cdh},\\
&\ell_{\mathcal{R}}(4,0) = \mathtt{afh}, \quad \ell_{\mathcal{R}}(4,1) = \mathtt{bdi}, \quad \ell_{\mathcal{R}}(4,2) = \mathtt{ceg}.
\end{aligned}$$

The resulting nine codewords $\mathrm{SZ}_\ell(\mathcal{R}) = \{\mathbf{x}(\mathtt{a}), \mathbf{x}(\mathtt{b}), \ldots, \mathbf{x}(\mathtt{i})\}$ are

$$\begin{aligned}
&\mathbf{x}(\mathtt{a}) = 0000, \quad \mathbf{x}(\mathtt{b}) = 0111, \quad \mathbf{x}(\mathtt{c}) = 0222, \quad \mathbf{x}(\mathtt{d}) = 1021, \quad \mathbf{x}(\mathtt{e}) = 1102,\\
&\mathbf{x}(\mathtt{f}) = 1210, \quad \mathbf{x}(\mathtt{g}) = 2012, \quad \mathbf{x}(\mathtt{h}) = 2120, \quad \mathbf{x}(\mathtt{i}) = 2201.
\end{aligned}$$

Note the dependence on the labeling – a different labeling gives a different code.

Example 3.68. We now extend the transformation SZ_ℓ to isomorphisms. Let $\mathrm{SZ}_\ell(\mathcal{R})$ and $\mathrm{SZ}_\ell(\mathcal{Q})$ be OE codes derived from resolutions $\mathcal{R}$ and $\mathcal{Q}$ using bijections $\ell_{\mathcal{R}}$ and $\ell_{\mathcal{Q}}$, and let $\mathcal{X}$ and $\mathcal{Y}$ be the 2-(qk, k, λ) designs underlying the resolutions $\mathcal{R}$ and $\mathcal{Q}$. For $f = (f_P, f_{\mathcal{B}}) \in \mathrm{Iso}(\mathcal{R}, \mathcal{Q})$, define

$$\mathrm{SZ}_\ell(f) = (k, h) \in S_q \wr S_r, \text{ where } \ell_{\mathcal{Q}}^{-1} f_{\mathcal{B}} \ell_{\mathcal{R}}(i, j) = (h(i), k_{h(i)}(j)) \tag{3.23}$$

for all $1 \le i \le r$ and $j \in Z_q$. Note that (k, h) is well-defined because $f_{\mathcal{B}}$ maps parallel classes in $\mathcal{R}$ onto parallel classes in $\mathcal{Q}$.

Theorem 3.69. *The transformation* SZ_ℓ *given by* (3.22) *and* (3.23) *defines a functor from the category of resolutions of* 2-(qk, k, λ) *designs with* $q > 1$ *into the category of OE codes with parameters* $(r, qk, r - \lambda)_q$.

Proof. Let $f = (f_P, f_{\mathcal{B}}) \in \mathrm{Iso}(\mathcal{R}, \mathcal{Q})$. To establish that $\mathrm{SZ}_\ell(f) = (k, h) \in \mathrm{Iso}(\mathrm{SZ}_\ell(\mathcal{R}), \mathrm{SZ}_\ell(\mathcal{Q}))$ it suffices to show that (k, h) transforms for every $p \in P(\mathcal{X})$ the codeword associated with p in $\mathrm{SZ}_\ell(\mathcal{R})$ into the codeword associated with $f_P(p)$ in $\mathrm{SZ}_\ell(\mathcal{Q})$. We write $\mathbf{x}(p) = (x(p)_1, x(p)_2, \ldots, x(p)_r)$ for the former codeword and $\mathbf{y}(f_P(p)) = (y(f_P(p))_1, y(f_P(p))_2, \ldots, y(f_P(p))_r)$ for the latter codeword; cf. (3.22). By definition of $\mathrm{SZ}_\ell(\mathcal{R})$, we have that $p \in P(\mathcal{X})$ is incident with the block $\ell_{\mathcal{R}}(i, x(p)_i)$ for all $1 \le i \le r$. Because $(f_P, f_{\mathcal{B}}) \in \mathrm{Iso}(\mathcal{X}, \mathcal{Y})$ and (3.23) holds, we have that $f_P(p) \in P(\mathcal{Y})$ is incident with $f_{\mathcal{B}} \ell_{\mathcal{R}}(i, x(p)_i) = \ell_{\mathcal{Q}}(h(i), k_{h(i)}(x(p)_i))$ for all $1 \le i \le r$. On the other hand, $f_P(p)$ is incident with $\ell_{\mathcal{Q}}(h(i), y(f_P(p))_{h(i)})$ for all $1 \le i \le r$. Thus, $y(f_P(p))_{h(i)} = k_{h(i)}(x(p)_i)$ for all $1 \le i \le r$. Consequently, (k, h) transforms $\mathbf{x}(p)$ into $\mathbf{y}(f_P(p))$ under (3.9).

To prove that SZ_ℓ is a functor, it remains to check (3.16) (obvious) and (3.17). To establish (3.17), let $\mathcal{T}$ be a resolution of a 2-(qk, k, λ) design $\mathcal{Z}$ and let $\mathrm{SZ}_\ell(\mathcal{T})$ be the OE code obtained from $\mathcal{T}$ via the labeling $\ell_{\mathcal{T}}$. For $g = (g_P, g_{\mathcal{B}}) \in \mathrm{Iso}(\mathcal{Q}, \mathcal{T})$, let $\mathrm{SZ}_\ell(g) = (t, s)$, where $\ell_{\mathcal{T}} g_{\mathcal{B}} \ell_{\mathcal{Q}}^{-1}(i, j) = (s(i), t_{s(i)}(j))$ for all $1 \le i \le r$ and $j \in Z_q$. We have $\mathrm{SZ}_\ell(g) \circ \mathrm{SZ}_\ell(f) = (t, s)(k, h)$, where the composition operation is given in (3.7). On the other hand, for $\mathrm{SZ}_\ell(g \circ f) = (v, u)$ we obtain for all $1 \le i \le r$ and $j \in Z_q$:

$$\begin{aligned}(u(i), v_{u(i)}(j)) &= \ell_{\mathcal{T}} g_{\mathcal{B}} f_{\mathcal{B}} \ell_{\mathcal{R}}^{-1}(i, j) \\ &= \ell_{\mathcal{T}} g_{\mathcal{B}} \ell_{\mathcal{Q}}^{-1} \ell_{\mathcal{Q}} f_{\mathcal{B}} \ell_{\mathcal{R}}^{-1}(i, j) \\ &= (sh(i), t_{sh(i)} k_{h(i)}(j)).\end{aligned}$$

By (3.7) thus $\mathrm{SZ}_\ell(g \circ f) = (v, u) = (t, s)(k, h) = \mathrm{SZ}_\ell(g) \circ \mathrm{SZ}_\ell(f)$. □

The following theorem removes the dependence on the labeling ℓ up to natural equivalence of functors (Definition 3.65). The techniques used in the proof – which we omit – are analogous to the previous proof.

Theorem 3.70. *Let* SZ_ℓ *and* $\mathrm{SZ}_{\tilde{\ell}}$ *be functors given by* (3.22) *and* (3.23) *subject to labelings* ℓ *and* $\tilde{\ell}$, *respectively. Then,* $t_{\mathcal{R}} = (k, h)$ *defined for all* $1 \le i \le r$ *and* $j \in Z_q$ *by* $(h(i), k_{h(i)}(j)) = \tilde{\ell}_{\mathcal{R}}^{-1} \ell_{\mathcal{R}}(i, j)$ *is a natural equivalence from* SZ_ℓ *to* $\mathrm{SZ}_{\tilde{\ell}}$.

Thus, up to natural equivalence we can speak of the functor SZ without the need to specify the particular labeling ℓ.

3.2.3 Reconstructibility and Equivalence of Categories

By Definition 3.57 it is clear that a functor F transforms isomorphic objects X and Y into isomorphic objects $F(X)$ and $F(Y)$. The converse property is captured in the following definition; cf. [18], [379, Chap. 15].

Definition 3.71. *A functor F from $\mathscr{A}$ into $\mathscr{B}$ is* reconstructible *if for all $X, Y \in \mathrm{Ob}(\mathscr{A})$ it holds that $F(X) \cong F(Y)$ implies $X \cong Y$.*

A reconstructible functor is a useful tool because it enables us to transform many problems concerning the objects in $\mathscr{A}$ into problems concerning the objects in $\mathscr{B}$. For example, $X \cong Y$ if and only if $F(X) \cong F(Y)$, so we can test X and Y for isomorphism in $\mathscr{A}$ by performing the test on $F(X)$ and $F(Y)$ in $\mathscr{B}$. Similarly, if we have a classification of the objects in $\mathscr{B}$ up to isomorphism, then it is usually straightforward to obtain a classification of the objects in $\mathscr{A}$.

Clearly, not all functors are reconstructible.

Example 3.72. By Theorem 2.44 the line graph (Example 3.59) of any square BIBD is isomorphic to the complete graph K_v, but in general more than one isomorphism class of square BIBDs with given parameters exists. Thus, the line graph functor is not reconstructible in the category of square BIBDs.

Reconstructible functors can further be divided into those that also give a one-to-one correspondence between $\mathrm{Mor}(X, Y)$ and $\mathrm{Mor}(F(X), F(Y))$ and those that do not.

Definition 3.73. *A functor F from $\mathscr{A}$ into $\mathscr{B}$ is* strongly reconstructible *if for all $X, Y \in \mathrm{Ob}(\mathscr{A})$ and $g \in \mathrm{Mor}(F(X), F(Y))$ there exists a unique $f \in \mathrm{Mor}(X, Y)$ such that $F(f) = g$.*

Alternatively, a strongly reconstructible functor is called a full and faithful functor [387].

It is perhaps not immediate that reconstructibility and strong reconstructibility are preserved by natural equivalence, so we will give a proof of this.

Theorem 3.74. *Let F and G be equivalent functors from $\mathscr{A}$ into $\mathscr{B}$. If F is (strongly) reconstructible, then so is G.*

Proof. Let t be a natural equivalence from F to G. Let $g \in \mathrm{Iso}(G(X), G(Y))$. By (3.21) we have $t_Y^{-1} g t_X \in \mathrm{Iso}(F(X), F(Y))$. Thus, reconstructibility of F implies reconstructibility of G.

Suppose now that F is strongly reconstructible. To establish injectivity of G on morphisms, let $G(f_1) = G(f_2)$ for some $f_1, f_2 \in \mathrm{Mor}(X, Y)$ and $X, Y \in \mathrm{Ob}(\mathscr{A})$. By (3.21) we obtain

$$F(f_1) = t_Y^{-1} G(f_1) t_X = t_Y^{-1} G(f_2) t_X = F(f_2),$$

which implies $f_1 = f_2$ by strong reconstructibility of F.

To establish surjectivity of G on morphisms, let $g \in \mathrm{Mor}(G(X), G(Y))$ and hence $t_Y^{-1} g t_X \in \mathrm{Mor}(F(X), F(Y))$. Because F is strongly reconstructible, there exists an $f \in \mathrm{Mor}(X, Y)$ such that $F(f) = t_Y^{-1} g t_X$. By (3.21) we have $F(f) = t_Y^{-1} G(f) t_X$. By combining the two equalities it follows that $G(f) = t_Y F(f) t_X^{-1} = g$, which shows that G is surjective on morphisms. □

An even stronger notion of reconstructibility is given in the following definition. We denote by $1_{\mathscr{A}}$ the identity functor in a category $\mathscr{A}$.

Definition 3.75. *A functor F from $\mathscr{A}$ into $\mathscr{B}$ is an* equivalence of categories *if there exists a functor G from $\mathscr{B}$ into $\mathscr{A}$ such that $GF \simeq 1_{\mathscr{A}}$ and $FG \simeq 1_{\mathscr{B}}$.*

We say that $\mathscr{A}$ and $\mathscr{B}$ are *equivalent* and write $\mathscr{A} \simeq \mathscr{B}$ if there exists an equivalence of categories from $\mathscr{A}$ into $\mathscr{B}$.

Theorem 3.76. *Let F be a functor from $\mathscr{A}$ to $\mathscr{B}$. If F is an equivalence of categories, then F is strongly reconstructible.*

Proof. Let G be a functor from $\mathscr{B}$ to $\mathscr{A}$ whose existence is guaranteed by Definition 3.75. To establish injectivity of F on morphisms, let $F(f_1) = F(f_2)$ for some $f_1, f_2 \in \mathrm{Mor}(X, Y)$, $X, Y \in \mathrm{Ob}(\mathscr{A})$. Then, $GF(f_1) = GF(f_2)$. Let t be a natural equivalence from $1_{\mathscr{A}}$ to GF. By (3.21) we obtain $f_1 = t_Y^{-1} GF(f_1) t_X = t_Y^{-1} GF(f_2) t_X = f_2$.

To establish surjectivity of F on morphisms, we start with a preliminary observation. Let $U, V \in \mathrm{Ob}(\mathscr{A})$ and $g \in \mathrm{Mor}(FGF(U), FGF(V))$. Because $1_{\mathscr{B}}$ and hence FG is strongly reconstructible, there exists a unique $f \in \mathrm{Iso}(F(U), F(V))$ such that $FG(f) = g$. Thus, $F : \mathrm{Mor}(GF(U), GF(V)) \to \mathrm{Mor}(FGF(U), FGF(V))$ is surjective for all $U, V \in \mathrm{Ob}(\mathscr{A})$.

Now let $X, Y \in \mathrm{Ob}(\mathscr{A})$ and $g \in \mathrm{Mor}(F(X), F(Y))$. Because $GF \simeq 1_{\mathscr{A}}$ there exist $U, V \in \mathrm{Ob}(\mathscr{A})$ and $r \in \mathrm{Iso}(X, GF(U))$, $s \in \mathrm{Iso}(Y, GF(V))$. Because $F(s) g F(r^{-1}) \in \mathrm{Mor}(FGF(U), FGF(V))$, the earlier observation implies that there exists an $f \in \mathrm{Mor}(GF(U), GF(V))$ such that $F(f) = F(s) g F(r^{-1})$. Consequently, $F(s^{-1} f r) = g$, which shows that F is surjective on morphisms. □

The following theorem demonstrates that a strongly reconstructible functor F is "almost" an equivalence of categories. The additional property that is required is that the image of F contains an object from every isomorphism class of objects in $\mathscr{B}$, cf. [387, IV.§4].

Theorem 3.77. *Let F be a strongly reconstructible functor from $\mathscr{A}$ to $\mathscr{B}$. If for every $Z \in \mathrm{Ob}(\mathscr{B})$ there exists an $X \in \mathrm{Ob}(\mathscr{A})$ such that $F(X) \cong Z$, then F is an equivalence of categories.*

Proof. By assumption we can associate with every $U \in \mathrm{Ob}(\mathscr{B})$ an object $G(U) \in \mathrm{Ob}(\mathscr{A})$ and an isomorphism $\ell_U \in \mathrm{Iso}(U, FG(U))$. For all $U, V \in \mathrm{Ob}(\mathscr{B})$ and $f \in \mathrm{Mor}(U,V)$, define $G(f) = F^{-1}(\ell_V f \ell_U^{-1})$. Note that $G(f)$ is well-defined because $\ell_V f \ell_U^{-1} \in \mathrm{Mor}(FG(U), FG(V))$ and F is strongly reconstructible.

First, we check that G defines a functor. Clearly $G(1_U) = 1_{G(U)}$. By (3.17) we have

$$\begin{aligned} F^{-1}(g \circ f) &= F^{-1}(FF^{-1}(g) \circ FF^{-1}(f)) \\ &= F^{-1}F(F^{-1}(g) \circ F^{-1}(f)) \\ &= F^{-1}(g) \circ F^{-1}(f) \end{aligned} \tag{3.24}$$

for every $X, Y, Z \in \mathrm{Ob}(\mathscr{A})$ and $f \in \mathrm{Mor}(F(X), F(Y))$, $g \in \mathrm{Mor}(F(Y), F(Z))$. Thus, for every $U, V, W \in \mathrm{Ob}(\mathscr{B})$ and $f \in \mathrm{Mor}(U,V)$, $g \in \mathrm{Mor}(V,W)$, we have by (3.24)

$$\begin{aligned} G(g \circ f) &= F^{-1}(\ell_W g f \ell_U^{-1}) \\ &= F^{-1}(\ell_W g \ell_V^{-1} \ell_V f \ell_U^{-1}) \\ &= F^{-1}(\ell_W g \ell_V^{-1}) F^{-1}(\ell_V f \ell_U^{-1}) \\ &= G(g) \circ G(f). \end{aligned}$$

This shows that G is a functor from $\mathscr{B}$ to $\mathscr{A}$.

We proceed to show that F is an equivalence of categories. Let $U, V \in \mathrm{Ob}(\mathscr{B})$. For all $f \in \mathrm{Mor}(U,V)$ we have $FG(f) = \ell_V f \ell_U^{-1}$, which shows that ℓ is a natural equivalence from $1_{\mathscr{B}}$ to FG.

Conversely, let $X, Y \in \mathrm{Ob}(\mathscr{A})$. For all $f \in \mathrm{Mor}(X,Y)$ we have

$$\begin{aligned} GF(f) &= F^{-1}(\ell_{F(Y)} F(f) F^{-1}(\ell_{F(X)}^{-1})) \\ &= F^{-1}(\ell_{F(Y)}) f F^{-1}(\ell_{F(X)}^{-1}) \\ &= F^{-1}(\ell_{F(Y)}) f F^{-1}(\ell_{F(X)})^{-1}. \end{aligned}$$

The second equality follows from (3.24) and the fact that both $\ell_{F(Y)}$ and $\ell_{F(X)}$ are isomorphisms between objects in the image of F in $\mathrm{Ob}(\mathscr{B})$. Thus, $t_X = F^{-1}(\ell_{F(X)})$ is a natural equivalence from $1_{\mathscr{A}}$ to GF. Since $GF \simeq 1_{\mathscr{A}}$ and $FG \simeq 1_{\mathscr{B}}$, we have that F is an equivalence of categories. □

Let us now proceed to combinatorial examples.

Example 3.78. The dual and complement functors in Example 3.58 both define an equivalence of categories because $F \circ F$ is the identity functor in both cases.

Example 3.79. Although the line graph functor in Example 3.59 is not reconstructible in general, there exist important special cases in which reconstructibility is obtained, this is discussed in more detail in Sect. 3.3.2.

Example 3.80. Consider the functor R in Example 3.60. We claim that R is an equivalence of categories.

Let $(\mathcal{X}, B)$ and $(\mathcal{Y}, C)$ be projective planes with blocks B and C individualized, respectively. First, by Theorem 2.62 an affine plane is always resolvable with a unique resolution. In particular, any isomorphism $g \in \text{Iso}(R(\mathcal{X}, B), R(\mathcal{Y}, C))$ maps parallel classes onto parallel classes. Furthermore, by the structure of a projective plane, the parallel classes are in a one-to-one correspondence with the points deleted in taking the residual. Thus, g defines a unique isomorphism $f \in \text{Iso}((\mathcal{X}, B), (\mathcal{Y}, C))$ with $R(f) = g$. This shows that R is strongly reconstructible.

By Theorem 2.54 any affine plane is a residual of a projective plane, so Theorem 3.77 implies that R is an equivalence of categories. Thus, isomorphism classes of affine planes of order n are in a one-to-one correspondence with isomorphism classes of projective planes of order n with one block individualized.

Example 3.81. Consider the Hadamard design functor HD and the categories $\mathscr{H}_1$ and $\mathscr{H}_2$ in Example 3.62. We claim that the functor is reconstructible.

To see this, let $\text{HD}(\mathbf{H}_1, \mathbf{e}_{a_1}, \mathbf{e}_{b_1}) \cong_{\mathscr{H}_2} \text{HD}(\mathbf{H}_2, \mathbf{e}_{a_2}, \mathbf{e}_{b_2})$, that is, there exists a pair $(\mathbf{P}, \mathbf{Q})$ of permutation matrices with

$$(\mathbf{P}, \mathbf{Q})\, \text{HD}(\mathbf{H}_1, \mathbf{e}_{a_1}, \mathbf{e}_{b_1}) = \text{HD}(\mathbf{H}_2, \mathbf{e}_{a_2}, \mathbf{e}_{b_2}).$$

Because only row a_1 and column b_1 in $\text{HD}(\mathbf{H}_1, \mathbf{e}_{a_1}, \mathbf{e}_{b_1})$ are filled with 1s, and similarly for row a_2 and column b_2 in $\text{HD}(\mathbf{H}_2, \mathbf{e}_{a_2}, \mathbf{e}_{b_2})$, we must have $\mathbf{P}\mathbf{e}_{a_1} = \mathbf{e}_{a_2}$ and $\mathbf{Q}\mathbf{e}_{b_1} = \mathbf{e}_{b_2}$. Thus, by (3.19) we have that $(\mathbf{P}, \mathbf{Q})$ is an isomorphism establishing

$$(\text{HD}(\mathbf{H}_1, \mathbf{e}_{a_1}, \mathbf{e}_{b_1}), \mathbf{e}_{a_1}, \mathbf{e}_{b_1}) \cong_{\mathscr{H}_1} (\text{HD}(\mathbf{H}_2, \mathbf{e}_{a_2}, \mathbf{e}_{b_2}), \mathbf{e}_{a_2}, \mathbf{e}_{b_2}). \tag{3.25}$$

The normalization operation (3.18) can be expressed as pre- and postmultiplication by a pair $(\mathbf{S}, \mathbf{T})$ of diagonal matrices with diagonal entries in $\{-1, 1\}$. Thus, by (3.19) we have

$$\begin{aligned} (\mathbf{H}_1, \mathbf{e}_{a_1}, \mathbf{e}_{b_1}) &\cong_{\mathscr{H}_1} (\text{HD}(\mathbf{H}_1, \mathbf{e}_{a_1}, \mathbf{e}_{b_1}), \mathbf{e}_{a_1}, \mathbf{e}_{b_1}), \\ (\mathbf{H}_2, \mathbf{e}_{a_2}, \mathbf{e}_{b_2}) &\cong_{\mathscr{H}_1} (\text{HD}(\mathbf{H}_2, \mathbf{e}_{a_2}, \mathbf{e}_{b_2}), \mathbf{e}_{a_2}, \mathbf{e}_{b_2}). \end{aligned} \tag{3.26}$$

Combining (3.25) and (3.26), we obtain $(\mathbf{H}_1, \mathbf{e}_{a_1}, \mathbf{e}_{b_1}) \cong_{\mathscr{H}_1} (\mathbf{H}_2, \mathbf{e}_{a_2}, \mathbf{e}_{b_2})$, and reconstructibility follows. (The functor HD is not strongly reconstructible because of the kernel of (3.19) generated by $(-\mathbf{I}, -\mathbf{I})$.)

Reconstructibility and the definition of $\mathscr{H}_2$ imply that the isomorphism classes in $\mathscr{H}_1$ and $\mathscr{H}_2$ are in a one-to-one correspondence. Based on the proof of Theorem 2.127 it is not difficult to check that $\mathscr{H}_2$ is equivalent to the category of Hadamard 2-$(4n-1, 2n-1, n-1)$ designs. Thus, isomorphism classes of Hadamard 2-$(4n-1, 2n-1, n-1)$ designs are in a one-to-one correspondence with equivalence classes of Hadamard matrices of order $4n$ with one row and column individualized.

The reader is encouraged to carry out the exercise of establishing analogous correspondences between the other categories of objects occurring in Theorems 2.120 and 2.127.

We conclude this section by showing that the functor SZ from resolutions of BIBDs into OE codes is an equivalence of categories.

Theorem 3.82. *The category of resolutions of* 2-(qk, k, λ) *designs is equivalent to the category of OE codes with parameters* $(r, qk, r-\lambda)_q$*, where* q, k, λ, r *are positive integers that satisfy* $q > 1$ *and* $r = \lambda(qk-1)/(k-1)$.

Proof. We apply Theorem 3.77 to the functor SZ defined by (3.22) and (3.23). By the generalized Plotkin bound (Theorem 2.82), every coordinate of an OE code with parameters $(r, qk, r-\lambda)_q$ contains exactly k occurrences of every coordinate value. Thus, for any such code $C \subseteq Z_q^r$, the incidence structure $\mathcal{X}$ defined by $P(\mathcal{X}) = C$, $\mathcal{B}(\mathcal{X}) = \{1, 2, \ldots, r\} \times Z_q$, and $I(\mathcal{X}) = \{(\mathbf{x}, (i,j)) : \mathbf{x} \in C,\ 1 \leq i \leq r,\ j \in Z_q,\ x_i = j\}$ is a 2-(qk, k, λ) design. Furthermore, $R = \{\{(i,j) : j \in Z_q\} : 1 \leq i \leq r\}$ defines a resolution $\mathcal{R} = (\mathcal{X}, R)$. Relative to the identity labeling $\ell_{\mathcal{R}}(i,j) = (i,j)$, we have $\mathrm{SZ}_\ell(\mathcal{R}) = C$. By Theorem 3.70 thus $\mathrm{SZ}(\mathcal{R}) \cong C$ independent of the labeling ℓ. Thus, every equivalence class of OE codes with parameters $(r, qk, r-\lambda)_q$ contains a code that lies in the image of SZ.

It remains to establish strong reconstructibility of SZ. By Theorems 3.70 and 3.74, strong reconstructibility is independent of the labeling ℓ employed. Let $\mathcal{R}$ and $\mathcal{Q}$ be resolutions of 2-(qk, k, λ) designs with underlying designs $\mathcal{X}$ and $\mathcal{Y}$, respectively.

To establish injectivity of $\mathrm{SZ}_\ell : \mathrm{Iso}(\mathcal{R}, \mathcal{Q}) \to \mathrm{Iso}(\mathrm{SZ}_\ell(\mathcal{R}), \mathrm{SZ}_\ell(\mathcal{Q}))$, observe that $\mathrm{SZ}_\ell(f)$ and (3.23) together uniquely determine $f_{\mathcal{B}}$. Furthermore, from $f_{\mathcal{B}}$ we can uniquely determine f_P since by $r > \lambda$ no two points are incident with the same set of blocks.

To establish surjectivity, let $(k, h) \in \mathrm{Iso}(\mathrm{SZ}_\ell(\mathcal{R}), \mathrm{SZ}_\ell(\mathcal{Q}))$. Define $f_P : P(\mathcal{X}) \to P(\mathcal{Y})$ so that $(k,h)(\mathbf{x}(p)) = \mathbf{y}(f_P(p))$ for all $p \in P(\mathcal{X})$. Define $f_{\mathcal{B}} : \mathcal{B}(\mathcal{X}) \to \mathcal{B}(\mathcal{Y})$ by setting $f_{\mathcal{B}}(\ell_{\mathcal{R}}(i,j)) = \ell_{\mathcal{Q}}(h(i), k_{h(i)}(j))$ for all $1 \leq i \leq r$ and $j \in Z_q$. Obviously $\mathrm{SZ}_\ell(f) = (k,h)$ for $f = (f_P, f_{\mathcal{B}})$ by (3.23). It is also immediate that $f_{\mathcal{B}}$ maps the parallel classes in $\mathcal{R}$ onto parallel classes in $\mathcal{Q}$. It remains to check that $f \in \mathrm{Iso}(\mathcal{X}, \mathcal{Y})$. Let $p \in P(\mathcal{X})$ and $B \in \mathcal{B}(\mathcal{X})$. Let i, j be determined from $\ell_{\mathcal{R}}(i,j) = B$. By (3.22), $(p, B) \in I(\mathcal{X})$ if and only if $x(p)_i = j$. By (3.9) and $(k,h)(\mathbf{x}(p)) = \mathbf{y}(f_P(p))$, we have $x(p)_i = j$ if and only if $y(f_P(p))_{h(i)} = k_{h(i)}(j)$. Again applying (3.22), the latter holds if and only if $(f_P(p), \ell_{\mathcal{Q}}(h(i), k_{h(i)}(j))) \in I(\mathcal{Y})$, that is, $(f_P(p), f_{\mathcal{B}}(B)) \in I(\mathcal{Y})$. It follows that $f \in \mathrm{Iso}(\mathcal{X}, \mathcal{Y})$ and hence $f \in \mathrm{Iso}(\mathcal{R}, \mathcal{Q})$. □

3.3 Isomorphism Computations

Isomorphism computations constitute an integral part of every classification algorithm. Not only is it important to detect and remove isomorphic objects from the final classification, but isomorphism computations are also in most cases required to eliminate redundant work when generating the objects of interest. The extent of isomorphism computations varies from algorithm to algorithm. In general, the best classification algorithms attempt to avoid expensive isomorphism computations as much as possible by taking advantage of the way in which the objects are constructed, and by replacing expensive computations by lighter computations by means of isomorphism invariants or group-theoretic techniques.

To understand and to implement classification algorithms in practice, we require a treatment of both the abstract concepts associated with isomorphism computations as well as the concrete details how to implement an algorithm that operates on a specific family of objects.

From an implementation perspective there are essentially two approaches to isomorphism computations on a family of combinatorial objects. Either the computations are carried out using a general-purpose algorithm operating on a category on which most types of objects can be represented, or a tailored algorithm specific to the objects of interest is used. Tailored algorithms have the best performance, but require time to implement, and they are arguably more prone to errors than a well-tested readily available general-purpose algorithm. Furthermore, the practical performance of a general-purpose algorithm can often be improved to a very acceptable level compared with tailored algorithms by the use of isomorphism invariants specific to the objects of interest.

These observations withstanding, we have chosen to base the present discussion of isomorphism computations on general-purpose algorithms operating on colored graphs. Once the algorithms operating on colored graphs are understood, it is possible to tailor the algorithms to the objects of interest if necessary.

We define colored graphs and develop representations for central types of combinatorial objects as colored graphs in Sect. 3.3.2. We then proceed to discuss isomorphism invariants and the use of invariants in expediting isomorphism computations in Sects. 3.3.3, 3.3.4, and 3.3.5. Once these tools are available, we are ready to discuss the types of isomorphism problems recurrently encountered and the tools for solving them in Sect. 3.3.6. For the subsequent treatment of isomorph rejection in classification algorithms, it will be convenient to define the problems and tools first in terms of an abstract group action; colored graphs are illustrated as a special case.

The implementation of general-purpose isomorphism algorithms is not discussed in detail until Sect. 5.6, where we have available further conceptual and algorithmic tools, namely, backtrack search, search trees, and certain permutation group algorithms. Also pointers to tailored isomorphism algorithms will be given in this context. What is convenient is that an appropriate isomorphism algorithm can in most cases be treated as a black-box subroutine in the context of classification algorithms, whereby a detailed understanding of the algorithm implementation is not required, and the black-box description given in Sect. 3.3.6 suffices. The computational complexity of isomorphism computations is discussed in Chap. 11.

3.3.1 Lexicographic Order

Before proceeding to the main subject matter, we will briefly discuss lexicographic order, which is frequently employed in the context of isomorphism computations because it is easy to define by extension to virtually any type of object constructed from elements of an ordered set. Lexicographic order also has other desirable properties in the context of isomorph rejection in classification algorithms. These will be discussed in Chap. 4.

An ordered set is *well-ordered* if every nonempty subset has a minimum element. In particular, all finite ordered sets are well-ordered.

Definition 3.83. *Let A be an ordered set, let J be a well-ordered set, and let* $\mathrm{Fun}(J, A)$ *be the set of all functions from J into A. The* lexicographic order $\prec$ *on* $\mathrm{Fun}(J, A)$ *is defined for all $f_1, f_2 \in \mathrm{Fun}(J, A)$ by $f_1 \prec f_2$ if and only if there exists an $i \in J$ such that $f_1(i) < f_2(i)$ and $f_1(j) = f_2(j)$ for all $j \in J$ with $j < i$.*

We use this basic order on functions to define a lexicographic order for other types of objects constructed from ordered sets. Let $(a_1, a_2, \ldots, a_m) \in A^m$ be an m-tuple with entries from an ordered set A. A lexicographic order for the m-tuples in A^m is defined by identifying each tuple with the function $i \mapsto a_i$ from $\{1, 2, \ldots, m\}$ into A, where $1 < 2 < \cdots < m$. In particular, for $A = \{0, 1, \ldots, q-1\}$ with $0 < 1 < \cdots < q-1$, this order corresponds to the usual ordering for m-digit q-ary integers.

Example 3.84. For $q = 2$ and $m = 3$, we have

$$000 \prec 001 \prec 010 \prec 011 \prec 100 \prec 101 \prec 110 \prec 111.$$

Given a well-ordered set S, a lexicographic order on the set of all subsets of S is defined by identifying a subset $T \subseteq S$ with the characteristic function $\chi_T : S \to \{0, 1\}$ defined for all $s \in S$ by

$$\chi_T(s) = \begin{cases} 1 & \text{if } s \in T, \\ 0 & \text{if } s \notin T. \end{cases}$$

An analogous lexicographic order for multisets with elements drawn from S is obtained by identifying a multiset T with the function χ_T associating to each $s \in S$ its multiplicity $\chi_T(s)$ in T.

Example 3.85. Let $S = \{1,2,3\}$ with $1 < 2 < 3$. The lexicographic order on the set of all subsets of S is

$$\emptyset \prec \{3\} \prec \{2\} \prec \{2,3\} \prec \{1\} \prec \{1,3\} \prec \{1,2\} \prec \{1,2,3\}.$$

Based on repeated application of these constructions we can extend lexicographic order to sets of subsets, tuples of subsets, tuples of tuples, and so forth. Similarly, we obtain by restriction a lexicographic order on the set of all k-subsets and partitions of a set. However, one has to be careful with well-ordering whenever a construction is applied to an infinite ordered set.

Example 3.86. Let $S = \{1,2,3\}$ with $1 < 2 < 3$. The lexicographic order on the set of all partitions of S is

$$\{\{1,2,3\}\} \prec \{\{1\},\{2,3\}\} \prec \{\{1,3\},\{2\}\} \prec \{\{1,2\},\{3\}\} \prec \{\{1\},\{2\},\{3\}\}.$$

3.3.2 Representing Objects as Colored Graphs

Most combinatorial objects occurring in practice can be succinctly represented as colored graphs for purposes of isomorphism computations.

For our subsequent discussion it will be convenient to view a coloring of the vertices as an ordered partition as follows.

Definition 3.87. *An* ordered partition *is a tuple* $\pi = (V_1, V_2, \ldots, V_m)$ *where* $\{V_1, V_2, \ldots, V_m\}$ *is a partition of a finite set* V. *The sets* $V_1, V_2, \ldots, V_m$ *are called the* cells *or* color classes *of the partition.*

Given $x \in V$, we write $\pi(x)$ for the index of the cell in which x occurs, that is, $x \in V_i$ if and only if $\pi(x) = i$. Intuitively, the coloring π assigns the color $\pi(x)$ to the vertex x. Accordingly, it is convenient to identify the ordered partition π with the function of V onto $\{1,2,\ldots,m\}$ defined by $x \mapsto \pi(x)$.

In the context of isomorphism computations we use the following precise notion of a colored graph.

Definition 3.88. *A* colored graph *is a pair* (G, π), *where* G *is a graph and* π *is an ordered partition of* $V(G)$. *For two colored graphs,* (G,π) *and* (H,σ), *the set of isomorphisms of* (G,π) *onto* (H,σ) *is defined by*

$$\begin{aligned}\mathrm{Iso}((G,\pi),(H,\sigma)) = \{&f \in \mathrm{Iso}(G,H) : \\ &\pi(u) = \sigma(f(u)) \text{ for all } u \in V(G)\}.\end{aligned} \tag{3.27}$$

In other words, an isomorphism of colored graphs must map vertices of each color onto vertices of the same color.

We will now present possible transformations into colored graphs from some of the main families of combinatorial objects discussed in this book, namely designs, codes, and Hadamard matrices. When the idea is understood, it is usually not difficult to develop transformations for other families. Note, however, that several different transformations are often possible. The transformations are presented mostly without a detailed proof; the concepts and tools suitable for checking the desired properties of the transformations in connection to associated notions of isomorphism can be found in Sect. 3.2.

The transformation for designs is a standard tool in design theory and occurs throughout the literature, cf. [126], [205, Remark 9.41], [342, Sect. 7.4.2], and [412].

Definition 3.89. *Let $\mathcal{X}$ be an incidence structure where $P(\mathcal{X})$ and $\mathcal{B}(\mathcal{X})$ are disjoint. The* incidence graph $\mathrm{IG}(\mathcal{X})$ *is the colored graph with vertex set $P(\mathcal{X}) \cup \mathcal{B}(\mathcal{X})$, edge set $\{\{p, B\} : (p, B) \in I(\mathcal{X})\}$, and vertex coloring $(P(\mathcal{X}), \mathcal{B}(\mathcal{X}))$.*

The incidence graph [55, 207] is also called the *Levi graph* [18, 139], especially in the context of projective planes.

Example 3.90. A colored graph that is obtained from the Fano plane in Example 2.34 is presented in Fig. 3.7. This particular graph without the coloring of the vertices is known as the *Heawood graph* [55].

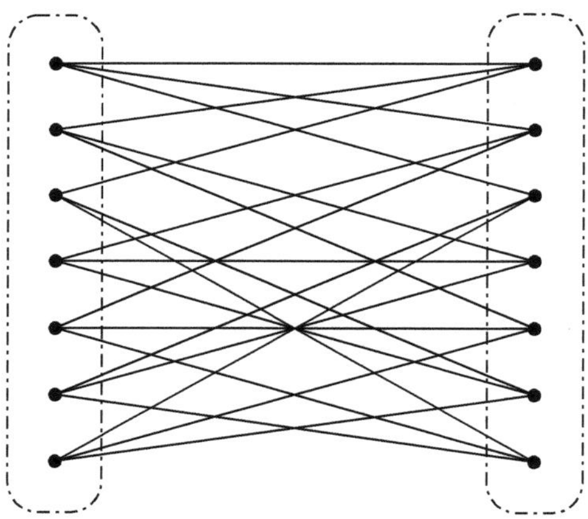

Fig. 3.7. Transforming an incidence structure into a colored graph

For illustration, let us review the steps that suffice to check that it is appropriate to use the incidence graph representation for isomorphism computations on incidence structures. First we extend to a functor the transformation IG from incidence structures into colored graphs. Let $f = (f_P, f_{\mathcal{B}}) \in \mathrm{Iso}(\mathcal{X}, \mathcal{Y})$ be an isomorphism of incidence structures. Associate with f the bijection $\mathrm{IG}(f)$ obtained by merging f_P and $f_{\mathcal{B}}$ into one bijection from $P(\mathcal{X}) \cup \mathcal{B}(\mathcal{X})$ onto $P(\mathcal{Y}) \cup \mathcal{B}(\mathcal{Y})$. Because f is an isomorphism of incidence structures, it follows by Definitions 2.22, 2.7, and 3.89 that $\mathrm{IG}(f)$ is an isomorphism of the colored graph $\mathrm{IG}(\mathcal{X})$ onto the colored graph $\mathrm{IG}(\mathcal{Y})$. Properties (3.16) and (3.17) are immediate, which allows us to conclude that IG now defines a functor. By the structure of the incidence graph and the coloring of the vertices, any isomorphism of incidence graphs defines a unique isomorphism of the original incidence structures; thus, the functor is strongly reconstructible, which enables us to alternate at will between the representations for purposes of isomorphism computations.

The reader who feels uncomfortable with categories and functors can alternatively consider the framework of group actions and Definition 3.16 in analyzing different representations. In this setting a functor corresponds to a homomorphism of group actions, to be discussed in Sect. 4.2.4. Group actions are employed to analyze a colored graph representation of Hadamard matrices in [410].

If the vertex coloring is removed from the incidence graph, the incidence graph functor is not reconstructible.

Example 3.91. Consider any incidence structure that is not isomorphic to its dual. One possible incidence structure is the Pasch configuration in Example 2.21. If we remove the vertex coloring from the incidence graph, then the Pasch configuration and its dual have isomorphic incidence graphs, see Fig. 3.8. Since the Pasch configuration is not isomorphic to its dual, the incidence graph functor without vertex coloring is not reconstructible.

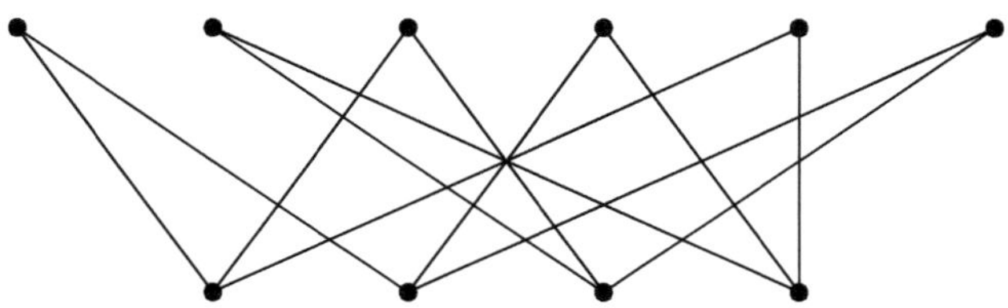

Fig. 3.8. An ambiguous incidence graph without vertex coloring

In addition to the incidence graph, sometimes also a more compact representation can be used for incidence structures. In particular, the line graph functor LG from Example 3.59. However, we have to be very careful with reconstructibility because in general an incidence structure is not reconstructible from its line graph.

Let us consider reconstructibility from line graphs in more detail; cf. [18]. It is easy to see that a point $p \in P(\mathcal{X})$ incident with r blocks defines an clique of size r in $\mathrm{LG}(\mathcal{X})$ that consists of the blocks incident with p. Thus, assuming that every point is incident with at least one block, we can reconstruct $\mathcal{X}$ up to isomorphism from $\mathrm{LG}(\mathcal{X})$ if we can identify the cliques in $\mathrm{LG}(\mathcal{X})$ that correspond to the points of $\mathcal{X}$. The following theorem of Deza [155, 156] allows us to sometimes identify such cliques based on their size.

Theorem 3.92. *Let $\mathcal{Z}$ be an incidence structure in which every block is incident with exactly k points and every pair of distinct blocks has exactly μ points in common. Then, either all the blocks have the same μ points in common, or* $|\mathcal{B}(\mathcal{Z})| \leq k^2 - k + 1$.

Corollary 3.93. *Let $\mathcal{X}$ be an incidence structure in which every block is incident with exactly k points and every pair of distinct blocks has at most one point in common. If every point of $\mathcal{X}$ is incident with more than $k^2 - k + 1$ blocks, then $\mathcal{X}$ is strongly reconstructible from* $\mathrm{LG}(\mathcal{X})$.

Proof. An clique of size r in $\mathrm{LG}(\mathcal{X})$ corresponds to a set of r blocks in $\mathcal{X}$ with pairwise exactly one point in common. By Theorem 3.92 such a clique is formed by blocks that have the same point of $\mathcal{X}$ in common if $r > k^2 - k + 1$. To reconstruct $\mathcal{X}$ up to isomorphism, let the blocks be the vertices of $\mathrm{LG}(\mathcal{X})$, and let the points be the maximal cliques of size greater than $k^2 - k + 1$; a block (vertex) is incident with a point (clique) if and only if the vertex occurs in the clique. Strong reconstructibility follows from the observation that any isomorphism of graphs must map maximal cliques onto maximal cliques. □

Corollary 3.94. *A Steiner system $S(2, k; v)$ is strongly reconstructible from its line graph if* $v > (k^2 - k + 1)(k - 1) + 1$.

In particular, a Steiner triple system of order v is strongly reconstructible from its line graph if $v > 15$.

Example 3.95. Strong reconstructibility from line graphs fails for Steiner triple systems of order 15. Namely, the STS(15) with automorphism group of order 20,160 has a line graph with automorphism group of order 40,320.

The transformation for codes appears in [459]. For a q-ary code $C \subseteq Z_q^n$, we construct a colored graph $\mathrm{CG}(C)$ with vertex set $C \cup (\{1, 2, \ldots, n\} \times Z_q)$, edge set $\{\{\mathbf{x}, (i, x_i)\} : \mathbf{x} \in C,\ i \in \{1, 2, \ldots, n\}\} \cup \{\{(j, a), (j, b)\} : j \in \{1, 2, \ldots, n\},\ a, b \in Z_q\}$, and vertex coloring $(C, \{1, 2, \ldots, n\} \times Z_q)$.

Example 3.96. The graph obtained from the code $\{000, 011, 221\} \subseteq Z_3^3$ is shown in Fig. 3.9.

It is not difficult to check that C is strongly reconstructible from $\mathrm{CG}(C)$. Namely, observe that the vertices of the second color induce a union of n disjoint copies of K_q in $\mathrm{CG}(C)$. The automorphism group of this graph is isomorphic to the group $S_q \wr S_n$ inducing equivalence for codes; vertices of the first color encode by adjacency the structure of the codewords.

By utilizing the equivalence of resolutions of BIBDs and OE codes (Theorem 3.82), the transformation for codes can be used for resolutions of BIBDs as well.

Also in the classification of linear codes, a transformation to colored graphs can be utilized. However, since such a transformation is not as straightforward

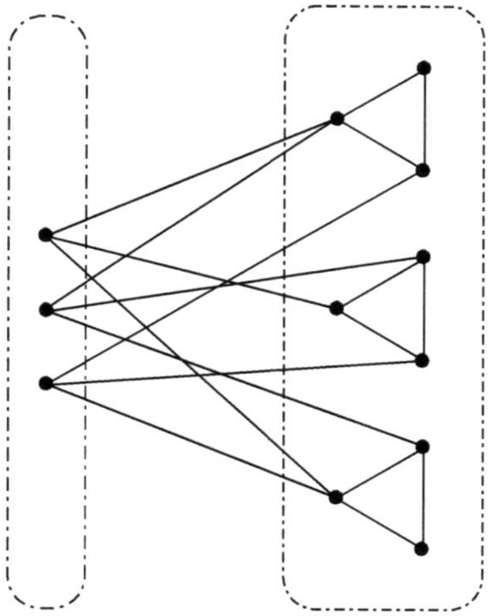

Fig. 3.9. Transforming an unrestricted code into a colored graph

as those presented here, it is postponed to the discussion of linear codes in Sect. 7.3.3.

If we consider Hadamard matrices as binary codes and use a transformation for codes, then we miss the transformations that complement codewords (cf. Definition 2.125). The following transformation, which applies to an Hadamard matrix $\mathbf{H} = (h_{ij})$ of order n, is a slight modification of that in [410]. Let $V = \{v_1, v_2, \ldots, v_n\}$, $V' = \{v'_1, v'_2, \ldots, v'_n\}$, $W = \{w_1, w_2, \ldots, w_n\}$, and $W' = \{w'_1, w'_2, \ldots, w'_n\}$ be pairwise disjoint sets. We now construct a colored graph $\mathrm{HG}(\mathbf{H})$ with vertex set $V \cup V' \cup W \cup W'$, edge set $\{\{v_i, w_j\} : h_{ij} = -1\} \cup \{\{v'_i, w'_j\} : h_{ij} = -1\} \cup \{\{v_i, w'_j\} : h_{ij} = 1\} \cup \{\{v'_i, w_j\} : h_{ij} = 1\} \cup \{\{v_i, v'_i\}, \{w_i, w'_i\} : i \in \{1, 2, \ldots, n\}\}$, and vertex coloring $(V \cup V', W \cup W')$.

Example 3.97. The colored graph obtained from the unique Hadamard matrix of order 2 (see Example 2.126) is shown in Fig. 3.10.

Again it is not difficult to check that $\mathbf{H}$ is strongly reconstructible from $\mathrm{HG}(\mathbf{H})$; cf. [410]. Observe that both color classes induce graphs with automorphism group isomorphic to $S_2 \wr S_n$; the edges between color classes encode the entries of the matrix.

The transformation described for Hadamard matrices can be generalized to any $n \times m$ matrices with the same notion of isomorphism.

As is implicit in the transformations for codes and Hadamard matrices, a transformation into a colored graph is typically based on representing the group whose action captures the isomorphisms for the objects of interest as the automorphism group of a colored graph. Once an appropriate colored graph representation for the acting group is available, it is usually straightforward to encode the structure in the objects of interest by adding vertices and edges.

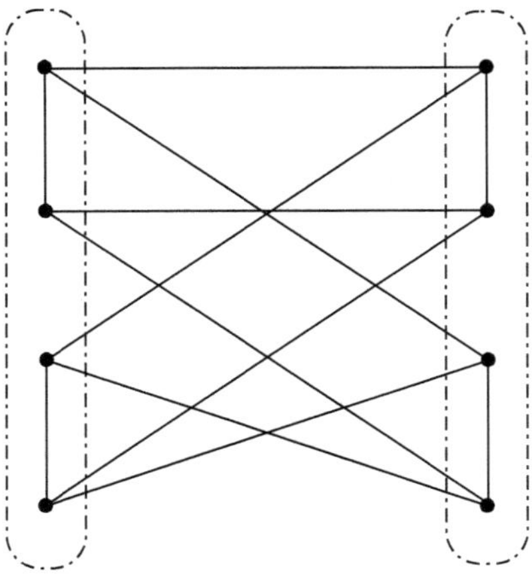

Fig. 3.10. Transforming an Hadamard matrix into a colored graph

3.3.3 Invariants and Certificates

Isomorphism computations are often laborious for combinatorial objects because of their regular structure. Appropriate isomorphism invariants can be used to expedite isomorphism computations in many difficult cases.

Definition 3.98. *An* (isomorphism) invariant *for a category* $\mathscr{A}$ *is a function* I *with domain* $\mathrm{Ob}(\mathscr{A})$ *such that for all* $X, Y \in \mathrm{Ob}(\mathscr{A})$ *it holds that* $X \cong Y$ *implies* $I(X) = I(Y)$.

In other words, an invariant is a property of the objects in $\mathscr{A}$ that is preserved by isomorphism. An obvious immediate application for an invariant is in showing that two objects are nonisomorphic: $I(X) \neq I(Y)$ implies $X \ncong Y$. In general, if $I(X) \neq I(Y)$, then we say that I *distinguishes* X from Y.

We begin by discussing graph invariants and then proceed to other types of objects. Some examples of graph invariants of varying computational difficulty are as follows.

Example 3.99. Any property of a graph that does not depend on the labels of the vertices is an invariant for the category of graphs. Such properties include the number of vertices, the number of edges, the multiset consisting of the degree of every vertex, the number of triangles (K_3 subgraphs) in the graph, the multiset of eigenvalues of an adjacency matrix, the order of the automorphism group, the multiset of lengths of automorphism orbits, and so forth.

Example 3.100. The graphs in Fig. 3.11 are nonisomorphic as witnessed by the number of triangles: the left-hand side graph has four triangles, the right-hand side graph has none.

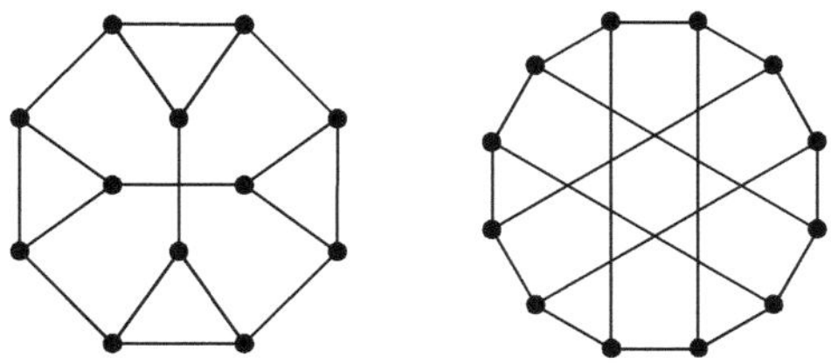

Fig. 3.11. Two nonisomorphic graphs distinguished by the number of triangles

Example 3.101. Graphs derived from combinatorial objects can be highly regular. For example, no invariant mentioned in Example 3.99 can distinguish between the nonisomorphic line graphs of the following two Steiner triple systems of order 15:

abc ade afg ahi ajk	abc ade afg ahi ajk
alm ano bdf beh bgj	alm ano bdf beh bgj
bik bln bmo cdg cel	bil bkn bmo cdg cef
cfh cij cmn cko dim	cio chl cjn ckm dij .
dkn hjn dho djl efn	dmn dho dkl hjm fim
gin eio ejm egk fil	egm ein eko ejl fjo
fjo fkm ghm glo hkl	fhk fln ghn gik glo

Invariants can of course be defined for categories other than graphs.

Example 3.102. Given a Steiner triple system, an invariant is obtained by counting the number of Pasch configurations (Example 2.21) occurring in the system. The systems in Example 3.101 both contain 12 Pasch configurations. A more powerful invariant is obtained from the multiset that contains for each block B the number of Pasch configurations in which B occurs. This invariant distinguishes the systems in Example 3.101.

Example 3.103. Given a code C, the multiset that contains the Hamming distances of every pair of codewords in C is an invariant for the category of codes under every notion of isomorphism considered for codes so far.

Example 3.104. Given a 1-factorization $\mathcal{F}$ of a graph, the union of every pair of distinct 1-factors is a disjoint union of cycles of even length at least 4, cf. Figs. 2.6 and 2.7. We obtain an invariant for 1-factorizations from the multiset consisting of the multiset of cycle lengths in each such union.

An invariant that can perfectly distinguish between different isomorphism classes in a category is called a certificate [427].

Definition 3.105. *An invariant I for $\mathscr{A}$ is a* certificate *if for all $X, Y \in \mathrm{Ob}(\mathscr{A})$ it holds that $I(X) = I(Y)$ implies $X \cong Y$.*

Alternative names for a certificate in the literature include a *code* [501] or a *complete set of invariants* (for a set of invariants) [254]. A certificate is particularly desirable in the context of isomorphism computations because it suffices to test certificate values for equality to test isomorphism: $X \cong Y$ if and only if $I(X) = I(Y)$.

Example 3.106. One possible certificate for graphs of order n is as follows. Given a graph G, let $\mathbf{A}$ be an adjacency matrix of G and put

$$I(G) = \max\{\mathbf{PAP}^{-1} : \mathbf{P} \text{ is an } n \times n \text{ permutation matrix}\}, \tag{3.28}$$

where the maximum is taken with respect to some order on the set of all $n \times n$ adjacency matrices of graphs.

In other words, $I(G)$ is the maximum adjacency matrix associated with G. It follows immediately from Definitions 2.5 and 2.7 that I is a certificate for graphs of order n.

A lexicographic order for adjacency matrices is obtained as follows. Associate with an adjacency matrix $\mathbf{A}$ the binary $n(n-1)/2$-tuple obtained by listing the entries above the diagonal column by column; that is,

$$(a_{12}, a_{13}, a_{23}, a_{14}, a_{24}, a_{34}, a_{15}, \ldots, a_{n-1,n}).$$

Each adjacency matrix is associated with a unique tuple because $\mathbf{A}$ is symmetric with zero diagonal. Now let the lexicographic order on adjacency matrices be defined from the lexicographic order on the associated tuples.

Example 3.107. The lexicographically maximum certificates for the graphs of order 4 in Fig. 2.3 are:

$$\begin{bmatrix}0000\\0000\\0000\\0000\end{bmatrix}, \begin{bmatrix}0100\\1000\\0000\\0000\end{bmatrix}, \begin{bmatrix}0100\\1000\\0001\\0010\end{bmatrix}, \begin{bmatrix}0110\\1000\\1000\\0000\end{bmatrix}, \begin{bmatrix}0110\\1001\\1000\\0100\end{bmatrix}, \begin{bmatrix}0111\\1000\\1000\\1000\end{bmatrix},$$

$$\begin{bmatrix}0110\\1010\\1100\\0000\end{bmatrix}, \begin{bmatrix}0110\\1001\\1001\\0110\end{bmatrix}, \begin{bmatrix}0111\\1010\\1100\\1000\end{bmatrix}, \begin{bmatrix}0111\\1011\\1100\\1100\end{bmatrix}, \begin{bmatrix}0111\\1011\\1101\\1110\end{bmatrix}.$$

It is not difficult to define certificates analogous to the certificate in Example 3.106 for colored graphs, incidence structures, and codes. In subsequent chapters, certificates of this type based on lexicographic order are employed in several classification algorithms. However, the practical applicability of these certificates is restricted to relatively small object sizes. Namely, the known methods for computing a certificate of this type quickly become very inefficient as the size of the objects increases. For example, a naïve implementation of the certificate in Example 3.106 searches among the $n!$ permutations for a permutation that produces the maximum incidence matrix. A somewhat more educated search strategy can of course do better than this in terms of the number of permutations searched, but nevertheless all the known approaches become impractical already for small values of n.

To obtain practical isomorphism algorithms also for larger objects, we must take advantage of the isomorphism-invariant structure within an object.

3.3.4 Subobject Invariants

Our main application for invariants will be in narrowing down the set of possible isomorphisms between objects to expedite isomorphism computations. In essentially all cases encountered so far, an isomorphism can be viewed as a bijection defined on certain subobjects of an object, say, the vertices of a graph, the points and blocks of an incidence structure, or the coordinate values in the coordinates of a code. If we can distinguish between different subobjects in an isomorphism-invariant manner, then we can restrict the possible isomorphisms between objects. This leads us to *subobject invariants*; that is, invariants for a category where each object is a pair consisting of an object and an individualized subobject.

For example, let I be an invariant for the category of graphs with one vertex individualized (Example 3.44), and let G and H be two graphs. For all $f \in \mathrm{Iso}(G, H)$ and $x \in V(G)$ we have by definition that $(G, x) \cong (H, f(x))$ in the category of graphs with one vertex individualized. Because I is an invariant for this category, $I(G, x) = I(H, f(x))$. It follows that any isomorphism from G to H must map a vertex $x \in V(G)$ into a vertex $y \in V(H)$ satisfying $I(G, x) = I(H, y)$. An invariant for the category of graphs with one vertex individualized is called a *vertex invariant*.

Before proceeding to discuss examples, we adopt the convention of writing multisets as sorted tuples consisting of the elements of a multiset.

Example 3.108. The degree $d(G, x)$ of a vertex $x \in V(G)$ is obviously a vertex invariant. For the graph in Fig. 2.1 we obtain

$$d(G,1) = d(G,5) = 1, \quad d(G,2) = d(G,3) = 3, \quad d(G,4) = d(G,6) = 2.$$

A more powerful vertex invariant is obtained from the multiset $d_\Gamma(G, x)$ of the degrees of vertices adjacent to $x \in V(G)$. For the graph in Fig. 2.1 we obtain

$$\begin{aligned} &d_\Gamma(G,1) = (3), && d_\Gamma(G,2) = (1,2,3), && d_\Gamma(G,3) = (2,2,3),\\ &d_\Gamma(G,4) = (1,3), && d_\Gamma(G,5) = (2), && d_\Gamma(G,6) = (3,3). \end{aligned}$$

Thus, in this particular situation we are able to distinguish all vertices from each other. Note that this leaves at most one possible isomorphism from G to any other graph H.

Invariants that individualize a subobject can obviously be generalized to categories other than graphs and subobjects other than vertices. For example, we can have a *point invariant* or a *block invariant* for the category of incidence structures. For codes we can have a *codeword invariant* or a *coordinate invariant*.

Example 3.109. Let $C \subseteq Z_q^n$ be a code. For a codeword $\mathbf{x} \in C$, let $I(C, \mathbf{x})$ be the multiset consisting of the Hamming distance from $\mathbf{x}$ to all the other codewords of C. This defines a codeword invariant for the category of codes. For example, for the $(5,3,3)_2$ code $C = \{00000, 00111, 11011\}$, we have

$$I(C, 00000) = (3,4), \quad I(C, 00111) = (3,3), \quad I(C, 11011) = (3,4).$$

Example 3.110. Let $C \subseteq Z_q^n$ be a code. For a coordinate $1 \le i \le n$, let $I(C, i)$ be the multiset consisting of the number of occurrences of each coordinate value $a \in Z_q$ in coordinate i of C. This defines a coordinate invariant for the category of codes. For example, for the $(3,3,2)_3$ code $C = \{000, 011, 112\}$, we have

$$I(C,1) = (0,1,2), \quad I(C,2) = (0,1,2), \quad I(C,3) = (1,1,1).$$

Example 3.111. Let $\mathcal{X}$ be an incidence structure. For a block $B \in \mathcal{B}(\mathcal{X})$, let $I(\mathcal{X}, B)$ be the multiset consisting of the number of points B and B' have in common, where B' ranges over all blocks in $\mathcal{B}(\mathcal{X})$. This defines a block invariant for the category of incidence structures. An illustrative example is obtained by applying the invariant to the blocks of each 2-$(8,4,3)$ design in Example 2.58.

A good invariant is both fast to compute and can distinguish well between different isomorphism classes in the category of objects under study. In essence, there are two main difficulties hindering the development of good invariants for combinatorial objects: *symmetry* and *regularity*.

Symmetry (as recorded by the automorphism group) presents a fundamental obstacle to subobject invariants. In particular, no subobject invariant can by definition distinguish between subobjects that are in the same orbit under the action of the automorphism group.

Example 3.112. Consider the graphs in Fig. 3.11. For both graphs the automorphism group acts transitively on the vertices. Thus, no vertex invariant can by definition distinguish between the vertices in such a graph. A similar situation occurs with the cube graph in Fig. 3.2.

Fortunately, even if the original object is highly symmetric, subobject invariants can often be employed to expedite isomorphism computations by first individualizing a subobject that breaks the symmetry, and then applying an invariant that takes into account the individualized subobject; we will return to this topic in Sects. 3.3.5 and 5.6.

Regularity of the objects being studied is another obstacle to invariants.

Example 3.113. The codeword and coordinate invariants in Examples 3.109 and 3.110 cannot distinguish between codewords and coordinates, respectively, in OE codes. Similarly, the block invariant in Example 3.111 fails to distinguish blocks in Steiner systems $S(2,k;v)$ and square designs.

Typically it is the case that the defining properties for a family of combinatorial objects imply regularity up to a threshold, after which the differences in structure start to appear. This is perhaps best illustrated in the case of designs. One possible source of invariants for designs is to look at small configurations occurring in the designs.

Example 3.114. Small configurations occurring in Steiner triple systems have been extensively studied, see [123, 219]; a more general treatment appears in [31] and in the references therein. Up to isomorphism there are two 2-line configurations, five 3-line configurations, and 16 4-line configurations that can occur in a Steiner triple system [220]. These configurations are depicted in Fig. 3.12.

A configuration is *constant* if its number of occurrences in an STS(v) depends only on v; otherwise the configuration is *variable.*

It can be shown that in an STS(v) all configurations with at most 3 lines are constant. For 4-line configurations, five of the 16 configurations are constant (C_4, C_7, C_8, C_{11}, and C_{15}), and the remaining 11 are variable in such a way that v and the number of occurrences of any one variable configuration together determine the number of occurrences of the other 10 variable configurations [220]. The Pasch configuration (C_{16}) is one of the variable 4-line configurations.

Once an appropriate configuration has been identified, then one possibility to obtain a point (respectively, block) invariant for a design is to count for each point (block) the number of occurrences in the configurations in the design – cf. Example 3.102.

Another common source of invariants for designs is to select a set of points (blocks), and look at invariant values for a substructure that is derived from the selection in an isomorphism-invariant manner.

Example 3.115. Select a point $p \in P$ in an STS(v). The configuration induced by the set of blocks incident with p is called the *neighborhood* of p. It is not difficult to verify that all one-point neighborhoods in an STS(v) are isomorphic to a $(v-1)/2$-*star configuration.* The 7-star is depicted in Fig. 3.13; configurations B_3 and C_7 in Fig. 3.12 are the 3-star and the 4-star, respectively.

Differences in structure start to appear if we select a pair of distinct points $p_1, p_2 \in P$ from an STS(v) and consider the configuration induced by the set of blocks incident with at least one selected point; this structure is a *double neighborhood* [123]. A double neighborhood consists of the unique block $\{p_1, p_2, p_3\}$ incident with both p_1 and p_2, and of the two sets of $(v-3)/2$

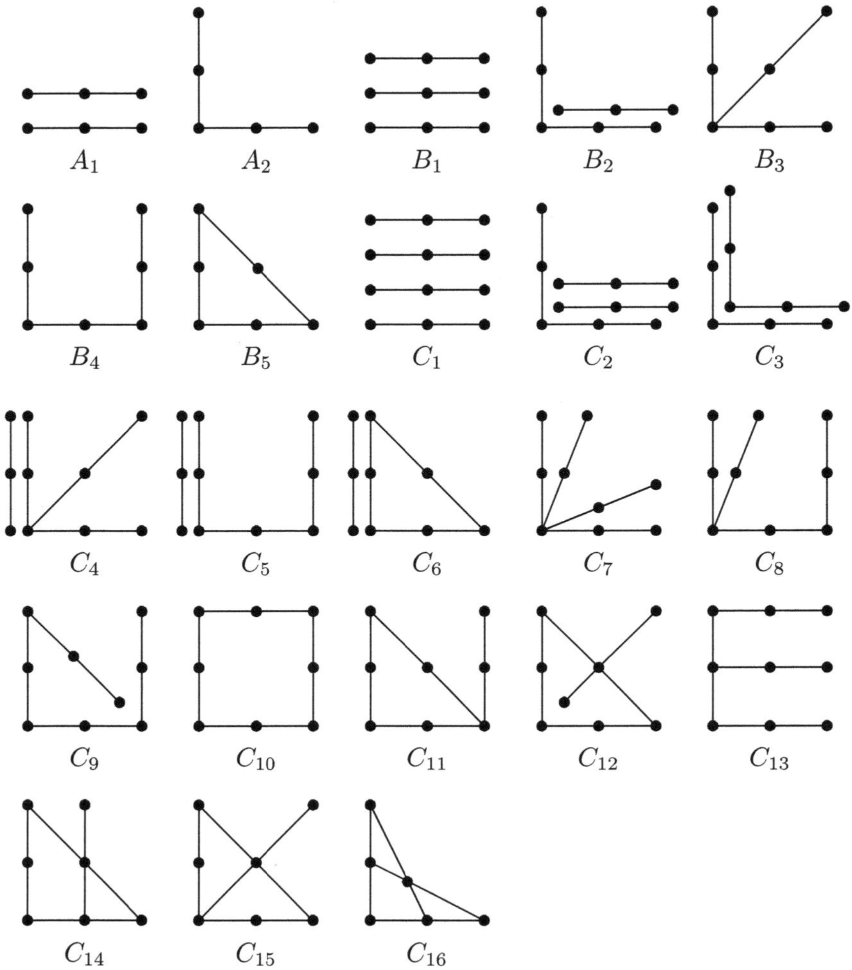

Fig. 3.12. Small configurations in Steiner triple systems

blocks incident with p_1 and p_2, respectively. The points $P \setminus \{p_1, p_2, p_3\}$ in a double neighborhood induce a 2-regular graph whose connected components are cycles of even length at least 4; this graph is frequently called the *cycle graph* defined by the pair p_1, p_2. It is easy to check that the multiset consisting of the cycle lengths in the cycle graph characterizes a double neighborhood up to isomorphism, and thus forms an invariant for double neighborhoods.

Example 3.116. Let us give a concrete example of double neighborhoods. Consider the left-hand side STS(15) in Example 3.101. The points `a` and `b` determine the double neighborhood

`abc ade afg ahi ajk alm ano bdf beh bgj bik bln bmo.`

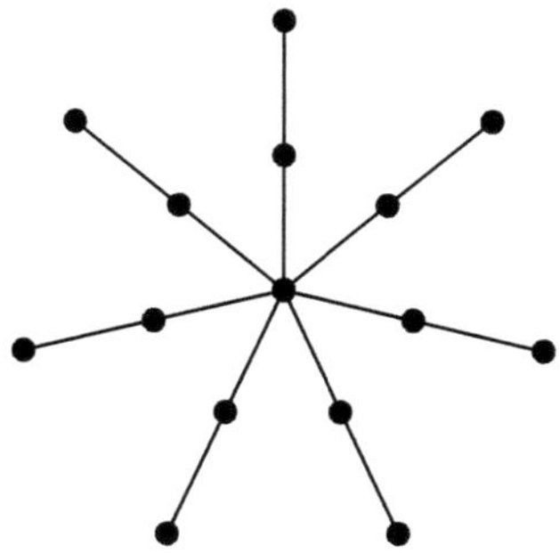

Fig. 3.13. A star configuration

The associated cycle graph consists of one 8-cycle and one 4-cycle with edges

`de eh hi ik kj jg gf fd,   lm mo on nl,`

respectively. On the other hand, the double neighborhood determined by the points `a` and `c` is

`abc ade afg ahi ajk alm ano cdg cel cfh cij cmn cko,`

where the associated cycle graph consists of one 12-cycle with edges

`de el lm mn no ok kj ji ih hf fg gd.`

Example 3.117. Point and block invariants for square designs can be obtained from invariants for the associated derived and residual designs.

Example 3.118. Any three points not on a common line in a projective plane of order n determine a configuration equivalent to a main class of Latin squares of order $n-1$ (see Definition 2.114 and Example 6.21). Thus, an invariant for main classes of Latin squares yields an invariant for projective planes.

Invariants applicable to designs are discussed in [8, 125, 204, 205, 206] and in the references therein.

For codes it appears that not many invariants have been published. By utilizing the equivalence of resolutions of BIBDs and OE codes (Theorem 3.82), invariants for the design underlying a corresponding resolution can be used as invariants for OE codes. Another possibility is to use block intersections between parallel classes, see [431, 432]. Invariants for Hadamard matrices are discussed in [184, 364] and the references therein. Many vertex invariants for colored graphs appear in [412].

3.3.5 Compounding and Iterative Refinement

A general strategy for increasing the distinguishing power of invariants is to compound multiple invariants into one.

Theorem 3.119. *Let $I_1, I_2, \ldots, I_m$ be invariants for $\mathscr{A}$. Then, the function I defined for all $X \in \mathrm{Ob}(\mathscr{A})$ by $I(X) = (I_1(X), I_2(X), \ldots, I_m(X))$ is an invariant for $\mathscr{A}$.*

Iterative refinement [137, 138, 601, 602] is another generic technique for increasing the distinguishing power of invariants. In essence, iterative refinement repeatedly applies a subobject invariant so that each new iteration takes advantage of the subobjects distinguished during the previous iteration. The iteration terminates when no further subobjects can be distinguished. To provide a more concrete discussion, we will work with colored graphs and vertex invariants for colored graphs. Once the basic ideas are understood, iterative refinement can be tailored to other categories and types of subobjects in a straightforward manner.

Let (G, π) be a colored graph, and let I be a vertex invariant for colored graphs. We assume that the range of I is an ordered set, for example, an appropriate lexicographic order can typically be used. To apply the invariant I iteratively, we construct from (G, π) a new colored graph, where the coloring π has been refined based on the vertices distinguished by I. In other words, we use the coloring of the vertices to keep track of the vertices distinguished during earlier iterations, whereby subsequent iterations can take advantage of the distinguished vertices. When the iteration terminates, the final color of each vertex constitutes a vertex invariant for the original colored graph.

Let us now make the previous ideas precise. We encode the vertices distinguished by I in (G, π) using an ordered partition $\bar{I}_{G,\pi}$ of the vertex set $V(G)$. The cells of $\bar{I}_{G,\pi}$ are maximal sets of vertices with constant value of I, and the cells are ordered by increasing value of I. Recalling the functional form for ordered partitions discussed in connection with Definition 3.87, an equivalent definition is that $\bar{I}_{G,\pi}$ is the ordered partition of $V(G)$ satisfying for all $u, v \in V(G)$ the requirement

$$\bar{I}_{G,\pi}(u) < \bar{I}_{G,\pi}(v) \quad \text{if and only if} \quad I(G, \pi, u) < I(G, \pi, v). \tag{3.29}$$

One recurrently encountered vertex invariant for colored graphs is as follows.

Definition 3.120. *Let D be the vertex invariant for colored graphs that associates to each vertex $u \in V(G)$ of a colored graph (G, π) the multiset $D(G, \pi, u)$ of colors of the vertices adjacent to u. This vertex invariant is called the* color degree *or* color valency *invariant.*

Example 3.121. Let G be the graph in Fig. 2.1 and let $\pi_1 = (\{1, 2, 3, 4, 5, 6\})$. The invariant values are

$$\begin{array}{lll} D(G, \pi_1, 1) = (1), & D(G, \pi_1, 2) = (1, 1, 1), & D(G, \pi_1, 3) = (1, 1, 1), \\ D(G, \pi_1, 4) = (1, 1), & D(G, \pi_1, 5) = (1), & D(G, \pi_1, 6) = (1, 1). \end{array}$$

Assuming lexicographic ordering for multisets of integers, we obtain the ordered partition $\bar{D}_{G,\pi_1} = (\{1, 5\}, \{4, 6\}, \{2, 3\})$.

We require some further terminology for ordered partitions to make precise how $\bar{I}_{G,\pi}$ is used to refine π. This terminology is also required in the discussion of isomorphism algorithms in Sect. 5.6.

Definition 3.122. *Let π and σ be two ordered partitions of a finite set V. We say that π is* finer *than σ if the following two properties hold:*

1. *every cell of π is a subset of a cell of σ,*
2. *σ can be obtained from π by repeatedly replacing two consecutive cells with their union.*

An equivalent definition in functional form is that π is finer than σ if for all $u, v \in V$ it holds that $\pi(u) < \pi(v)$ implies $\sigma(u) \leq \sigma(v)$. Note that an ordered partition is by definition finer than itself. Furthermore, every ordered partition of V is finer than the *unit* partition whose only cell is V. An ordered partition is *discrete* if all of its cells are singleton sets. The only partition finer than a discrete partition π is π itself.

Definition 3.123. *Let $\pi = (V_1, V_2, \ldots, V_m)$ and $\sigma = (W_1, W_2, \ldots, W_n)$ be two ordered partitions of a finite set V. The* intersection *of π and σ is the ordered partition $\pi \wedge \sigma$ of V whose cells are all nonempty sets of the form $V_i \cap W_j$, where $1 \leq i \leq m$ and $1 \leq j \leq n$; the cells are lexicographically ordered so that $V_{i_1} \cap W_{j_1}$ precedes $V_{i_2} \cap W_{j_2}$ if and only if $(i_1, j_1) \prec (i_2, j_2)$.*

An equivalent definition in functional form is that $\pi \wedge \sigma$ is the ordered partition that satisfies for all $u, v \in V$ the requirement $(\pi \wedge \sigma)(u) < (\pi \wedge \sigma)(v)$ if and only if $(\pi(u), \sigma(u)) \prec (\pi(v), \sigma(v))$. Note that the intersection operation is in general not commutative. Furthermore, because of the lexicographic order employed, $\pi \wedge \sigma$ is always finer than π.

Example 3.124. Let

$$\pi = (\{3, 6\}, \{1, 5\}, \{2, 4\}), \quad \sigma = (\{3, 5\}, \{1, 2, 4, 6\}).$$

We have

$$\pi \wedge \sigma = (\{3\}, \{6\}, \{5\}, \{1\}, \{2, 4\})$$

and

$$\sigma \wedge \pi = (\{3\}, \{5\}, \{6\}, \{1\}, \{2, 4\}).$$

Definition 3.125. *Let I be a vertex invariant for colored graphs. The* refinement transformation *R_I from colored graphs into colored graphs is defined for a colored graph (G, π) by $R_I(G, \pi) = (G, \pi \wedge \bar{I}_{G,\pi})$.*

A colored graph (G, π) is *stable* with respect to I if $R_I(G, \pi) = (G, \pi)$. If the graph G is clear from the context, then it is a convenient abuse of terminology to speak only of the ordered partition in the context of refinement and stability with respect to I.

Example 3.126. Let us continue Example 3.121 and refine π_1 to a stable partition with respect to D. From π_1 and $\bar{D}_{G,\pi_1}$ computed in Example 3.121, we obtain the refined partition

$$\pi_2 = \pi_1 \wedge \bar{D}_{G,\pi_1} = (\{1,5\},\{4,6\},\{2,3\}).$$

Evaluating D for (G,π_2), we obtain

$$\begin{aligned} &D(G,\pi_2,1) = (3), && D(G,\pi_2,2) = (1,2,3), && D(G,\pi_2,3) = (2,2,3), \\ &D(G,\pi_2,4) = (1,3), && D(G,\pi_2,5) = (2), && D(G,\pi_2,6) = (3,3). \end{aligned}$$

Thus,

$$\bar{D}_{G,\pi_2} = (\{1\},\{6\},\{5\},\{3\},\{4\},\{2\})$$

and

$$\pi_3 = \pi_2 \wedge \bar{D}_{G,\pi_2} = (\{1\},\{5\},\{6\},\{4\},\{3\},\{2\}).$$

Since π_3 is a discrete partition, it is obviously stable.

Ordered partitions that are stable with respect to the color degree invariant D are often called *equitable* [342, Sect. 7.3.2], [411]. Let us continue with another example.

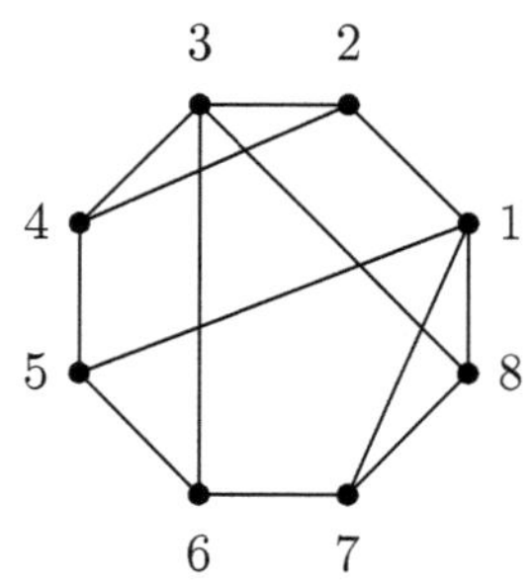

Fig. 3.14. The graph for Example 3.127

Example 3.127. Consider the graph G in Fig. 3.14. Let us refine the ordered partition $\pi_1 = (\{1,2,3,4,5,6,7,8\})$ using the color degree invariant D until an equitable partition is reached. As in previous examples, we assume lexicographic order for the invariant values when constructing $\bar{D}_{G,\pi}$, we obtain

$$\begin{aligned} \pi_2 &= \pi_1 \wedge \bar{D}_{G,\pi_1} = (\{2,4,5,6,7,8\},\{1,3\}), \\ \pi_3 &= \pi_2 \wedge \bar{D}_{G,\pi_2} = (\{2,8\},\{4,5,6,7\},\{1,3\}), \\ \pi_4 &= \pi_3 \wedge \bar{D}_{G,\pi_3} = (\{2,8\},\{5,6\},\{4,7\},\{1,3\}), \\ \pi_5 &= \pi_4 \wedge \bar{D}_{G,\pi_4} = (\{2,8\},\{5,6\},\{4,7\},\{1,3\}). \end{aligned}$$

Thus, π_4 is stable with respect to D.

The following example illustrates refinement and symmetry. The relevant observation here is that once symmetry is broken, invariants can again be used to distinguish isomorphism-invariant structure and thereby expedite isomorphism computations.

Example 3.128. The automorphism group of the cube Q_3 (see Example 3.11) acts transitively on the vertices. Thus, no vertex invariant can distinguish between the vertices. However, we can break the symmetry by individualizing a vertex and then apply refinement. In the setting of colored graphs, individualization is carried out by assigning a unique color to a vertex. If we individualize the vertex 1, we obtain

$$\begin{aligned}\pi_1 &= (\{1\},\{2,3,4,5,6,7,8\}),\\ \pi_2 = \pi_1 \wedge \bar{D}_{Q_3,\pi_1} &= (\{1\},\{4,6,7,8\},\{2,3,5\}),\\ \pi_3 = \pi_2 \wedge \bar{D}_{Q_3,\pi_2} &= (\{1\},\{4,6,7\},\{8\},\{2,3,5\}),\\ \pi_4 = \pi_3 \wedge \bar{D}_{Q_3,\pi_3} &= (\{1\},\{4,6,7\},\{8\},\{2,3,5\}).\end{aligned}$$

This stable partition cannot be refined further because of symmetry – it is easy to check based on Example 3.11 that $\mathrm{Aut}(Q_3,\pi_1)$ has the orbit partition $\{\{1\},\{2,3,5\},\{4,6,7\},\{8\}\}$ on its natural action on the vertices. Further symmetry breaking can be carried out by individualizing additional vertices.

After these examples, we proceed to study in detail the isomorphism-preserving properties of the refinement transformation $R_I(G,\pi) = (G,\pi \wedge \bar{I}_{G,\pi})$ and refinement to a stable colored graph.

Theorem 3.129. *For any two isomorphic colored graphs* (G,π) *and* (H,σ) *it holds that* $\mathrm{Iso}((G,\pi),(H,\sigma)) = \mathrm{Iso}(R_I(G,\pi),R_I(H,\sigma))$.

Proof. ($\subseteq$): Let $f \in \mathrm{Iso}((G,\pi),(H,\sigma))$ be an arbitrary isomorphism. We proceed to show that $f \in \mathrm{Iso}((G,\pi\wedge\bar{I}_{G,\pi}),(H,\sigma\wedge\bar{I}_{H,\sigma}))$. Because $f \in \mathrm{Iso}(G,H)$ by (3.27), it suffices to check for all $u \in V(G)$ that $(\pi \wedge \bar{I}_{G,\pi})(u) = (\sigma \wedge \bar{I}_{H,\sigma})(f(u))$. Because $\pi(u) = \sigma(f(u))$ for all $u \in V(G)$ by (3.27), it follows from Definition 3.123 that it suffices to check that $\bar{I}_{G,\pi}(u) = \bar{I}_{H,\sigma}(f(u))$ for all $u \in V(G)$. Let $u \in V(G)$ be arbitrary. As witnessed by f, we have $(G,\pi,u) \cong (H,\sigma,f(u))$ in the category of colored graphs with one vertex individualized. Thus, $I(G,\pi,u) = I(H,\sigma,f(u))$ because I is a vertex invariant for colored graphs. It thus follows from (3.29) that $\bar{I}_{G,\pi}(u) = \bar{I}_{H,\sigma}(f(u))$.

($\supseteq$): Clearly, $\mathrm{Aut}(G,\pi) \supseteq \mathrm{Aut}(G,\pi\wedge\mu)$ for any ordered partition μ of $V(G)$. In particular, the ($\subseteq$)-part of the proof thus gives that $\mathrm{Aut}(G,\pi) = \mathrm{Aut}(G,\pi\wedge\bar{I}_{G,\pi})$. Because (G,π) and (H,σ) are isomorphic, there exists an $h \in \mathrm{Iso}((G,\pi),(H,\sigma)) \subseteq \mathrm{Iso}((G,\pi\wedge\bar{I}_{G,\pi}),(H,\sigma\wedge\bar{I}_{H,\sigma}))$. Let $f \in \mathrm{Iso}((G,\pi\wedge\bar{I}_{G,\pi}),(H,\sigma\wedge\bar{I}_{H,\sigma}))$ be arbitrary. Because $a = h^{-1}f \in \mathrm{Aut}(G,\pi\wedge\bar{I}_{G,\pi}) = \mathrm{Aut}(G,\pi)$, it follows that $f = ha \in \mathrm{Iso}((G,\pi),(H,\sigma))$. □

It follows that R_I extends to a functor if we define the transformation on isomorphisms by $R_I(f) = f$. Note, however, that R_I is not reconstructible in general – consider, for example, the color degree invariant D and the colored graphs (G, π_3) and (G, π_4) in Example 3.127.

For a colored graph (G, π), we write $R_I^+(G, \pi)$ for the stable colored graph obtained by iteratively applying R_I.

Theorem 3.130. *For any two isomorphic colored graphs (G, π) and (H, σ) it holds that* $\text{Iso}((G, \pi), (H, \sigma)) = \text{Iso}(R_I^+(G, \pi), R_I^+(H, \sigma))$.

Proof. A discrete ordered partition is always stable. Because each iteration of R_I either has already produced a stable partition or increases the number of cells by at least one, we always have $R_I^+(G, \pi) = R_I^{n-1}(G, \pi)$ and $R_I^+(H, \sigma) = R_I^{n-1}(G, \sigma)$, where $n = |V(G)| = |V(H)|$. Applying $n-1$ times Theorem 3.129 completes the proof. □

Theorem 3.131. *Let (G, π) be a colored graph, and let $R_I^+(G, \pi) = (G, \pi^+)$. Then, I^+ defined for all $u \in V(G)$ by $I^+(G, \pi, u) = \pi^+(u)$ is a vertex invariant for colored graphs.*

Proof. Let $(G, \pi, u) \cong (H, \sigma, v)$ in the category of colored graphs with one vertex individualized. We prove that $I^+(G, \pi, u) = I^+(H, \sigma, v)$. Because $(G, \pi, u) \cong (H, \sigma, v)$, there exists an $f \in \text{Iso}((G, \pi), (H, \sigma))$ with $f(u) = v$. Theorem 3.130 implies $f \in \text{Iso}(R_I^+(G, \pi), R_I^+(H, \sigma))$; in particular, $\pi^+(u) = \sigma^+(f(u)) = \sigma^+(v)$. □

Invariants and refinement are central to practical isomorphism algorithms on combinatorial objects. Typically it is the case that a significant fraction of the running time of an algorithm is spent in evaluating subobject invariants and refining ordered partitions. Thus, a very fast implementation is required for good performance. A more detailed discussion and algorithms appear in [325], [342, Chap. 7], and [411]; also consulting the source code of *nauty* [412] is recommended to the reader interested in low-level implementation.

Invariants obviously exhibit varying computational difficulty. For example, the color degree invariant D is cheap to evaluate, whereas an invariant counting the occurrences of a nontrivial subgraph or configuration is more expensive. Performancewise it is thus prudent to employ a hierarchy of invariants ordered from cheap to expensive, whereby an expensive invariant is applied only when the cheaper invariants fail to refine a given ordered partition. As long as the order of application for the invariants is invariant under isomorphism, results analogous to Theorems 3.130 and 3.131 hold for a stable colored graph obtained using a hierarchy of invariants.

For reasons of performance it is typically the case that invariants are best evaluated using a natural representation of the objects, even in the case where a colored-graph representation is eventually used to carry out the isomorphism computations. For example, a block invariant for Steiner triple systems is typically best evaluated in a triple system representation compared with the line graph representation. After a subobject invariant has been evaluated in a natural representation, most colored-graph representations allow us to encode the distinctions made by the invariant by refining the coloring. For example, in the line graph representation we can color the vertices based on the invariant values obtained from the block invariant evaluated in the triple system representation.

3.3.6 Isomorphism Problems and Tools

To sum up the previous development, we now have available the conceptual and practical machinery enabling the transformation of questions concerning isomorphism in the central categories of combinatorial objects into questions concerning isomorphism on colored graphs. Given an object in a category of interest, we can represent the object as a colored graph, and carry out isomorphism computations on the colored graphs representing the objects. The results of a computation can then be translated back into the original representation by means of the (strongly) reconstructible functor mediating the change of representation. Isomorphism invariants can be employed in the process to distinguish isomorphism-invariant features (if any) within the objects, thereby narrowing down the set of possible isomorphisms and expediting isomorphism computations.

We proceed to discuss the basic computational problems associated with isomorphism and the tools for solving them. To obtain generality and to hide away unnecessary detail, it will be convenient to work with a group G that acts on a finite set Ω with isomorphisms given by Definition 3.16. Colored graphs are an immediate special case.

By varying the group action, we can capture the notion of isomorphism on graphs, colored graphs, designs, codes, Hadamard matrices, and so forth. Furthermore, the abstract setup also captures more advanced applications where the acting group ranges from the automorphism group of an object to the normalizer $N_G(H) = \{g \in G : gHg^{-1} = H\}$ of a prescribed group H of automorphisms in the isomorphism-inducing group G, cf. Chap. 9.

Example 3.132. To capture isomorphism on colored graphs (Definition 3.88), let $V = \{1, 2, \ldots, n\}$ and let Ω be the set of all colored graphs with vertex set V. For a colored graph $X \in \Omega$, we write $V(X) = V$ for the vertex set, $E(X)$ for the edge set, and $\pi(X)$ for the ordered partition of V into color classes. A permutation $g \in G = \mathrm{Sym}(V)$ acts on $X \in \Omega$ to produce the colored graph $g * X \in \Omega$ defined by

$$\begin{aligned} V(g * X) &= V, \\ E(g * X) &= \{\{g(u), g(v)\} : \{u, v\} \in E(X)\}, \\ \pi(g * X) &= (g(V_1), g(V_2), \ldots, g(V_m)), \end{aligned} \tag{3.30}$$

where $\pi(X) = (V_1, V_2, \ldots, V_m)$ and $g(V_i) = \{g(u) : u \in V_i\}$ for all $1 \le i \le m$.

Example 3.133. Let X be the colored graph on $V = \{1, 2, 3, 4, 5, 6\}$ with

$$\begin{aligned} V(X) &= V, \\ E(X) &= \{\{1,2\},\{2,3\},\{2,6\},\{3,4\},\{3,6\},\{4,5\}\}, \\ \pi(X) &= (\{1,5\},\{4,6\},\{2,3\}). \end{aligned}$$

For $g = (1\ 2\ 3\ 4\ 5\ 6) \in \mathrm{Sym}(V)$, we obtain the colored graph gX with

$$\begin{aligned} V(gX) &= V, \\ E(gX) &= \{\{1,3\},\{1,4\},\{2,3\},\{3,4\},\{4,5\},\{5,6\}\}, \\ \pi(gX) &= (\{2,6\},\{1,5\},\{3,4\}). \end{aligned}$$

The most basic computational problem associated with isomorphism is the problem of testing whether two given objects are isomorphic.

Problem 3.134. (ISOMORPHISM) Given $X, Y \in \Omega$, decide whether $X \cong Y$.

In the case $X \cong Y$ a variant of the isomorphism problem also asks for an isomorphism $g \in G$ with $gX = Y$. Another basic problem is that of computing a set of generators for the automorphism group of an object.

Problem 3.135. (AUTOMORPHISM GROUP GENERATORS) Given $X \in \Omega$, compute a set of generators for the automorphism group $\mathrm{Aut}(X)$.

Of these two basic problems, the automorphism group generator problem occurs frequently in practice, whereas the isomorphism problem is rarely applied in practice in this direct form of testing two given objects for isomorphism. What is more useful is a special type of certificate (Definition 3.105).

Definition 3.136. *A* canonical representative map *for the action of G on Ω is a function $\rho : \Omega \to \Omega$ that satisfies the following two properties:*

1. *for all $X \in \Omega$ it holds that $\rho(X) \cong X$,*
2. *for all $X, Y \in \Omega$ it holds that $X \cong Y$ implies $\rho(X) = \rho(Y)$.*

For an object $X \in \Omega$, the object $\rho(X)$ is the *canonical form* of X with respect to ρ. Analogously, X is *in canonical form* if $\rho(X) = X$. For an orbit $GX \subseteq \Omega$, the object $\rho(X)$ is the *canonical representative* of the orbit with respect to ρ.

A common way to define a canonical representative map is to order the set Ω, and to declare an order-extremal element of each orbit to be the canonical representative of that orbit.

Example 3.137. To order the set of all colored graphs with vertex set $V = \{1, 2, \ldots, n\}$, one possibility is to use a lexicographic order analogous to the order in Example 3.106. For example, given a colored graph X, define the corresponding ordered tuple by first listing the color of every vertex, followed by a list of the entries in the upper triangular part of the associated adjacency matrix. Note that since the vertex set is $\{1, 2, \ldots, n\}$, we have a natural correspondence between graphs and adjacency matrices. For example, the ordered tuple that corresponds to the colored graph X in Example 3.133 is

$$(1, 3, 3, 2, 1, 2, 1, 0, 1, 0, 0, 1, 0, 0, 0, 1, 0, 1, 1, 0, 0).$$

The lexicographically maximum colored graph in Ω isomorphic to X corresponds to the ordered tuple

$$(3, 3, 2, 2, 1, 1, 1, 1, 1, 1, 0, 0, 0, 1, 0, 0, 0, 0, 0, 1, 0).$$

As with Example 3.106, this canonical representative map for colored graphs is easy to define, but practical to compute only for relatively small n.

The computational problem associated with a canonical representative map is to place a given object into canonical form.

Problem 3.138. (Canonical Form) Given $X \in \Omega$, compute the canonical form of X with respect to a canonical representative map ρ.

Obviously, if we have an algorithm that places objects into canonical form, then isomorphism testing is easy: $X \cong Y$ if and only if $\rho(X) = \rho(Y)$. Furthermore, to detect isomorphs from a set of objects, instead of pairwise testing it suffices to place each object into canonical form, and then test the objects for equality, for example, by sorting the objects in canonical form.

Associated with a canonical representative map are the isomorphisms that transform an object into canonical form. Such isomorphisms will be useful in subsequent applications in classification algorithms. We give the following definition in a form that is independent of the associated canonical representative map.

Definition 3.139. *A* canonical labeling map *for the action of G on Ω is a function $\kappa : \Omega \to G$ such that for all $g \in G$ and $X \in \Omega$ it holds that*

$$\kappa(gX)gX = \kappa(X)X. \tag{3.31}$$

For an object $X \in \Omega$, the isomorphism $\kappa(X)$ is a *canonical labeling* of X with respect to κ.

A canonical labeling map determines a unique canonical representative map.

Lemma 3.140. *Let κ be a canonical labeling map for the action of G on Ω. Then, the function $\rho : \Omega \to \Omega$ defined for all $X \in \Omega$ by $\rho(X) = \kappa(X)X$ is a canonical representative map.*

Proof. Clearly, $\rho(X) \cong X$ for all $X \in \Omega$. Let $X, Y \in \Omega$ be two isomorphic objects. Thus, there exists a $g \in G$ with $gX = Y$. By (3.31) we thus obtain $\rho(Y) = \kappa(Y)Y = \kappa(gX)gX = \kappa(X)X = \rho(X)$. □

Conversely, a canonical representative map ρ determines a canonical labeling map κ up to $\kappa(X) \in \mathrm{Iso}(X, \rho(X))$ for all $X \in \Omega$.

The computational problem associated with a canonical labeling map is (obviously) that of computing $\kappa(X)$ for a given object X.

Problem 3.141. (CANONICAL LABELING) Given $X \in \Omega$, compute a canonical labeling of X with respect to a canonical labeling map κ.

Example 3.142. The software package *nauty* [412] computes canonical labeling of a colored graph X given as input. As a side-effect, *nauty* also produces a set of generators for the automorphism group $\mathrm{Aut}(X)$. The algorithm used by *nauty* is discussed in Sect. 5.6.

Once the canonical labeling of X is available, it is straightforward to bring X into canonical form.

A recurrent approach to isomorphism computations on combinatorial objects is to transform an object into a colored graph and apply *nauty*, after which the canonical labeling and automorphism group generators can be used to solve the relevant problem.

If better performance is required, in many cases using one of the built-in vertex invariants of *nauty* will improve performance; however, finding the right invariant and arguments may require experimentation. Also a custom invariant may be used. The following lemma shows that a refinement transformation (cf. Theorem 3.129) based on a subobject invariant can be used as a preprocessing step before evaluating a canonical labeling map.

Lemma 3.143. *Let κ be a canonical labeling map for the action of G on Ω, and let $R : \Omega \to \Omega$ be a function such that for all $X, Y \in \Omega$ it holds that $X \cong Y$ implies $\mathrm{Iso}(X, Y) = \mathrm{Iso}(R(X), R(Y))$. Then, $\kappa' : \Omega \to G$ defined for all $X \in \Omega$ by $\kappa'(X) = \kappa(R(X))$ is a canonical labeling map for the action of G on Ω.*

Proof. First observe that for all $g \in G$ and $X \in \Omega$ we have $gR(X) = R(gX)$ because $X \cong gX$ and hence by assumption $g \in \mathrm{Iso}(X, gX) = \mathrm{Iso}(R(X), R(gX))$. Now let $g \in G$ and $X \in \Omega$ be arbitrary. We proceed along the following sequence of equalities:

$$\begin{aligned} \kappa'(gX)R(gX) &= \kappa(R(gX))R(gX) = \kappa(gR(X))gR(X) \\ &= \kappa(R(X))R(X) = \kappa'(X)R(X) = R(\kappa'(X)X). \end{aligned} \tag{3.32}$$

The first and fourth equality follow immediately from definition of κ'. The second and fifth equality hold by the initial observation. The third equality holds because κ is a canonical labeling map. From (3.32) we thus have $\kappa'(gX) \in \mathrm{Iso}(R(gX), R(\kappa'(X)X))$. Because clearly $gX \cong \kappa'(X)X$, we have by assumption $\kappa'(gX) \in \mathrm{Iso}(gX, \kappa'(X)X)$, that is, $\kappa'(gX)gX = \kappa'(X)X$. □

4

Isomorph-Free Exhaustive Generation

Central to classification is the problem of exhaustively generating, without isomorphs, all objects meeting a collection of constraints. This chapter looks at general techniques for solving such problems. There are two main topics to consider: exhaustive generation and isomorph rejection.

Exhaustive generation, discussed in Sect. 4.1, is inevitably specific to the objects at hand, whereby different types of objects have relatively little in common. The aim is thus to cover only the basic techniques encountered in essentially every algorithm, namely, backtrack search and associated search trees. A related recurrent topic is the estimation of the resource requirements of a search.

Isomorph rejection techniques, discussed in Sect. 4.2, serve two purposes in exhaustive generation. First, it is necessary to eliminate isomorphs among the generated objects if isomorph-free generation is desired. Second, isomorph rejection is in most cases required to eliminate excessive redundant work when generating the objects. A number of general-purpose techniques exist for isomorph rejection. Section 4.2.1 discusses the basic strategy of keeping a record of the objects encountered. Sections 4.2.2 and 4.2.3 discuss the more advanced strategies of orderly generation and generation by canonical augmentation, which is followed by a discussion of group-theoretic techniques based on homomorphisms of group actions and localization in Sect. 4.2.4.

4.1 Exhaustive Generation

From a practical perspective, generation problems for combinatorial objects range roughly between two extremes. At one extreme, there are objects with a straightforward inductive structure.

Example 4.1. A permutation of $\{1, 2, \ldots, n\}$ can be viewed as a list containing each element $1, 2, \ldots, n$ exactly once. Deleting the element n from such a list, we obtain a list representing a permutation of $\{1, 2, \ldots, n-1\}$. Reversing the process, we obtain a method for exhaustive generation of permutations.

Further examples of such objects with a straightforward inductive structure include subsets, tuples, set and integer partitions, Catalan families, labeled trees, and so forth (see [342, 505, 516, 608]). A characteristic of such objects is that algorithmic generation is easy based on an inductive structure, and limits to generation are placed only by the often explosive growth in the number of objects as a function of object size, and possibly by a nontrivial isomorphism relation.

At the other extreme, there are objects with nontrivial regularity properties, such as designs, and objects with extremal structural properties, such as codes optimal with respect to minimum distance or covering radius. A characteristic of such objects is that due to the lack of – or lack of knowledge of – structure expediting generation, we must *search* for the objects through a larger class of partial objects, whereby the limits to generation are typically placed by the structure of the search. In particular, not all partial objects considered lead to the discovery of an object to be generated, whereby potentially many more partial objects are considered during the search than there are objects to be generated; a case in point being when the nonexistence of an object is established by considering all possible ways to construct it. With objects of this type, most of the effort in algorithm design is expended in experimenting with different ways to structure the search to constrain the partial objects considered. In some cases this requires considerable insight into the properties of the objects being investigated; the material in Chap. 12 is an excellent example in this respect.

Example 4.2. Looking at the examples encountered in Chap. 2, it is certainly not immediate from Definitions 2.16 and 2.32 that the 1-factorizations of K_8 and the 2-$(8,4,3)$ designs are as displayed in Fig. 2.7 and Example 2.58, respectively. In particular, it is not immediate from the defining properties how one should proceed to generate such objects.

4.1.1 Searching and Search Trees

How should one proceed to generate "nontrivial" combinatorial objects? The answer to this question obviously depends on the objects themselves and the knowledge that we have on their structure, that is, as implied by the defining properties. In practice there are two features that are present in essentially every algorithm design. First, the objects are generated step by step, exhaustively covering all possibilities. Second, in most cases we are forced to search through a strictly larger set of partial objects to find the objects that we want to generate.

The purpose of this section is to illustrate and develop these ideas in a more formally precise setting to enable subsequent discussion. It is convenient to work with a running example.

Example 4.3. Consider the following generation task (cf. [347]): Generate all 0-1 matrices of size 4×4 with exactly two 1s in every row and in every column.

Now, this is obviously not a very difficult problem to solve, even by hand calculation. There are six possible 0-1 rows subject to a row containing two 1s:

$$1100, \quad 1010, \quad 1001, \quad 0110, \quad 0101, \quad 0011. \tag{4.1}$$

We can thus proceed one row at a time, making sure that no column contains more than two 1s:

$$\begin{bmatrix} ???? \\ ???? \\ ???? \\ ???? \end{bmatrix}, \begin{bmatrix} 1100 \\ ???? \\ ???? \\ ???? \end{bmatrix}, \begin{bmatrix} 1100 \\ 1100 \\ ???? \\ ???? \end{bmatrix}, \begin{bmatrix} 1100 \\ 1100 \\ 0011 \\ ???? \end{bmatrix}, \begin{bmatrix} 1100 \\ 1100 \\ 0011 \\ 0011 \end{bmatrix}.$$

The last matrix is clearly one of the matrices that needs to be generated.

So far so good, but suppose we take a different path:

$$\begin{bmatrix} ???? \\ ???? \\ ???? \\ ???? \end{bmatrix}, \begin{bmatrix} 1100 \\ ???? \\ ???? \\ ???? \end{bmatrix}, \begin{bmatrix} 1100 \\ 1010 \\ ???? \\ ???? \end{bmatrix}, \begin{bmatrix} 1100 \\ 1010 \\ 0110 \\ ???? \end{bmatrix}.$$

Now we are stuck, because none of the six possible rows can be placed into the last row without introducing a column with three 1s. This illustrates a partial object that does not lead to the discovery of an object to be generated. Accepting defeat, we can always proceed to consider the remaining possibilities.

If we carry out this procedure systematically, it is obvious that we will obtain all 0-1 matrices that were to be generated. In doing so, we end up considering a tree of matrices that looks like the tree in Fig. 4.1. Parts of the tree are not displayed because of space limitations; the truncated subtrees are marked with dots.

Let us now place this example in a more abstract framework. First, it is convenient to assume that a search has a domain on which it operates.

Definition 4.4. *The* domain *of a search is a finite set Ω that contains all objects considered by the search.*

Example 4.5. Let the domain be the set of all 0-1 matrices of size 4×4 with entries in zero or more rows set to the undefined value "?".

Second, the structure of a search is conveniently modeled by a rooted tree.

Definition 4.6. *A* search tree *is a rooted tree whose vertices – or* nodes *– are objects in the domain Ω. Two nodes are joined by an edge if and only if they are related by one search step. The root node is the starting point of the search.*

Alternatively a search tree is called a *state space tree* [342].

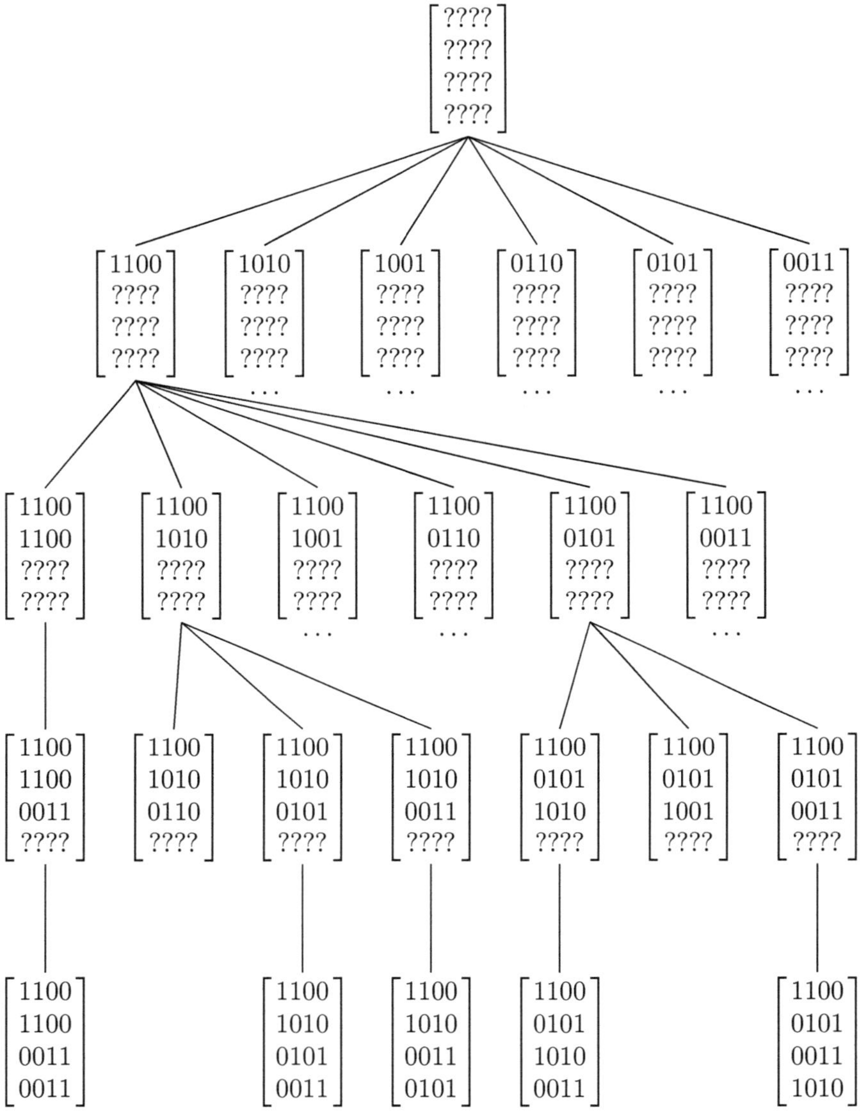

Fig. 4.1. A search tree (truncated in part)

Definition 4.7. *For a node X in a search tree we write $C(X)$ for the set of all child nodes of X. For a nonroot node X we write $p(X)$ for the parent node of X.*

In practice a search tree is defined only implicitly through the domain Ω, the root node $R \in \Omega$, and the rule $X \mapsto C(X)$ for forming the child nodes of a node X.

Example 4.8. Let the domain Ω of the search be as in Example 4.5, and let the root node $\mathbf{R}$ be the matrix with all entries undefined. Given a node $\mathbf{X} \in \Omega$, the set $C(\mathbf{X})$ of child nodes is formed by placing each of the six possible rows (4.1) into the first undefined row of $\mathbf{X}$ (if any), and removing any resulting matrices that contain more than two 1s in some column. The resulting search tree is depicted in Fig. 4.1.

In Example 4.8 the rule for forming the child nodes is straightforward. In most cases occurring in practice we are actually facing a search whose implementation involves multiple layers of searching, where in order to generate the child nodes in an upper level search it is necessary to carry out a lower level search to actually generate the children. However, it is typically possible to study the upper level search by treating the lower level search as a subroutine that produces the child nodes one at a time.

The execution of a search algorithm can be viewed as *traversing* a search tree by systematically following its edges to visit all the nodes. The operation of a typical search algorithm can be modeled as a *depth-first* traversal of the search tree; that is, starting from the root node R, the algorithm recursively visits the children of a node before returning from the node. Pseudocode for a depth-first traversal is given in Algorithm 4.1.

Algorithm 4.1 Depth-first traversal of a search tree

```
procedure DEPTH-FIRST-TRAVERSE(X: node)
1: report X (if applicable)
2: for all Y ∈ C(X) do
3:     DEPTH-FIRST-TRAVERSE(Y)
4: end for
end procedure
```

We do not prescribe any order in which the children of a node are visited during traversal – unless explicitly indicated otherwise, any order will do.

4.1.2 Backtrack Search

The previous section can be considered as providing a high-level model of exhaustive generation and search. To implement a search in practice one has to adopt a more low-level perspective.

Backtrack search or *backtracking* [215, 596] is an algorithmic principle that formalizes the intuitive "step by step, try out all the possibilities" approach to finding all solutions to a finite problem. Textbooks that discuss backtrack search include [342, 495, 505].

Again it is convenient to develop an example in parallel with the more abstract treatment.

Example 4.9. Many low-level problems encountered in practice involve a finite number of variables $x_1, x_2, \ldots, x_m$ that assume values in a finite set, say $x_j \in \{0, 1\}$. To look at a concrete example, we can represent a 0-1 matrix of size 4×4 using one 0-1 variable for each entry:

$$\mathbf{X} = \begin{bmatrix} x_1 & x_2 & x_3 & x_4 \\ x_5 & x_6 & x_7 & x_8 \\ x_9 & x_{10} & x_{11} & x_{12} \\ x_{13} & x_{14} & x_{15} & x_{16} \end{bmatrix}.$$

A *partial solution* in a backtrack search is a tuple $(a_1, a_2, \ldots, a_\ell)$, where the a_i are elements of a finite set U. The intuition is that each a_i encodes a choice made in one search step. A partial solution $(a_1, a_2, \ldots, a_\ell)$ that constitutes a solution to the problem at hand is called a *solution*. It is assumed that every solution to the problem at hand can be represented by such a finite tuple.

Example 4.10. It is natural to regard as one search step the act of setting the value of one variable. A choice made in one search step can be represented as a pair $a_i = (j, v)$ encoding the fact that $x_j = v$, where $j \in \{1, 2, \ldots, 16\}$ and $v \in \{0, 1\}$. A solution is a tuple that assigns a unique value to each of the 16 variables so that they encode a 0-1 matrix of size 4×4 with exactly two 1s in every row and in every column.

Backtrack search operates by recursively extending a partial solution one step at a time as dictated by the constraints of the problem at hand. More formally, given a partial solution $(a_1, a_2, \ldots, a_\ell)$ as input, a backtrack search procedure computes a *choice set* $A_{\ell+1} = A_{\ell+1}(a_1, a_2, \ldots, a_\ell) \subseteq U$, and for each $a_{\ell+1} \in A_{\ell+1}$ recursively invokes itself with the input $(a_1, a_2, \ldots, a_\ell, a_{\ell+1})$. A choice set can be empty, implying that no extension of the current partial solution can lead to a solution. After all alternatives $a_{\ell+1}$ have been considered, the procedure returns control – or backtracks – to the invoking procedure. The initial invocation is made with the empty tuple (), and solutions are reported as they are discovered.

A pseudocode description of backtrack search is given as Algorithm 4.2.

To exhaustively generate all solutions, it is required that the choice sets satisfy the following consistency property: if $(a_1, a_2, \ldots, a_n)$ is a solution, then for all $\ell = 0, 1, \ldots, n-1$ we must have $a_{\ell+1} \in A_{\ell+1}(a_1, a_2, \ldots, a_\ell)$. The essence of backtrack search is to exclude alternatives from the choice set in a consistent manner based on the known properties of a solution and the current partial solution.

Algorithm 4.2 Backtrack search

```
procedure BACKTRACK((a_1, a_2, ..., a_ℓ): partial solution)
  if (a_1, a_2, ..., a_ℓ) is a solution then
      report (a_1, a_2, ..., a_ℓ)
  end if
  compute the choice set A_{ℓ+1}(a_1, a_2, ..., a_ℓ)
  for all a_{ℓ+1} ∈ A_{ℓ+1} do
      BACKTRACK((a_1, a_2, ..., a_{ℓ+1}))
  end for
end procedure
```

Example 4.11. One search strategy is to select the lowest-numbered variable x_j lacking a value in the current partial solution, and put $A_{\ell+1} \subseteq \{(j,0),(j,1)\}$ so that an alternative (j,v) is excluded if and only if it results in a row or column in the 4×4 matrix with more than two occurrences of the value v in it. A more sophisticated search strategy could select a variable x_j with the minimum number of alternatives in the previous sense.

Backtrack search is obviously a general principle rather than a detailed technique for solving a problem requiring exhaustive search. Existing backtrack algorithms for well-known combinatorial problems can often be employed in solving an exhaustive generation problem or a related subproblem. Recurrently encountered problems and algorithms in this respect are discussed in Chap. 5.

The practicality of a backtrack algorithm is determined by the design choices made when transforming the abstract problem into the backtrack search framework. Design principles for backtrack algorithms together with examples can be found in [49, 416, 505]. The following basic principles are adapted from [416].

Incorporate what you know into the algorithm. In many cases a fair amount of combinatorial information is available on the structure of a solution. Incorporating this information into the algorithm design can significantly decrease the number of partial solutions considered. In essence, searching should start only when combinatorial arguments yield no further information. However, a detailed case by case analysis is usually best left to a computer.

Minimize the number of alternatives in a choice set. There are usually a number of ways to extend a given partial solution. Selecting a way that minimizes the number of alternatives in the choice set is usually worth the computational effort expended in finding it. One manifestation of this principle is *constraint propagation* or *forcing*, where priority is placed on extensions of a partial solution that have only one alternative in the associated choice set. Thus, the constraints of the problem "force" the choice – cf. Example 4.11.

Abort early if possible. Often it is possible to detect that a partial solution cannot lead to a solution although it is still possible to extend the partial solution further. Detecting such partial solutions and backtracking immediately – this is called *pruning* – can avoid a lot of work compared with the cost of detection. One manifestation of this principle is the *branch-and-bound* technique, where the solutions are associated with a target value – for example, the size of a clique in a graph – and a *bounding function* provides for each partial solution an upper bound on the value of any partial solution obtained by extending it. If the bounding function indicates that the target value cannot be reached by extending the current partial solution, the algorithm can backtrack immediately.

Minimize the amount of work at each recursive call. Even a small computational overhead at each recursive call can require a lot of time if there are many calls. A careful design of the data structures can significantly decrease the time overhead. There are three basic operations. First, when a choice is made, the data structure must be updated to reflect the implications of the choice – for example, in the form of constraint propagation – and to facilitate subsequent choices. Second, upon backtracking it is necessary to rewind the data structure back to its original state. Third, when the search returns from a recursive call it is necessary to determine the next alternative in the choice set. Copying of data should be avoided whenever possible.

Keep it simple. A small loss in efficiency is worthwhile for a gain in simplicity. Complex algorithms contain errors more often than simple ones.

4.1.3 Estimating Resource Requirements

Compared with a traditional text on algorithms, an obvious omission in the present book is that almost no attempt has been made to formally analyze the algorithms employed in terms of resource requirements such as running time or storage space. There are essentially two reasons for this omission. First, in most cases an analysis accurately reflecting practical performance is lacking, and is challenging to conduct because it must take into account in a nontrivial way the objects being generated and the method of generation. Second, due to the often explosive growth in the number of objects as a function of object size, algorithm analysis in the traditional asymptotic setting is of somewhat limited use – in practice we are restricted to small object sizes, where practical performance is obtained through tailoring the algorithms to specific instances rather than to the general case.

Because of the challenges in applying analytical tools, estimation is applied as a replacement. Especially with large searches, one recurrently encountered problem is to estimate the feasibility of the search in terms of the required computer time. Other quantities of interest for which rough estimates are sometimes useful include the number of nodes in a search tree, and the number of nodes of a given type. In this section we discuss two basic techniques for obtaining such estimates. A further discussion can be found in [321, 414, 494, 505, 508].

One estimation technique is described by Knuth [321]. Given a search tree (T, R), associate with every node $X \in V(T)$ a *weight* $w(X)$. For example, we may take $w(X)$ to be the time required to visit X, or we may take $w(X) = 1$ to estimate the size of T. The goal is now to estimate the total weight of the tree T:

$$w(T) = \sum_{X \in V(T)} w(X).$$

The estimation technique is based on repeatedly sampling random paths in (T, R). Define a random path from the root of (T, R) to a leaf node as follows. Initially, let $X_0 = R$. For $i = 0, 1, \ldots$, if X_i is a leaf node, then put $n = i$ and stop; otherwise, select uniformly at random one of the children of X_i as X_{i+1} and continue. Denote by d_i the number of children of X_i; that is, $d_i = |C(X_i)|$ for all $i = 0, 1, \ldots, n$. The estimate for $w(T)$ obtained from the path $X_0, X_1, \ldots, X_n$ is now

$$W = w(X_0) + d_0 \cdot w(X_1) + d_0 d_1 \cdot w(X_2) + \cdots + d_0 d_1 \cdots d_{n-1} \cdot w(X_n).$$

In other words, the estimate assumes that all nodes at each level $i = 0, 1, \ldots, n$ in (T, R) have the same weight and number of children as X_i. Of course this is not true for a typical search tree; however, we can prove that W always has the correct expected value:

Theorem 4.12. *The expected value of* W *is* $w(T)$.

Proof. An equivalent way to define W is to view W as the sum of random variables W_X, one for each node $X \in V(T)$, such that

$$W_X = \begin{cases} 0 & \text{if } X \notin \{X_0, X_1, \ldots, X_n\}, \\ d_0 d_1 \cdots d_{\ell-1} \cdot w(X) & \text{if } X \in \{X_0, X_1, \ldots, X_n\}, \end{cases}$$

where ℓ is the level of X in (T, R) and d_j is the number of children of the ancestor of X at level $j = 0, 1, \ldots, \ell - 1$. Because the probability for the event $X \in \{X_0, X_1, \ldots, X_n\}$ is $1/d_0 \cdot 1/d_1 \cdot \cdots \cdot 1/d_{\ell-1}$, the expected value of W_X is $w(X)$ for all $X \in V(T)$. The claim follows by linearity of expectation. □

Thus, the mean of repeated samples of W approaches $w(T)$ by the law of large numbers. This approach is formalized in Algorithm 4.3, where the parameter m specifies the number of samples of W used.

Theorem 4.12 does not convey any information about the speed of convergence towards $w(T)$, which is contingent on the structure of the search tree; for a discussion, see [321]. Although a very good estimate is often obtained with a small number of samples, there are (not only artificial) examples of trees for which an impractically large number of estimates is needed to arrive at an accurate estimate. An extension of the technique is described in [494].

Another estimation technique, described by McKay [414], is to employ probabilistic pruning during a depth-first traversal of (T, R). Associated with

Algorithm 4.3 Estimating the weight of a search tree via random paths

function PATH-ESTIM(X: node, d: integer): cost estimate
select uniformly at random an $Y \in C(X)$
return $d \cdot w(X)+$ PATH-ESTIM(Y, $d \cdot |C(X)|$)
function TREE-COST(R: root node, m: integer): cost estimate
$W_{\text{tot}} \leftarrow 0$
for all $i \in \{1, 2, \ldots, m\}$ **do**
    $W_{\text{tot}} \leftarrow W_{\text{tot}}+$ PATH-ESTIM($R, 1$)
end for
return W_{tot}/m

each level ℓ in T there is an acceptance probability $0 \leq p_\ell \leq 1$. Whenever a node X at level ℓ is encountered, the subtree rooted at X is pruned with probability $1 - p_\ell$. Thus, the probability that any given node at level ℓ is accepted is $p_0 p_1 \cdots p_\ell$. By linearity of expectation, the expected number of nodes accepted at level ℓ is $p_0 p_1 \cdots p_\ell \cdot N_\ell$, where N_ℓ is the total number of nodes at level ℓ in (T, R). Consequently, the number of nodes accepted at level ℓ divided by $p_0 p_1 \cdots p_\ell$ is a random variable with expectation N_ℓ. The mean of repeated samples thus approaches N_ℓ as the number of samples is increased, but again the speed of convergence depends on the structure of the search tree. Nevertheless, a testament to the practical power of the technique is the correct estimation (between $1.1 \cdot 10^{10}$ and $1.2 \cdot 10^{10}$) of the number of isomorphism classes of STS(19) in [414]. Compared with the previous technique, an advantage of the present technique is that it is very straightforward to implement on top of an existing search algorithm with the help of a pseudorandom number generator.

4.2 Techniques for Isomorph Rejection

The previous section developed a basic framework for algorithms for exhaustive generation. In this section we proceed to consider isomorph-free exhaustive generation. Again an abstract framework is appropriate to cover the subsequent applications to codes and designs, but it is convenient to start with an example.

Example 4.13. Consider the following classification version of the generation task in Example 4.3: Generate exactly one representative from every isomorphism class of 0-1 matrices of size 4×4 with exactly two 1s in every row and in every column. Two matrices are regarded as isomorphic if one can be obtained from the other by an independent permutation of the rows and the columns.

In the previous examples we have developed an exhaustive generation approach for the 0-1 matrices in question – now we have to ensure isomorph-free generation.

A straightforward solution is to generate all the matrices row by row, as described by the search tree in Example 4.8, keep a record of the complete matrices encountered, and output a complete matrix only if it is not isomorphic to a complete matrix appearing on record.

Disregarding implementation details, this solution is obviously correct, but not particularly efficient. If we look at the illustration of the search tree in Fig. 4.1, it is obvious that the same isomorphism classes are encountered several times, although we would be perfectly content with only one matrix from every isomorphism class of complete matrices.

Example 4.13 serves to illustrates a more general situation. Namely, it is typically not known how to operate directly on the level of isomorphism classes, which forces one to work with labeled objects for purposes of representing and generating the isomorphism classes. Working with labeled objects essentially always introduces *redundancy* into exhaustive generation; that is, (up to isomorphism) the same object is encountered several times. This redundancy must be explicitly addressed. In the minimum, isomorphic objects must be filtered from the output of the algorithm to ensure isomorph-free generation. In most cases redundancy must also be addressed on the level of partial objects. We quote Swift [567]:

> In many problems involving exhaustive searches the limiting factor with respect to speed or completion may not be the efficiency with which the search as such is conducted but rather the number of times the same basic problem is investigated. That is, the search routine may be effective in rejecting impossible cases in large blocks and still fail to accomplish its purpose in that cases which must be investigated are looked at too frequently.

The term *isomorph rejection* was introduced by Swift [567] for the techniques for eliminating redundancy in exhaustive search.

To discuss isomorph rejection in a general setting, we introduce an isomorphism relation to the search tree model. It is natural to work with a group action because the isomorphism relations occurring in practice can be captured in this manner.

Let G be a group that acts on the search domain Ω with isomorphisms given by Definition 3.16.

Example 4.14. Consider Example 4.13. Let the domain of the search be as in Example 4.5. To capture the notion of isomorphism in Example 4.13, let the acting group be $S_4 \times S_4$, where a pair of permutations $(h, k) \in S_4 \times S_4$ acts on a matrix $\mathbf{X} = (x_{ij})$ of size 4×4 so that h permutes the rows and k permutes the columns; more precisely, $\mathbf{Y} = (h, k) * \mathbf{X}$, where $y_{ij} = x_{h^{-1}(i),k^{-1}(j)}$ for all $i, j = 1, 2, 3, 4$. For example, the matrices

$$\begin{bmatrix} 1100 \\ 1010 \\ ???? \\ ???? \end{bmatrix}, \quad \begin{bmatrix} 1100 \\ 0101 \\ ???? \\ ???? \end{bmatrix} \tag{4.2}$$

are isomorphic under this action – an isomorphism is $(h, k) = (\epsilon, (1\ 2)(3\ 4))$.

Isomorph rejection techniques can now be described in terms of this model by making assumptions about the process that generates the objects – that is, assumptions about the structure of an associated search tree – in relation to the group action inducing isomorphism on the search domain.

Example 4.15. Armed with the isomorphism relation from the previous example, let us take a new look at the search tree in Fig. 4.1. It is not difficult to observe from the figure – and if necessary, formally prove based on Example 4.8 – that isomorphic nodes have isomorphic children. This is an example of an assumption about an abstract search tree that we will make in the context of isomorph rejection.

If the search tree meets the assumptions made, then redundancy can be detected via appropriate isomorphism computations on search tree nodes. The precise form of the assumptions, the isomorphism computations, and what is considered redundant are specific to a technique.

Example 4.16. Consider the assumption that isomorphic nodes have isomorphic children. A corollary – which we will prove in detail later – is that two rooted subtrees with isomorphic root nodes consist of (up to isomorphism) the same nodes. Thus, traversing only one such subtree suffices for isomorph-free exhaustive generation, making any other subtrees redundant. For a concrete illustration, consider the subtrees with roots (4.2) in Fig. 4.1.

From a practical perspective the present search tree model for isomorph rejection has two main applications. First, the assumptions about the search tree dictate how an algorithm should be structured in practice so that an isomorph rejection technique can be applied. Thus, each isomorph rejection technique essentially provides a framework for algorithm design. Second, provided that the operation of an actual algorithm is accurately described by a traversal of a search tree meeting the assumptions, the correctness of isomorph rejection is established as an immediate corollary of a correctness proof in the search tree model.

The main design goal for isomorph rejection in the search tree model is to traverse a subtree of the abstract search tree as efficiently as possible subject to the correctness constraint that exactly one object is output from every isomorphism class that we wish to generate. The best isomorph rejection techniques in general gain efficiency by taking advantage of the way in which the objects are generated, whereby potentially expensive explicit isomorphism computations (such as canonical labeling) are either traded for group-theoretic techniques relying on prescribed groups of automorphisms or replaced with less expensive computation by means of invariants in the majority of cases. Other design goals for isomorph rejection include the ability to carry out the search in parallel – that is, the ability to traverse disjoint subtrees independently of each other – and space-efficiency in terms of objects that need to be stored in memory during traversal.

In what follows we proceed to discuss different isomorph rejection techniques in detail, but before this let us state explicitly one basic principle that supplements the principles in Sect. 4.1.2.

Carry out isomorph rejection on partial objects only if there is redundancy. Regardless of which technique is employed, isomorph rejection on partial objects is a waste of effort if the investment is not returned in gains in eliminated redundancy. Typically a good strategy is to carry out isomorph rejection on the first few levels of a search tree up to a level where most of the redundancy is eliminated, and subsequent search can proceed with no isomorph rejection or only fast checks. Often a good measure of the remaining redundancy is the order of the automorphism group of a partial object.

The subsequent treatment roughly follows [69, 414] in the division of the techniques into different types.

4.2.1 Recorded Objects

Among the "folklore" techniques for isomorph rejection is the approach of keeping a global record $\mathscr{R}$ of the objects seen so far during traversal of a search tree. Whenever an object X is encountered, it is tested for isomorphism against the recorded objects in $\mathscr{R}$. If X is isomorphic to a recorded object, then the subtree rooted at X is pruned as redundant. This approach is presented in Algorithm 4.4.

Algorithm 4.4 Isomorph rejection via recorded objects

procedure RECORD-TRAVERSE(X: node)
if there exists an $Y \in \mathscr{R}$ such that $X \cong Y$ **then**
    return
end if
$\mathscr{R} \leftarrow \mathscr{R} \cup \{X\}$
report X (if applicable)
for all $Z \in C(X)$ **do**
    RECORD-TRAVERSE(Z)
end for
end procedure

The fundamental assumption with this technique is that what is regarded as redundant really should be redundant for purposes of exhaustive generation up to isomorphism. Namely, it is assumed that isomorphic nodes in the search tree have isomorphic children in the following precise sense:

for all nodes X, Y it holds that if $X \cong Y$, then for every $Z \in C(X)$ there exists a $W \in C(Y)$ with $Z \cong W$. (4.3)

Example 4.17. It is not difficult to check that the search tree in Example 4.8 satisfies (4.3) – the rule for producing the children of a node yields isomorphic children when applied to isomorphic nodes. Figure 4.2 shows a subtree of the search tree traversed using Algorithm 4.4. Nodes marked with "×" are isomorphic to nodes encountered earlier. Note that the subtree traversed depends on the order of traversal for the children of a node; here we assume that the children of each node are traversed in decreasing lexicographic order of the augmenting rows.

We now prove the correctness of isomorph rejection based on recorded objects. We first require a technical consequence of (4.3) that will also be useful later.

Lemma 4.18. *Every search tree that satisfies* (4.3) *admits a total order relation $<$ defined on the isomorphism classes of its nodes such that for every nonroot node X it holds that $p(X) < X$.*

Proof. Consider a directed graph with the isomorphism classes of nodes as vertices. Let the edges be defined so that for every nonroot node X, there is a directed edge from the isomorphism class of $p(X)$ into the isomorphism class of X. We claim that this directed graph is loopless and acyclic. To reach a contradiction, suppose that this is not the case. Let the isomorphism classes in the cycle be $\mathcal{X}_0, \mathcal{X}_1, \ldots, \mathcal{X}_{m-1}$, where there is a directed edge from $\mathcal{X}_i$ to $\mathcal{X}_{(i+1) \bmod m}$ for all $i = 0, 1, \ldots, m-1$. (In the case of a loop, we have $m = 1$.) For $j = 0, 1, 2, \ldots$, we define a sequence of nodes X_j with $X_j \in \mathcal{X}_{j \bmod m}$ as follows. Let X_0 be a node in $\mathcal{X}_0$. Given X_j, by definition of the directed graph there exist nodes $Y \in \mathcal{X}_j$ and $Z \in \mathcal{X}_{(j+1) \bmod m}$ with $X_j \cong Y$ and $p(Z) = Y$, that is, $Z \in C(Y)$. Applying (4.3), we thus conclude that there exists a node $X_{j+1} \in C(X_j)$ such that $X_{j+1} \cong Z$, that is $X_{j+1} \in \mathcal{X}_{(j+1) \bmod m}$. In this way we can produce an arbitrarily long path induced by $X_0, X_1, \ldots$ in the search tree. Because a search tree is finite, this is a contradiction. Thus, the directed graph is acyclic and loopless, whereby a topological sorting of its vertices gives a desired total order relation. □

Theorem 4.19. *When implemented on a search tree satisfying* (4.3), *Algorithm* 4.4 *reports exactly one node from every isomorphism class of nodes.*

Proof. By the structure of the algorithm it is obvious that the record $\mathscr{R}$ contains at most one node from every isomorphism class of nodes. Furthermore, a node X is reported if and only if $X \in \mathscr{R}$. Let us say that a node X *appears* in $\mathscr{R}$ if there exists a $Y \in \mathscr{R}$ with $X \cong Y$. We claim that every node appears in $\mathscr{R}$ when the invocation RECORD-TRAVERSE(R) returns, where R is the root node of the search tree.

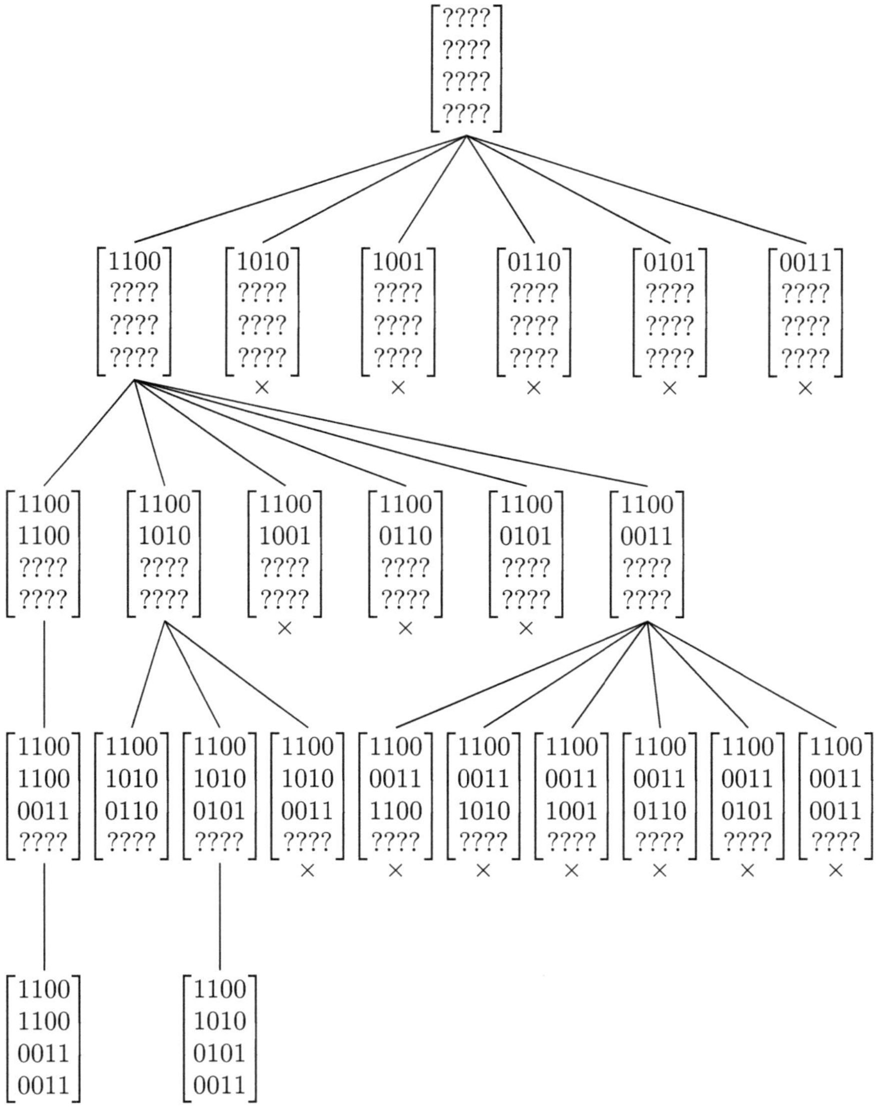

Fig. 4.2. A search tree traversed with isomorph rejection

To reach a contradiction, suppose that some node does not appear in $\mathscr{R}$. Then, with respect to the order given by Lemma 4.18, there exists a minimum isomorphism class of nodes not appearing in $\mathscr{R}$. Let X be any node in this isomorphism class. Because R appears in $\mathscr{R}$ and X is minimum, $p(X)$ appears in $\mathscr{R}$. But then by the structure of the algorithm and (4.3) also X must appear in $\mathscr{R}$, a contradiction. □

In practice lines 1–4 in Algorithm 4.4 are implemented using a certificate and a hash table – or some other data structure that allows fast searching from a large collection of objects; see [136, 322]. First, the certificate of X is computed. Then, it is checked whether the certificate occurs in the hash table. If so, the object and the associated subtree are pruned; otherwise the certificate is inserted into the hash table.

Isomorph rejection via recorded objects is sufficient for generating many families of combinatorial objects. Indeed, because a certificate for the objects occurring in the search is often easily obtainable by transforming into a graph, this approach is fast to implement and less error-prone compared with the more advanced techniques. Furthermore, it is easy to experiment with isomorph rejection on partial objects by omitting isomorph rejection on some of the partial objects – the technique remains correct for classification purposes as long as lines 1–4 are executed whenever X is an object that we want to classify.

There are at least three difficulties with isomorph rejection via recorded objects. Perhaps the most fundamental difficulty is the need to store the objects encountered. Especially when the number of nonisomorphic partial objects is large, the available storage space can quickly run out. The second difficulty is that the search does not parallelize easily because a search process must somehow communicate with the other search processes to find out whether an object has been already encountered. (Note, however, that different subtrees can be searched independently in parallel, but then we have to carry out isomorph rejection among the objects reported by each search process. If the number of nonisomorphic objects is small this is not a major difficulty.) The third difficulty is that computing the certificate of every object encountered can be computationally expensive compared with the use of cheaper invariants in the more advanced techniques.

4.2.2 Orderly Generation

One possibility to perform isomorph rejection is to carefully select a canonical representative from every isomorphism class of nodes in the search tree, and then consider only nodes in canonical form in the search. This general approach is formulated in Algorithm 4.5, where ρ is the canonical representative map for the action G on the search domain Ω that is used to decide whether a node is in canonical form or not.

Algorithm 4.5 Orderly generation

procedure CANREP-TRAVERSE(X: node)
if $X \neq \rho(X)$ **then**
  return
end if
report X (if applicable)
for all $Y \in C(X)$ **do**
  CANREP-TRAVERSE(Y)
end for
end procedure

Of course this isomorph rejection strategy requires assumptions on the search tree in connection with the canonical representatives to function properly. The following two assumptions are simple to state but nontrivial to implement:

$$\text{the canonical form } \rho(X) \text{ of every node } X \text{ is a node.} \tag{4.4}$$

Furthermore:

$$\text{for every nonroot node } X \text{ in canonical form, it holds that the parent node } p(X) \text{ is also in canonical form.} \tag{4.5}$$

Theorem 4.20. *When implemented on a search tree satisfying* (4.4) *and* (4.5), *Algorithm* 4.5 *reports exactly one node from every isomorphism class of nodes.*

Proof. Because only canonical representatives are reported, it is obvious that at most one node is reported from every isomorphism class of nodes. Consider an arbitrary node X. By (4.4) the canonical form $\rho(X)$ is a node. An induction on (4.5) shows that all ancestors of $\rho(X)$ up to and including the root R are in canonical form. Thus, $\rho(X)$ is reported. □

Although the basic idea of generating only canonical representatives is simple, finding appropriate canonical representatives and structuring the search accordingly are nontrivial tasks in practice. Typically the canonical representatives are extremal elements of orbits relative to a lexicographic order on the search domain Ω, whereby the desired structure for the search tree is obtained by constructing the objects in steps so that the lexicographically most significant part is completed first, followed by the lexicographically less significant parts. Thus, this strategy of generation is commonly called *orderly generation* (cf. [500]), although the term is occasionally used for a larger family of algorithms (cf. [414, 510]). Orderly generation was introduced independently by Faradžev [185] and Read [500].

The following framework illustrates how lexicographic order is applied to achieve property (4.5).

In many cases we can view objects in the search domain as n-tuples $\mathbf{x} = (x_1, x_2, \ldots, x_n) \in \Sigma^n$ over an alphabet Σ consisting of q symbols and an additional symbol "?" for the purpose of indicating that an entry is undefined. Let Σ^n be lexicographically ordered with respect to an order on Σ such that "?" is the minimum symbol.

Example 4.21. To cast the search tree in Example 4.8 into this framework, let $\Sigma = \{0, 1, ?\}$ be ordered by $? < 0 < 1$, and identify each 4×4 matrix in the tree with the 16-tuple over Σ obtained by concatenating the rows of the matrix. For example,

$$\begin{bmatrix} 1100 \\ 1010 \\ ???? \\ ???? \end{bmatrix}$$

is identified with the 16-tuple

$$(1, 1, 0, 0, 1, 0, 1, 0, ?, ?, ?, ?, ?, ?, ?, ?).$$

Similarly, we can in many cases view the notion of isomorphism on the search domain as being induced by a permutation group $G \le S_n$ that acts on Σ^n by permuting the entries in a tuple, see (3.10).

Example 4.22. To cast the group action in Example 4.14 into this framework, observe that a permutation of the rows and columns of a 4×4 matrix $\mathbf{X}$ defines a corresponding permutation of the entries in the 16-tuple $\mathbf{x}$ obtained from $\mathbf{X}$. Accordingly, we can represent the acting group in Example 4.14 as a subgroup $G \le S_{16}$ whose action is equivalent to the action in Example 4.14.

A search tree over the domain Σ^n constructs the n-tuples in order of lexicographic significance if every nonroot node $\mathbf{x}$ and its parent $p(\mathbf{x})$ have the form

$$\begin{aligned} p(\mathbf{x}) &= (y_1, y_2, \ldots, y_j, ?, ?, \ldots, ?), \\ \mathbf{x} &= (x_1, x_2, \ldots, x_k, ?, ?, \ldots, ?), \end{aligned}$$

for some $0 \le j < k \le n$ such that $x_i = y_i$ holds for all $1 \le i \le j$ and $x_i \ne ?$ holds for all $1 \le i \le k$.

Example 4.23. The search tree in Example 4.21 has this property: whenever a new row becomes defined in a 4×4 matrix, the four lexicographically most significant undefined entries in the associated 16-tuple become defined.

Theorem 4.24. *If a search tree over the domain Σ^n constructs the n-tuples in order of lexicographic significance, and the lexicographically maximum n-tuple of every orbit of G on Σ^n is in canonical form, then* (4.5) *holds.*

Proof. We prove the contrapositive claim; that is, for every nonroot node $\mathbf{x}$ it holds that if $p(\mathbf{x})$ is not in canonical form, then neither is $\mathbf{x}$. Clearly, if $p(\mathbf{x})$ is not in canonical form, then there exists a $g \in G$ such that $g * p(\mathbf{x}) \succ p(\mathbf{x})$. We claim that $g * \mathbf{x} \succ \mathbf{x}$, which implies that $\mathbf{x}$ is not in canonical form. Because the n-tuples are constructed in order of lexicographic significance, any undefined symbols "?" in $\mathbf{x}$ and $p(\mathbf{x})$ occur in the lexicographically least significant entries. Furthermore, since the undefined symbol is the minimum symbol in the alphabet Σ, the lexicographically most significant entries in $g * p(\mathbf{x})$ up to and including the entry where $g * p(\mathbf{x})$ and $p(\mathbf{x})$ differ cannot contain an undefined symbol. Since g acts by permuting the coordinates, and $p(\mathbf{x})$ is equal to $\mathbf{x}$ in the entries not containing an undefined symbol, we must have $g * \mathbf{x} \succ \mathbf{x}$. Thus, $\mathbf{x}$ is not in canonical form. □

Example 4.25. Under the assumption that the lexicographically maximum 16-tuple of every orbit is in canonical form, it is easy to see that the search tree in Example 4.21 satisfies (4.4). Also (4.5) holds by Theorem 4.24. Thus, we can apply Algorithm 4.5.

Viewed in terms of 4×4 matrices instead of 16-tuples, the subtree traversed by Algorithm 4.5 is identical to the tree depicted in Fig. 4.2. The symbol "×" now marks nodes that are not in canonical form.

Compared with isomorph rejection via recorded objects, orderly generation has the convenient property that no isomorphism tests between different nodes of the search tree are required. The decision whether to accept or reject a node can be made locally, based on a procedure that determines whether the current node X is in canonical form. Thus, the search can be efficiently parallelized because disjoint subtrees can be searched independently of each other. Furthermore, no objects need to be stored in memory for purposes of isomorph rejection.

A further key advantage obtainable with orderly generation is that it is often possible to exploit the properties of order-extremal objects that we want to classify in pruning subtrees that cannot contain such an object.

Example 4.26. Let $\mathbf{X}$ be a 0-1 matrix of size 4×4 with exactly two 1s in every row and every column. Furthermore, suppose that $\mathbf{X}$ is the lexicographically maximum matrix relative to permutation of the rows and the columns. It follows from the properties of lexicographic order that the matrix $\mathbf{X}$ must have the form

$$\begin{bmatrix} 1 & 1 & 0 & 0 \\ 1 & x_6 & x_7 & x_8 \\ 0 & x_{10} & x_{11} & x_{12} \\ 0 & x_{14} & x_{15} & x_{16} \end{bmatrix}.$$

Namely, a matrix not of this form can be transformed by permutation of the rows and columns to a lexicographically greater matrix of this form. This observation can now be applied to prune the search tree in Example 4.8. For example, no descendant of the lexicographically maximum matrix

$$\begin{bmatrix} 1100 \\ 0011 \\ ???? \\ ???? \end{bmatrix}$$

is a lexicographically maximum matrix that we want to classify. Thus, the associated subtree can be pruned.

Further examples of this type of order-based constraints relying on the structure of the objects that we want to classify can be found in subsequent chapters and [154, 162, 238, 302, 415, 424, 425, 522, 524, 549]. It should be noted that such constraints on partial objects can to some extent be implemented through the use of invariants in the other isomorph rejection techniques, but this is rather more tedious.

The main drawback with orderly generation is that testing whether an object is in canonical form relative to a lexicographic order is often computationally expensive. The typical approach for testing canonicity is to represent the acting group G as a permutation group and employ backtrack search on cosets of a point stabilizer chain in G to verify that $gX \preceq X$ for all $g \in G$. Lexicographic order and discovered automorphisms can be employed to prune the associated search tree on cosets. Also the fact that $p(X)$ is in canonical form can be exploited to restrict the search. In many cases a useful heuristic observation is that a $g \in G$ with $gX \succ X$ is likely to establish $gY \succ Y$ for a sibling Y of X as well (cf. [415]) – in [152, 153] this observation is developed into a backjumping strategy for the backtrack search that generates the children of a node.

4.2.3 Canonical Augmentation

Introduced by McKay [414], generation by canonical augmentation requires that an object is generated "in a canonical way", as opposed to orderly generation, which requires that the object itself be canonical. The presentation that follows differs somewhat from the original presentation in [414], but the central ideas are the same.

We begin with a simplified version of generation by canonical augmentation that is already sufficient for many purposes. This technique is perhaps most appropriately called generation by *weak* canonical augmentation.

Consider a search tree. Every node X in the tree has a finite sequence of ancestors from which it has been constructed:

$$X,\ p(X),\ p(p(X)),\ p(p(p(X))),\ \ldots. \tag{4.6}$$

Due to redundancy, an isomorphism class of nodes in general occurs multiple times in the search tree. In other words, for a given node X there in general exists a node Y with $X \cong Y$ and $X \neq Y$. Such a node Y also has a sequence of ancestors from which it has been constructed:

$$Y,\ p(Y),\ p(p(Y)),\ p(p(p(Y))),\ \ldots. \tag{4.7}$$

A fundamental observation is now that even though $X \cong Y$, the associated ancestor sequences (4.6) and (4.7) need not consist of the same nodes up to isomorphism. That is, on the level of isomorphism classes of nodes, the sequences (4.6) and (4.7) can be distinct even if $X \cong Y$. The main idea is now to exploit such differences among isomorphic nodes in rejecting isomorphs. Before proceeding further, let us look at an example of ancestor sequences.

Example 4.27. Consider the search tree in Example 4.8. One sequence of ancestors is

$$\begin{bmatrix}1100\\1010\\0101\\0011\end{bmatrix},\quad \begin{bmatrix}1100\\1010\\0101\\????\end{bmatrix},\quad \begin{bmatrix}1100\\1010\\????\\????\end{bmatrix},\quad \begin{bmatrix}1100\\????\\????\\????\end{bmatrix},\quad \begin{bmatrix}????\\????\\????\\????\end{bmatrix}.$$

Another sequence of ancestors is

$$\begin{bmatrix}1100\\0011\\0110\\1001\end{bmatrix},\quad \begin{bmatrix}1100\\0011\\0110\\????\end{bmatrix},\quad \begin{bmatrix}1100\\0011\\????\\????\end{bmatrix},\quad \begin{bmatrix}1100\\????\\????\\????\end{bmatrix},\quad \begin{bmatrix}????\\????\\????\\????\end{bmatrix}.$$

Note that the matrices in respective sequence positions 1, 2, 4, and 5 are isomorphic, but the matrices in position 3 are nonisomorphic. Thus, the ancestor sequences are distinct on the level of isomorphism classes of matrices.

To arrive at an isomorph rejection strategy, let us now specify on the level of isomorphism classes a "canonical way" to generate each nonroot node of the search tree. More formally, let Ω_{nr} be the union of all orbits of G on Ω that contain a nonroot node of the search tree. Associate with every object $X \in \Omega_{\mathrm{nr}}$ a *weak canonical parent* $w(X) \in \Omega$ such that the following property holds:

$$\text{for all } X, Y \in \Omega_{\mathrm{nr}} \text{ it holds that } X \cong Y \text{ implies } w(X) \cong w(Y). \tag{4.8}$$

The function w defines for every object $X \in \Omega_{\mathrm{nr}}$ a sequence of objects analogous to (4.6):

$$X,\ w(X),\ w(w(X)),\ w(w(w(X))),\ \ldots. \tag{4.9}$$

Because of (4.8), any two isomorphic objects have identical sequences (4.9) if we look at the sequences on the level of isomorphism classes of objects.

In rejecting isomorphs, the sequence (4.9) now constitutes the "canonical way" to generate objects in the isomorphism class of X. When the search tree is traversed, a node X is considered further only if it has been constructed in the canonical way specified by (4.9); that is, every node in the ancestor sequence (4.6) should be isomorphic to the object in the corresponding position in the sequence (4.9). In practice this property is tested one sequence position at a time. Accordingly, we say that a node X is *generated by weak canonical augmentation* if

$$p(X) \cong w(X). \tag{4.10}$$

Before proceeding with further technical details, let us look at the implications for isomorph rejection. Suppose that X, Y are isomorphic nodes encountered when traversing the search tree. Furthermore, suppose that both nodes are generated by weak canonical augmentation. By (4.8) we obtain

$$p(X) \cong w(X) \cong w(Y) \cong p(Y).$$

In other words, isomorphic nodes generated by weak canonical augmentation have isomorphic parent nodes. Assuming that isomorph rejection has been performed on parent nodes, we obtain that isomorphic nodes generated by weak canonical augmentation must be siblings; that is, $p(X) = p(Y)$. Thus, if generation by weak canonical augmentation is employed, then it suffices to perform isomorph rejection only among siblings when traversing a search tree.

Algorithm 4.6 traverses a search tree using generation by weak canonical augmentation reinforced with isomorph rejection on siblings.

To achieve isomorph-free exhaustive generation, we make the natural assumption (4.3) that isomorphic nodes have isomorphic children. Furthermore, we assume that the weak canonical parent function w is compatible with the search tree in the following sense:

> for every nonroot node X, there exists a nonroot node Y such that $X \cong Y$ and $p(Y) \cong w(Y)$. (4.11)

In essence, (4.3) and (4.11) together imply that, for every node X, the canonical parent sequence (4.9) is realized on the level of isomorphism classes by an actual sequence (4.7) of nodes occurring in the search tree.

Algorithm 4.6 Generation by weak canonical augmentation

procedure WEAK-CANAUG-TRAVERSE(X: node)
- report X (if applicable)
- $\mathcal{Z} \leftarrow \emptyset$
- **for all** $Z \in C(X)$ **do**
  - **if** $p(Z) \cong w(Z)$ **then**
    - $\mathcal{Z} \leftarrow \mathcal{Z} \cup \{Z\}$
  - **end if**
- **end for**
- remove isomorphs from $\mathcal{Z}$
- **for all** $Z \in \mathcal{Z}$ **do**
  - WEAK-CANAUG-TRAVERSE(Z)
- **end for**

end procedure

Let us prove correctness of Algorithm 4.6 subject to these assumptions.

Theorem 4.28. *When implemented on a search tree satisfying* (4.3) *and* (4.11), *Algorithm* 4.6 *reports exactly one node from every isomorphism class of nodes.*

Proof. Order the isomorphism classes of nodes in the search tree using Lemma 4.18. We proceed by induction in the order of isomorphism classes. Note that the only node in the first isomorphism class is the root node R. Thus, the induction base follows from the initial invocation WEAK-CANAUG-TRAVERSE(R).

To establish uniqueness, suppose uniqueness holds for the first ℓ isomorphism classes of nodes. Let Z_1, Z_2 be nodes in isomorphism class number $\ell+1$ such that WEAK-CANAUG-TRAVERSE is invoked with input Z_1, Z_2. Thus, we must have $p(Z_i) \cong w(Z_i)$ for $i = 1, 2$. Consequently, $Z_1 \cong Z_2$ and (4.8) imply $p(Z_1) \cong p(Z_2)$. The order on isomorphism classes implies that we can apply the inductive hypothesis (uniqueness) on $p(Z_1)$ and $p(Z_2)$. Thus, $p(Z_1) = p(Z_2)$. Because isomorph rejection has been performed on siblings, $Z_1 = Z_2$.

It remains to establish existence; suppose existence holds for the first ℓ isomorphism classes of nodes. Let W be a node in isomorphism class number $\ell + 1$. We show that there exists a node Z such that $Z \cong W$ and WEAK-CANAUG-TRAVERSE is invoked with Z. By (4.11) there exists a node $Y \cong W$ such that $w(Y) \cong p(Y)$. The order on isomorphism classes implies that we can apply the inductive hypothesis (existence) and conclude that WEAK-CANAUG-TRAVERSE is invoked at least once with input X such that $X \cong p(Y)$. It follows from (4.3) that there exists a $Z \in C(X)$ such that $Z \cong Y$. Now (4.8) implies $p(Z) = X \cong p(Y) \cong w(Y) \cong w(Z)$. Thus, WEAK-CANAUG-TRAVERSE is invoked either with Z or with a sibling isomorphic to Z. □

Let us now abandon the simplified framework and proceed to discuss generation by canonical augmentation. The essential extension to the simplified framework is that, in addition to requiring that a node X has a specific parent node with $p(X) \cong w(X)$, we also require that X must be generated by augmenting $p(X)$ in a specific "way". To describe the framework in an abstract setting, we must first model the process in which an object is generated by augmenting another object. In terms of a search tree, the ordered pair $(X, p(X))$ characterizes the augmentation that is performed to generate a node X from $p(X)$ during the search.

Example 4.29. Consider the search tree in Example 4.8. The matrix

$$\mathbf{X} = \begin{bmatrix} 1100 \\ 1010 \\ 0101 \\ ???? \end{bmatrix}$$

is generated by augmenting the matrix

$$p(\mathbf{X}) = \begin{bmatrix} 1100 \\ 1010 \\ ???? \\ ???? \end{bmatrix}.$$

The ordered pair $(\mathbf{X}, p(\mathbf{X}))$ now contains the information how $\mathbf{X}$ was generated by augmenting $p(\mathbf{X})$ in the search.

Formally, an *augmentation* is an ordered pair $(X, Z) \in \Omega \times \Omega$ of objects from the search domain. Define isomorphism for augmentations by extending the isomorphism-inducing action of G on Ω to the elementwise action on $\Omega \times \Omega$. In particular, $(X, Z) \cong (Y, W)$ if and only if there exists a $g \in G$ with $gX = Y$ and $gZ = W$.

Example 4.30. Recall the group action in Example 4.14. The augmentations

$$\left(\begin{bmatrix} 1100 \\ 1010 \\ 0101 \\ ???? \end{bmatrix}, \begin{bmatrix} 1100 \\ 1010 \\ ???? \\ ???? \end{bmatrix} \right), \quad \left(\begin{bmatrix} 0101 \\ 0110 \\ 1010 \\ ???? \end{bmatrix}, \begin{bmatrix} 0101 \\ 0110 \\ ???? \\ ???? \end{bmatrix} \right)$$

are isomorphic with respect to the induced action on augmentations; an isomorphism is $(h, k) = ((1\ 2), (1\ 2\ 3\ 4))$.

To define what a canonical augmentation is, associate with every object $X \in \Omega_{\mathrm{nr}}$ a *canonical parent* $m(X) \in \Omega$ satisfying the following property:

$$\text{for all } X, Y \in \Omega_{\mathrm{nr}} \text{ it holds that } X \cong Y \text{ implies } (X, m(X)) \cong (Y, m(Y)). \tag{4.12}$$

The augmentation $(X, m(X))$ is the *canonical augmentation* associated with X. Note that (4.12) implies that m satisfies (4.8), but the converse implication does not hold in general. Now, analogously to the ancestor sequence (4.6) and the canonical parent sequence (4.9) in the simplified framework, we look at the augmentation sequence

$$(X, p(X)),\ (p(X), p(p(X))),\ (p(p(X)), p(p(p(X)))),\ \ldots$$

and the canonical augmentation sequence

$$(X, m(X)),\ (m(X), m(m(X))),\ (m(m(X)), m(m(m(X)))),\ \ldots$$

in performing isomorph rejection. Again this is done one sequence position at a time. Accordingly, we say that a node Z occurring in the search tree is *generated by canonical augmentation* if

$$(Z, p(Z)) \cong (Z, m(Z)). \tag{4.13}$$

Note that (4.13) implies (4.10) but the converse need not hold.

The main observation to be exploited in isomorph rejection is now as follows. Suppose that Z and W are isomorphic nodes encountered when traversing the search tree. Furthermore, suppose that both nodes are generated by canonical augmentation. By (4.12) and (4.13), we obtain

$$(Z, p(Z)) \cong (Z, m(Z)) \cong (W, m(W)) \cong (W, p(W)). \tag{4.14}$$

In particular, $p(Z) \cong p(W)$. Assuming that isomorph rejection has already been performed on parent nodes, we obtain $p(Z) = p(W)$. Let $X = p(Z) = p(W)$ and observe that (4.14) implies $(Z, X) \cong (W, X)$, that is, there exists an $a \in \operatorname{Aut}(X)$ with $aZ = W$. Thus, we conclude that any two isomorphic nodes Z, W generated by canonical augmentation must be related by an automorphism of their common parent node, X.

Compared with the simplified framework, this observation often results in more efficient isomorph rejection among siblings. In particular, if $\operatorname{Aut}(X)$ is trivial, then the test (4.13) suffices for complete isomorph rejection among the children of X. Another advantage is that cheap isomorphism invariants can often be used to perform the test (4.13); this will be discussed later.

Algorithm 4.7 gives pseudocode for generation by canonical augmentation. Isomorph rejection on $C(X)$ based on automorphisms of X is performed in lines 2 and 3.

Algorithm 4.7 Generation by canonical augmentation

procedure CANAUG-TRAVERSE(X: node)
1: report X (if applicable)
2: **for all** $\mathcal{Z} \in \{C(X) \cap \{aZ : a \in \operatorname{Aut}(X)\} : Z \in C(X)\}$ **do**
3:    select any $Z \in \mathcal{Z}$
4:    **if** $(Z, p(Z)) \cong (Z, m(Z))$ **then**
5:       CANAUG-TRAVERSE(Z)
6:    **end if**
7: **end for**
end procedure

We make the following assumptions to obtain isomorph-free exhaustive generation. First, isomorphic nodes in the search tree must have isomorphic children such that an isomorphism applies also to the parent nodes:

for all nodes X, Y it holds that if $X \cong Y$, then for every $Z \in C(X)$ there exists a $W \in C(Y)$ such that $(Z, X) \cong (W, Y)$. (4.15)

Second, for every nonroot node, there must exist an isomorphic node that is generated by canonical augmentation:

for every nonroot node X, there exists a node Y such that $X \cong Y$ and $(Y, m(Y)) \cong (Y, p(Y))$. (4.16)

Note that (4.15) and (4.16) are strengthened versions of (4.3) and (4.11), respectively. The following theorem is analogous to Theorem 4.28; nevertheless we give a full proof for completeness.

Theorem 4.31. *When implemented on a search tree satisfying* (4.15) *and* (4.16), *Algorithm* 4.7 *reports exactly one node from every isomorphism class of nodes.*

Proof. Because (4.15) implies (4.3), we can order the isomorphism classes of nodes in the search tree using Lemma 4.18. We proceed by induction in the order of isomorphism classes. Note that the only node in the first isomorphism class is the root node R. Thus, the induction base follows from the initial invocation CANAUG-TRAVERSE(R).

To establish uniqueness, suppose uniqueness holds for the first ℓ isomorphism classes of nodes. Let Z_1, Z_2 be nodes in isomorphism class number $\ell + 1$ such that CANAUG-TRAVERSE is invoked with input Z_1, Z_2. Thus, we must have $(Z_i, p(Z_i)) \cong (Z_i, m(Z_i))$ for $i = 1, 2$. Consequently, $Z_1 \cong Z_2$ and (4.12) imply $(Z_1, p(Z_1)) \cong (\mathbb{Z}_2, p(Z_2))$. In particular, $p(Z_1) \cong p(Z_2)$. The order on isomorphism classes implies that we can apply the inductive hypothesis (uniqueness) on $p(Z_1)$ and $p(Z_2)$. Thus, $X = p(Z_1) = p(Z_2)$. By $(Z_1, p(Z_1)) \cong (Z_2, p(Z_2))$ there exists an automorphism $a \in \mathrm{Aut}(X)$ such that $aZ_1 = Z_2$. Thus, $Z_1 = Z_2$ by the structure of the algorithm.

It remains to establish existence; suppose existence holds for the first ℓ isomorphism classes of nodes. Let W be a node in isomorphism class number $\ell + 1$. We show that there exists a node Z such that $Z \cong W$ and CANAUG-TRAVERSE is invoked with input Z. By (4.16) there exists a node $Y \cong W$ such that $(Y, m(Y)) \cong (Y, p(Y))$. The order on isomorphism classes implies that we can apply the inductive hypothesis (existence) and conclude that CANAUG-TRAVERSE is invoked at least once with node X such that $X \cong p(Y)$. It follows from (4.15) that there exists a $Z \in C(X)$ such that $(Z, X) \cong (Y, p(Y))$. Let $a \in \mathrm{Aut}(X)$ such that $aZ \in C(X)$. Clearly, $p(aZ) = X = p(Z)$ and $(aZ, X) \cong (Z, X)$. Furthermore, (4.12) implies $(aZ, m(aZ)) \cong (Z, m(Z)) \cong (Y, m(Y)) \cong (Y, p(Y)) \cong (Z, p(Z)) \cong (aZ, p(aZ))$. Thus, aZ passes the test (4.13) for all applicable choices of $a \in \mathrm{Aut}(X)$. It follows that CANAUG-TRAVERSE is invoked with an input isomorphic to $Z \cong Y \cong W$. □

Observe that as a corollary to the previous proof it is also possible to interchange the order in which the rejection of nodes isomorphic under $\mathrm{Aut}(X)$ and the test (4.13) are performed so that isomorph rejection is carried out only among those objects $Z \in C(X)$ that satisfy (4.13); cf. Algorithm 4.6. Also recall from the preceding discussion that nodes in $C(X)$ generated by canonical augmentation are isomorphic under $\mathrm{Aut}(X)$ if and only if they are isomorphic under G.

Let us now discuss how to implement generation by canonical augmentation in practice. The essential prerequisite is that we have a generation strategy corresponding to a search tree satisfying (4.15). In most cases occurring in practice a search tree where isomorphic nodes have isomorphic children in the sense of (4.3) also satisfies the stronger requirement (4.15).

Example 4.32. The search tree in Example 4.8 satisfies (4.15). Observe that for any two nodes $\mathbf{X}, \mathbf{Y}$ in the search tree, $\mathbf{X} \cong \mathbf{Y}$ implies that there exists an isomorphism $(h, k) \in \mathrm{Iso}(\mathbf{X}, \mathbf{Y})$ that fixes the rows that are undefined – such rows always occur as the last rows when nodes of the search tree are considered. For any child $\mathbf{Z} \in C(\mathbf{X})$ put $\mathbf{W} = (h, k) * \mathbf{Z}$ and observe that $\mathbf{W} \in C(\mathbf{Y})$. Furthermore, $(\mathbf{Z}, \mathbf{X}) \cong (\mathbf{W}, \mathbf{Y})$ as witnessed by (h, k). A more concrete illustration may be obtained by looking at Fig. 4.1: for any two isomorphic nodes there exists a permutation of the rows and columns that takes one node and any selected child onto the isomorphic node and child, respectively.

Once the search tree is available, we must implement the canonical parent function m in a manner that meets (4.12) and (4.16). The intuition is that we have to specify for every isomorphism class of nodes an augmentation that occurs in the search tree. In most cases a generation approach proceeds by extending subobjects of some kind – for example, graphs are constructed from subgraphs, codes are constructed from subcodes with smaller parameters, designs are constructed from smaller designs or substructures of some kind, and so forth. Thus, by looking at a generated object, we can identify every possible subobject from which the object is generated in the search, and select one of these as the canonical parent. This is the intuition that we will pursue in the following abstract treatment.

We assume that we can associate to every object $Z \in \Omega_{\mathrm{nr}}$ a nonempty set $S(Z) \subseteq \Omega$ of *subobjects* such that

$$\text{for all } Z \in \Omega_{\mathrm{nr}} \text{ and } X \in S(Z) \text{ it holds that } (Z, X) \cong (Y, p(Y)) \text{ for some nonroot node } Y \text{ in the search tree.} \tag{4.17}$$

Example 4.33. Considering the search tree in Example 4.8, every child node is generated by introducing one more defined row into a node. Thus, for a matrix $\mathbf{Z}$ isomorphic to a nonroot node, we can let $S(\mathbf{Z})$ consist of all matrices obtained from $\mathbf{Z}$ by transforming one defined row into an undefined row in all possible ways. For example,

$$S\left(\begin{bmatrix} 1100 \\ 0011 \\ 0110 \\ ???? \end{bmatrix}\right) = \left\{ \begin{bmatrix} ???? \\ 0011 \\ 0110 \\ ???? \end{bmatrix}, \begin{bmatrix} 1100 \\ ???? \\ 0110 \\ ???? \end{bmatrix}, \begin{bmatrix} 1100 \\ 0011 \\ ???? \\ ???? \end{bmatrix} \right\}.$$

It is immediate that (4.17) holds in this case.

We also assume that any isomorphism of objects also maps associated subobjects onto subobjects:

$$\text{for all } g \in G \text{ and } Z \in \Omega_{\text{nr}} \text{ it holds that } gS(Z) = S(gZ). \tag{4.18}$$

In practice this property holds for most natural notions of a subobject, including the subobjects in Example 4.33.

The necessary assumptions on subobjects are now in place. In this setting the typical way to define a canonical parent function m is to rely on a canonical labeling map in selecting a subobject so that (4.12) will hold. Let κ be a canonical labeling map for the action of G on Ω. Given an object $Z \in \Omega_{\text{nr}}$, we compute the canonical parent $m(Z)$ as follows. First, we compute the canonical labeling $\kappa(Z) \in G$ and the canonical form $\hat{Z} = \kappa(Z)Z$ of Z. Then, we select any subobject $\hat{Z}_0 \in S(\hat{Z})$ so that the selection depends only on the canonical form $\hat{Z}$ and not on Z. Finally, we set $m(Z) = \kappa(Z)^{-1}\hat{Z}_0$.

To establish (4.12), observe that the selection of the subobject is made on the canonical form of the input object. Thus, for any two isomorphic objects $Z, W \in \Omega_{\text{nr}}$ we have $\hat{Z} = \hat{W}$, $\hat{Z}_0 = \hat{W}_0$, and therefore $(Z, m(Z)) \cong (W, m(W))$ as witnessed by the isomorphism $\kappa(W)^{-1}\kappa(Z)$. Thus, (4.12) holds.

To establish (4.16), observe that by (4.18) we have that $m(Z) \in S(Z)$ for all $Z \in \Omega_{\text{nr}}$. Thus, (4.17) implies that for any $Z \in \Omega_{\text{nr}}$ there exists a nonroot node Y in the search tree with $(Z, m(Z)) \cong (Y, p(Y))$. Because $Y \cong Z$, we have $(Y, m(Y)) \cong (Z, m(Z)) \cong (Y, p(Y))$ by (4.12). Thus, (4.16) holds.

Example 4.34. Order matrices lexicographically as in Example 4.21. Let a matrix be in canonical form if it is the lexicographic maximum of its orbit under the action in Example 4.14, and define $m(\mathbf{Z})$ by selecting the lexicographically maximum subobject $\hat{\mathbf{Z}}_0 \in S(\hat{\mathbf{Z}})$. (The canonical labeling $\kappa(\mathbf{Z}) \in \text{Iso}(\mathbf{Z}, \hat{\mathbf{Z}})$ may be selected arbitrarily.)

Example 4.35. Consider the search tree in Example 4.8 and Algorithm 4.7. Suppose the function m from Example 4.34 is used, and suppose that the lexicographically maximum matrix is always selected in line 3 of the algorithm. Then, the subtree traversed by Algorithm 4.7 is identical to the earlier tree depicted in Fig. 4.2, where "×" now marks nodes that fail the test (4.13) or are not the maximum in their respective orbits on $C(\mathbf{X})$ under the action of $\text{Aut}(\mathbf{X})$.

In an algorithm implementation occurring in practice the canonical parent $m(Z)$ is rarely explicitly computed in the form just described, because we are essentially only interested in testing whether $(Z, m(Z)) \cong (Z, p(Z))$ holds for a nonroot node $Z \in \Omega_{\text{nr}}$ encountered in the search. Equivalently, assuming the subobject framework, we are only interested in testing whether $m(Z)$ and $p(Z)$ are in the same orbit of the action of $\text{Aut}(Z)$ on the set of subobjects $S(Z)$. This observation can often be exploited to great efficiency.

First, it is frequently the case that the subobjects in $S(Z)$ are in a one-to-one correspondence with certain elementary objects that make up Z – for example, the codewords of a code when the search augments a code one codeword at a time, the blocks of a design when generating designs block by block, and so forth – in which case the automorphism orbits of subobjects are immediately obtained by computing a canonical labeling and generators for $\mathrm{Aut}(Z)$ in a permutation representation that involves the elementary objects.

Example 4.36. In Example 4.33 there is a natural one-to-one correspondence between the defined rows of a matrix and the subobjects associated with the matrix. The automorphism orbits of subobjects correspond to the automorphism orbits of rows.

Second, it is possible to use subobject invariants to test whether $m(Z)$ and $p(Z)$ are in the same orbit of $\mathrm{Aut}(Z)$ on $S(Z)$. For example, suppose we select $m(Z)$ in such a manner that it always has the maximum invariant value in $S(Z)$. If $p(Z)$ is the unique subobject in $S(Z)$ with the maximum invariant value, then we must have $p(Z) = m(Z)$, and this can be decided by invariant computations only without ever computing $m(Z)$ explicitly via canonical labeling. Similarly, if $p(Z)$ does not have the maximum invariant value in $S(Z)$, then $p(Z)$ and $m(Z)$ must occur in different orbits of $\mathrm{Aut}(Z)$ on $S(Z)$.

Example 4.37. One subobject invariant that can be applied in the setting of Examples 4.33 and 4.36 is to associate with every defined row i of a matrix the total number of 1s that occur in the columns that contain a 1 in row i.

Further examples on the use of subobject invariants in connection with generation by canonical augmentation can be found in Sect. 6.1.4 and [70, 304, 414, 415].

Generation by canonical augmentation can be efficiently parallelized because the test (4.13) depends only on the current node Z and its parent $p(Z)$. Furthermore, the isomorph rejection reinforcing (4.13) needs to be performed only among siblings, so knowledge of nodes encountered elsewhere in the search tree is not required. Efficiency in terms of required storage space depends in general on the strategy chosen to reject isomorphs among siblings.

4.2.4 Homomorphisms of Group Actions and Localization

In this section we take a more algebraic approach to isomorph rejection – or more accurately, to isomorph-free exhaustive generation in general – in terms of homomorphisms of group actions and group-theoretic localization. In this connection it is natural to abandon the search tree model and work only with the relevant algebraic concepts, that is, groups, group actions, and associated homomorphisms.

In the algebraic setting, the orbits of a group action correspond to the isomorphism classes of objects. Accordingly, isomorph-free exhaustive generation corresponds to producing a set of objects that contains exactly one object from every orbit. Such a set is called an *orbit transversal.* Thus, in algebraic terms the fundamental problem to be considered is that we have a group G

that acts on a finite set Ω that is implicitly defined, and we are asked to produce an orbit transversal $T_\Omega \subseteq \Omega$ for the action.

Example 4.38. Let us consider as a running example the problem of constructing an orbit transversal for the induced action of $C_6 = \langle(1\ 2\ 3\ 4\ 5\ 6)\rangle$ on 2-subsets of $\{1,2,3,4,5,6\}$; cf. Example 3.19.

Kerber and Laue [308, 309, 356, 357] together with collaborators have extensively studied the use of homomorphisms of group actions in solving orbit transversal problems. In essence, the idea is to first solve an orbit transversal problem for a secondary action, and then use this transversal in solving the primary transversal problem, where the primary and secondary action are connected by means of a homomorphism of group actions. For simplicity, we assume that the acting group G is fixed.

Definition 4.39. *Let G be a group that acts on two finite sets Ω and Π. A* homomorphism of group actions *is a map $\varphi : \Omega \to \Pi$ such that $\varphi(gX) = g\varphi(X)$ for all $g \in G$ and $X \in \Omega$.*

A bijective homomorphism of group actions is called an *isomorphism*. Two actions are *equivalent* if they are related by an isomorphism of group actions.

Basic homomorphisms of group actions are the map $X \mapsto N_G(X)$ taking an object $X \in \Omega$ to its stabilizer (G acts on its subgroups by conjugation), the bijection $gX \mapsto gN_G(X)$ taking an orbit element $gX \in GX$ to a left coset of the stabilizer (G acts on left cosets of its subgroups by left multiplication), and the map $gH \mapsto gK$ taking a left coset into a (larger) left coset, $H \leq K \leq G$. In many cases these homomorphisms can be used to transform a problem involving an "external" action of a group on a finite set Ω into a problem involving a "local" action of the group on its subgroups or cosets of subgroups.

Example 4.40. A local description of the orbit transversal problem in Example 4.38 can be obtained in terms of the group $S_6 = \mathrm{Sym}(\{1,2,3,4,5,6\})$ and the setwise stabilizer $N_{S_6}(\{1,2\})$ of $\{1,2\}$ in S_6. Namely, the induced action of S_6 on 2-subsets of $\{1,2,3,4,5,6\}$ is easily checked – cf. Theorem 3.20 – to be equivalent to the action of S_6 on the left cosets $S_6/N_{S_6}(\{1,2\})$ by left multiplication. Accordingly, for the group $C_6 = \langle(1\ 2\ 3\ 4\ 5\ 6)\rangle \leq S_6$, a "localized" orbit transversal problem equivalent to Example 4.38 is to determine an orbit transversal for the action of C_6 on $S_6/N_{S_6}(\{1,2\})$ by left multiplication.

Another basic family of homomorphisms is obtained from maps that "forget" some structure in an object or make the object "coarser" in such a way that the group action is respected. An important special case are projection maps of various types. For example, let G act on finite sets Ψ_1 and Ψ_2, and let $\Lambda \subseteq \Psi_1 \times \Psi_2$, where G acts elementwise on ordered pairs in Λ, that is, for $g \in G$ and $(Y_1, Y_2) \in \Lambda$, define $g * (Y_1, Y_2) = (gY_1, gY_2)$. In this case the projection maps $\Psi_1 \overset{\psi_1}{\leftarrow} \Lambda \overset{\psi_2}{\rightarrow} \Psi_2$ defined for all $(Y_1, Y_2) \in \Lambda$ by $Y_1 \overset{\psi_1}{\leftarrow\!\shortmid} (Y_1, Y_2) \overset{\psi_2}{\mapsto} Y_2$ are homomorphisms of group actions.

Provided that suitable homomorphisms are available, an orbit transversal problem can be solved step by step along a sequence of homomorphisms. In more precise terms, we have a sequence of group actions – G being the acting group in each case – connected by homomorphisms of group actions:

$$\Omega_1 \xrightarrow{\varphi_1} \Omega_2 \xrightarrow{\varphi_2} \Omega_3 \xrightarrow{\varphi_3} \cdots \xrightarrow{\varphi_{n-2}} \Omega_{n-1} \xrightarrow{\varphi_{n-1}} \Omega_n. \qquad (4.19)$$

An orbit transversal for the desired action on Ω_1 is now obtained by starting with an orbit transversal for the action on Ω_n, which is assumed to be trivially available or obtained through other means. Then, for each $i = n-1, n-2, \ldots, 1$ the orbit transversal for Ω_{i+1} and the homomorphism φ_i are used to produce an orbit transversal for Ω_i. Depending on whether $\varphi_i : \Omega_i \to \Omega_{i+1}$ or $\varphi_i : \Omega_{i+1} \to \Omega_i$, each step either *lifts* an orbit transversal from the image Ω_{i+1} to the domain Ω_i, or *projects* surjectively from the domain Ω_{i+1} to the image Ω_i. This general approach to classification is often called classification by the *homomorphism principle* [356, 357].

Let us consider examples of this situation before looking at the lifting and projecting steps in more detail.

Example 4.41. A sequence of homomorphisms appropriate for the localized problem in Example 4.40 can be obtained by varying the right-hand side group in $S_6/N_{S_6}(\{1,2\})$; cf. [518, 519, 520]. Let

$$H_1 = N_{S_6}(\{1,2\}), \quad H_2 = N_{S_6}((1,2)), \quad H_3 = N_{S_6}(1), \quad H_4 = S_6.$$

Note that $H_1 \geq H_2$, $H_2 \leq H_3$, and $H_3 \leq H_4$. We obtain a sequence of homomorphisms of group actions:

$$\begin{aligned} \varphi_1 &: S_6/H_1 \leftarrow S_6/H_2, & sH_1 &\leftarrow\!\shortmid\, sH_2; \\ \varphi_2 &: S_6/H_2 \to S_6/H_3, & sH_2 &\mapsto sH_3; \\ \varphi_3 &: S_6/H_3 \to S_6/H_4, & sH_3 &\mapsto sH_4. \end{aligned} \qquad (4.20)$$

In each case the relevant action is the action of $C_6 = \langle(1\ 2\ 3\ 4\ 5\ 6)\rangle$ by left multiplication, that is, $g * sH_i = (gs)H_i$ for all $g \in C_6$, $s \in S_6$, and $1 \leq i \leq 4$. The solution approach is now to start with an orbit transversal for the action of C_6 on $S_6/H_4 = S_6/S_6$ – which is trivially $\{S_6\}$ – and proceed along the sequence (4.20) in the reverse direction until we have an orbit transversal for the action of C_6 on S_6/H_1. That is, in terms of (4.19) we have

$$S_6/H_1 \xleftarrow{\varphi_1} S_6/H_2 \xrightarrow{\varphi_2} S_6/H_3 \xrightarrow{\varphi_3} S_6/H_4.$$

The homomorphisms φ_3 and φ_2 induce lifting steps, whereas φ_1 induces a projecting step. Note that all of the homomorphisms are surjective.

Example 4.42. Let us illustrate a sequence of projection homomorphisms in terms of our other running example on 0-1 matrices. For $k = 0, 1, 2, 3, 4$, let Γ_k be the set of all 4×4 matrices with entries from $\{0, 1, ?\}$ such that $4 - k$

rows are 0-1 rows with two 1s, k rows consist of the undefined symbol "?", and every column contains at most two 1s. For $k = 0, 1, 2, 3$, let $\Sigma_k \subseteq \Gamma_k \times \Gamma_{k+1}$ consist of all ordered pairs $(\mathbf{X}_k, \mathbf{X}_{k+1})$ such that $\mathbf{X}_{k+1}$ is obtained from $\mathbf{X}_k$ by transforming one 0-1 row into an undefined row. Let $S_4 \times S_4$ act on each of the sets $\Gamma_0, \Gamma_1, \Gamma_2, \Gamma_3, \Gamma_4$ by row and column permutation – see Example 4.14 – and let $S_4 \times S_4$ act elementwise on the ordered pairs in $\Sigma_0, \Sigma_1, \ldots, \Sigma_3$. For $k = 0, 1, 2, 3$, define projection homomorphisms π_k and λ_k for all $(\mathbf{X}_k, \mathbf{X}_{k+1}) \in \Sigma_k$ by $\pi_k : (\mathbf{X}_k, \mathbf{X}_{k+1}) \mapsto \mathbf{X}_k$ and $\lambda_k : (\mathbf{X}_k, \mathbf{X}_{k+1}) \mapsto \mathbf{X}_{k+1}$, respectively. In terms of (4.19), we have the sequence

$$\Gamma_0 \xleftarrow{\pi_0} \Sigma_0 \xrightarrow{\lambda_0} \Gamma_1 \xleftarrow{\pi_1} \Sigma_1 \xrightarrow{\lambda_1} \Gamma_2 \xleftarrow{\pi_2} \Sigma_2 \xrightarrow{\lambda_2} \Gamma_3 \xleftarrow{\pi_3} \Sigma_3 \xrightarrow{\lambda_3} \Gamma_4.$$

The homomorphisms λ_k induce lifting steps and the surjective homomorphisms π_k induce projecting steps.

The following theorem contains the fundamental structural information to be exploited in the lifting and projecting steps. Define the *preimage* of $Z \in \Pi$ with respect to $\varphi : \Omega \to \Pi$ by $\varphi^{-1}(Z) = \{X \in \Omega : \varphi(X) = Z\}$. Note that a preimage may be empty if φ is not surjective.

Theorem 4.43. *Let G be a group that acts on finite sets Ω and Π, and let $\varphi : \Omega \to \Pi$ be a homomorphism of group actions. Then,*

1. *for all $Z, W \in \varphi(\Omega)$ the sets of orbits on Ω intersected by the preimages $\varphi^{-1}(Z)$ and $\varphi^{-1}(W)$ are either equal or disjoint, where equality holds if and only if $Z \cong W$,*
2. *for all $Z \in \varphi(\Omega)$ and $X \in \varphi^{-1}(Z)$ it holds that*

$$\mathrm{Aut}(Z) = \bigcup_{Y \in \varphi^{-1}(Z)} \mathrm{Iso}(X, Y). \tag{4.21}$$

Proof. We first prove that if $\varphi^{-1}(Z)$ and $\varphi^{-1}(W)$ intersect the same orbit, say $GX \in G\backslash\Omega$, then $Z \cong W$. Let $g_1, g_2 \in G$ such that $g_1 X \in \varphi^{-1}(Z)$ and $g_2 X \in \varphi^{-1}(W)$. Thus,

$$g_2 g_1^{-1} Z = g_2 g_1^{-1} \varphi(g_1 X) = g_2 g_1^{-1} g_1 \varphi(X) = g_2 \varphi(X) = \varphi(g_2 X) = W.$$

Conversely, we prove that $Z \cong W$ implies that $\varphi^{-1}(Z)$ and $\varphi^{-1}(W)$ intersect the same orbits in $G\backslash\Omega$. Note that both $\varphi^{-1}(Z)$ and $\varphi^{-1}(W)$ are nonempty because of the assumption $Z, W \in \varphi(\Omega)$. Let $g_0 \in G$ satisfy $g_0 Z = W$ and let $GX \in G\backslash\Omega$ be an arbitrary orbit. By symmetry it suffices to prove that if $\varphi^{-1}(Z)$ intersects GX, then $\varphi^{-1}(W)$ intersects GX. Let $g \in G$ such that $gX \in \varphi^{-1}(Z)$. We obtain

$$W = g_0 Z = g_0 \varphi(gX) = \varphi(g_0 g X).$$

Thus, $g_0 g X \in \varphi^{-1}(W)$. This completes the proof of the first claim.

To establish the second claim, let $Z \in \varphi(\Omega)$ and $X \in \varphi^{-1}(Z)$. For an arbitrary $g \in \mathrm{Aut}(Z)$ we have $\varphi(gX) = g\varphi(X) = gZ = Z$. Thus, for $Y = gX$ it holds that $Y \in \varphi^{-1}(Z)$ and $g \in \mathrm{Iso}(X,Y)$. Conversely, for arbitrary $X, Y \in \varphi^{-1}(Z)$ and $g \in \mathrm{Iso}(X,Y)$, we have $gZ = g\varphi(X) = \varphi(gX) = \varphi(Y) = Z$. Thus, $g \in \mathrm{Aut}(Z)$. □

Theorem 4.43 contains a wealth of useful information. For example, φ maps orbits on Ω into orbits on Π, and for all $X \in \Omega$ we have $\mathrm{Aut}(X) \leq \mathrm{Aut}(\varphi(X))$. However, from a classification point of view the fundamental consequence is that Theorem 4.43 enables divide-and-conquer solutions to orbit transversal problems. In particular, instead of considering the "large" sets Ω and Π, it suffices to consider many "small" sets in the form of preimages of transversal elements. Also, instead of considering the "large" acting group G in isomorphism computations, it suffices to restrict the acting group to the "small" automorphism groups of transversal elements.

Let us first consider the structure of a lifting step in more detail. In this case we are given a homomorphism $\varphi : \Omega \to \Pi$ and an orbit transversal $T_\Pi = \{Z_1, Z_2, \ldots, Z_m\} \subseteq \Pi$. The task is to determine an orbit transversal for the action of G on Ω. The first claim of Theorem 4.43 immediately gives that the preimages $\varphi^{-1}(Z_1), \varphi^{-1}(Z_2), \ldots, \varphi^{-1}(Z_m)$ intersect all orbits of the action of G on Ω, where different preimages intersect different orbits. Thus, in computing an orbit transversal for the action of G on Ω, it suffices to consider only the preimages of transversal elements Z_i, and each preimage $\varphi^{-1}(Z_i)$ can be considered independently of the other preimages. Furthermore, the second claim of Theorem 4.43 implies that it suffices to consider only the group $\mathrm{Aut}(Z_i)$ in rejecting elements in $\varphi^{-1}(Z_i)$ belonging to the same orbit; indeed, because $\mathrm{Iso}(X,Y) \subseteq \mathrm{Aut}(Z_i)$ for all $X, Y \in \varphi^{-1}(Z_i)$, we have $X \cong Y$ if and only if there exists an $a \in \mathrm{Aut}(Z_i)$ with $aX = Y$. Also note that $\mathrm{Aut}(X) \leq \mathrm{Aut}(Z_i)$ for all $X \in \varphi^{-1}(Z_i)$. Thus, the homomorphism $\varphi : \Omega \to \Pi$ has effectively reduced the orbit transversal problem for the action of G on Ω into the problem of first determining an orbit transversal for the action of G on Π, followed by the problem of determining, independently for each transversal element $Z_i \in \Pi$, an orbit transversal for the action of $\mathrm{Aut}(Z_i)$ on the preimage $\varphi^{-1}(Z_i)$.

Example 4.44. Let us illustrate lifting steps in the situation of Example 4.41. The orbit transversal for the action of C_6 on $S_6/H_4 = S_6/S_6 = \{S_6\}$ is trivially $\{S_6\}$. Accordingly, the preimage $\varphi_3^{-1}(S_6)$ consists of all the cosets in $S_6/H_3 = S_6/N_{S_6}(1)$, namely

$$\epsilon H_3, \quad (1\ 2)H_3, \quad (1\ 3)H_3, \quad (1\ 4)H_3, \quad (1\ 5)H_3, \quad (1\ 6)H_3.$$

Note that because $H_3 = N_{S_6}(1)$ a left coset sH_3 is uniquely determined by the image $s(1) \in \{1,2,3,4,5,6\}$, where $s \in S_6$. Thus, because C_6 is transitive on $\{1,2,3,4,5,6\}$, the action of C_6 by left multiplication on S_6/H_3 is also transitive. Let us select $\{\epsilon H_3\}$ as an orbit transversal for the action of C_6

on S_6/H_3. Note that the associated automorphism group $\mathrm{Aut}(\epsilon H_3) = \{\epsilon\}$. Proceeding to the next lifting step and the homomorphism $\varphi_2 : S_6/H_2 \to S_6/H_3$, the preimage $\varphi_2^{-1}(\epsilon H_3)$ consists of five cosets:

$$\epsilon H_2, \quad (2\ 3)H_2, \quad (2\ 4)H_2, \quad (2\ 5)H_2, \quad (2\ 6)H_2. \tag{4.22}$$

Because $\mathrm{Aut}(\epsilon H_3) = \{\epsilon\}$, Theorem 4.43 implies that these cosets lie on pairwise different orbits of C_6 on S_6/H_2. Furthermore, each coset has a trivial automorphism group. These cosets form an orbit transversal for the action of C_6 on S_6/H_2.

Example 4.45. In the situation of Example 4.42, an orbit transversal for the action of $S_4 \times S_4$ on Γ_2 consists of the three matrices

$$\begin{bmatrix} 1100 \\ 1100 \\ ???? \\ ???? \end{bmatrix}, \quad \begin{bmatrix} 1100 \\ 1010 \\ ???? \\ ???? \end{bmatrix}, \quad \begin{bmatrix} 1100 \\ 0011 \\ ???? \\ ???? \end{bmatrix}.$$

Lifting from Γ_2 to Σ_1 using λ_1, the preimage of the first matrix intersects one orbit on Σ_1 represented by

$$\left(\begin{bmatrix} 1100 \\ 1100 \\ 0011 \\ ???? \end{bmatrix}, \begin{bmatrix} 1100 \\ 1100 \\ ???? \\ ???? \end{bmatrix} \right),$$

the preimage of the second matrix intersects two orbits represented by

$$\left(\begin{bmatrix} 1100 \\ 1010 \\ 0110 \\ ???? \end{bmatrix}, \begin{bmatrix} 1100 \\ 1010 \\ ???? \\ ???? \end{bmatrix} \right), \quad \left(\begin{bmatrix} 1100 \\ 1010 \\ 0101 \\ ???? \end{bmatrix}, \begin{bmatrix} 1100 \\ 1010 \\ ???? \\ ???? \end{bmatrix} \right),$$

and the preimage of the third matrix intersects two orbits represented by

$$\left(\begin{bmatrix} 1100 \\ 0011 \\ 1100 \\ ???? \end{bmatrix}, \begin{bmatrix} 1100 \\ 0011 \\ ???? \\ ???? \end{bmatrix} \right), \quad \left(\begin{bmatrix} 1100 \\ 0011 \\ 1010 \\ ???? \end{bmatrix}, \begin{bmatrix} 1100 \\ 0011 \\ ???? \\ ???? \end{bmatrix} \right).$$

These five ordered pairs of matrices together form an orbit transversal for the action of $S_4 \times S_4$ on Σ_1.

For a projecting step we are given a surjective homomorphism $\varphi : \Omega \to \Pi$ and an orbit transversal $\{X_1, X_2, \ldots, X_m\}$ for the action of G on Ω. The task is to produce an orbit transversal for the action of G on Π. Because φ is surjective, the set $\{\varphi(X_1), \varphi(X_2), \ldots, \varphi(X_m)\}$ intersects every orbit on Π. It thus suffices to remove from this set all but one element from every orbit.

Example 4.46. Consider the orbit transversal (4.22) for the action of C_6 on S_6/H_2 in Example 4.44 and the homomorphism $\varphi_1 : S_6/H_2 \to S_6/H_1$ in Example 4.41. From Theorem 4.43 we know that two images of transversal elements are isomorphic if and only if the respective preimages intersect the same orbits. We obtain the following preimages of images of the transversal elements:

$$\begin{aligned}
\varphi_1^{-1}(\epsilon H_1) &= \{\epsilon H_2, (1\ 2)H_2\}, \\
\varphi_1^{-1}((2\ 3)H_1) &= \{(2\ 3)H_2, (1\ 3\ 2)H_2\}, \\
\varphi_1^{-1}((2\ 4)H_1) &= \{(2\ 4)H_2, (1\ 4\ 2)H_2\}, \\
\varphi_1^{-1}((2\ 5)H_1) &= \{(2\ 5)H_2, (1\ 5\ 2)H_2\}, \\
\varphi_1^{-1}((2\ 6)H_1) &= \{(2\ 6)H_2, (1\ 6\ 2)H_2\}.
\end{aligned}$$

Because

$$(1\ 2\ 3\ 4\ 5\ 6) * (1\ 6\ 2)H_2 = (3\ 4\ 5\ 6)H_2 = \epsilon H_2$$

we obtain $\epsilon H_1 \cong (2\ 6)H_1$. Similarly,

$$(1\ 3\ 5)(2\ 4\ 6) * (1\ 5\ 2)H_2 = (2\ 3\ 5\ 4\ 6)H_2 = (2\ 3)H_2$$

implies $(2\ 3)H_1 \cong (2\ 5)H_1$. With some further effort it can be checked no other images of transversal elements are isomorphic under the action of C_6. Thus, $\{\epsilon H_1, (2\ 3)H_1, (2\ 4)H_1\}$ is an orbit transversal for the action of C_6 on S_6/H_1. The associated automorphism groups are also straightforward to determine using Theorem 4.43. Namely, for $\varphi_1^{-1}((2\ 4)H_1)$ we have $(1\ 4)(2\ 5)(3\ 6) * (2\ 4)H_2 = (2\ 1\ 4\ 5)(3\ 6)H_2 = (1\ 4\ 2)H_2$ and $\mathrm{Aut}((2\ 4)H_2) = \{\epsilon\}$. Thus, using Theorem 4.43 we obtain

$$\begin{aligned}
\mathrm{Aut}((2\ 4)H_1) &= \mathrm{Aut}((2\ 4)H_2) \cup \mathrm{Iso}((2\ 4)H_2, (1\ 4\ 2)H_2) \\
&= \{\epsilon, (1\ 4)(2\ 5)(3\ 6)\}.
\end{aligned}$$

On the other hand, $\varphi_1^{-1}(\epsilon H_1)$ and $\varphi_1^{-1}((2\ 3)H_1)$ contain only pairwise nonisomorphic cosets under the action of C_6, each with a trivial automorphism group; therefore $\mathrm{Aut}(\epsilon H_1) = \mathrm{Aut}((2\ 3)H_1) = \{\epsilon\}$.

Example 4.47. Consider the orbit transversal for the action of $S_4 \times S_4$ on Σ_1 in Example 4.45. Projecting this transversal into a transversal for the action on Γ_1 using π_1, we obtain three orbits with representatives

$$\begin{bmatrix} 1100 \\ 1100 \\ 0011 \\ ???? \end{bmatrix}, \quad \begin{bmatrix} 1100 \\ 1010 \\ 0110 \\ ???? \end{bmatrix}, \quad \begin{bmatrix} 1100 \\ 1010 \\ 0101 \\ ???? \end{bmatrix}.$$

In particular, the first and fourth orbit on Σ_1 (in order of representatives in Example 4.47) project to the first orbit on Γ_1, the second orbit on Σ_1 projects to the second orbit on Γ_1, and the third and fifth orbit on Σ_1 project to the third orbit on Γ_1.

The implementation of the lifting and projecting steps into a practical algorithm greatly depends on the group actions and homomorphisms in question. Indeed, in this connection generation by homomorphisms should be regarded more as a general principle – such as backtrack search – rather than a specific implementation technique for isomorph rejection. Thus, our discussion here will be limited to some rather general implementation principles.

Typically it is the case that an implementation proceeding step by step along a sequence of group actions requires also information on the automorphism groups of transversal elements; cf. Examples 4.44 and 4.46. Furthermore, it is convenient to have available a canonical labeling map for each action in the sequence.

Definition 4.48. *An* extended orbit transversal *consists of an orbit transversal $T_\Omega \subseteq \Omega$, a set S_X of generators for the automorphism group* $\mathrm{Aut}(X)$ *for each $X \in T_\Omega$, and a practical way to evaluate a canonical labeling map $\kappa_{G,\Omega} : \Omega \to G$ satisfying $\kappa_{G,\Omega}(X)X \in T_\Omega$ for all $X \in \Omega$.*

The implementation of a lifting step from Π to Ω via a homomorphism $\varphi : \Omega \to \Pi$ clearly requires a way to construct objects in the preimage $\varphi^{-1}(Z)$ for every $Z \in T_\Pi$. Also required is a way to reject objects isomorphic under the action of $\mathrm{Aut}(Z)$ in $\varphi^{-1}(Z)$. If $\mathrm{Aut}(Z)$ is small, then this can be done by exhaustive search through elements of $\mathrm{Aut}(Z)$ – in particular, no search is required if $\mathrm{Aut}(Z)$ is trivial. If $\varphi^{-1}(Z)$ is small, then the orbit algorithms in Sect. 5.5 can be employed. In both cases it is possible to determine generators for $\mathrm{Aut}(X)$ and canonical labeling for any $X \in \varphi^{-1}(Z)$ under the action of $\mathrm{Aut}(Z)$; in the former case the entire group $\mathrm{Aut}(Z)$ is obtained as a side effect of executing the search, in the latter case a left transversal L of $\mathrm{Aut}(X)$ in $\mathrm{Aut}(Z)$ is easily obtained as a side effect of orbit computations, after which Theorem 5.26 and sifting yield a small set of generators for $\mathrm{Aut}(X)$; see Sect. 5.5. In general, a lifting step requires for every $Z \in T_\Pi$ the solution of an (extended) orbit transversal problem for the action of $\mathrm{Aut}(Z)$ on $\varphi^{-1}(Z)$. By Theorem 4.43 from such a set of solutions we obtain directly an orbit transversal T_Ω and associated automorphism group generators. Canonical labeling for the action of G on Ω can be defined for all $X \in \Omega$ by

$$\kappa_{G,\Omega}(X) = \kappa_{\mathrm{Aut}(g\varphi(X)),\varphi^{-1}(g\varphi(X))}(gX)g, \quad g = \kappa_{G,\Pi}(\varphi(X)).$$

To verify this claim, observe that we first canonically label the image $\varphi(X)$ using $\kappa_{G,\Pi}$ – which is available by assumption that we have an extended orbit transversal for the action of G on Π – and then canonically label gX in the preimage $\varphi^{-1}(g\varphi(X)) \subseteq \Omega$ using a canonical labeling map given by the extended orbit transversals on the preimages.

Let us now proceed to consider a projecting step from Ω to Π via a surjective homomorphism $\varphi : \Omega \to \Pi$. There are two basic approaches to implement a projecting step. The first approach is to employ Theorem 4.43 and dynamic programming (cf. [356]). It is assumed that the preimages $\varphi^{-1}(Z)$ are small and easily computable; furthermore, it is assumed that sufficient storage space is available to store entire orbit transversals together with auxiliary information. The idea is to maintain a work set $\Delta \subseteq \Omega$, initially consisting of an orbit transversal T_Ω, from which orbit representatives are removed in steps by selecting $X \in \Delta$, computing $Z = \varphi(X)$, and deleting all orbit representatives of orbits intersecting $\varphi^{-1}(Z)$ from Δ. The process stops when Δ is empty, at which point all the objects Z computed in the process form an orbit transversal T_Π by surjectivity of φ and Theorem 4.43. As a side-effect, it is possible to obtain automorphism group generators and a look-up array facilitating rapid evaluation of a canonical labeling map. A pseudocode implementation of the procedure is given in Algorithm 4.8.

Algorithm 4.8 Projecting an orbit transversal using dynamic programming

```
procedure TRANS-PROJECT( (T_Ω, {S_X}, κ_{G,Ω}): extended orbit transversal)
T_Π ← ∅
τ[·] ← empty array
Δ ← T_Ω
while Δ ≠ ∅ do
   select any X ∈ Δ
   Z ← φ(X)
   τ[X] ← 1_G
   T_Π ← T_Π ∪ {Z}
   S_Z ← S_X
   for all Y ∈ φ^{-1}(X) do
      g_0 ← κ_{G,Ω}(Y)
      Y_0 ← g_0 Y
      Δ ← Δ \ {Y_0}
      if Y_0 = X then
         S_Z ← S_Z ∪ {g_0^{-1}}
      else
         τ[Y_0] ← g_0^{-1}
      end if
   end for
end while
end procedure
```

Theorem 4.49. *The following claims hold when Algorithm 4.8 terminates:*

1. T_Π *is an orbit transversal for the action of* G *on* Π*,*
2. $\mathrm{Aut}(Z) = \langle S_Z \rangle$ *for all* $Z \in T_\Pi$*,*
3. *a canonical labeling map* $\kappa_{G,\Pi}$ *can be defined for every* $Z \in \Pi$ *by using the look-up array* $\tau[\cdot]$ *as follows: select any* $X \in \varphi^{-1}(Z)$*, and then put*

$$\kappa_{G,\Pi}(Z) = \tau[gX]g, \quad g = \kappa_{G,\Omega}(X).$$

Proof. The first claim holds by surjectivity of φ and Theorem 4.43. To establish the second claim, observe that $X \cong Y$ implies $\text{Iso}(X,Y) = g_0^{-1}\text{Aut}(X)$ and apply Theorem 4.43. To verify the third claim, observe that for each $Y_0 \in T_\Omega$ the array entry $\tau[Y_0]$ contains a group element taking Y_0 to an object $Y \in \Omega$ with $\varphi(Y) \in T_\Pi$. Thus, because $gZ = g\varphi(X) = \varphi(gX)$ and $gX \in T_\Omega$, we have $\kappa_{G,\Pi}(Z)Z = \tau[gX]\varphi(gX) = \varphi(\tau[gX]gX) \in T_\Pi$. □

Note that the look-up array $\tau[\cdot]$ is defined for all $X \in T_\Omega$. Thus, the amount of storage space required by the dynamic programming approach is proportional to the number of orbits on Ω and Π.

The second approach to implement a projecting step is to assume the existence of practical auxiliary algorithms for evaluating a canonical labeling map $\kappa_{G,\Pi} : \Pi \to G$ and a set S_Z generators for $\text{Aut}(Z)$, $Z \in \Pi$. In this way it is possible to obtain a parallelizable and more memory-efficient approach analogous to generation by canonical augmentation. The idea is to associate with every orbit in $GZ \in G\backslash\Pi$ a canonical orbit in the preimage $\varphi^{-1}(GZ)$ from which the orbit GZ should be generated by projection. In practice it is convenient to define the association with respect to a preimage $\varphi^{-1}(Z)$, whereby the connection to generation by canonical augmentation becomes more apparent (cf. [357]).

Theorem 4.50. *Let $\varphi : \Omega \to \Pi$ be a surjective homomorphism of group actions and let G be the acting group for both actions. Let μ be a function that associates to every $Z \in \Pi$ a nonempty* $\text{Aut}(Z)$*-orbit $\mu(Z) \subseteq \varphi^{-1}(Z)$ such that $g\mu(Z) = \mu(gZ)$ holds for all $g \in G$. Then,*

1. *for all $X, Y \in \Omega$ it holds that $X \in \mu(\varphi(X))$, $Y \in \mu(\varphi(Y))$, and $\varphi(X) \cong \varphi(Y)$ together imply $X \cong Y$,*
2. *for every orbit $GZ \in G\backslash\Pi$ there exists an orbit $GX \in G\backslash\Omega$ with $\varphi(GX) = GZ$ and $X_0 \in \mu(\varphi(X_0))$ for all $X_0 \in GX$.*

Proof. To establish the first claim, let $g \in G$ satisfy $g\varphi(X) = \varphi(Y)$. From $X \in \mu(\varphi(X))$ we obtain $gX \in g\mu(\varphi(X)) = \mu(g\varphi(X)) = \mu(\varphi(Y))$. Because $\mu(\varphi(Y))$ is an $\text{Aut}(\varphi(Y))$-orbit and $Y \in \mu(\varphi(Y))$, there exists an $a \in \text{Aut}(\varphi(Y))$ with $agX = Y$. Thus, $X \cong Y$. To establish the second claim, select an arbitrary $Z \in \Pi$ and let $X \in \mu(Z)$. Clearly, $\varphi(X) = Z$ and hence $\varphi(GX) = GZ$. Select an arbitrary $X_0 \in GX$ and let $g \in G$ satisfy $X_0 = gX$. From $X \in \mu(Z) = \mu(\varphi(X))$ we obtain

$$X_0 = gX \in g\mu(\varphi(X)) = \mu(g\varphi(X)) = \mu(\varphi(gX)) = \mu(\varphi(X_0)).$$

The claim follows because Z and X_0 were arbitrary. □

Corollary 4.51. *Let T_Ω be an orbit transversal for the action on Ω. Then,*

$$T_\Pi = \{\varphi(X) : X \in T_\Omega,\ X \in \mu(\varphi(X))\}$$

is an orbit transversal for the action on Π.

and easily computable; furthermore, it is assumed that sufficient storage space is available to store entire orbit transversals together with auxiliary information. The idea is to maintain a work set $\Delta \subseteq \Omega$, initially consisting of an orbit transversal T_Ω, from which orbit representatives are removed in steps by selecting $X \in \Delta$, computing $Z = \varphi(X)$, and deleting all orbit representatives of orbits intersecting $\varphi^{-1}(Z)$ from Δ. The process stops when Δ is empty, at which point all the objects Z computed in the process form an orbit transversal T_Π by surjectivity of φ and Theorem 4.43. As a side-effect, it is possible to obtain automorphism group generators and a look-up array facilitating rapid evaluation of a canonical labeling map. A pseudocode implementation of the procedure is given in Algorithm 4.8.

Algorithm 4.8 Projecting an orbit transversal using dynamic programming

---

procedure TRANS-PROJECT($(T_\Omega, \{S_X\}, \kappa_{G,\Omega})$: extended orbit transversal)
$T_\Pi \leftarrow \emptyset$
$\tau[\cdot] \leftarrow$ empty array
$\Delta \leftarrow T_\Omega$
while $\Delta \neq \emptyset$ **do**
    select any $X \in \Delta$
    $Z \leftarrow \varphi(X)$
    $\tau[X] \leftarrow 1_G$
    $T_\Pi \leftarrow T_\Pi \cup \{Z\}$
    $S_Z \leftarrow S_X$
    for all $Y \in \varphi^{-1}(X)$ **do**
        $g_0 \leftarrow \kappa_{G,\Omega}(Y)$
        $Y_0 \leftarrow g_0 Y$
        $\Delta \leftarrow \Delta \setminus \{Y_0\}$
        if $Y_0 = X$ **then**
            $S_Z \leftarrow S_Z \cup \{g_0^{-1}\}$
        else
            $\tau[Y_0] \leftarrow g_0^{-1}$
        end if
    end for
end while
end procedure

Theorem 4.49. *The following claims hold when Algorithm 4.8 terminates:*

1. *T_Π is an orbit transversal for the action of G on Π,*
2. *$\mathrm{Aut}(Z) = \langle S_Z \rangle$ for all $Z \in T_\Pi$,*
3. *a canonical labeling map $\kappa_{G,\Pi}$ can be defined for every $Z \in \Pi$ by using the look-up array $\tau[\cdot]$ as follows: select any $X \in \varphi^{-1}(Z)$, and then put*

$$\kappa_{G,\Pi}(Z) = \tau[gX]g, \quad g = \kappa_{G,\Omega}(X).$$

Proof. The first claim holds by surjectivity of φ and Theorem 4.43. To establish the second claim, observe that $X \cong Y$ implies $\mathrm{Iso}(X,Y) = g_0^{-1}\mathrm{Aut}(X)$ and apply Theorem 4.43. To verify the third claim, observe that for each $Y_0 \in T_\Omega$ the array entry $\tau[Y_0]$ contains a group element taking Y_0 to an object $Y \in \Omega$ with $\varphi(Y) \in T_\Pi$. Thus, because $gZ = g\varphi(X) = \varphi(gX)$ and $gX \in T_\Omega$, we have $\kappa_{G,\Pi}(Z)Z = \tau[gX]\varphi(gX) = \varphi(\tau[gX]gX) \in T_\Pi$. □

Note that the look-up array $\tau[\cdot]$ is defined for all $X \in T_\Omega$. Thus, the amount of storage space required by the dynamic programming approach is proportional to the number of orbits on Ω and Π.

The second approach to implement a projecting step is to assume the existence of practical auxiliary algorithms for evaluating a canonical labeling map $\kappa_{G,\Pi} : \Pi \to G$ and a set S_Z generators for $\mathrm{Aut}(Z)$, $Z \in \Pi$. In this way it is possible to obtain a parallelizable and more memory-efficient approach analogous to generation by canonical augmentation. The idea is to associate with every orbit in $GZ \in G\backslash\Pi$ a canonical orbit in the preimage $\varphi^{-1}(GZ)$ from which the orbit GZ should be generated by projection. In practice it is convenient to define the association with respect to a preimage $\varphi^{-1}(Z)$, whereby the connection to generation by canonical augmentation becomes more apparent (cf. [357]).

Theorem 4.50. *Let $\varphi : \Omega \to \Pi$ be a surjective homomorphism of group actions and let G be the acting group for both actions. Let μ be a function that associates to every $Z \in \Pi$ a nonempty* $\mathrm{Aut}(Z)$*-orbit $\mu(Z) \subseteq \varphi^{-1}(Z)$ such that $g\mu(Z) = \mu(gZ)$ holds for all $g \in G$. Then,*

1. *for all $X, Y \in \Omega$ it holds that $X \in \mu(\varphi(X))$, $Y \in \mu(\varphi(Y))$, and $\varphi(X) \cong \varphi(Y)$ together imply $X \cong Y$,*
2. *for every orbit $GZ \in G\backslash\Pi$ there exists an orbit $GX \in G\backslash\Omega$ with $\varphi(GX) = GZ$ and $X_0 \in \mu(\varphi(X_0))$ for all $X_0 \in GX$.*

Proof. To establish the first claim, let $g \in G$ satisfy $g\varphi(X) = \varphi(Y)$. From $X \in \mu(\varphi(X))$ we obtain $gX \in g\mu(\varphi(X)) = \mu(g\varphi(X)) = \mu(\varphi(Y))$. Because $\mu(\varphi(Y))$ is an $\mathrm{Aut}(\varphi(Y))$-orbit and $Y \in \mu(\varphi(Y))$, there exists an $a \in \mathrm{Aut}(\varphi(Y))$ with $agX = Y$. Thus, $X \cong Y$. To establish the second claim, select an arbitrary $Z \in \Pi$ and let $X \in \mu(Z)$. Clearly, $\varphi(X) = Z$ and hence $\varphi(GX) = GZ$. Select an arbitrary $X_0 \in GX$ and let $g \in G$ satisfy $X_0 = gX$. From $X \in \mu(Z) = \mu(\varphi(X))$ we obtain

$$X_0 = gX \in g\mu(\varphi(X)) = \mu(g\varphi(X)) = \mu(\varphi(gX)) = \mu(\varphi(X_0)).$$

The claim follows because Z and X_0 were arbitrary. □

Corollary 4.51. *Let T_Ω be an orbit transversal for the action on Ω. Then,*

$$T_\Pi = \{\varphi(X) : X \in T_\Omega,\ X \in \mu(\varphi(X))\}$$

is an orbit transversal for the action on Π.

Proof. Because T_Ω contains exactly one object from every orbit on Ω, the first claim of Theorem 4.50 implies that T_Π contains at most one object from every orbit on Π; conversely, the second claim implies that T_Π contains at least one object from every orbit on Π. □

In particular, given a function μ satisfying the requirements of Theorem 4.50, the decision whether to include the image $\varphi(X)$ in an orbit transversal can be made without relying on stored transversal elements or other auxiliary information.

One possibility to implement the test $X \in \mu(\varphi(X))$ is to rely on a canonical labeling map $\kappa_{G,\Pi}$ as follows. Given $X \in \Omega$, first compute $Z = \varphi(X)$, $g = \kappa_{G,\Pi}(Z)$, and $Z_0 = gZ$. Then, select any $X_0 \in \varphi^{-1}(Z_0)$ so that the selection depends only on Z_0 and $\varphi^{-1}(Z_0)$. Finally, put $m(Z) = g^{-1}X_0$ let $\mu(Z)$ be the $\mathrm{Aut}(Z)$-orbit of $m(Z)$. The test $X \in \mu(\varphi(X))$ is equivalent to checking whether $m(Z)$ and X are in the same $\mathrm{Aut}(Z)$-orbit on $\varphi^{-1}(Z)$; cf. Sect. 4.2.3.

A further discussion on generation by homomorphisms and more detailed algorithm implementations can be found in [230, 231, 309, 356, 357, 518, 520, 521].

5

Auxiliary Algorithms

In developing classification algorithms, one recurrently encounters subproblems belonging to a few common problem classes. Algorithms for solving such problems are discussed in this chapter. The reader who is not interested in implementational details of classification algorithms may well skip this chapter and assume that there is a “black box” that takes care of these tasks. Those who wish to implement classification algorithms, however, should notice that these auxiliary algorithms often constitute the core of the whole search and thereby affect the overall efficiency in a direct way. On the other hand, the chapter can also be studied separately to get an insight into some contemporary algorithms for several important hard problems (for which no polynomial time algorithms are known).

Almost without exception, we want to develop algorithms that find *all* solutions fulfilling given criteria. Most algorithms for finding just one solution of a certain kind utilize heuristics for recognizing the part of the search space where it most probably can be found. In an exhaustive exploration, however, we have to traverse the whole search space and need not implement heuristics of this kind.

In Sect. 5.1, algorithms for finding cliques in graphs are discussed. Algorithms for finding exact covers and covers of a set with given subsets are considered in Sects. 5.2 and 5.3, respectively. All these problems can be formulated as instances of Diophantine linear systems of equations, algorithms for which are discussed in Sect. 5.4. Section 5.5 provides a treatment of the basic permutation group algorithms often required in isomorphism computations. Isomorphism algorithms based on refinement are considered in Sect. 5.6. The chapter is ended in Sect. 5.7 with a discussion of the practical issue of distributing an extensive computation to a network of computers.

5.1 Clique Algorithms

Clique search plays a central role for many classification problems. Actually, clique search can be used for any problem where one wants to construct objects that are sets of smaller subobjects (of any kind) that must fulfill certain pairwise requirements. By mapping all possible subobjects into vertices of a graph and inserting an edge between two vertices exactly when the requirements are fulfilled, the final objects correspond to cliques in this graph. For a survey of application of clique algorithms to constructing and classifying objects of this type, see [458].

Whereas most clique algorithms in the literature focus on finding a maximum clique – or a maximum-weight clique in a weighted graph – our interest is in finding all cliques of a given size or weight (although, if they exist, these are in most cases maximum or maximum-weight cliques). The two types of problems for unweighted graphs are presented as Problems 5.1 and 5.2.

Problem 5.1. (MAXIMUM CLIQUE) Given a graph G, find (the size of) a maximum clique in G.

Problem 5.2. (CLIQUES) Given a graph G and an integer k, find all cliques of size k in G.

We will here restrict our consideration to unweighted graphs. There is an obvious backtrack algorithm for Problem 5.2, where one simply adds one vertex at a time to increase the size of a clique and uses a simple bounding function to prune the search [95]. We will here present another algorithm, which performs well for instances related to classification problems for codes and designs; the algorithm is from [456] and is here slightly modified to get an algorithm that finds all cliques of a given size. A search for all maximum cliques requires two passes: one that finds the size of a maximum clique and one that finds all cliques of that size. These tasks are accomplished by the algorithm from [456] and Algorithm 5.1, respectively.

In calling Algorithm 5.1, the order of the graph is given by n and the vertices are labeled $1, 2, \ldots, n$ in some order. The ordering of the vertices has a significant impact on the efficiency of the algorithm; appropriate orderings are mentioned later in the discussion of classification algorithms. The size of the cliques that we search for is given by k. We use the standard notation $\Gamma(i)$ for the neighborhood of i.

Algorithm 5.1 proceeds by repeatedly calculating entries of the array $c[i]$, the size of a maximum clique in the graph induced by the vertices in $S_i = \{i, i+1, \ldots, n\}$. However, it turns out that the value of $c[i]$ is useful only when it is smaller than $k-1$ so if the size of a maximum clique in S_i is greater than $k-1$, then we let $c[i] = k-1$. The value of $c[i]$ is used for pruning the backtrack search. Algorithms that utilize functions of this type for pruning are known as *Russian doll search* algorithms [595].

Algorithm 5.1 All cliques of size k

```
procedure CLQ(U: set, s: integer)
if s = k then
    report the current solution y_1, y_2, ..., y_k
    return
end if
if s > max then
    max ← s
    found ← TRUE
    return
end if
while U ≠ ∅ do
    if s + |U| ≤ max then
        return
    end if
    i ← min{j : j ∈ U}
    if s + c[i] ≤ max then
        return
    end if
    U ← U \ {i}
    y_{s+1} ← i
    CLQ(U ∩ Γ(i), s + 1)
    if found then
        return
    end if
end while
end procedure
procedure CLIQUE(G: graph, k: integer)
max ← 0
for i ← n, n − 1, ..., 1 do
    found ← FALSE
    y_1 ← i
    CLQ(S_i ∩ Γ(i), 1)
    c[i] ← max
end for
end procedure
```

For certain hard instances, when no cliques of size k exist, a speed-up may be achieved by first calculating values of $d[i]$, the size of a maximum clique the graph induced by the vertices in $T_i = \{1, 2, \ldots, i\}$, but only for $i \leq n/2$ [456, 459]. Then $d[i] + c[i+1]$ gives an upper bound on the size of a clique in the graph.

Alternative approaches for finding cliques include algorithms that use vertex coloring for pruning [529] and branch-and-bound algorithms; we will have a brief look at an approach of the latter type.

We may formulate the problem of finding a clique of a given size as an integer linear system of inequalities. For $1 \leq i \leq n$, let $x_i \in \{0,1\}$ take value 1 exactly when the vertex $i \in V$ is in a particular set. A clique is a set of vertices no two of which are nonadjacent, and thereby corresponds to a solution of the system

$$x_i + x_j \leq 1 \quad \text{for all } \{i,j\} \notin E,\ i \neq j. \tag{5.1}$$

The search for a maximum clique may be viewed as an optimization problem, where we want to find

$$\max \sum_{i=1}^{n} x_i$$

over all solutions of (5.1). To find one or all cliques of size k, we instead add to (5.1) the equation

$$\sum_{i=1}^{n} x_i = k.$$

Since no two vertices of an independent set can occur in a clique, inequalities of the form

$$\sum_{i \in S} x_i \leq 1,$$

where S is an independent set, may be added to the system (5.1). Such inequalities have no effect on the solutions of the system, but they do have an impact on the pruning strategy that we now present.

In a backtrack search for cliques of size k – and analogously in the search for maximum cliques – a relaxation of the integer constraint on the variables may be used for bounding. The relaxed variables are real numbers $0 \leq x_i \leq 1$, and linear programming methods can then be used to find the maximum of $\sum_{i=1}^{n} x_i$ under the given constraints and values of x_i known from the partial solution of the backtrack search. If the maximum is smaller than k, then the search tree can be pruned. The solution of the relaxed problem can also provide useful information for the choice of the next variable to branch on in the backtrack algorithm.

If an instance has many more inequalities than there are variables, then one should think about solving the linear programming relaxation via its *dual* problem instead; see [81], which also contains several other ideas for solving clique problems in this manner.

The outlined methods can be further generalized to weighted graphs with integer weights on the vertices. The algorithm published in [453] – which generalizes the one in [456] to weighted graphs – has been implemented as a set of C routines for finding cliques in arbitrary weighted graphs, called Cliquer [443].

Problem 5.3. (Weighted Cliques) Given a weighted graph G and an integer k, find all cliques of weight k in G.

5.2 Exact Cover Algorithms

For most of the problems encountered in this book, the performance of the known algorithms depends on the type of instance, and the algorithm to use should be carefully chosen on a case-by-case basis. For the problem to be discussed in this section, however, there is an algorithm that lacks serious competitors.

Definition 5.4. *A* set cover, *or simply* cover, *of a finite set P is a set of nonempty subsets of P whose union is P. A cover consisting of pairwise disjoint subsets is an* exact cover.

Problem 5.5. (EXACT COVERS) Given a set $\mathcal{S}$ of subsets of a finite set P, find all exact covers of P consisting of sets from $\mathcal{S}$.

The algorithm to be discussed can be traced back to the early paper on backtrack search by Golomb and Baumert [215]; its application to the construction of combinatorial objects is discussed, for example, in [401, 458].

An instance of EXACT COVERS can be solved by a transformation into an instance of WEIGHTED CLIQUES. Add one vertex for each subset in the given set $\mathcal{S}$, let the weight of a vertex be the size of the corresponding subset, and insert an edge between two vertices exactly when the corresponding subsets are disjoint. The cliques of weight $|P|$ in this graph give the solutions of the original instance of EXACT COVERS. If all weights of the vertices are the same, that is, if all subsets have the same size, then we get an instance of CLIQUES.

A simple, but yet remarkably efficient, algorithm for EXACT COVERS is displayed as Algorithm 5.2, which is invoked with EXACT$(\mathcal{S}, P, 1)$. Following one of the main principles for designing efficient backtrack algorithms listed in Sect. 4.1.2, on each level of the search tree we pick an element to be covered so as to minimize the number of children of the current node of the search tree. For more advanced strategies – with different kinds of *lookahead* – which minimize the size of the search tree even further, the overhead of the strategies often leads to slower overall performance.

Research Problem 5.6. Determine under what conditions, and how, lookahead can improve the performance of Algorithm 5.2.

Knuth [323] observed that an idea from [265] can speed up a direct implementation of Algorithm 5.2 by a factor of about two with the following designated data structure.

Let $\mathbf{A}$ be a 0-1 matrix of size $n \times m$ with one column for each element of P, one row for each set in $\mathcal{S}$, and a 1 indicating that an element occurs in a subset. This means that the exact covers correspond to the solutions of $\mathbf{A}^T\mathbf{x} = \mathbf{1}$, where $\mathbf{x}$ is a 0-1 column vector.

The employed data structure consists of multiple circular doubly linked lists. Each 1 in the matrix $\mathbf{A}$ corresponds to a list entry i with five fields:

Algorithm 5.2 All exact covers

procedure EXACT($\mathcal{S}$: set of sets, P: set, s: integer)
if $P = \emptyset$ **then**
  report the current solution $Q_1, Q_2, \ldots, Q_{s-1}$
  return
end if
find a $p \in P$ that minimizes $|\{S \in \mathcal{S} : p \in S\}|$
$\mathcal{S}' \leftarrow \{S \in \mathcal{S} : p \in S\}$
for all $S' \in \mathcal{S}'$ **do**
  $Q_s \leftarrow S'$
  $\mathcal{S}'' \leftarrow \{S \in \mathcal{S} : S \cap S' = \emptyset\}$
  EXACT($\mathcal{S}'', P \setminus S', s+1$)
end for
end procedure

$L[i]$, $R[i]$, $U[i]$, $D[i]$, and $C[i]$. The rows of the matrix are doubly linked as circular lists via the L and R fields ("left" and "right"), and the columns are doubly linked as circular lists via the U and D fields ("up" and "down"). Each column list includes a special entry called the column header. The C field of each list entry points to the column header of the column in which the entry lies. A column header contains an additional field, $S[c]$ ("size"), which is used to keep track of the number of rows linked to the column list.

The uncovered columns are linked to a circular list via the L and R fields of their column headers. The column header list also contains a special entry, h, which is used to access the column header list.

Example 5.7. Let $P = \{1, 2, 3, 4, 5, 6\}$ and $\mathcal{S}$ consist of

$$S_1 = \{3, 5, 6\}, \quad S_2 = \{1, 4\}, \quad S_3 = \{2, 3, 6\},$$
$$S_4 = \{1, 4\}, \quad S_5 = \{2\}, \quad S_6 = \{4, 5\}.$$

The data structure initialized with this instance is shown in Fig. 5.1. The arrows represent the L, R, U, D links in the list entries depicted by boxes. The column headers appear in the topmost row.

Algorithm 5.3, to be invoked with SEARCH(1), shows in detail the implementation of Algorithm 5.2 using the data structure with linked lists.

Consider the first two lines of the procedure COVER in Algorithm 5.3. The operations

$$L[R[c]] \leftarrow L[c], \quad R[L[c]] \leftarrow R[c]$$

clearly remove the column header c from the list of uncovered columns. The crucial observation for backtracking is that the operations

$$L[R[c]] \leftarrow c, \quad R[L[c]] \leftarrow c$$

suffice to insert c back into the list. Thus, it suffices to keep track of the elements deleted from a list to enable their insertion back into the list. This

Algorithm 5.3 All exact covers with dancing links

```
procedure COVER(c: column header)
  L[R[c]] ← L[c]
  R[L[c]] ← R[c]
  for i ← D[c], D[D[c]], ..., while i ≠ c do
     for j ← R[i], R[R[i]], ..., while j ≠ i do
        U[D[j]] ← U[j]
        D[U[j]] ← D[j]
        S[C[j]] ← S[C[j]] − 1
     end for
  end for
end procedure
procedure UNCOVER(c: column header)
  for i ← U[c], U[U[c]], ..., while i ≠ c do
     for j ← L[i], L[L[i]], ..., while j ≠ i do
        S[C[j]] ← S[C[j]] + 1
        U[D[j]] ← j
        D[U[j]] ← j
     end for
  end for
  L[R[c]] ← c
  R[L[c]] ← c
end procedure
procedure SEARCH(s: integer)
  if R[h] = h then
     report the current solution Q_1, ..., Q_{s−1}
     return
  end if
  pick an uncovered column c that minimizes S[c]
  COVER(c)
  for r ← D[c], D[D[c]], ..., while r ≠ c do
     Q_s ← r
     for j ← R[r], R[R[r]], ..., while j ≠ r do
        COVER(C[j])
     end for
     SEARCH(s + 1)
     for j ← L[r], L[L[r]], ..., while j ≠ r do
        UNCOVER(C[j])
     end for
  end for
  UNCOVER(c)
  return
end procedure
```

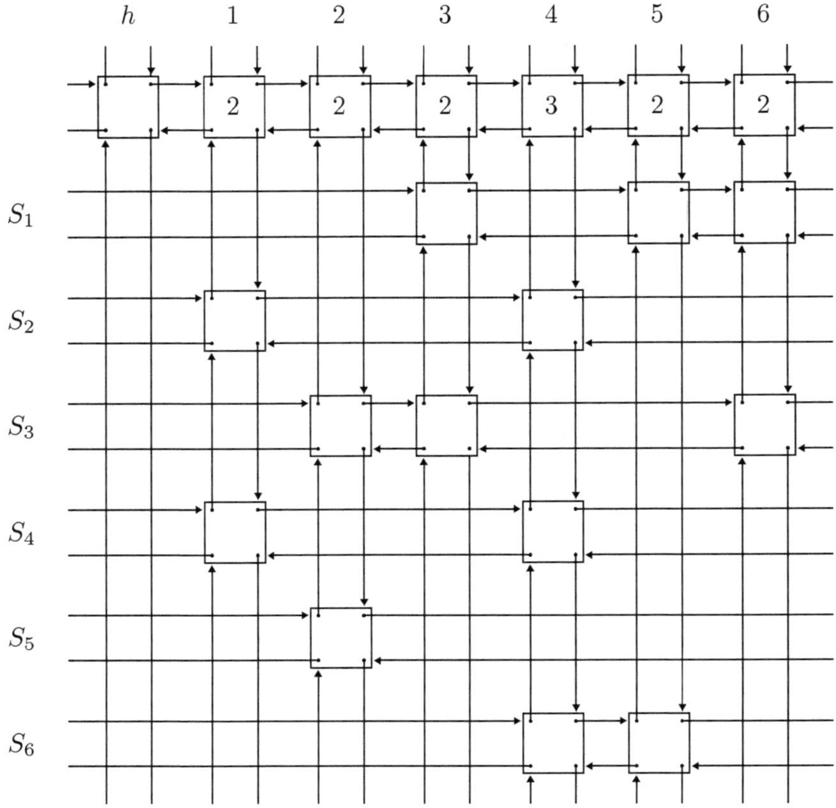

Fig. 5.1. Data structure for exact cover algorithm

eliminates essentially all of the bookkeeping usually required in backtracking since all information required to "rewind" the global data structure to the earlier state is present in the L, R, U, D fields of the deleted list entries. Knuth [323] describes the behaviour of the linked lists as dancing – *dancing links*.

One may generalize the EXACT COVERS problem and search for sets from a given collection $\mathcal{S}$ of subsets of P with the requirement that all elements of P occur in exactly λ sets. Algorithm 5.2 be applied to such a problem after a slight modification, but its efficiency decreases rapidly with increasing λ.

5.3 Set Cover Algorithms

In the set cover problem we do not require that the subsets be disjoint, so it is more general than EXACT COVERS. Obviously, we could then just take all given subsets (forming $\mathcal{S}$) of a finite set P and check whether their union is P, so we need another parameter: weights are associated with the subsets and we want to find set covers of given weight. The case when all weights are 1 is recurrent.

Problem 5.8. (SET COVERS) Given a set $\mathcal{S}$ of subsets of a finite set P and an integer k, find all covers of P consisting of k sets from $\mathcal{S}$.

Problem 5.9. (WEIGHTED SET COVERS) Given a set $\mathcal{S}$ of weighted subsets of a finite set P and an integer k, find all covers of P consisting of sets from $\mathcal{S}$ with total weight k.

A set cover with the property that by removing any subset it is no longer a cover is said to be *minimal*. It is often reasonable to restrict the search to minimal covers, since it is not difficult to construct all covers given all minimal covers. The unweighted case, SET COVERS, is considered in the rest of this section.

An algorithm for SET COVERS restricted to minimal covers can be obtained by slightly modifying Algorithm 5.2 and is presented as Algorithm 5.4. The elements in the union of the sets of a partial solution are said to be *covered* and the elements in P that are not covered are said to be *uncovered*.

Algorithm 5.4 All minimal set covers of size k

```
procedure cvr(𝒮: set of sets, P: set, s: integer)
if s = k + 1 then
   if P = ∅ then
      report the current solution Q_1, Q_2, ..., Q_k
   end if
   return
end if
if P = ∅ or a bounding function shows that no solution is possible then
   return
end if
use a heuristic to pick p ∈ P
𝒮' ← {S ∈ 𝒮 : p ∈ S}
𝒮'' ← 𝒮
for all S' ∈ 𝒮' do
   Q_s ← S'
   𝒮'' ← 𝒮'' \ {S'}
   cvr(𝒮'', P \ S', s + 1)
end for
end procedure
```

To improve the performance of a set cover algorithm, we need a bounding function that identifies situations where it is not possible to find a (minimal) set cover with k subsets. If all subsets have the same size, then an obvious lower bound on the number of subsets needed for a cover is obtained by dividing the number of uncovered elements with the size of the sets. The following results from [10] give more effective bounding functions.

Theorem 5.10. *Let P be a finite set of elements and $\mathcal{S}$ a set of subsets of P, and let $w : P \to \mathbb{R}$, where $\mathbb{R}$ is the set of real numbers, be a function satisfying*

$$\sum_{x \in S} w(x) \leq 1$$

for all $S \in \mathcal{S}$. Then any set cover of P using sets from $\mathcal{S}$ has size at least

$$\sum_{x \in P} w(x).$$

Proof. For a set cover $\mathcal{C} \subseteq \mathcal{S}$, we have

$$|\mathcal{C}| = \sum_{Y \in \mathcal{C}} 1 \geq \sum_{Y \in \mathcal{C}} \sum_{x \in Y} w(x) \geq \sum_{x \in P} w(x).$$

□

One possible choice for the function $w(x)$ is

$$w(x) = \frac{1}{\max\{|S| : x \in S \in \mathcal{S}\}}.$$

Corollary 5.11. *The size of a set cover of a finite set P using sets from $\mathcal{S}$ is bounded from below by*

$$\sum_{x \in P} \frac{1}{\max\{|S| : x \in S \in \mathcal{S}\}}.$$

There are several possible heuristics for choosing the next element to cover; we list two possible criteria:

1. Minimize the number of children of a node of the search tree.
2. Maximize the average number of uncovered elements for the children of a node of the search tree.

Roughly, these criteria attempt to minimize, respectively, the width and the height of a search tree. The first criterion is used in [10]. Other possible strategies include choosing the smallest – according to some total order such as lexicographic order – uncovered element.

An alternative approach, used in [10], is to focus on the subsets with which we are covering instead of the elements in P to be covered. The covering problem can be formulated as a Diophantine linear system of equations (cf. Sect. 5.4) with 0-1 variables associated with the sets and telling whether they are used or not. A branch-and-bound algorithm can then be used to find a minimum covering, cf. Sect. 5.1. Several conditions for pruning a search tree are listed in [10]. A branch-and-bound algorithm is also used in [395] to solve a covering problem of this type.

5.4 Diophantine Linear Systems of Equations

The problems in Sects. 5.1 to 5.3 can all be transformed into Diophantine linear systems of equations; we have already seen examples in this direction.

Problem 5.12. (DIOPHANTINE) Given an $m \times n$ integer matrix $\mathbf{A} = (a_{ij})$, an $m \times 1$ integer vector $\mathbf{b}$, and an $n \times 1$ integer vector $\mathbf{u}$, find all integer solutions $\mathbf{x} = [x_1\ x_2\ \cdots\ x_n]^T$ to $\mathbf{Ax} = \mathbf{b}$ subject to $0 \le x_i \le u_i$ for all $1 \le i \le n$.

If inequalities occur in a problem instance, these can be eliminated through the standard method of introducing *slack variables*. The requirement that $x_i \ge 0$ is not restricting, since any interval $l_i \le x_i \le u_i$ with $l_i \ne 0$ can be replaced by $0 \le x'_i \le u_i - l_i$, simultaneously replacing $\mathbf{b}$ by $\mathbf{b} - l_i\mathbf{a}_i$ with $\mathbf{A} = [\mathbf{a}_1\ \mathbf{a}_2\ \cdots\ \mathbf{a}_n]$.

In Sect. 5.2 it is shown that an instance of EXACT COVERS can be transformed into an instance of DIOPHANTINE with $\mathbf{b} = \mathbf{1}$. This transformation is in fact reversible: when $\mathbf{b} = \mathbf{1}$ in DIOPHANTINE, we have an instance of EXACT COVERS. It is pointed out in [205] that Mathon and Magliveras have applied this idea to solve DIOPHANTINE with $\mathbf{b}$ being the all-λ vector and $\lambda > 1$ using an algorithm for a generalization of EXACT COVERS (see also [401] and the end of Sect. 5.2).

Even if many problems can be formulated in the framework of DIOPHANTINE, for reasons of efficiency it is not desirable to use one general algorithm, but tailored algorithms should be used whenever possible. Obviously, there are instances of DIOPHANTINE that do not fall within any of the considered classes of subproblems, so we need an algorithm for the general case as well.

Commercial software available for solving problems involving linear equations and inequalities are generally designed to solve optimization problems where one optimal solution suffices. For classification problems, however, we need *all* solutions satisfying the given (in)equalities. Moreover, since it is often desirable to have full control of the program – for example, to be able to estimate the magnitude of the search – a program with access to the source code is often preferable.

In the classification of combinatorial objects, one meets different types of instances of DIOPHANTINE. Some subproblems of this type have a negligible impact on the overall computing time, but in some cases such subproblems are even the performance bottleneck of a classification. The type of subproblem determines the amount of effort one should put on developing and tuning algorithms for the particular instances.

Algorithm 5.5 uses backtracking to find all solutions to a Diophantine linear system of equations. Here we assume further that all entries a_{ij} of the matrix $\mathbf{A}$ are nonnegative. This assumption actually holds for virtually all instances that one meets in the subsequent classification algorithms.

In Algorithm 5.5 the values of x_i are fixed for increasing value of the parameter i. The purpose of the auxiliary vectors $\mathbf{s}$ and $\mathbf{m}$ is to guide the search. The vector $\mathbf{s}$ is a linear combination of the columns of $\mathbf{A}$ given by the

Algorithm 5.5 All solutions of a Diophantine linear system of equations

```
procedure LNQ(s: vector, j: integer)
if j = n + 1 then
    report the current solution x_1, x_2, ..., x_n
    return
end if
for x_j ← 0, 1, ..., u_j do
    ok ← TRUE
    for i ← 1, 2, ..., m do
        s'_i ← s_i + a_ij x_j
        if s'_i > b_i then
            return
        end if
        if j = m_i and s'_i ≠ b_i then
            ok ← FALSE
            break
        end if
    end for
    if ok then
        LNQ(s', j + 1)
    end if
end for
end procedure
procedure DIOPHANTINE(A: matrix, b: vector, u: vector)
for i ← 1, 2, ..., m do
    s_i ← 0
    m_i ← max{j : a_ij ≠ 0}
end for
LNQ(s, 1)
end procedure
```

partial solution, and the value of m_i gives the last level that is able to change the value of s_i, that is, $a_{i,m_i} \neq 0$ and $a_{ij} = 0$ for $j > m_i$.

Like with clique algorithms, the ordering of the input data affects the speed of the algorithm. It is, for example, desirable that the values of m_i be small to get early pruning of the search. Therefore one may sort the rows of $\mathbf{A}$ in order of increasing number of nonzero entries. Thereafter, the columns may be sorted in reverse lexicographic order, with the entries of the first row being the most significant.

If the matrix $\mathbf{A}$ contains identical columns – say columns number j_1, $j_2, \ldots, j_k$ are identical – one should remove all of these but column number j_1, and replace u_{j_1} by $\sum_{i=1}^{k} u_{j_i}$. With many identical columns, this may have a significant impact on the size of the search tree. Actually, the preprocessing in the beginning of [333, Algorithm 6] has the same goal (but the framework is different). There are various ways of further improving Algorithm 5.5, including constraint propagation: if $x_1 + x_2 + x_5 = 2$ and the values of x_1 and x_2 are known, then we can calculate the value of x_5.

Various pruning techniques applicable to instances of DIOPHANTINE related to construction of designs appear in [204, 206] and [400, Sect. 4].

One possibility to structure an algorithm for DIOPHANTINE is to employ the following meet-in-the-middle strategy suggested by H. Haanpää. First, a look-up table of all partial solutions for the last $n - k$ variables $x_{k+1}, x_{k+2}, \ldots, x_n$ is compiled. Each such solution is indexed by the vector $\mathbf{b}' = (b'_1, b'_2, \ldots, b'_m)$, where $b'_i = b_i - \sum_{j=k+1}^{n} a_{ij}x_j$. Clearly, the maximum number of partial solutions in the table is $\prod_{j=k+1}^{n}(u_j + 1)$. A partial solution can be eliminated from the table if $b'_i < 0$ for some i; the pruning techniques cited above can also be used to eliminate partial solutions. The look-up table can then be employed to complete the search in one step after the values of the first k variables have been set. Namely, all completions of the partial solution $(x_1, \ldots, x_k)$ appear in the look-up table under the index $\mathbf{s} = (s_1, s_2, \ldots, s_m)$, where $s_i = \sum_{j=1}^{k} a_{ij}x_j$.

Other algorithms for DIOPHANTINE include those by Kramer, Leavitt, and Magliveras [333]; and Schmalz [521]. Actually, Schmalz's algorithm is not far from Algorithm 5.5. Ivanov [278] observed that in solving instances coming from extending partial incidence matrices of BIBDs row by row (Sect. 6.1.1), one may make use of the recursive structure of the problem in developing efficient algorithms (see also [480]); an analogous idea works for constructing codes codeword by codeword (Sect. 7.1.2).

There are other ways of transforming a system of equations than just reordering the variables and the equations. In particular, any linear combination of the given equations can be added to the system.

Example 5.13. If two of the equations are $x_1 + x_2 + x_3 + x_4 = 2$ and $x_2 + x_3 + x_4 + x_5 = 2$, then it follows that $x_1 - x_5 = 0$. This implies that the value of x_5 can be calculated if the value of x_1 is known (constraint propagation).

Example 5.13 shows that even if $\mathbf{A}$ contains only nonnegative entries – which is the case for most instances related to problems in this book – one should not ignore the possibility of adding equations leading to negative entries. A systematic procedure for adding equations for instances related to construction of designs is described in [384], where also a pruning technique for instances where $\mathbf{A}$ contains both positive and negative entries is presented.

In general it is desirable to have equations with as few variables as possible, that is, the Hamming weight (or, more generally, norm) of the rows should preferably be small. We will now have a brief look at a family of methods that explicitly aim at transforming an instance into one dealing with vectors of small norm. This work has led to some of the best known algorithms for solving certain hard instances of DIOPHANTINE.

A solution to $\mathbf{Ax} = \mathbf{b}$ is obviously a solution to

$$\begin{bmatrix} \mathbf{A} & -\mathbf{b} \\ \mathbf{I}_n & \mathbf{0} \end{bmatrix} \begin{bmatrix} \mathbf{x} \\ 1 \end{bmatrix} = \begin{bmatrix} \mathbf{0} \\ \mathbf{x} \end{bmatrix}.$$

Consequently, to get a solution to $\mathbf{Ax} = \mathbf{b}$, we must find a vector that is an integer linear combination of the column vectors of

$$\begin{bmatrix} \mathbf{A} & -\mathbf{b} \\ \mathbf{I}_n & \mathbf{0} \end{bmatrix} \tag{5.2}$$

and that has 0s in the m first coordinates followed by values from the ranges of the respective variables x_i (and check that it is indeed a solution). The vectors that can be obtained as an integer linear combination of a given set of vectors form a *lattice*. Accordingly, we now want to find a vector with the given properties in the lattice spanned by the columns of (5.2). One property of such a solution vector that leads to the success of this approach is that the Euclidean norm

$$\sqrt{x_1^2 + x_2^2 + \cdots + x_n^2}$$

is rather small. For example, with $x_i \in \{0, 1\}$, the Euclidean norm is bounded from above by $\sqrt{n}$.

The columns of (5.2) are linearly independent and thereby form a *basis* of the lattice that they span. A lattice has many bases, and we are interested in transforming a basis into another basis with vectors of smaller norm; this procedure is called *lattice basis reduction*. The seminal lattice basis reduction algorithm is the Lenstra–Lenstra–Lovász (LLL) algorithm [363]. Subsequently, a large number of other algorithms have been presented; different methods can also be combined. For an introduction to the theory of basis reduction, see [342, Chap. 8].

Much of the work in this area has considered existence problems, where one hopes to find a solution to the original problem among the vectors (or the vectors negated) in the bases encountered in the reduction process. However, Wassermann [598, 599] – inspired by results of Kaib and Ritter [289] – developed an exhaustive algorithm for solving DIOPHANTINE that starts from a reduced basis, and applied this to classifying designs in [598]. It turns out that for optimal performance other lattices than that of (5.2) are to be preferred. The best choice depends on the type of problem, for example, whether it has only 0-1 variables or not. One lattice proposed in [599] for general instances with $0 \le x_i \le u_i$ is the lattice spanned by the columns of

$$\left[\begin{array}{cccc|c} \multicolumn{4}{c|}{N\mathbf{A}} & -N\mathbf{b} \\ \hline 2c_1 & 0 & \cdots & 0 & -u \\ 0 & 2c_2 & & 0 & -u \\ \vdots & & \ddots & \vdots & \vdots \\ 0 & 0 & \cdots & 2c_n & -u \\ \hline 0 & 0 & \cdots & 0 & u \end{array}\right],$$

where N is a large positive integer, $u = \operatorname{lcm}\{u_1, u_2, \ldots, u_n\}$, and $c_i = u/u_i$ for $i = 1, 2, \ldots, n$. The reader is referred to [598, 599] for further details on using lattice basis reduction for solving DIOPHANTINE.

Techniques developed by the integer linear programming community are worth considering in particular when the ranges of the variables are (much) larger than $\{0, 1\}$.

5.5 Permutation Group Algorithms

For purposes of isomorph rejection it is often necessary to solve problems involving permutation groups. This section makes a small excursion into elementary algorithms and data structures associated with permutation groups. For an extensive treatment of permutation group algorithms, see [82, 266, 528].

Throughout this section we assume that $G \leq \mathrm{Sym}(\Sigma)$ is a permutation group that acts on a finite set Ω. Furthermore, we assume that an efficient procedure exists for computing the image $gX \in \Omega$ given $g \in G$ and $X \in \Omega$ as input.

A recurrently occurring problem is computing the orbit of a given object. A typical situation is that we have the generators for the automorphism group of some object, and we must determine the automorphism orbit of some subobject. For example, it may be that we must determine the automorphism orbit of a block in an incidence structure, given generators for the automorphism group.

Problem 5.14. (Orbit of Object) Given a set S of generators for G and an object $X \in \Omega$, output all objects in the orbit GX.

In many cases we also require for every $Y \in GX$ a group element $l(Y) \in G$ such that $l(Y)X = Y$; in other words, the set $L = \{l(Y) : Y \in GX\}$ constitutes a left transversal for the stabilizer $N_G(X)$ in G.

The object orbit problem can be solved by computing closure under the generators in S. We maintain a set $\mathcal{O}$ that contains the objects in the orbit GX discovered so far, and a work list $\mathcal{W}$ that contains those objects in $\mathcal{O}$ that have not been processed yet. An object $Y \in \mathcal{W}$ is processed by evaluating the image gY for all generators $g \in S$ and updating $\mathcal{O}, \mathcal{W}$ accordingly. A pseudocode description appears in Algorithm 5.6.

When Algorithm 5.6 returns, $\mathcal{O} = GX$. The algorithm requires $|GX| \cdot |S|$ image computations, $|GX| \cdot |S|$ set membership tests, and $|GX|$ set insertions. The set operations can be implemented using a data structure that supports fast searching and updating, such as a hash table [136, 322].

Algorithm 5.6 can be enhanced to compute a left transversal L for the stabilizer $N_G(X)$ in G. During the execution of the algorithm, we record for every object $Y \in \mathcal{O}$ a group element $l(Y) \in G$ such that $l(Y)X = Y$. Initially we assign $l(X) = \epsilon$. Subsequently we put $l(Z) = g \cdot l(Y)$ whenever an object Z is inserted into $\mathcal{O}$. The resulting set $L = \{l(Y) : Y \in \mathcal{O}\}$ is a left transversal for $N_G(X)$ in G.

Algorithm 5.6 Computing the orbit of an object

```
function OBJECT-ORBIT(S: generators for G, X: object): orbit
  O ← {X}
  initialize W to a list consisting of the object X
  while W is not empty do
    select and remove any object Y from W
    for all g ∈ S do
      Z ← gY
      if Z ∉ O then
        O ← O ∪ {Z}
        append Z into the list W
      end if
    end for
  end while
  return O
```

A slightly different variant of the object orbit problem is to maintain an orbit partition when generators are added into the group. This problem occurs frequently during isomorphism computations when we have to update an existing automorphism orbit partition to take into account newly discovered automorphisms.

Problem 5.15. (ORBIT PARTITION UPDATE) Given the orbit partition $\langle S\rangle\backslash\Omega$ for some $S \subseteq \mathrm{Sym}(\Sigma)$, and a permutation $g \in \mathrm{Sym}(\Sigma)$, determine the orbit partition $\langle S \cup \{g\}\rangle\backslash\Omega$.

To solve the partition update problem, it obviously suffices to determine the sets of orbits (if any) in $\langle S\rangle\backslash\Omega$ that fuse under the action of g. More precisely, $\langle S \cup \{g\}\rangle\backslash\Omega$ is the join of the partitions $\langle S\rangle\backslash\Omega$ and $\langle\{g\}\rangle\backslash\Omega$ in the lattice of partitions of Ω. The partition $\langle\{g\}\rangle\backslash\Omega$ can be determined by decomposing g (viewed as a permutation of Ω) into disjoint cycles, which requires $|\Omega|$ image computations. To compute the join of two partitions, we can proceed as follows. For a partition P of Ω, let $F(P)$ be a forest with vertex set Ω such that each cell in P is the vertex set of a connected component of $F(P)$. For two partitions P, Q of Ω, the *join* $P \vee Q$ is the partition whose cells are the vertex sets of the connected components in $F(P) \cup F(Q)$. Thus, the join can be computed in time linear in $|\Omega|$ using depth-first search on the graph $F(P) \cup F(Q)$. Alternatively, data structures for disjoint sets [136] can be used to compute the join of two partitions.

In some cases we must work directly with a permutation group $G \leq \mathrm{Sym}(\Sigma)$ given by a set of generators S. Examples of such situations include testing membership in G, computing the order of G, and computing the stabilizer of a point $p \in \Sigma$ in G. A straightforward solution is to compute closure for the generators or, equivalently, the orbit of the identity permutation under the action of G on itself by left multiplication (cf. Algorithm 5.6). However, it is obvious that this is practical only for groups of small order, because all group elements are generated and stored in memory.

Sims [537, 538] defined the fundamental concepts of a base and strong generating set that enable the efficient solution of many permutation group problems.

Definition 5.16. *Let $G \le \mathrm{Sym}(\Sigma)$. A sequence $B = (b_1, b_2, \ldots, b_m)$ of distinct elements of Σ is a* base *for G if $N_G(B) = \{\epsilon\}$.*

A base defines a chain of point stabilizer subgroups

$$G = G_1 \ge G_2 \ge \cdots \ge G_m \ge G_{m+1} = \{\epsilon\}, \tag{5.3}$$

where $G_i = N_G((b_1, \ldots, b_{i-1}))$, $1 \le i \le m+1$. A base is *nonredundant* if G_{i+1} is a proper subgroup of G_i for all $1 \le i \le m$.

Example 5.17. Recall the automorphism group of the cube Q_3 from Example 3.11. The sequence $B = (1, 2, 3)$ is a base for $G = \mathrm{Aut}(Q_3)$. The associated stabilizer chain is

$$\begin{aligned} G_1 &= \mathrm{Aut}(Q_3), \\ G_2 &= \{\epsilon,\ (3\ 5)(4\ 6),\ (2\ 3)(6\ 7),\ (2\ 3\ 5)(4\ 7\ 6), \\ &\qquad (2\ 5\ 3)(4\ 6\ 7),\ (2\ 5)(4\ 7)\}, \\ G_3 &= \{\epsilon,\ (3\ 5)(4\ 6)\}, \\ G_4 &= \{\epsilon\}. \end{aligned}$$

Example 5.18. A nonredundant base is not necessarily unique. For example, $B_1 = (1, 3, 7)$, $B_2 = (4, 8)$, and $B_3 = (7)$ are nonredundant bases of

$$G = \langle (1\ 2)(3\ 4\ 5\ 6)(7\ 8\ 9\ 10\ 11\ 12\ 13\ 14) \rangle.$$

A strong generating set contains generators for every group G_i in the stabilizer chain (5.3) in a conveniently accessible form.

Definition 5.19. *Let $G \le \mathrm{Sym}(\Sigma)$ and let $B = (b_1, b_2, \ldots, b_m)$ be a base for G. A set $S \subseteq \mathrm{Sym}(\Sigma)$ is a* strong generating set (SGS) *for G relative to B if*

$$\langle S \cap G_i \rangle = G_i \quad \textit{for all } 1 \le i \le m. \tag{5.4}$$

In other words, G_i is generated by those permutations in S that stabilize the base points $b_1, b_2, \ldots, b_{i-1}$.

Example 5.20. An SGS for $\mathrm{Aut}(Q_3)$ relative to $B = (1, 2, 3)$ is

$$S = \{(1\ 2)(3\ 4)(5\ 6)(7\ 8),\ (2\ 3\ 5)(4\ 7\ 6),\ (3\ 5)(4\ 6)\}.$$

Once an SGS is available it can be used in a number of ways. Associated with every G_i is the *fundamental orbit* $\Delta_i = G_i b_i$ and – by the orbit-stabilizer theorem (Theorem 3.20) – a corresponding left transversal L_i for G_{i+1} in G_i. These can be determined from generators for G_i using Algorithm 5.6.

Example 5.21. For $G = \mathrm{Aut}(Q_3)$ and $B = (1, 2, 3)$ the fundamental orbits are

$$\Delta_1 = \{1,2,3,4,5,6,7,8\}, \quad \Delta_2 = \{2,3,5\}, \quad \Delta_3 = \{3,5\}.$$

A corresponding sequence of left transversals is

$$\begin{aligned} L_1 = \{&\epsilon,\ (1\ 2)(3\ 4)(5\ 6)(7\ 8),\ (1\ 3\ 4\ 2)(5\ 7\ 8\ 6),\ (1\ 4)(5\ 8), \\ &(1\ 5\ 7\ 8\ 4\ 2)(3\ 6),\ (1\ 6\ 4)(3\ 5\ 8), \\ &(1\ 7\ 6\ 4)(2\ 3\ 5\ 8),\ (1\ 8)(2\ 4)(3\ 6)(5\ 7)\}, \\ L_2 = \{&\epsilon,\ (2\ 3\ 5)(4\ 7\ 6),\ (2\ 5\ 3)(4\ 6\ 7)\}, \\ L_3 = \{&\epsilon,\ (3\ 5)(4\ 6)\}. \end{aligned}$$

Because two cosets in G_i/G_{i+1} are either equal or disjoint, it follows that every $g \in G$ can be written uniquely as a product

$$g = l_1 l_2 \cdots l_m, \quad l_i \in L_i \text{ for all } 1 \le i \le m. \tag{5.5}$$

In particular, by Lagrange's theorem (Theorem 3.7) and the orbit-stabilizer theorem (Theorem 3.20), we have

$$|G| = \prod_{i=1}^{m} \frac{|G_i|}{|G_{i+1}|} = \prod_{i=1}^{m} |L_i| = \prod_{i=1}^{m} |\Delta_i|.$$

Thus, we can easily compute the order of G based on an SGS. Furthermore, we can backtrack through the elements of G by using the product representation (5.5). Compared with Algorithm 5.6, this enables a space-efficient solution for the problem of listing the elements of G.

Problem 5.22. (Group Elements) Given a set S of generators for G, output all the elements in G.

The product representation (5.5) can be computed iteratively for a given $g \in G$ as follows. Initially, we put $g_1 = g$. Given $g_i \in G_i$ for $1 \le i \le m$, we find the unique transversal element $l_i \in L_i$ that satisfies $g_i(b_i) = l_i(b_i)$, and put $g_{i+1} = l_i^{-1} g_i$. This procedure is called *sifting* g.

Example 5.23. Let us sift $g = (1\ 8)(2\ 6\ 5\ 7\ 3\ 4) \in \mathrm{Aut}(Q_3)$ using the left transversals in Example 5.21:

$$\begin{aligned} g_1 &= (1\ 8)(2\ 6\ 5\ 7\ 3\ 4), & l_1 &= (1\ 8)(2\ 4)(3\ 6)(5\ 7), \\ g_2 &= (2\ 3)(6\ 7), & l_2 &= (2\ 3\ 5)(4\ 7\ 6), \\ g_3 &= (3\ 5)(4\ 6), & l_3 &= (3\ 5)(4\ 6), \\ g_4 &= \epsilon. \end{aligned}$$

Sifting enables us to solve the membership problem for G.

Algorithm 5.7 Sifting a permutation

```
function SIFT(h: permutation of Σ): permutation of Σ
for i ← 1, 2, ..., m do
    if h(b_i) ∉ Δ_i then
        return h
    end if
    select the transversal element l_i ∈ L_i such that l_i(b_i) = h(b_i)
    h ← l_i^{-1} h
end for
return h
```

Problem 5.24. (Group Membership) Given a set S of generators for G, and a permutation $h \in \mathrm{Sym}(\Sigma)$, decide whether $h \in G$.

Algorithm 5.7 attempts to sift a given permutation $h \in \mathrm{Sym}(\Sigma)$ through the stabilizer chain (5.3). It is easy to check that $\textsc{sift}(h) = \epsilon$ if and only if $h \in G$.

It remains to describe how to construct a base and an SGS from an arbitrary set T of generators for G.

Problem 5.25. (Strong Generating Set) Given a set T of generators for G, determine a nonredundant base B and an associated SGS S for G.

The following description of the Schreier–Sims algorithm for SGS construction owes a lot to [528, Chap. 4]. The main difficulty in constructing an SGS is in obtaining a small set of generators for G_{i+1}, given generators for G_i. The following theorem is due to O. Schreier.

Theorem 5.26. *Let K be a group and let $H \leq K$. If S is a set of generators for K and L is a left transversal for H in K with $1 \in L$, then*

$$R = \{u^{-1}sv : u, v \in L,\ s \in S,\ u^{-1}sv \in H\} \tag{5.6}$$

is a set of generators for H.

Proof. Let $h \in H$. Because K is finite and S generates K, there exist $s_1, s_2, \ldots, s_\ell \in S$ such that $h = s_\ell s_{\ell-1} \cdots s_1$. Put $v_1 = 1$ and $r_1 = u_1^{-1}s_1v_1$, where $u_1 \in L$ is the unique transversal element such that $u_1^{-1}s_1v_1 \in H$. Clearly, $r_1 \in R$. For $i = 1, 2, \ldots, \ell - 1$, put $r_{i+1} = u_{i+1}^{-1}s_{i+1}v_{i+1}$, where $v_{i+1} = u_i$ and $u_{i+1} \in L$ is the unique transversal element such that $u_{i+1}^{-1}s_{i+1}v_{i+1} \in H$. Thus, $r_{i+1} \in R$. It follows that

$$h = s_\ell s_{\ell-1} \cdots s_1 = u_\ell^{-1} r_\ell r_{\ell-1} \cdots r_1.$$

Since $s_\ell s_{\ell-1} \cdots s_1 \in H$ and $r_\ell r_{\ell-1} \cdots r_1 \in H$, we must have $u_\ell \in H$; that is, $u_\ell = 1$ because $L \cap H = \{1\}$. Thus, $h = r_\ell r_{\ell-1} \cdots r_1$ and R generates H. □

The generator set (5.6) is called a set of *Schreier generators* for H.

Given $S_i \subseteq \mathrm{Sym}(\Sigma)$ such that $\langle S_i \rangle = G_i$, we obtain a set of Schreier generators for $G_{i+1} = N_{G_i}(b_i) = N_{\langle S_i \rangle}(b_i)$ as follows. First, we compute from S_i the fundamental orbit $\Delta_i = G_i b_i$ and a left transversal L_i for $G_{i+1} = N_{G_i}(b_i)$ in G_i using Algorithm 5.6. Then, we construct a set S_{i+1} of Schreier generators for G_{i+1} from S_i and L_i using (5.6); note that $u^{-1}sv \in G_{i+1}$ if and only if $u^{-1}sv(b_i) = b_i$.

We shall now prove that this approach, iterated for $i = 1, 2, \ldots, m-1$, produces an SGS relative to a given base B. However, from an algorithmic perspective the approach is still unsatisfactory because the number of generators in S_i can get very large as i grows.

Theorem 5.27. *Let $S_1, S_2, \ldots, S_m \subseteq \mathrm{Sym}(\Sigma)$ satisfy $S_j \subseteq G_j$ for all $1 \leq j \leq m$. If $\langle S_1 \rangle = G$ and*

$$N_{\langle S_j \rangle}(b_j) = \langle S_{j+1} \rangle \tag{5.7}$$

for all $1 \leq j \leq m-1$, then $S = \cup_{j=1}^{m} S_j$ is an SGS for G relative to B.

Proof. We show by induction on i that $\langle G_i \cap S \rangle = \langle S_i \rangle = G_i$ for all $1 \leq i \leq m$. The base case $i = 1$ holds because $S_1 \subseteq S \subseteq G_1$ and $\langle S_1 \rangle = G_1$ by assumption. For the inductive step, suppose the claim holds for i. Clearly, $\langle G_{i+1} \cap S \rangle \leq G_{i+1}$. Conversely, by (5.7) and the inductive hypothesis, we have $\langle G_{i+1} \cap S \rangle \geq \langle S_{i+1} \rangle = N_{\langle S_i \rangle}(b_i) = N_{G_i}(b_i) = G_{i+1}$. □

A natural solution to overcome the growth in the number of generators is to detect and disregard redundant generators by sifting. This approach leads to the Schreier–Sims algorithm, given as Algorithm 5.8. The algorithm constructs a base and an SGS for G by sifting the Schreier generators for $N_{\langle S_i \rangle}(b_i)$ relative to the stabilizer chain $\langle S_{i+1} \rangle \geq \langle S_{i+2} \rangle \geq \cdots \geq \langle S_m \rangle$. Sifting is always possible because either $i = m$ or (5.7) holds for all $i+1 \leq j \leq m-1$ whenever UPDATE is invoked with input i. Furthermore, after the invocation of UPDATE, (5.7) holds for all $i \leq j \leq m-1$.

The set S_{i+1} is augmented in Algorithm 5.8 only when a Schreier generator is not in $\langle S_{i+1} \rangle$. Thus, $|S_{i+1}|$ is bounded from above by the length of the longest subgroup chain in $\mathrm{Sym}(\Sigma)$, which is in turn bounded by $3|\Sigma|/2$ (see [90]), or by $|\Sigma| \log_2 |\Sigma|$ using Lagrange's theorem (Theorem 3.7). Because UPDATE is invoked only when one of the sets S_i is augmented with new generators, and $|R| \leq |\Sigma||S_i|$ by (5.6), the worst-case running time and the memory requirement of the algorithm are bounded by a polynomial in $|\Sigma|$ and $|T|$. A more accurate worst-case analysis can be found in [528, Sect. 4.2].

Algorithm 5.8 is sufficient for the applications in this book, but it should be noted that more efficient algorithms exist both in terms of time and memory usage, see [20, 88, 281, 528].

5.6 Isomorphism Algorithms

Although isomorphism computations can be carried out in a black-box fashion, it is often advantageous to have a basic understanding of how an isomorphism

Algorithm 5.8 Constructing a base and an associated SGS

```
procedure UPDATE(i: integer)
  initialize/augment Δ_i and L_i using S_i
  R ← a set of Schreier generators for N_⟨S_i⟩(b_i)
  for all g ∈ R do
     if i < m then
        if SIFT(g) ≠ ϵ then
           S_{i+1} ← S_{i+1} ∪ {g}
           UPDATE(i + 1)
        end if
     else
        if g ≠ ϵ then
           m ← m + 1
           b_{i+1} ← a point in Σ moved by g
           S_{i+1} ← {g}
           UPDATE(i + 1)
        end if
     end if
  end for
end procedure
function BASE-SGS(T: set of generators for G ≤ Sym(Σ)): base and SGS for G
  if T ⊆ {ϵ} then
     return (), ∅
  else
     m ← 1
     b_1 ← a point in Σ moved by T
     S_1 ← T
     UPDATE(1)
     B ← (b_1, b_2, ..., b_m)
     S ← ∪_{j=1}^{m} S_j
     return B, S
  end if
```

algorithm operates. This section provides a high-level discussion on the structure of algorithms based on iterative refinement via invariants that are employed in practice, the primary example being *nauty* [412]. A more extensive treatment including algorithm pseudocode can be found in [83, 325, 366, 411] and [342, Chap. 7]; see also [22]. Early work on refinement-based techniques is discussed in [85, 501, 601]. Refinement techniques tailored for specific families of objects are studied in [73, 114, 121, 364, 560, 426]. The limitations of refinement-based techniques are studied in [85, 430].

It is convenient to develop the basic ideas using the abstract framework from Sect. 3.3.6. Once the basic ideas have been treated, we illustrate practical implementation using ordered partitions. Let G be a group that acts on a finite set Ω with isomorphisms given by Definition 3.16. An example setting to keep in mind is to let Ω be the set of all graphs with vertex set $\{1, 2, \dots, n\}$, and let $G = \mathrm{Sym}(\{1, 2, \dots, n\})$ act on Ω by permuting the vertices as in (3.2).

Given an object $X \in \Omega$ as input, a refinement-based algorithm carries out a backtrack search on the group G to find an object Y isomorphic to X, where it is required that the same object Y is obtained for all inputs isomorphic to X. As a side-effect of the search we obtain an isomorphism $g \in \mathrm{Iso}(X, Y)$ and a set of generators for the automorphism group $\mathrm{Aut}(X) = \mathrm{Iso}(X, X)$. The structure of the search is easiest to describe using an associated search tree $T(X, G)$. The nodes in the tree are right cosets of the form Hg, where $H \leq G$ and $g \in G$. The tree is defined inductively using the following two basic operations.

The *partitioning rule* takes a coset $Hg \neq \{g\}$ and partitions it into a set of subcosets $\{Kh_1g, Kh_2g, \dots, Kh_ug\}$, where K is a proper subgroup of H and $\{h_1, h_2, \dots, h_u\}$ is a right transversal for K in H. In essence, the partitioning rule takes a subproblem and partitions it into a set of smaller subproblems in the hope that they would be easier to solve.

The *refinement transformation* takes a coset Hg and refines it to a subcoset Khg, where $K \leq H$ and $h \in H$. We say that Khg is the *refinement* of Hg.

Definition 5.28. *Let $X \in \Omega$. The search tree $T(X, G)$ is defined inductively as follows:*

1. *the root node is the refinement of G,*
2. *if $Hg \neq \{g\}$ is a node, then the child nodes of Hg are the refinements of cosets into which Hg is partitioned by the partitioning rule,*
3. *no other nodes except those forced by (1) and (2) occur in the tree.*

Example 5.29. Let $G = \mathrm{Sym}(\{1, 2, 3\})$. An elementary partitioning rule for a coset Hg in G is obtained by letting K be the stabilizer of the least point moved by H. If the refinement transformation is the identity mapping, then we obtain the search tree depicted in Fig. 5.2.

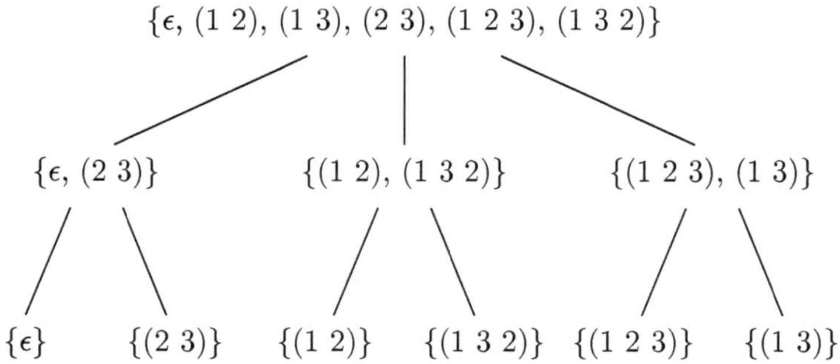

Fig. 5.2. A search tree on right cosets

In practice both the partitioning rule and the refinement transformation depend on the input X. We require any dependence on X to be such that the following two requirements are met. First, if on input X the coset $Hg \neq \{g\}$ partitions into $\{Kh_1g, Kh_2g, \ldots, Kh_ug\}$, then on input g_0X the coset Hgg_0^{-1} partitions into $\{Kh_1gg_0^{-1}, Kh_2gg_0^{-1}, \ldots, Kh_ugg_0^{-1}\}$ for all $g_0 \in G$. Second, if on input X the refinement of Hg is Khg, then on input g_0X the refinement of Hgg_0^{-1} is $Khgg_0^{-1}$ for all $g_0 \in G$. These two requirements and an induction on Definition 5.28 immediately imply that search trees associated with isomorphic inputs are isomorphic in the following sense.

Theorem 5.30. *Let $X \in \Omega$ and $g_0 \in G$. Then, the mapping $Hg \mapsto Hgg_0^{-1}$ on right cosets of subgroups of G induces an isomorphism of the rooted tree $T(X, G)$ onto the rooted tree $T(g_0X, G)$.*

In the language of group actions, Theorem 5.30 states that the action $g_0 * Hg = Hgg_0^{-1}$ of G on right cosets of its subgroups by right multiplication induces an action of G on the set of all rooted trees $\{T(X, G) : X \in \Omega\}$ so that

$$g_0 * T(X, G) = T(g_0X, G) \quad \text{for all } X \in \Omega \text{ and } g_0 \in G. \tag{5.8}$$

Figure 5.3 illustrates the situation.

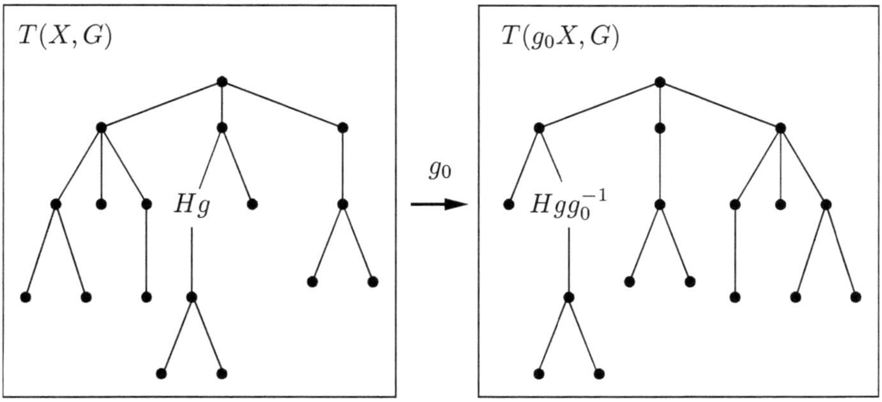

Fig. 5.3. Illustration of Theorem 5.30 and (5.8)

A set of search trees can be used to define a canonical representative map as follows. Observe that every leaf node in a search tree $T(X, G)$ is a singleton set $\{g\}$. Associate with every leaf node $\{g\}$ in $T(X, G)$ the *leaf object* $gX \in \Omega$. By Theorem 5.30 we have that $\{g\}$ is a leaf node of $T(X, G)$ if and only if $\{gg_0^{-1}\}$ is a leaf node of $T(g_0X, G)$. Because $gX = (gg_0^{-1})g_0X$, the search trees $T(X, G)$ and $T(g_0X, G)$ have the same set of leaf objects. A leaf object $Y \in \Omega$ selected in a manner that depends only on the set of leaf objects is a canonical form of X. Assuming that Ω is ordered, a simple leaf selection rule is to keep track of the minimum leaf object encountered so far while traversing $T(X,G)$. We obtain a canonical labeling map by setting $\kappa(X) = g$ for any leaf node $\{g\}$ with $Y = gX$, where Y is the minimum leaf object.

The structure of a refinement-based isomorphism algorithm is now evident. On input $X \in \Omega$ the algorithm traverses the search tree $T(X,G)$ and returns a canonical labeling $\kappa(X) \in G$ that takes X to the leaf object indicated by the leaf selection rule. A set of generators for the automorphism group $\mathrm{Aut}(X)$ is computed during the tree traversal, and discovered automorphisms are used to prune redundant parts of the tree – this will be discussed in more detail later.

To implement an algorithm in practice, we require first a representation for the cosets that occur in the search trees. If G is a permutation group, then a direct representation for a coset $Hg \subseteq G$ is obtained from a pair consisting of a permutation in Hg and a set of permutations that generate $H \leq G$. Permutation group algorithms (Sect. 5.5) can then be used to compute with cosets in this representation. The permutation representation is typically only employed in special cases, however. In most cases the acting group G is a symmetric group or a direct product thereof, whereby ordered partitions offer a very efficient representation for a certain family of cosets of subgroups.

We proceed to develop the connection between ordered partitions and cosets. At this point it may be useful to recall the terminology and notation for ordered partitions introduced in Sects. 3.3.2 and 3.3.5. For brevity we write S_n for the symmetric group $\mathrm{Sym}(\{1,2,\ldots,n\})$ and Π_n for the set of all ordered partitions of $\{1,2,\ldots,n\}$. Let $g \in S_n$ act on an ordered partition $\pi = (V_1, V_2, \ldots, V_m) \in \Pi_n$ by $g * \pi = (g(V_1), g(V_2), \ldots, g(V_m))$. Associate every ordered partition $\pi \in (V_1, V_2, \ldots, V_m) \in \Pi_n$ the ordered partition $\overrightarrow{\pi} = (W_1, W_2, \ldots, W_m)$ defined for all $i = 1, 2, \ldots, m$ by $t_0 = 0$, $t_i = t_{i-1} + |V_i|$, and $W_i = \{t_{i-1}+1, t_{i-1}+2, \ldots, t_{i-1}+|V_i|\}$. In other words, $\overrightarrow{\pi}$ has the same sequence of cell sizes as π, and the elements within each cell and on successive cells are consecutively numbered. For $\pi \in \Pi_n$, let $Q(\pi) = \{g \in S_n : g*\pi = \overrightarrow{\pi}\}$.

Example 5.31. For $\pi = (\{2,3\},\{1,5\},\{4\})$, we have $\overrightarrow{\pi} = (\{1,2\},\{3,4\},\{5\})$ and

$$Q(\pi) = \{(1\ 3\ 2)(4\ 5),\ (1\ 3)(4\ 5),\ (1\ 4\ 5\ 3\ 2),\ (1\ 4\ 5\ 3)\}.$$

The following observations are straightforward to verify. First, for every $\pi \in \Pi_n$, the set $Q(\pi)$ is a right coset of the stabilizer $N_{S_n}(\overrightarrow{\pi}) = \{g \in S_n : g * \overrightarrow{\pi} = \overrightarrow{\pi}\}$; that is, $Q(\pi) = N_{S_n}(\overrightarrow{\pi})g$, where $g \in Q(\pi)$ is arbitrary. Second, the set $Q(\pi)$ uniquely determines π. Third, let $g_0 \in S_n$ act on the coset $Q(\pi)$ by $g_0 * Q(\pi) = Q(\pi)g_0^{-1}$. Then $g_0 * Q(\pi) = Q(g_0 * \pi)$, and thus C defines an isomorphism of the group actions of S_n on Π_n and on $Q(\Pi_n)$. Fourth, for $\pi_1, \pi_2 \in \Pi_n$ we have that π_1 is finer than π_2 if and only if $Q(\pi_1) \subseteq Q(\pi_2)$. These observations together imply that the ordered partitions in Π_n provide an alternative representation for the cosets $Q(\Pi_n)$. An algorithm implementation can thus in practice operate on ordered partitions, with the background provided by the cosets $Q(\Pi_n)$ in the group-theoretic framework.

Example 5.32. Figure 5.4 shows the search tree in Fig. 5.2 (rotated 90 degrees counterclockwise) where each right coset is now represented using the corresponding ordered partition. Note, however, that in general not every coset of a subgroup of a symmetric group can be represented using an ordered partition.

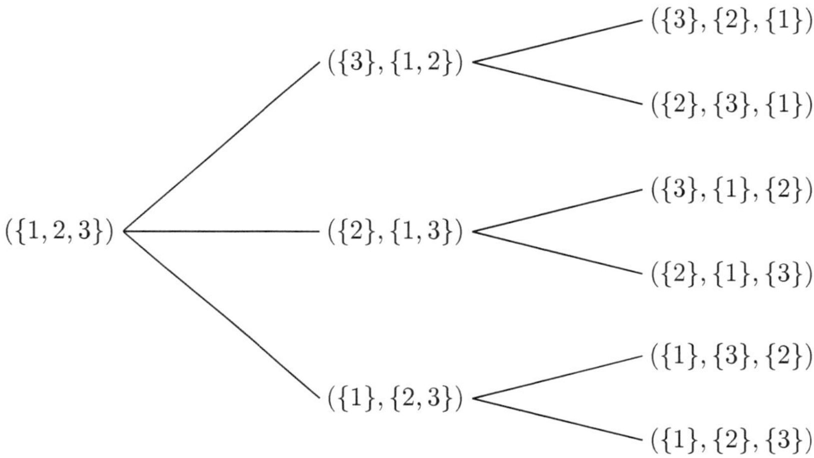

Fig. 5.4. A search tree on ordered partitions

Let us now briefly develop the structure of a partitioning rule and a refinement transformation in the setting of ordered partitions.

Given $X \in \Omega$ and a nondiscrete $\pi = (V_1, V_2, \ldots, V_m) \in \Pi_n$ as input, the typical partitioning rule *splits* a cell V_i with $|V_i| > 1$ by individualizing an element $x \in V_i$ in all possible ways. More precisely, the partitioning rule produces from π the set $\{\pi \wedge (\{x\}, \{1, 2, \ldots, n\} \setminus \{x\}) : x \in V_i\}$ of ordered partitions. Looking at the corresponding right cosets, we have

$$Q(\pi) = \bigcup_{x \in V_i} Q(\pi \wedge (\{x\}, \{1, 2, \ldots, n\} \setminus \{x\})),$$

where the union consists of pairwise disjoint proper subcosets of $Q(\pi)$. A simple strategy is to split the first cell of the minimum size greater than 1. More complex rules for selecting the cell to be split can be used, but to ensure that the requirement on cosets preceding Theorem 5.30 is met, the index i of the selected cell must be the same for all isomorphic inputs X, π and $g_0 X, g_0 \pi$, where $g_0 \in S_n$.

Example 5.33. Splitting the second cell of $\pi = (\{2, 3\}, \{1, 5\}, \{4\})$, we obtain the ordered partitions

$$\pi_1 = (\{2, 3\}, \{1\}, \{5\}, \{4\}), \quad \pi_2 = (\{2, 3\}, \{5\}, \{1\}, \{4\}).$$

Comparing

$$Q(\pi_1) = \{(1\ 3\ 2)(4\ 5),\ (1\ 3)(4\ 5)\},$$
$$Q(\pi_2) = \{(1\ 4\ 5\ 3\ 2),\ (1\ 4\ 5\ 3)\}$$

with $Q(\pi)$ in Example 5.31, we see that $Q(\pi) = Q(\pi_1) \cup Q(\pi_2)$ holds. Figures 5.4 and 5.2 provide a further illustration.

A refinement transformation in the setting of ordered partitions takes as input $X \in \Omega$ and $\pi \in \Pi_n$, and outputs a $\pi^+ \in \Pi_n$ that is finer than π. For all $g_0 \in G$ it is required that on input $g_0 X, g_0 \pi$ the output is $g_0 \pi^+$. Because $Q(g_0 \pi^+) = Q(\pi^+) g_0^{-1}$, this requirement guarantees that the requirement on cosets preceding Theorem 5.30 is met. Theorem 3.130 establishes that the refinement transformations studied in Sect. 3.3.5 can be applied on graphs and colored graphs in the present context – the reader is referred to Sect. 3.3.5 for examples of refinement of ordered partition. Most algorithm implementations rely on the color degree invariant (Definition 3.120) to provide the default refinement transformation.

In addition to the direct permutation representation of cosets and the representation in terms of ordered partitions, a third possibility is to combine the two representations; that is, to use a permutation representation for the current coset together with an ordered partition for keeping track of invariant values. This technique is particularly useful in the context of more general computation with permutation groups; see [367, 368].

We have here provided a general outline of how a refinement-based isomorphism algorithm operates. It still remains to discuss how automorphisms are discovered while traversing the search tree, and how discovered automorphisms are used to prune the search tree.

First, it is useful to observe that automorphism pruning is mandatory for practical performance if $\mathrm{Aut}(X)$ has large order. Namely, observe that if $\{g\}$ is a leaf node in $T(X, G)$, then $\{ga^{-1}\}$ is a leaf node in $T(X, G)$ for all $a \in \mathrm{Aut}(X)$ by Theorem 5.30.

A typical algorithm implementation discovers automorphisms by keeping track of one or more leaf nodes and the associated leaf objects. If $\{g\}$ is a leaf node that we keep track of, and during the search we encounter a leaf node $\{g_1\}$ such that the leaf objects agree – that is, $gX = g_1 X$ – then clearly $a = g_1^{-1} g$ is an automorphism of X. Moreover, every automorphism of X can be discovered in this manner.

The more leaf nodes we keep track of, the sooner we discover automorphisms, but at the cost of bookkeeping and comparing leaf objects. Typically most objects have a trivial or small automorphism group, whereby most algorithms rely only on one or two leaf objects for discovering automorphisms. The current candidate object for the canonical form of X indicated by the leaf selection rule is usually employed in discovering automorphisms – additional tracking of leaf nodes is motivated depending on the typical order of the automorphism group. A more detailed discussion and additional strategies for discovering automorphisms can be found in [411].

Once we have discovered automorphisms, we can start using them to prune the search tree $T(X,G)$. Let $A \leq G$ be the group generated by the automorphisms discovered so far. If we have traversed the subtree rooted at a node Hg in $T(X,G)$, and upon traversing $T(X,G)$ encounter a node Hg_1 such that $Hga_0^{-1} = Hg_1$ for some $a_0 \in A$, then we can prune the subtree rooted at Hg_1. To see this, observe that by Theorem 5.30 we have that $\{hg\}$ is a leaf node in the subtree rooted at Hg if and only if $\{hga_0^{-1}\}$ is a leaf node in the subtree rooted at Hg_1. Thus, the two subtrees have the same set of leaf objects – if we have traversed the subtree rooted at Hg, then we will not discover any new leaf objects by traversing the subtree rooted at Hg_1. This observation also applies to automorphisms. Namely, if $\{a \in G : aX = X,\ Hga^{-1} = Hg\} \subseteq A$, then also $\{a \in G : aX = X,\ Hg_1a^{-1} = Hg_1\} \subseteq A$. Indeed, an arbitrary $a \in G$ with $aX = X$ and $Hg_1a^{-1} = Hg_1$ can be expressed as a product $a = a_0a_1a_0^{-1}$, where $a_1 = a_0^{-1}aa_0$ satisfies $a_1X = X$ and $Hga_1^{-1} = Hg$. Because both $a_0 \in A$ and $a_1 \in A$, we have $a \in A$.

A typical algorithm implementation exploits the previous pruning strategy in two places: upon discovery of a new automorphism, and upon traversing new child nodes of a node where at least one child has already been traversed. Here we assume that a depth-first traversal strategy is used for the search tree.

When an automorphism $a = g_1^{-1}g$ is discovered, let Hg be the minimal node in $T(X,G)$ that contains both leaf nodes $\{g\}$ and $\{g_1\}$. The child of Hg that contains g has been traversed because of depth-first traversal order; moreover, a takes the child of Hg that contains $\{g\}$ onto the child containing $\{g_1\}$, which shows that we can prune the subtree rooted at the child containing $\{g_1\}$ after inserting a into a set of generators for A.

When traversing the children of a node Hg, we can compute the stabilizer $N_A(Hg) = \{a \in A : Hga^{-1} = Hg\}$, and look at the orbits of $N_A(Hg)$ on the children of Hg. It suffices to traverse only one child and the associated subtree from each such orbit. However, computing the stabilizer can be expensive, and should always be weighed against the gains obtained. For example, *nauty* – as described in [411] – does not employ this type of pruning to the fullest extent possible. Algorithm implementations where this pruning strategy is employed are described in more detail in [83, 325, 342].

5.7 Distributing Computer Search

There are two aspects of dividing a computation into parts for distribution among several computers: methodology and implementation. For most classification algorithms in this work, both of these issues are easily dealt with.

If a computation is divided into parts that need no mutual communication whatsoever, the implementation of such a computation is technically straightforward. Fortunately, many backtrack algorithms have this desirable property.

A common approach is to pick out a level, say L, of the search tree and distribute the subtrees rooted at that level among the computers. In Fig. 5.5, $L = 1$, and the complete search is divided into 4 parts (subtrees). The appropriate value of L depends on many factors, for example, the number of computers available, and should be decided on a case-by-case basis. With n computers numbered $0, 1, \dots, n-1$, the computer numbered m may now process the subtrees whose roots are the nodes number $m + ni$ on level L for $i = 0, 1, \dots$. The tasks are then fairly evenly distributed, but it may still be the case that some subtree is considerably larger than the other ones, or that a computer involved is relatively slow, whereby the total computation time is not reduced by a factor close to n. A solution to both of these issues is that the computers involved process one subtree (or a few) at a time, repeatedly consulting a database for the next job – that is, the next unprocessed subtree. For problems with extremely biased search trees, more sophisticated methods of partitioning a search tree into subtrees are necessary.

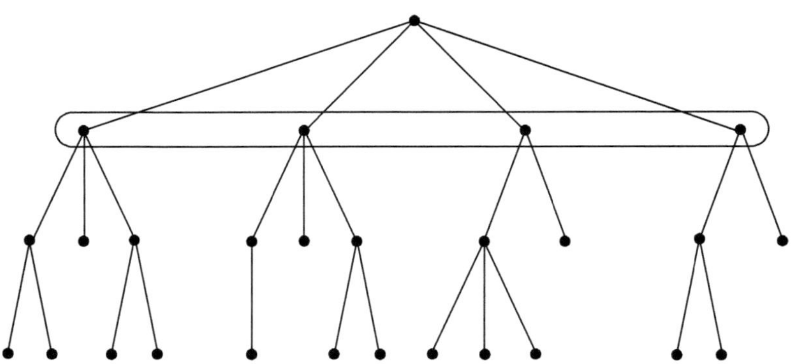

Fig. 5.5. Distributing parts of a search tree

There are two possibilities for a participating computer to determine the state of the root node of an allotted subtree. If the levels up to L can be processed very fast, it is possible to carry out a breadth-first search or a truncated depth-first search to find the desired node on level L. However, if the overhead due to traversing the first few levels of the search tree is considerable, then the nodes on level L may be precomputed and tabulated. Each subprocess then fetches the appropriate starting point for its computation from this table.

With a network of computers, the use of a batch system, such as `autoson` [413], is highly recommended. Most backtrack search programs need only minor modifications to be able to utilize `autoson`. For example, if we have a program `search` with one argument n, $1 \leq n \leq 500$, which indicates the starting point of the partial search, and the program writes to standard output, then the whole search is simply carried out with

```
auadd -cyc 1 -lim 500 -log #.out search #
```

The macro character `#` takes the value of the incremental parameter, which goes through all values from 1 to 500. The output of the first process is written to the file `1.out` and the executed program is `search 1`. The `autoson` system comprises extensive possibilities for administrating the processes involved, which may be affected by various situations such as computer shutdown. Other software systems for batch control and parallel search that have been used for traversing large search trees include Condor [130], PVM [42], and ZRAM [77].

6

Classification of Designs

In this chapter we turn to the first main topic of this book, classification of designs. The designs to be discussed are divided into three classes: balanced incomplete block designs (BIBDs) – that is, 2-designs – are considered in Sect. 6.1, t-designs with $t \geq 3$ in Sect. 6.2, and resolutions of designs in Sect. 6.3. In the discussion of resolutions of designs, the main focus is on resolutions of BIBDs. Extensive tables of classification results are provided at the end of each section. Finally, in Sect. 6.4, classification of designs with certain additional properties is briefly discussed.

6.1 Balanced Incomplete Block Designs

The two main approaches for classifying BIBDs proceed point by point and block by block; such algorithms are discussed in Sects. 6.1.1 and 6.1.3, respectively. With a few exceptions, these two approaches suffice to reproduce the known classification results for BIBDs. Some methods for getting at instances that cannot be handled with the basic techniques are discussed in Sect. 6.1.5.

6.1.1 Classification Point by Point

In terms of an incidence matrix of a BIBD, a point-by-point construction completes the matrix one row at a time. A 0-1 matrix $\mathbf{N} = (n_{ij})$ is an incidence matrix of a BIBD with parameters v, k, λ, b, r if and only if the matrix has size $v \times b$ and

$$\mathbf{N1} = r\mathbf{1}, \quad \mathbf{NN}^T = (r - \lambda)\mathbf{I} + \lambda\mathbf{J}, \quad \mathbf{1}^T\mathbf{N} = k\mathbf{1}^T. \tag{6.1}$$

In other words, by (6.1) the number of 1s in every row of such an incidence matrix $\mathbf{N}$ is r, the inner product (over the integers) of every two distinct rows is λ, and the number of 1s in every column is k.

An approach to constructing BIBDs row by row via incidence matrices is now evident. Given a $v' \times b$ partial incidence matrix $\mathbf{N}$ with $v' < v$, we first find all 0-1 vectors $\mathbf{y}$ of size $b \times 1$ such that

$$\mathbf{1}^T\mathbf{y} = r, \quad \mathbf{N}\mathbf{y} = \lambda\mathbf{1}, \quad \mathbf{1}^T\mathbf{N} + \mathbf{y}^T \le k\mathbf{1}^T. \tag{6.2}$$

For each $\mathbf{y}$ that is a solution to (6.2), we augment $\mathbf{N}$ by the row defined by this vector, and proceed recursively to augment the new matrix. This process can be described by a search tree whose nodes are the possible partial incidence matrices. The root of the tree is the $0 \times b$ empty matrix, and the children of a $v' \times b$ matrix $\mathbf{N}$ are all the $(v'+1) \times b$ matrices obtainable by appending a solution $\mathbf{y}$ of (6.2) to $\mathbf{N}$.

Isomorph rejection is obviously required for this approach to be practical. Here we will develop an isomorph rejection strategy based on orderly generation and lexicographically maximum canonical representatives.

We begin by defining the relevant lexicographic order and associated notation. Let $\mathbf{A} = (a_{ij})$ be a 0-1 matrix of size $s \times t$. We write $\mathbf{A}_{[i,\cdot]} = [a_{i1}\ a_{i2}\ \cdots\ a_{it}]$ and $\mathbf{A}_{[\cdot,j]} = [a_{1j}\ a_{2j}\ \cdots\ a_{sj}]^T$, and generalize this notation in the obvious way to sets of rows and columns by writing $\mathbf{A}_{[S,\cdot]}$ and $\mathbf{A}_{[\cdot,S]}$, for a set S of row (column) indices, respectively. We associate with $\mathbf{A}$ the st-tuple $w(\mathbf{A})$ formed by concatenating the rows of $\mathbf{A}$; more precisely,

$$w(\mathbf{A}) = (a_{11}, a_{12}, \ldots, a_{1t}, a_{21}, a_{22}, \ldots, a_{2t}, \ldots, a_{s1}, a_{s2}, \ldots, a_{st}).$$

This induces a lexicographic order $\prec$ on the set of all 0-1 matrices of size $s \times t$ by $\mathbf{A} \prec \mathbf{B}$ if and only if $w(\mathbf{A}) \prec w(\mathbf{B})$, where the latter order relation refers to the lexicographic order on tuples (Sect. 3.3.1).

We say that two 0-1 matrices of size $s \times t$ are *isomorphic* if one can be obtained from the other by permuting the rows and the columns. Note that if we view the matrices as incidence structures, this is equivalent to saying that the two incidence structures are isomorphic (Definition 2.22). This notion of isomorphism partitions the set of all 0-1 matrices of size $s \times t$ into isomorphism classes. We say that a matrix is *canonical* – that is, the canonical representative of its isomorphism class – if it is the lexicographic maximum of the matrices in its isomorphism class.

The next theorem gives the following fundamental property required by orderly generation: a canonical matrix is obtained only by augmenting a canonical matrix with a new row. (Here we assume that the augmenting row always becomes the last row.) Thus, it suffices to consider only canonical matrices while traversing the search tree.

Theorem 6.1. *Let* $\mathbf{A}$ *be a canonical* 0-1 *matrix of size* $s \times t$. *Then, the submatrix* $\mathbf{A}_{[\{1,2,\ldots,i\},\cdot]}$ *is canonical for any* $1 \le i \le s$.

Proof. Assume that $\mathbf{A}_{[\{1,2,\ldots,i\},\cdot]}$ is not canonical. Then we can permute the rows and columns of $\mathbf{A}_{[\{1,2,\ldots,i\},\cdot]}$ to obtain a lexicographically greater matrix.

Extend such a pair of permutations to permute the rows and columns of **A** by keeping the rows $i+1, i+2, \ldots, s$ fixed. The matrix obtained from **A** using this pair of permutations is lexicographically greater than **A**, which shows that **A** is not canonical, a contradiction. □

Example 6.2. Figure 6.1 shows the first five levels of the search tree for 2-(7,3,2) designs. Only canonical matrices are displayed.

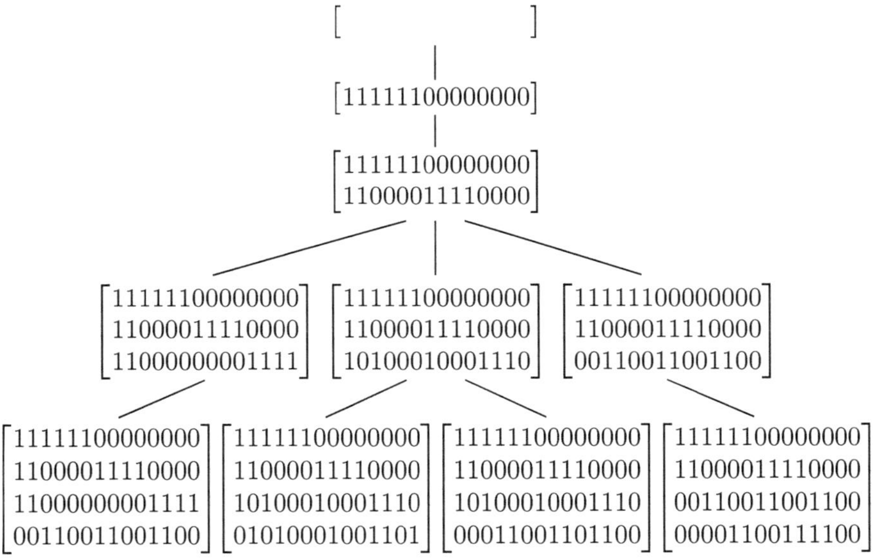

Fig. 6.1. A search tree for 2-(7,3,2) designs (truncated to first five levels)

We continue with some observations that enable a more efficient traversal of the search tree and achieve further pruning.

Because testing whether a matrix is canonical is computationally expensive, executing the full canonicity test should be avoided whenever possible. The following observation can be exploited in this regard.

Theorem 6.3. *Let* **A** *be a canonical* 0-1 *matrix of size* $s \times t$. *Then,*

$$\mathbf{A}_{[\cdot,1]} \succeq \mathbf{A}_{[\cdot,2]} \succeq \cdots \succeq \mathbf{A}_{[\cdot,t]}, \tag{6.3}$$

$$\mathbf{A}_{[1,\cdot]} \succeq \mathbf{A}_{[2,\cdot]} \succeq \cdots \succeq \mathbf{A}_{[s,\cdot]}. \tag{6.4}$$

Proof. If (6.3) does not hold, then we can sort the columns of **A** into decreasing lexicographic order so that (6.3) holds. This column-sorted matrix is lexicographically greater than **A**, which is impossible if **A** is canonical. The reasoning for (6.4) is similar. □

The result of Theorem 6.3 can be exploited as follows. When traversing a node $\mathbf{N}$, all solutions $\mathbf{y}$ of (6.2) that do not satisfy (6.3) can be disregarded. More precisely, $\mathbf{y} = [y_1\ y_2\ \cdots\ y_b]^T$ can be disregarded if there exist $1 \le j_1 < j_2 \le b$ such that $y_{j_1} < y_{j_2}$ and $\mathbf{N}_{[\cdot,j_1]} = \mathbf{N}_{[\cdot,j_2]}$. This idea can be extended further by considering the automorphisms of $\mathbf{N}$ restricted to act on the columns. Namely, $\mathbf{y}$ can be discarded if there exists such an automorphism that transforms $\mathbf{y}$ to a lexicographically greater vector by permuting the entries in $\mathbf{y}$. It is also possible to consider the column automorphisms of the submatrices $\mathbf{N}_{[\{1,2,\ldots,i\},\cdot]}$, $1 \le i < v'$; in this case $\mathbf{y}$ can be disregarded if a column automorphism transforms it to a vector that is lexicographically greater than $\mathbf{N}_{[i+1,\cdot]}{}^T$. A more detailed treatment of these automorphism pruning strategies can be found in [152, 153].

Further pruning can be achieved by exploiting the property that a complete canonical incidence matrix satisfies $\mathbf{1}^T\mathbf{N} = k\mathbf{1}$. Namely, let $\mathbf{N}$ be a $v' \times b$ incidence matrix with $\mathbf{1}^T\mathbf{N} \le k\mathbf{1}^T$ and $v' < v$. Let

$$j_0 = \min\{j : \mathbf{1}^T\mathbf{N}_{[\cdot,j]} < k\}.$$

To extend $\mathbf{N}$ to a complete incidence matrix, the column j_0 must eventually be completed so that the number of 1s is k. Furthermore, to obtain a canonical complete incidence matrix, the rows that contain a 1 in column j_0 must be added before any rows that contain a 0 in column j_0 – otherwise the complete incidence matrix violates (6.4). Thus, $\mathbf{y}$ can be disregarded if $y_{j_0} = 0$.

The previous observation is more powerful than it first appears. This is because it restricts the number of canonical matrices that the search traverses. Without the requirement $y_{j_0} = 1$, the search would traverse every canonical matrix isomorphic to a matrix of the form $\mathbf{N}_{[I,\cdot]}$, where $I \subseteq \{1, 2, \ldots, v\}$ and $\mathbf{N}$ is a complete incidence matrix of size $v \times b$. Requiring $y_{j_0} = 1$ prunes many of these intermediate solutions that would otherwise be traversed. For example, requiring $y_{j_0} = 1$ in Fig. 6.1 already prunes the entire rightmost branch and the right child of the middle branch.

It is also useful to note that the full canonicity test need not be performed at every level in the search tree. Indeed, the algorithm remains correct even if we perform the test only for $v \times b$ incidence matrices. The standard approach is to employ the canonicity test for the first few levels of the search tree, after which the algorithm is executed without the canonicity test until the matrix is complete and subjected to a final canonicity test.

For performance reasons it is often advantageous to structure a phase of the search where no canonicity tests are performed as a clique search. For a $v' \times b$ incidence matrix $\mathbf{N}$, let G be the graph whose vertices are all the solutions $\mathbf{y}$ of (6.2) satisfying $\mathbf{y} \preceq \mathbf{N}_{[i,\cdot]}{}^T$ for all $1 \le i \le v'$. Two vertices in G are connected by an edge if and only if the inner product of the associated vectors is λ. The graph G is the *compatibility graph* for $\mathbf{N}$. A clique of size s in G now corresponds to a set of s rows extending $\mathbf{N}$, and vice versa. If $s = v - v'$, then the clique completes $\mathbf{N}$ to an incidence matrix satisfying (6.1); see [458, Theorem 3.3].

Example 6.4. Figure 6.2 shows the compatibility graph for the canonical matrix

$$\mathbf{N} = \begin{bmatrix} 1111110000000 \\ 1100001111000 0 \\ 1010001000111 0 \\ 0101000100110 1 \end{bmatrix}$$

that occurs in the search tree for 2-$(7,3,2)$ designs in Fig. 6.1.

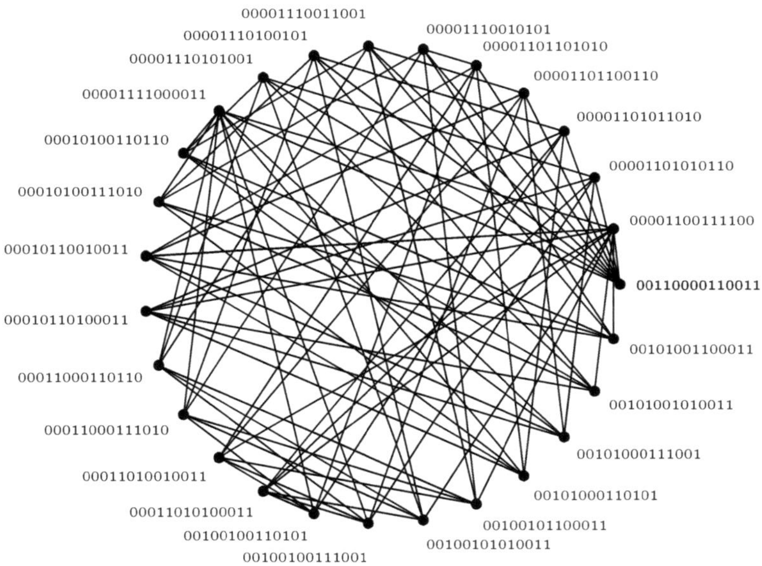

Fig. 6.2. Compatibility graph for the matrix in Example 6.4

In principle, the entire search could be carried out in this manner as a plain clique search. However, in practice this is not possible because of the size and symmetry of the initial compatibility graph. Typically the fastest solution is to first proceed row by row with isomorph rejection via canonicity testing up to a level where the compatibility graph is not too large – currently at most a few thousand vertices, depending on the size of the cliques – and then proceed using clique search. See, for example, [303, 545, 549].

This completes the outline of an orderly algorithm. Before discussing approaches for solving (6.2), we list a few work points. Canonicity testing of incidence matrices is considered separately in Sect. 6.1.2.

Research Problem 6.5. Develop techniques for detecting a partial incidence matrix that cannot be completed. The bounding functions in clique search constitute one such technique – this is one of the reasons why clique search is more efficient than the basic algorithm in some situations – however, this is surely not the only possibility to prune the search tree. Bounding functions based on relaxations of integer linear programs could provide useful in this respect; cf. [395].

Research Problem 6.6. Develop algorithms based on generation by canonical augmentation for classifying BIBDs point by point. What are the possibilities of implementing requirements similar to the requirement in orderly generation that $y_{j_0} = 1$?

Let a canonical 0-1 matrix $\mathbf{N}$ of $v' \times b$ be given. The equations and inequalities in (6.2) clearly constitute an integer linear programming problem, which can be solved using the tools from Sect. 5.4.

Example 6.7. Consider the canonical matrix

$$\mathbf{N} = \begin{bmatrix} 11111100000000 \\ 11000011110000 \\ 10100010001110 \\ 01010001001101 \end{bmatrix}$$

that occurs in the search tree for 2-$(7,3,2)$ designs in Fig. 6.1. The equations and inequalities in (6.2) give the following Diophantine linear system:

$$\begin{array}{llllllllllllll}
y_1+y_2+y_3+y_4+y_5+y_6+y_7+y_8+y_9+y_{10}+y_{11}+y_{12}+y_{13}+y_{14} = 6, \\
y_1+y_2+y_3+y_4+y_5+y_6 = 2, \\
y_1+y_2 \quad +y_7+y_8+y_9+y_{10} = 2, \\
y_1 \quad +y_3 \quad +y_7 \quad +y_{11}+y_{12}+y_{13} = 2, \\
\quad y_2 \quad +y_4 \quad +y_8 \quad +y_{11}+y_{12} \quad +y_{14} = 2
\end{array} \tag{6.5}$$

subject to $y_1 = y_2 = 0$ and $y_3, y_4, \ldots, y_{14} \in \{0, 1\}$.

By the previous discussion we have the additional constraints

$$\mathbf{y} \preceq \mathbf{N}_{[i,\cdot]}{}^T \quad \text{for all } 1 \le i \le v' \tag{6.6}$$

and

$$y_{j_0} = 1 \quad \text{for } j_0 = \min\{j : \mathbf{1}^T \mathbf{N}_{[\cdot,j]} < k\}. \tag{6.7}$$

Example 6.8. Property (6.6) gives no additional constraints to (6.5). Property (6.7) gives the constraint $y_3 = 1$.

To eliminate redundancy in the search space for DIOPHANTINE – especially in the early stages of the search – it is useful to partition $\mathbf{N}$ into regions of identical columns, and replace the variables associated with a region by a single variable.

Example 6.9. In Example 6.7, the partition into regions of identical columns and the associated variables are as follows.

x_1	x_2	x_3	x_4	x_5	x_6	x_7	x_8	x_9	x_{10}	x_{11}
1	1	1	1	1 1	0	0	0 0	0 0	0	0
1	1	0	0	0 0	1	1	1 1	0 0	0	0
1	0	1	0	0 0	1	0	0 0	1 1	1	0
0	1	0	1	0 0	0	1	0 0	1 1	0	1

Taking into account the constraint $y_3 = 1$ from Example 6.8, we obtain the reduced system

$$\begin{aligned} x_1 + x_2 + x_3 + x_4 + x_5 + x_6 + x_7 + x_8 + x_9 + x_{10} + x_{11} &= 6, \\ x_1 + x_2 + x_3 + x_4 + x_5 &= 2, \\ x_1 + x_2 + x_6 + x_7 + x_8 &= 2, \\ x_1 + x_3 + x_6 + x_9 + x_{10} &= 2, \\ x_2 + x_4 + x_7 + x_9 + x_{11} &= 2 \end{aligned} \tag{6.8}$$

subject to $x_1 = x_2 = 0$, $x_3 = 1$, $x_4, x_6, x_7, x_{10}, x_{11} \in \{0,1\}$, and $x_5, x_8, x_9 \in \{0,1,2\}$.

The system (6.8) has three solutions.

x_1	x_2	x_3	x_4	x_5	x_6	x_7	x_8	x_9	x_{10}	x_{11}
0	0	1	1	0	0	0	2	0	1	1
0	0	1	0	1	0	1	1	0	1	1
0	0	1	0	1	0	0	2	1	0	1

(6.9)

A solution to the reduced system corresponds to a set of solutions of the original system, where a reduced variable x_ℓ specifies the number of original variables with $y_j = 1$ in the corresponding region.

Example 6.10. From (6.9) we see that there are $1 + 2 \cdot 2 + 2 \cdot 2 = 9$ solutions to (6.5) with $y_3 = 1$.

y_1	y_2	y_3	y_4	y_5	y_6	y_7	y_8	y_9	y_{10}	y_{11}	y_{12}	y_{13}	y_{14}
0	0	1	1	0	0	0	0	1	1	0	0	1	1
0	0	1	0	1	0	0	1	1	0	0	0	1	1
0	0	1	0	1	0	0	1	0	1	0	0	1	1
0	0	1	0	0	1	0	1	1	0	0	0	1	1
0	0	1	0	0	1	0	1	0	1	0	0	1	1
0	0	1	0	1	0	0	0	1	1	1	0	0	1
0	0	1	0	1	0	0	0	1	1	0	1	0	1
0	0	1	0	0	1	0	0	1	1	1	0	0	1
0	0	1	0	0	1	0	0	1	1	0	1	0	1

(6.10)

Note that for a reduced solution $\mathbf{x}$ there is exactly one original solution $\mathbf{y}$ that can produce a canonical matrix by augmenting $\mathbf{N}$. Namely, the variable values must form a nonincreasing sequence in each region of identical columns; more precisely, $\mathbf{N}$ augmented by $\mathbf{y}$ violates (6.3) unless we have

$$y_j \geq y_{j'} \quad \text{for all } j < j' \text{ such that } \mathbf{N}_{[\cdot,j]} = \mathbf{N}_{[\cdot,j']}. \tag{6.11}$$

Example 6.11. The first, second, and sixth solution in Example 6.10 satisfy (6.11).

On the other hand, if we are looking for all solutions to (6.2) subject to (6.6) – for example, to build the compatibility graph for $\mathbf{N}$ – then neither (6.7) nor (6.11) can be used to restrict the solutions; cf. Example 6.4 and Fig. 6.2.

6.1.2 Testing Canonicity of Incidence Matrices

A canonicity test determines whether a given matrix $\mathbf{A}$ of size $s \times t$ is the lexicographic maximum of its isomorphism class determined by permutation of the rows and the columns. The only known way to perform this test is via backtrack search. Fortunately, extensive pruning is possible, which makes backtracking practical even for some relatively large instances.

An important observation is that we need not search through all the row and column permutations. Once a permutation of the rows is fixed, the maximum matrix subject to this ordering of the rows is obtained by sorting the columns into decreasing lexicographic order. For a matrix $\mathbf{A}$, we write $\overrightarrow{\mathbf{A}}$ for the matrix obtained by sorting the columns of $\mathbf{A}$ into decreasing lexicographic order. More precisely, $\overrightarrow{\mathbf{A}}$ is the matrix obtained by permuting the columns in $\mathbf{A}$ such that

$$\overrightarrow{\mathbf{A}}_{[\cdot,1]} \succeq \overrightarrow{\mathbf{A}}_{[\cdot,2]} \succeq \cdots \succeq \overrightarrow{\mathbf{A}}_{[\cdot,t]}.$$

Example 6.12. We show an example matrix and the corresponding column-sorted matrix:

$$\mathbf{A} = \begin{bmatrix} 11000011110000 \\ 01010001001101 \\ 10100010001110 \\ 11111100000000 \end{bmatrix}, \quad \overrightarrow{\mathbf{A}} = \begin{bmatrix} 11111100000000 \\ 11000011110000 \\ 00110011001100 \\ 10100000101011 \end{bmatrix}.$$

Thus, if $\mathbf{A}$ is not canonical, then a counterexample to the canonicity of $\mathbf{A}$ can be found by considering all possible row permutations, and testing for each matrix $\mathbf{B}$ obtained by permuting the rows of $\mathbf{A}$ whether $\overrightarrow{\mathbf{B}} \succ \mathbf{A}$. If so, we have a counterexample.

We can construct a row permutation one row at a time, which allows us to effectively disregard many row permutations that cannot yield a counterexample. An intuitive description of this situation is as follows. Let $\mathbf{B}$ be a matrix that is obtained by permuting the rows of the input matrix $\mathbf{A}$. Now suppose that the submatrices consisting of the first i rows of the matrices satisfy

$\overrightarrow{\mathbf{B}}_{[\{1,2,\ldots,i\},\cdot]} \prec \overrightarrow{\mathbf{A}}_{[\{1,2,\ldots,i\},\cdot]}$. Then $\overrightarrow{\mathbf{B}} \prec \overrightarrow{\mathbf{A}}$ by the definition of lexicographic order. Thus, a counterexample cannot be found by extending $\mathbf{B}_{[\{1,2,\ldots,i\},\cdot]}$ if $\overrightarrow{\mathbf{B}}_{[\{1,2,\ldots,i\},\cdot]} \prec \overrightarrow{\mathbf{A}}_{[\{1,2,\ldots,i\},\cdot]}$.

To facilitate pruning with discovered automorphisms, it is convenient to formulate the search over possible permutations of the rows using group-theoretic terminology. Let $G \leq \mathrm{Sym}(\{1,2,\ldots,s\})$ be a permutation group that acts on a matrix $\mathbf{A}$ by permuting the rows. More precisely, a group element $g \in G$ acts on $\mathbf{A}$ so that the resulting matrix $g\mathbf{A}$ is defined by

$$(g\mathbf{A})_{[i,\cdot]} = \mathbf{A}_{[g^{-1}(i),\cdot]} \quad \text{for all } 1 \leq i \leq s. \tag{6.12}$$

In other words, row i in $\mathbf{A}$ becomes row $g(i)$ in $g\mathbf{A}$ for all $1 \leq i \leq s$. In practice we take $G = \mathrm{Sym}(\{1,2,\ldots,s\})$ when working with BIBDs, but for other types of designs – such as GDDs, cf. [480] – a smaller group is often motivated. Thus, we will develop the algorithm for a generic permutation group G.

Let G_i be the pointwise stabilizer of $1,2,\ldots,i-1$ in G, that is, $G_i = N_G((1,2,\ldots,i-1))$, $1 \leq i \leq s$. Let R_i be a right transversal for G_{i+1} in G_i, $1 \leq i \leq s-1$. For BIBDs and $G = \mathrm{Sym}(\{1,2,\ldots,s\})$, we clearly have $G_i \cong \mathrm{Sym}(\{i,i+1,\ldots,s\})$ for all $1 \leq i \leq s$. In this setting it is convenient to use transpositions for the right transversals:

$$R_i = \{(i\ \ i+1),(i\ \ i+2),\ldots,(i\ \ s)\}, \quad 1 \leq i < s.$$

Example 6.13. For $G = \mathrm{Sym}(\{1,2,3\})$ we have

$$\begin{aligned}
G_1 &= \{\epsilon,(2\ 3),(1\ 2),(1\ 3),(1\ 2\ 3),(1\ 3\ 2)\}, & R_1 &= \{\epsilon,(1\ 2),(1\ 3)\},\\
G_2 &= \{\epsilon,(2\ 3)\}, & R_2 &= \{\epsilon,(2\ 3)\},\\
G_3 &= \{\epsilon\}.
\end{aligned}$$

When G is an arbitrary permutation group, the right transversals of point stabilizers can be determined using the tools from Sect. 5.5. Observe that if L_i is a left transversal for G_i in G_{i+1}, then $R_i = L_i^{-1}$ is a right transversal for G_i in G_{i+1}.

The search now constructs a row permutation $g \in G$ one point at a time. First, the value $g^{-1}(1)$ is fixed, then the value $g^{-1}(2)$, and so forth. Relative to (6.12), this corresponds to selecting the first row of $g\mathbf{A}$, then the second row, and so forth. In this way we can check the pruning condition $\overrightarrow{g\mathbf{A}}_{[\{1,2,\ldots,i\},\cdot]} \prec \mathbf{A}_{[\{1,2,\ldots,i\},\cdot]}$ immediately after $g^{-1}(i)$ is fixed. In group-theoretic language, the permutation construction amounts to first selecting a right coset $G_2r_1 \in G_2\backslash G_1$, $r_1 \in R_1$, then a right coset $G_3r_2r_1 \in G_3\backslash G_1$, $r_2 \in R_2$, and so forth. Eventually $g = r_{s-1}r_{s-2}\cdots r_1$. Because $r_j(i) = i$ for all $i < j < s$, the eventual value $g^{-1}(i) = r_1^{-1}r_2^{-1}\cdots r_s^{-1}(i)$ is determined as soon as $r_1, r_2, \ldots, r_i$ are fixed.

Algorithm 6.1 is a backtrack search implementation of the canonicity predicate. The nodes of the associated search tree are best viewed as right cosets

Algorithm 6.1 Canonicity predicate for a 0-1 matrix

```
function COSET(i: integer, g: permutation): integer
if $\overrightarrow{g\mathbf{A}}_{[\{1,2,\ldots,i\},\cdot]} \succ \mathbf{A}_{[\{1,2,\ldots,i\},\cdot]}$ then
    return $-1$
end if
if $\overrightarrow{g\mathbf{A}}_{[\{1,2,\ldots,i\},\cdot]} \prec \mathbf{A}_{[\{1,2,\ldots,i\},\cdot]}$ then
    return 1
end if
if $i < s$ then
    for all $r_i \in R_i$ do
        $z \leftarrow$ COSET$(i+1, r_i g)$
        if $z \le 0$ then
            return $z$
        end if
    end for
    return 1
else
    $\Theta \leftarrow \Theta \vee (\langle\{g\}\rangle \backslash \{1,2,\ldots,s\})$
    $S \leftarrow S \cup \{g\}$
    return 0
end if
function FIRST(i: integer): integer
if $i < s$ then
    if FIRST$(i+1) < 0$ then
        return $-1$
    end if
    $U \leftarrow \emptyset$
    for all $r_i \in R_i \setminus G_{i+1}$ do
        if $\Theta(r_i^{-1}(i)) \cap U = \emptyset$ then
            if COSET$(i+1, r_i) < 0$ then
                return $-1$
            end if
        end if
        $U \leftarrow U \cup \{r_i^{-1}(i)\}$
    end for
else
    if $\overrightarrow{\mathbf{A}} \succ \mathbf{A}$ then
        return $-1$
    end if
end if
return 0
function CANONICAL(A: matrix of size $s \times t$): boolean
$\Theta \leftarrow \{\{i\} : 1 \le i \le s\}$
$S \leftarrow \emptyset$
if FIRST$(1) < 0$ then
    return FALSE
end if
report $S$ as an SGS for $\mathrm{Aut_r}(\mathbf{A})$ relative to the base $(1,2,\ldots,s)$
return TRUE
```

of the form $G_i g$, where $g \in G$ and $1 \le i \le s$. For $1 \le i < s$, the children of a node $G_i g$ are all nodes of the form $G_{i+1} r g$ for $r \in R_i$. The search tree is traversed in depth-first order starting from the root node $G = G_1$. More precisely, the invocation FIRST(i) traverses the subtree rooted at G_i, and an invocation COSET(i, g) traverses the subtree rooted at $G_i g$, where $g \in G \setminus G_i$. By the structure of the algorithm, the invocation FIRST(i) returns before any invocations COSET(i, g) are made for a given i. Thus, the traversal is ordered so that for every i, the coset of the identity G_{i+1} is traversed before all the other cosets in G_i.

Example 6.14. The search tree on cosets for the right transversals in Example 6.13 is depicted in Fig. 6.3. Note that the tree is identical to the tree in Fig. 5.2.

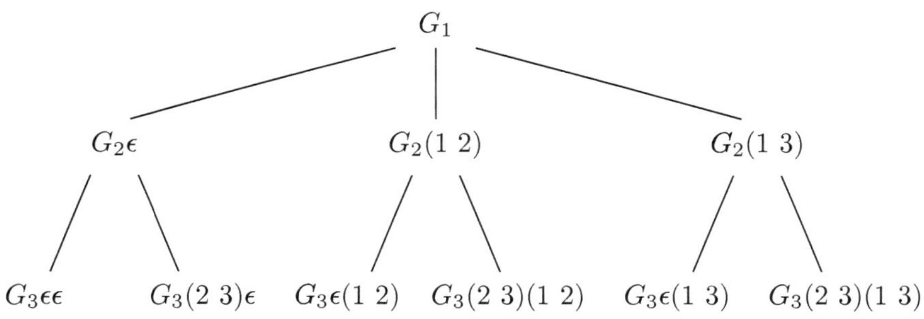

Fig. 6.3. Search tree for $G = \mathrm{Sym}(\{1, 2, 3\})$ and Example 6.13

Algorithm 6.1 records the discovered automorphisms into the set S and uses these to prune the search tree. The relevant notion of an automorphism is captured by the group $\mathrm{Aut_r}(\mathbf{A}) = \{g \in G : g\overrightarrow{\mathbf{A}} = \overrightarrow{\mathbf{A}}\}$. Equivalently, $\mathrm{Aut_r}(\mathbf{A})$ is the group consisting of all row permutations (in G) that can be extended to an automorphism of $\mathbf{A}$ by permuting the columns.

By the structure of the algorithm it is obvious that $\langle S \rangle \le \mathrm{Aut_r}(\mathbf{A})$ at all times. Associated with S is Θ, the partition of $\{1, 2, \ldots, s\}$ into $\langle S \rangle$-orbits. Whenever a new automorphism is discovered, the orbit partition Θ is updated via the partition join operation $\vee$ discussed in Sect. 5.5. For $\ell \in \{1, 2, \ldots, s\}$, we write $\Theta(\ell)$ for the cell of Θ that contains ℓ.

Automorphisms are employed in two different ways in pruning the search tree (cf. Sect. 5.6). The first observation is that we can prune the current coset $G_i g$ whenever a new automorphism is discovered. Here i is the smallest value for which the invocation FIRST(i) is complete when the automorphism is discovered. In Algorithm 6.1 this pruning strategy is implemented in the function COSET, which returns 0 when an automorphism has been discovered, and equality in line 10 takes care of the pruning. The following lemma shows that this pruning strategy is correct, provided that all automorphisms have been recorded – and no counterexample to the canonicity of $\mathbf{A}$ was found – while traversing the subtree rooted at G_i.

For a set of permutations $T \subseteq G$ we write $\overrightarrow{T\mathbf{A}}$ for the set $\{\overrightarrow{t\mathbf{A}} : t \in T\}$. By $\overrightarrow{T\mathbf{A}} \preceq \mathbf{A}$ we indicate that the relation holds for all matrices in the set $\overrightarrow{T\mathbf{A}}$.

Lemma 6.15. *Let $\overrightarrow{G_i\mathbf{A}} \preceq \mathbf{A}$ and $\mathrm{Aut_r}(\mathbf{A}) \cap G_i = \langle S\rangle \cap G_i$. If $\langle S\rangle \cap G_i g$ is nonempty for some $g \in G$, then $\overrightarrow{G_i g\mathbf{A}} \preceq \mathbf{A}$ and $\mathrm{Aut_r}(\mathbf{A}) \cap G_i g = \langle S\rangle \cap G_i g$.*

Proof. Let $a \in \langle S\rangle \cap G_i g$. Because $a \in \mathrm{Aut_r}(\mathbf{A})$, we have $\overrightarrow{G_i g\mathbf{A}} = \overrightarrow{G_i g a^{-1}\mathbf{A}} = \overrightarrow{G_i\mathbf{A}} \preceq \mathbf{A}$. Furthermore, let $a' \in \mathrm{Aut_r}(\mathbf{A}) \cap G_i g$ be arbitrary. Because $a'a^{-1} \in \mathrm{Aut_r}(\mathbf{A}) \cap G_i \le \langle S\rangle$ and $a \in \langle S\rangle$, we have $a' \in \langle S\rangle$. It follows that $\mathrm{Aut_r}(\mathbf{A}) \cap G_i g = \langle S\rangle \cap G_i g$. □

The second observation – implemented in line 26 – is as follows. Having traversed the subtree rooted at $G_{i+1}g_0$, where $g_0 \in G_i$, we record the image $g_0^{-1}(i)$ into the set U. In considering a subtree rooted at $G_{i+1}g$ with $g \in G_i$, we check whether there exists an automorphism $a \in \langle S\rangle \cap G_i$ taking the coset $G_{i+1}g_0$ into $G_{i+1}g$, that is, $G_{i+1}g_0 a^{-1} = G_{i+1}g$. If this is the case, then we can prune the subtree rooted at $G_{i+1}g$; this is justified by the following lemma. Provided that Θ is the orbit partition of $\langle S\rangle \cap G_i$ (which is the case with the function FIRST because $\langle S\rangle \le G_i$), then it is easy to see that a required automorphism a exists if and only if $\Theta(g^{-1}(i)) \cap U \neq \emptyset$.

Lemma 6.16. *Let $g_0 \in G_i$, $\overrightarrow{G_{i+1}g_0\mathbf{A}} \preceq \mathbf{A}$, and $\mathrm{Aut_r}(\mathbf{A}) \cap G_{i+1}g_0 = \langle S\rangle \cap G_{i+1}g_0$. If $g \in G_i$ such that there exists an $a \in \langle S\rangle \cap G_i$ with $ag^{-1}(i) = g_0^{-1}(i)$, then $\overrightarrow{G_{i+1}g\mathbf{A}} \preceq \mathbf{A}$, and $\mathrm{Aut_r}(\mathbf{A}) \cap G_{i+1}g = \langle S\rangle \cap G_{i+1}g$.*

Proof. We have $g_0 a g^{-1}(i) = i$ and $g, g_0, a \in G_i$. Thus, $G_{i+1}ga^{-1} = G_{i+1}g_0$. It follows that $\overrightarrow{G_{i+1}g\mathbf{A}} = \overrightarrow{G_{i+1}ga^{-1}\mathbf{A}} = \overrightarrow{G_{i+1}g_0\mathbf{A}} \preceq \mathbf{A}$. Let $a' \in \mathrm{Aut_r}(\mathbf{A}) \cap G_{i+1}g$. Because $a'a^{-1} \in \mathrm{Aut_r}(\mathbf{A}) \cap G_{i+1}g_0 = \langle S\rangle \cap G_{i+1}g_0$ and $a \in \langle S\rangle$, we have $a' \in \langle S\rangle$. Thus, $\langle S\rangle \cap G_{i+1}g = \mathrm{Aut_r}(\mathbf{A}) \cap G_{i+1}g$. □

If no pruning at all is performed, the algorithm will obviously find a counterexample if one exists. It is clear that the pruning based on lexicographic order in line 4 does not lead the search to miss an automorphism or a counterexample to canonicity. When pruning based on automorphisms is performed in lines 10 and 26, Lemmata 6.15 and 6.16 provide the inductive step to an induction on the number of pruning operations establishing that neither a counterexample nor a necessary generator for $\mathrm{Aut_r}(\mathbf{A})$ is missed. These observations are summarized in the next theorem.

Theorem 6.17. *The following two claims hold for Algorithm* 6.1:

1. *if $\mathbf{A}$ is not canonical, then the algorithm returns* FALSE,
2. *if $\mathbf{A}$ is canonical, then the algorithm returns* TRUE. *Furthermore, whenever the invocation* FIRST(i) *returns, S is a strong generating set for* $\mathrm{Aut_r}(\mathbf{A}) \cap G_i$ *relative to the base $(i, i+1, \ldots, s)$ with $|S| \le s - i$.*

To establish the inequality in the second claim of Theorem 6.17, observe that some cells in Θ merge whenever S is updated, and there can be at most $s - i$ such merges.

Automorphism pruning analogous to line 26 could also be performed within the function COSET by changing the base of $\langle S \rangle$ with permutation group techniques (see [528]), but this is worth the effort only for large automorphism groups. For isomorphism algorithms employing base change, see the concluding references in Sect. 5.6.

6.1.3 Classification Block by Block

Another possible way to approach the construction of BIBDs is to view a BIBD as a solution of the following Diophantine linear system of equations (cf. [222, 609]).

For parameters v, k, λ, let V be a set of v points, and let $\mathbf{A} = (a_{TK})$ be the 0-1 matrix whose rows and columns are indexed by the 2-subsets $T \subseteq V$ and the k-subsets $K \subseteq V$, respectively, such that

$$a_{TK} = \begin{cases} 1 & \text{if } T \subseteq K, \\ 0 & \text{if } T \not\subseteq K. \end{cases} \tag{6.13}$$

A solution $\mathbf{x}$ to the system

$$\mathbf{A}\mathbf{x} = \lambda \mathbf{1}, \quad x_K \in \{0, 1, \ldots, \lambda\} \text{ for all } K \subseteq V, \ |K| = k \tag{6.14}$$

clearly corresponds to a 2-(v, k, λ) design (represented as a set system over the point set V), where x_K specifies the multiplicity of the block K in the design. In particular, if $\mathbf{x}$ is a 0-1 vector, then the corresponding design is simple.

The construction of BIBDs block by block is equivalent to solving (6.14) by setting the value of one variable at a time. General methods for solving (6.14) are discussed in Chap. 5; we now consider tailored approaches for instances related to the construction of designs. Historically, block-by-block classification of designs can be traced back at least to Cole, Cummings, and White [129].

The system (6.14) is highly symmetric, so isomorph rejection is again required. To illustrate the symmetry, let $\mathbf{z}$ be a vector whose components are indexed by the 2-subsets or the k-subsets of V. We obtain a group action on the sets of all such vectors by letting $g \in \mathrm{Sym}(V)$ permute the components, that is, $g * \mathbf{z}$ is defined by $(g * z)_{g(W)} = z_W$ for all appropriate $W \subseteq V$. As an immediate consequence of (6.13), we obtain the following theorem.

Theorem 6.18. *The matrix* $\mathbf{A}$ *satisfies* $g(\mathbf{A}\mathbf{x}) = \mathbf{A}(g\mathbf{x})$ *for all* $g \in \mathrm{Sym}(V)$ *and all integer vectors* $\mathbf{x}$ *with the components indexed by the* k*-subsets of* V.

In particular, if $\mathbf{x}$ is a solution (or a partial solution) to (6.14), then so is $g\mathbf{x}$ for all $g \in \mathrm{Sym}(V)$.

In developing an isomorph rejection strategy it is more natural to work with a set system $\mathcal{X} = (V, \mathcal{B})$ whose blocks consist of k-subsets rather than a nonnegative integer vector $\mathbf{x}$ with components indexed by the k-subsets. Throughout this section we assume that all set systems have a fixed point set V. Isomorphism of set systems is defined by the induced action of $\mathrm{Sym}(V)$ permuting the points in the blocks.

The following general strategy for isomorph rejection has proven to be successful in many instances. First, a set of *seed* subsystems occurring in the designs is identified; it is required that every design to be classified contains at least one seed as a subsystem. Second, the seeds are classified up to isomorphism. Third, by extending the seeds in all possible ways and rejecting isomorphs, a classification of the designs is obtained.

Note that the concept of seeds is an abstract concept that can be used also for other approaches and other types of objects, when using the aforementioned three-step approach.

For two set systems, $\mathcal{S}$ and $\mathcal{X}$, we say that $\mathcal{X}$ *extends* $\mathcal{S}$ – or conversely, $\mathcal{S}$ *occurs* in $\mathcal{X}$ – if every block occurring in $\mathcal{S}$ occurs in $\mathcal{X}$ with greater or equal multiplicity. For two sets of set systems, $\mathscr{S}$ and $\mathscr{D}$, we say that $\mathscr{S}$ is a set of *seeds* for $\mathscr{D}$ if for every $\mathcal{X} \in \mathscr{D}$ there exists an $\mathcal{S} \in \mathscr{S}$ such that $\mathcal{X}$ extends a set system isomorphic to $\mathcal{S}$.

To study the search space for seed extension it is convenient to view a set system $\mathcal{X}$ constructed by extending a seed $\mathcal{S}$ as an ordered pair $(\mathcal{X}, \mathcal{S})$. Isomorphism of such ordered pairs – or augmentations, cf. Sect. 4.2.3 – is induced by the action of $\mathrm{Sym}(V)$ on set systems over V. In particular, $(\mathcal{X}_1, \mathcal{S}_1) \cong (\mathcal{X}_2, \mathcal{S}_2)$ indicates that there exists a $g \in \mathrm{Sym}(V)$ such that $g\mathcal{X}_1 = \mathcal{X}_2$ and $g\mathcal{S}_1 = \mathcal{S}_2$.

The practicality of the seed-based classification approach depends on a number of factors. To begin with, it must be feasible to classify the seeds up to isomorphism. This is a potentially difficult classification problem on its own. Furthermore, the structure of the seeds must be such that it is feasible to extend the seeds. There are two main points to consider from the perspective of isomorph rejection. First, the seeds should have a small automorphism group. The smaller automorphism group a seed has, the fewer isomorphic set systems are encountered during seed extension. This observation is justified by the following lemma, which follows immediately from the orbit-stabilizer theorem (Theorem 3.20).

Lemma 6.19. *Let $\mathcal{X}$ extend $\mathcal{S}$. Then, the number of pairs of the form $(\mathcal{X}', \mathcal{S})$ isomorphic to $(\mathcal{X}, \mathcal{S})$ is $|\mathrm{Aut}(\mathcal{S})|/|\mathrm{Aut}(\mathcal{X}, \mathcal{S})|$.*

Second, a design should not contain too many subsystems isomorphic to a seed, because the design is constructed as an extension of every automorphism orbit of such subsystems occurring in it. More precisely, for a given design $\mathcal{X}$, let $\mathcal{S}_1, \mathcal{S}_2, \ldots, \mathcal{S}_k$ consist of exactly one representative from each $\mathrm{Aut}(\mathcal{X})$-orbit of set systems occurring in $\mathcal{X}$ that are isomorphic to a seeds. Assuming that the seeds considered for extension are pairwise nonisomorphic, the total number of times a design isomorphic to $\mathcal{X}$ is generated is

$$\sum_{i=1}^{k} \frac{|\mathrm{Aut}(\mathcal{S}_i)|}{|\mathrm{Aut}(\mathcal{X}, \mathcal{S}_i)|}. \tag{6.15}$$

It is of course desirable to have a collection of seeds where this number is as small as possible for the majority of designs.

Finally, isomorphic designs must be eliminated from the output of the algorithm. This is straightforward if representatives of the designs encountered can be stored in memory; otherwise a more elaborate approach such as generation by canonical augmentation is required.

For BIBDs, a good set of seeds is often obtained by considering the blocks that have nonempty intersection with a small subset of points. We use as running examples the classification algorithms for Steiner triple systems of order 19 (2-$(19, 3, 1)$ designs) in [304] and projective planes of orders $n = 8, 9$ (2-$(73, 9, 1)$ and 2-$(91, 10, 1)$ designs) in [249, 346].

Example 6.20. Any block $\{x, y, z\}$ of an STS(19) has a nonempty intersection with $1 + 3(r - 1) = 25$ blocks. Disregarding the block $\{x, y, z\}$, the blocks incident with each of the points x, y, z form edge sets of three pairwise edge-disjoint 1-factors on the remaining 16 points. Thus, up to isomorphism such a set of 25 blocks can be described by an incidence matrix of the form in Fig. 6.4, where the matrices $\mathbf{F}_1$ and $\mathbf{F}_2$ specify two 1-factors.

1	11111111	00000000	00000000
1	00000000	11111111	00000000
1	00000000	00000000	11111111
0	10000000		
0	10000000		
0	01000000		
0	01000000		
0	00100000		
0	00100000		
0	00010000		
0	00010000	$\mathbf{F}_1$	$\mathbf{F}_2$
0	00001000		
0	00001000		
0	00000100		
0	00000100		
0	00000010		
0	00000010		
0	00000001		
0	00000001		

Fig. 6.4. Structure of seeds for STS(19)

There are up to isomorphism seven different ways to complete the matrix $\mathbf{F}_1$, corresponding to the seven partitions of 16 into even integers at least 4:

$$4+4+4+4, \quad 6+6+4, \quad 8+4+4, \quad 8+8, \quad 10+6, \quad 12+4, \quad 16.$$

A corresponding set of choices for $\mathbf{F}_1$ is depicted in Fig. 6.5.

10000000	10000000	10000000	10000000	10000000	10000000	10000000
01000000	01000000	01000000	01000000	01000000	01000000	01000000
10000000	10000000	10000000	10000000	10000000	10000000	10000000
01000000	01000000	01000000	01000000	00100000	00100000	00100000
00100000	00100000	00100000	00100000	01000000	01000000	01000000
00010000	00010000	00010000	00010000	00100000	00010000	00010000
00100000	00100000	00100000	00100000	00010000	00100000	00100000
00010000	00010000	00001000	00001000	00001000	00010000	00001000
00001000	00001000	00010000	00010000	00010000	00001000	00010000
00000100	00000100	00001000	00000100	00000100	00000100	00000100
00001000	00001000	00000100	00001000	00001000	00001000	00001000
00000100	00000010	00000010	00000010	00000010	00000010	00000010
00000010	00000100	00000100	00000100	00000100	00000100	00000100
00000001	00000001	00000001	00000001	00000001	00000001	00000001
00000010	00000010	00000010	00000010	00000010	00000010	00000010
00000001	00000001	00000001	00000001	00000001	00000001	00000001

Fig. 6.5. Possible choices for the $\mathbf{F}_1$ matrix in Fig. 6.4

To complete the matrix $\mathbf{F}_2$, a computer search is required. In this case it suffices (although this is clearly very inefficient) to perform an exhaustive search over all choices of $\mathbf{F}_1$ and all $16!/(2^8 \times 8!) = 2{,}027{,}025$ distinct 1-factors of K_{16}, with isomorph rejection via recorded objects. An alternative approach is to classify up to equivalence the equireplicate $(3, 16, 2)_8$ codes, whose equivalence classes are in a one-to-one correspondence with the isomorphism classes of sets of three pairwise edge-disjoint 1-factors of K_{16}.

Up to isomorphism there are 14,648 25-block seeds with the structure in Fig. 6.4. The order of the automorphism group for each of these seeds is displayed in Table 6.1.

As can be seen from Table 6.1, most of the seeds have a trivial or small automorphism group. Also, it is easy to check that two different blocks in an STS(v) cannot intersect the same set of blocks unless $v = 7$. Thus, every STS(19) contains 57 seeds, one for each block. These observations suggest that most isomorphism classes of STS(19) will not be encountered significantly more than 57 times with this collection of seeds, although in principle it could still be the case that most STS(19) arise as extensions of the seeds with a large automorphism group; in practice this does not happen. In fact, the total number of STS(19) that extend the 14,648 seeds is 710,930,186,096. Compared with the 11,084,874,829 isomorphism classes of STS(19), an isomorphism class of STS(19) is encountered about 64 times on the average.

Table 6.1. Automorphism group order for the 25-block seeds for STS(19)

$\lvert\mathrm{Aut}(\mathcal{S})\rvert$	Seeds	$\lvert\mathrm{Aut}(\mathcal{S})\rvert$	Seeds	$\lvert\mathrm{Aut}(\mathcal{S})\rvert$	Seeds
1	11,706	32	25	192	2
2	2,218	36	3	256	4
3	14	40	1	288	1
4	412	48	5	512	2
6	20	64	9	768	1
8	127	72	2	1,536	1
12	13	96	7	1,728	1
16	50	120	1	36,864	1
24	16	128	6		

Example 6.21. We will now discuss the type of seed is used in the classification of projective planes of orders $n = 8$ [249] and $n = 9$ [346].

A *triangle* in a projective plane is a set of three points not on a common line. A triangle induces a seed with the structure depicted in Fig. 6.6.

Each of the three pairs of distinct points in the triangle occurs on a unique line; these lines constitute the first three columns in Fig. 6.6. Because the lines of a projective plane have pairwise exactly one point in common, there are $3(n-1)$ additional lines incident with exactly one point of the triangle; these lines form the remaining columns in Fig. 6.6, which are partitioned into three groups based on their incidence with the triangle.

Because every pair of distinct points must occur on exactly one line, and every two lines must have exactly one point in common, it is not difficult to check that the matrix $\mathbf{T}$ must be a $(n-1)^2 \times 3(n-1)$ transposed incidence matrix of a transversal design $\mathrm{TD}(3, n-1)$. Up to isomorphism these are equivalent to the main classes of Latin squares of side $n-1$. By Table 8.2, there are 147 Latin squares of side 7 and 283,657 Latin squares of side 8, giving the number of such objects for $n = 8$ and $n = 9$, respectively.

Let us now proceed to discuss extension of classified seeds to designs. For a seed $\mathcal{S}$, let $\mathbf{s}$ be the corresponding integer vector with components indexed by the k-subsets $K \subseteq V$ such that s_K is the multiplicity of K in $\mathcal{S}$. Then, the designs $\mathcal{X}$ that extend $\mathcal{S}$ are in a one-to-one correspondence with the nonnegative integer solutions $\mathbf{y}$ to the system

$$\mathbf{A}\mathbf{y} = \lambda\mathbf{1} - \mathbf{A}\mathbf{s}. \tag{6.16}$$

The properties of the system (6.16) vary depending on the parameters v, k, λ and the choice of seeds. An effective algorithm is best determined by experimentation and by studying the instances at hand.

	$n-1$	n 1	n 1
011	111 1	000 0	000 0
101	000 ··· 0	111 ··· 1	000 ··· 0
110	000 0	000 0	111 1
100	100 0	000 0	000 0
100	010 0	000 ··· 0	000 ··· 0
100	001 0	000 0	000 0
⋮	⋱	⋮ ⋮	⋮ ⋮
100	000 1	000 ··· 0	000 ··· 0
010	000 0	100 0	000 0
010	000 ··· 0	010 0	000 ··· 0
010	000 0	001 0	000 0
⋮	⋮ ⋮	⋱	⋮ ⋮
010	000 ··· 0	000 1	000 ··· 0
001	000 0	000 0	100 0
001	000 ··· 0	000 ··· 0	010 0
001	000 0	000 0	001 0
⋮	⋮ ⋮	⋮ ⋮	⋱
001	000 ··· 0	000 ··· 0	000 1
000			
000			
000		**T**	
⋮			
000			

Fig. 6.6. Structure of triangle-based seeds for projective planes of order n

Example 6.22. The extension problems for STS(19) originating from the 25-block seeds in Example 6.20 can be efficiently solved by formulating (6.16) as an instance of EXACT COVERS – see Sects. 5.2 and 5.4 – and applying Algorithm 5.2. See [304].

Example 6.23. The extension problems for projective planes of order $n = 8$ and $n = 9$ relative to the triangle-based seeds in Example 6.21 can be solved as follows. First, working with the entire system (6.16) is not practical because $\binom{v}{k} = \binom{n^2+n+1}{n+1}$. To reduce the size of the system, observe that every line in a projective plane has exactly one point in common with every other line. Thus, it suffices to consider only candidate lines that satisfy this condition relative to lines in the seed. In fact, it is easy to check that the problem of generating the candidate lines is an instance of EXACT COVERS, where the task is to cover the lines in the seed using points occurring in at most n lines of the seed. Each such cover consisting of $n + 1$ points is a candidate line.

It remains to find sets of candidate lines that are compatible with each other so as to complete the seed to a projective plane. This is obviously an instance of CLIQUES, where any two candidate lines are connected by an edge if and only if they have exactly one point in common; cf. [346, 350, 572].

Compared with CLIQUES, a more efficient approach is obtained by completing the points to full incidence (point occurs in $n+1$ lines) one at a time using backtrack search and EXACT COVERS as follows. Disregarding the points already at full incidence, let x be a point incident to the maximum number of lines. Apply EXACT COVERS to find all the candidate lines through x. Next, apply EXACT COVERS to find all sets of candidate lines that – together with the existing lines through x – cover all the other points once.

6.1.4 Isomorph Rejection for Designs Extending a Seed

Provided that the number of nonisomorphic designs is not too large – such as with projective planes of order 8 and 9 – and evaluating a certificate for the generated designs is not too expensive, isomorph rejection via recorded objects suffices. If this is not the case – such as in the case of STS(19) – more advanced isomorph rejection techniques must be employed for the generated designs.

The seed-based approach is well-suited for generation by canonical augmentation, assuming that it is easy to identify from a given design $\mathcal{X}$ all the subsystems in $\mathcal{X}$ that are isomorphic to seeds.

To apply generation by canonical augmentation, the seed-based construction of designs can be described by the following search tree. The tree consists of three levels of nodes: the root node, the set systems isomorphic to seeds, and the designs. The root node has as children all the set systems isomorphic to seeds, and the children of such a set system $\mathcal{S}$ are all the designs that extend $\mathcal{S}$.

In addition to the search tree, we require a canonical parent function m that associates with every design $\mathcal{X}$ a subsystem $m(\mathcal{X})$ occurring in $\mathcal{X}$. It is required that $m(\mathcal{X})$ is isomorphic to a seed, and, in conformance with (4.12), that

> for all designs $\mathcal{X}, \mathcal{Y}$ it holds that $\mathcal{X} \cong \mathcal{Y}$ implies $(\mathcal{X}, m(\mathcal{X})) \cong (\mathcal{Y}, m(\mathcal{Y}))$.

Using the subobject framework in Sect. 4.2.3, a canonical parent function m can be constructed with the help of a canonical labeling map κ for the induced action of $\mathrm{Sym}(V)$ on set systems with point set V. For a design $\mathcal{X}$, let $\hat{\mathcal{X}} = \kappa(\mathcal{X})\mathcal{X}$ and select a set system $\hat{\mathcal{S}}$ occurring in $\hat{\mathcal{X}}$ such that $\hat{\mathcal{S}}$ is isomorphic to a seed, the selection depending only on the canonical form $\hat{\mathcal{X}}$. Then, $m(\mathcal{X}) = \kappa(\mathcal{X})^{-1}\hat{\mathcal{S}}$ defines a canonical parent function.

Example 6.24. In the case of STS(19) and the seeds from Example 6.20, selecting a set system $\hat{\mathcal{S}}$ isomorphic to a seed amounts to selecting a block $\hat{B}_m$ from $\hat{\mathcal{X}}$, and letting $\hat{\mathcal{S}}$ consist of all blocks in $\hat{\mathcal{X}}$ intersecting $\hat{B}_m$.

Let us sketch a proof that the search tree and the canonical parent function satisfy the requirements of generation by canonical augmentation in Sect. 4.2.3. By construction, every design $\mathcal{X}$ extends its canonical parent $m(\mathcal{X})$, which is isomorphic to a seed and hence occurs in the search tree. Thus, (4.16) holds. To verify (4.15) it suffices to consider only set systems $\mathcal{S}, \mathcal{S}'$ isomorphic to seeds. Suppose $\mathcal{S} \cong \mathcal{S}'$ and that $\mathcal{X}$ is a design that extends $\mathcal{S}$. Let $g \in G$ such that $g\mathcal{S} = \mathcal{S}'$. Then, $g\mathcal{X}$ extends $\mathcal{S}'$ and $g(\mathcal{X}, \mathcal{S}) = (g\mathcal{X}, \mathcal{S}')$. This shows that (4.15) holds.

Assuming that the seeds considered for extension are pairwise nonisomorphic, we can now apply generation by canonical augmentation. According to the framework, the parent test (4.13) rejects a design $\mathcal{X}$ unless $(\mathcal{X}, m(\mathcal{X})) \cong (\mathcal{X}, p(\mathcal{X}))$, that is, $m(\mathcal{X}) \cong_{\mathrm{Aut}(\mathcal{X})} p(\mathcal{X})$.

For performance reasons the parent test is often carried out in a different representation than the direct set system representation.

Example 6.25. To implement the test $m(\mathcal{X}) \cong_{\mathrm{Aut}(\mathcal{X})} p(\mathcal{X})$ for STS(19) and the seeds in Example 6.20, it is useful to observe that it suffices to consider the blocks B_m and B_p in $\mathcal{X}$ that define $m(\mathcal{X})$ and $p(\mathcal{X})$, respectively. An equivalent test is then $B_m \cong_{\mathrm{Aut}(\mathcal{X})} B_p$.

Because an STS(19) is strongly reconstructible from its line graph (Corollary 3.94), the test $B_m \cong_{\mathrm{Aut}(\mathcal{X})} B_p$ is equivalent to testing whether the vertices corresponding to B_m and B_p are in the same automorphism orbit of the line graph of $\mathcal{X}$. This approach is employed in [304].

A subobject invariant can be used to further improve the performance.

Example 6.26. For STS(19) and the seeds in Example 6.20, a significant performance improvement is obtainable through the following block invariant [304]. Associate with every block B in $\mathcal{X}$ the number $P(\mathcal{X}, B)$ of Pasch configurations in which B occurs in $\mathcal{X}$.

Now, in selecting the block $\hat{B}_m$ in Example 6.24, we always select $\hat{B}_m$ so that it has the maximum invariant value among the blocks in $\hat{\mathcal{X}}$. Because P is a block invariant, it follows that B_m has the maximum invariant value among the blocks in $\mathcal{X}$. Thus, we have $B_p \cong_{\mathrm{Aut}(\mathcal{X})} B_m$ only if B_p has the maximum invariant value among the blocks in $\mathcal{X}$, so we can reject $\mathcal{X}$ if B_p does not have the maximum invariant value. Similarly, if B_p is the unique block with the maximum invariant value, then we can accept $\mathcal{X}$ based on the invariant values only.

A further performance gain is obtained from the fact that the ordered partition of blocks defined by the invariant values can be used to expedite the evaluation of a canonical labeling for the line graph in the case it is necessary to compute B_m explicitly – see Lemma 3.143.

If $\mathcal{X}$ is accepted in the parent test, then a second test is required to reject possible remaining isomorphs. Assuming the seeds considered for extension are pairwise nonisomorphic, it follows from the general theory in Sect. 4.2.3 that any two isomorphic designs $\mathcal{X}, \mathcal{Y}$ accepted in the parent test must both extend the same seed $\mathcal{S} = p(\mathcal{X}) = p(\mathcal{Y})$. Furthermore, there exists a $g \in \text{Aut}(\mathcal{S})$ such that $g\mathcal{X} = \mathcal{Y}$. Thus, in the second test it suffices to reject Aut($\mathcal{S}$)-isomorphs only, and no further testing is required if Aut($\mathcal{S}$) is trivial.

A canonical labeling map $\kappa_{\text{Aut}(\mathcal{S})}$ for the action of Aut($\mathcal{S}$) on set systems can be used to reject Aut($\mathcal{S}$)-isomorphs. Namely, for every design $\mathcal{X}$ that is accepted in the parent test, it holds that all designs in the Aut($\mathcal{S}$)-orbit of $\mathcal{X}$ are generated as extensions of $\mathcal{S}$ and accepted in the parent test. Thus, rejection of Aut($\mathcal{S}$)-isomorphs is achieved by rejecting a generated design $\mathcal{X}$ unless $\mathcal{X} = \kappa_{\text{Aut}(\mathcal{S})}(\mathcal{X})\mathcal{X}$.

Example 6.27. For STS(19), most of the seeds in Example 6.20 have a small automorphism group. For such seeds $\mathcal{S}$, a canonical labeling map is easily obtained by exhaustive search on Aut($\mathcal{S}$). For example, we can set $\kappa_{\text{Aut}(\mathcal{S})}(\mathcal{X})$ equal to the first group element $g \in \text{Aut}(\mathcal{S})$ encountered that takes $\mathcal{X}$ to the lexicographically minimum set system in its Aut($\mathcal{S}$)-orbit. This is the approach employed in [304] for seeds with a small automorphism group.

Alternatively, a recorded certificates of designs can be used for rejecting isomorphs that are accepted in the parent test. The disadvantage with this approach is that the certificates need to be stored in memory.

Example 6.28. In [304], a hash table is used to reject isomorphs among designs accepted in the parent test for the 11 seeds whose automorphism group order exceeds 200. For these seeds, the maximum number of certificates that has to be stored in a hash table is 100,813.

6.1.5 Tailored Approaches

We now proceed to discuss some of the more tailored classification approaches for BIBDs that have been used in the literature.

In terms of the extent of the combinatorial analysis required, the most elaborate classification result for BIBDs to date is the proof of nonexistence of projective planes of order 10, to be discussed in Chap. 12.

A family of designs that has attracted a lot of interest is the family of 2-$(22, 8, 4)$ designs, where existence – conjecturally, nonexistence – is still open. Studies focusing on these designs include [29, 48, 246, 418, 419, 457]; see also the survey article [503]. Of the tailored approaches developed in this setting, at least two approaches should be mentioned because of their potentially wider applicability.

The first approach is based on studying the binary linear code generated by the rows of an incidence matrix – this approach is discussed in detail in Chap. 12, so we will not go into details here. In the context of 2-$(22, 8, 4)$ designs, the fundamental idea in [48] is to first classify a set of doubly-even self-orthogonal $[33, 16, 4]_2$ codes (see Sect. 7.3.6) that must contain the rows of a putative incidence matrix as weight $r = 12$ codewords if a 2-$(22, 8, 4)$ design is to exist. Then such codes are eliminated one at a time using combinatorial arguments and computer search.

The second approach, described in [418] and employed in [418, 419], is based on step by step refinement of pd-systems. Perhaps the best intuition of a pd-system is obtained by thinking of an incidence matrix of a design where the operation of aggregating a set of rows into one row consisting of their sum has been applied zero or more times. Classification via pd-systems proceeds in the reverse direction whereby aggregated rows are refined step by step.

Definition 6.29. *A* pd-system *of a BIBD with parameters v, k, λ, b, r is an ordered pair $(\mathbf{s}, \mathbf{T})$, where $\mathbf{s} = (s_i)$ is an $m \times 1$ positive integer vector with $\sum_{i=1}^{m} s_i = v$ and $\mathbf{T} = (t_{ij})$ is an $m \times b$ matrix with nonnegative integer entries satisfying the following five properties:*

1. *$t_{ij} \le s_i$ for all $1 \le i \le m$ and $1 \le j \le b$,*
2. *$\sum_{i=1}^{m} t_{ij} = k$ for all $1 \le j \le b$,*
3. *$\sum_{j=1}^{b} t_{ij} = s_i r$ for all $1 \le i \le m$,*
4. *$\sum_{j=1}^{b} t_{i_1 j} t_{i_2 j} = \lambda s_{i_1} s_{i_2}$ for all $1 \le i_1 < i_2 \le m$,*
5. *$\sum_{j=1}^{b} \binom{t_{ij}}{2} = \lambda \binom{s_i}{2}$ for all $1 \le i \le m$.*

The intuition here is that s_i specifies the number of rows of a putative incidence matrix of a design aggregated to obtain row i of $\mathbf{T}$. For $m = v$ the matrix $\mathbf{T}$ is an incidence matrix of a BIBD by (6.1).

A pd-system $(\mathbf{s}_1, \mathbf{T}_1)$ is *finer* than $(\mathbf{s}_2, \mathbf{T}_2)$ if $(\mathbf{s}_2, \mathbf{T}_2)$ can be obtained from $(\mathbf{s}_1, \mathbf{T}_1)$ by aggregating rows. Two pd-systems are *isomorphic* if $\mathbf{T}_1$ can be transformed into $\mathbf{T}_2$ by permuting the rows and columns so that the permutation of the rows transforms $\mathbf{s}_1$ simultaneously into $\mathbf{s}_2$.

Example 6.30. A sequence of progressively finer pd-systems leading to a Fano plane. The vertical line separates the $\mathbf{s}$ vector from the matrix $\mathbf{T}$.

$$
\left[\begin{array}{c|c} 1 & 1110000 \\ 3 & 2102121 \\ 3 & 0121212 \end{array}\right],
\left[\begin{array}{c|c} 1 & 1110000 \\ 1 & 1001100 \\ 2 & 1101021 \\ 3 & 0121212 \end{array}\right],
\left[\begin{array}{c|c} 1 & 1110000 \\ 1 & 1001100 \\ 1 & 1000011 \\ 1 & 0101010 \\ 3 & 0121212 \end{array}\right],
\left[\begin{array}{c|c} 1 & 1110000 \\ 1 & 1001100 \\ 1 & 1000011 \\ 1 & 0101010 \\ 1 & 0100101 \\ 2 & 0021111 \end{array}\right],
\left[\begin{array}{c|c} 1 & 1110000 \\ 1 & 1001100 \\ 1 & 1000011 \\ 1 & 0101010 \\ 1 & 0100101 \\ 1 & 0011001 \\ 1 & 0010110 \end{array}\right]
$$

A classification approach relying on pd-systems begins with a set of starter pd-systems classified up to isomorphism, and proceeds to refine these into an incidence matrix of a BIBD. The task of refining one aggregate row ($s_i > 1$) into a singleton row ($s'_i = 1$) and an aggregate row ($s'_{i+1} = s_i - 1$) in all possible ways is an instance of DIOPHANTINE, where the variables are the 0-1 entries in the singleton row. Isomorphism computations can be carried out by transformation into a colored graph or by a tailored approach. For details in the context of 2-$(22, 8, 4)$ designs, see [418, 419].

One approach for restricting the search space of an algorithm is to carry out an analysis of small configurations in a design and dichotomise the search space based on the presence or absence of the particular configurations. An approach of this type is employed to settle the nonexistence of 2-$(46, 6, 1)$ designs in [271].

In some cases a clique search used to complete a design can be tailored by taking into account the properties of the canonical designs. The classification of 2-$(25, 4, 1)$ designs is carried out in this manner [549]. In [545] a hybrid of the point-by-point and block-by-block classification approaches is used to classify the 2-$(31, 10, 3)$ designs. Due to improvements in clique algorithms and computer performance, the latter classification can currently be carried out by relying only on the basic point-by-point algorithm combined with clique searching.

Also results relating specific parameter families and different types of objects – cf. Theorems 2.120 and 2.127 – can be used to arrive at a classification. For example, from the projective planes of order n we obtain a classification of the affine planes of order n by taking the residual design with respect to every automorphism orbit of blocks in each isomorphism class of projective planes of order n (Example 3.80). More generally, given a family of square designs whose associated quasi-residual designs are embeddable, a classification of the quasi-residual designs can be obtained by taking the residual with respect to every block in every isomorphism class of square designs, and rejecting isomorphs. For example, by Theorem 2.54 the 2-$(28, 7, 2)$ quasi-residual designs can be classified in this manner based on a classification of the 2-$(37, 9, 2)$ square designs. In an analogous manner one can obtain from a classification of the Hadamard matrices of order $4n$ a classification of the objects listed in Theorem 2.127, for example, the 2-$(4n-1, 2n-1, n-1)$ square designs (Example 3.81).

6.1.6 Results

The standard reference for tables of classified BIBDs is [407], whose earlier versions are [405] and [406]. The entries in those tables are ordered lexicographically by r, k, and λ (in this order), which effectively gives parameter sets for which classification is feasible an early position in the total order imposed. However, for the sake of readability, we have chosen a different order in collecting the results in Tables 6.2 to 6.10. There is one table for each value of k, and the rows are ordered lexicographically by λ and v.

The entries of the tables contain the five central parameters of a design, the number of nonisomorphic designs (N), a reference to the result, and finally its entry number in the table of [407]. For each pair of k and λ, we stop at the smallest v for which there is no complete classification and indicate this situation by omitting the N value. An exception to this rule is that such an entry is completely omitted if there is an open case listed with the same k, smaller λ, and smaller or equal v. As in [407], we stop at $r = 41$ at the latest.

For $k > 11$, the only known classification result – apart from nonexistence results that follow from Theorems 2.47 and 2.54 – is that there are 208,310 2-$(27, 13, 6)$ Hadamard designs [547].

The following conventions are adhered to throughout the book. We have tried to trace the original reference for each result. If more than one reference is given, this indicates either that the result has been obtained in independent studies – then the form is [A], [B] – or that a result builds on several papers, possibly including corrigenda, shown by [A,B]. However, if a result builds on a corpus of papers by the same researcher or team, like for 2-$(111, 11, 1)$ designs, only the last paper in the series is referenced. If a reference in Tables 6.2 to 6.10 is [407], then the result is unpublished and the originator is listed within parentheses; throughout this book, results that are not obtained by the cited authors are presented in the same way. The reference [H] means that the result has been obtained while preparing this book.

Obviously, it is not always clear whether a result should be interpreted as a reinvention (when the author should have been able to find out that the result is not novel). The importance of reproved results should, however, not be underestimated as they strengthen the confidence in correctness of the original results. The following list is an extraction of (not too old) references containing among other things either such classification results or related surveys: [36, 80, 152, 204, 225, 226, 227, 229, 244, 278, 298, 404, 405, 406, 407, 480, 556]. Several of these studies also contain results on t-designs with $t \geq 3$.

A formula for determining the number of 2-$(6, 3, \lambda)$ designs (cf. Table 6.2) has been obtained by Mathon [400, 404]. Classification results for affine planes can be obtained from classified projective planes by Theorem 2.120 and the observation in Example 3.80 (uniqueness of the projective planes of orders 2, 3, and 4 is easily proved by hand; no references are given for these cases). In general, classification results for quasi-residual designs with $\lambda = 1$ or $\lambda = 2$ can be produced by Theorem 2.54. References to the literature are omitted when the Bruck–Ryser–Chowla theorem (Theorem 2.47) can be applied, and when Theorem 2.54 or 2.120 is applicable to settled cases.

Hadamard designs are indicated by HD; the reference(s) listed for these occasionally address some of the other families in Theorem 2.127.

Some classification results related to simple BIBDs have been tabulated in [227].

Table 6.2. Classification of 2-$(v, 3, \lambda)$ designs

v	k	λ	b	r	N	References	No
7	3	1	7	3	1	HD	1
9	3	1	12	4	1	Theorem 2.120	2
13	3	1	26	6	2	[143], [76]	8
15	3	1	35	7	80	[129]	14
19	3	1	57	9	11,084,874,829	[304]	29
21	3	1	70	10			42
6	3	2	10	5	1	Theorem 2.54	4
7	3	2	14	6	4	[439]	9
9	3	2	24	8	36	[434, 403]	21
10	3	2	30	9	960	[115], [278]	30
12	3	2	44	11	242,995,846	[452]	55
13	3	2	52	12			65
7	3	3	21	9	10	[434]	31
9	3	3	36	12	22,521	[402]	66
11	3	3	55	15			116
6	3	4	20	10	4	[228]	43
7	3	4	28	12	35	[228]	67
9	3	4	48	16	16,585,031	[407] (Spence)	145
10	3	4	60	18			190
7	3	5	35	15	109	[278, 407] (Spence)	117
9	3	5	60	20	5,862,121,434	[463]	235
6	3	6	30	15	6	[228]	118
7	3	6	42	18	418	[278, 481]	191
8	3	6	56	21	3,077,244	[407] (Spence)	275
9	3	6	72	24			356
7	3	7	49	21	1,508	[481]	276
6	3	8	40	20	13	[278], [404, 400]	236
7	3	8	56	24	5,413	[481]	357
7	3	9	63	27	17,785	[481]	477
6	3	10	50	25	19	[404, 400]	409
7	3	10	70	30	54,613	[407] (Mathon & Pietsch)	595
7	3	11	77	33	155,118	[407] (Mathon & Pietsch)	735
6	3	12	60	30	34	[404, 400]	596
7	3	12	84	36	412,991	[407] (Mathon & Pietsch)	881
7	3	13	91	39	1,033,129	[407, Addendum] (Mathon & Pietsch)	1030
6	3	14	70	35	48	[404, 400]	816
6	3	16	80	40	76	[404, 400]	1078

Table 6.3. Classification of 2-$(v, 4, \lambda)$ designs

v	k	λ	b	r	N	References	No
13	4	1	13	4	1		3
16	4	1	20	5	1	Theorem 2.120	5
25	4	1	50	8	18	[549]	22
28	4	1	63	9			32
10	4	2	15	6	3	Theorem 2.54	10
13	4	2	26	8	2,461	[480]	23
16	4	2	40	10			44
8	4	3	14	7	4	[439]	15
9	4	3	18	8	11	[206], [557], [375], [66]	24
12	4	3	33	11			56
10	4	4	30	12	13,769,944	[154]	71
8	4	6	28	14	2,310	[480]	101
9	4	6	36	16	270,474,142	[454]	150
10	4	6	45	18			193
8	4	9	42	21	8,360,901	[153]	278
9	4	9	54	24			364
8	4	12	56	28			524

Table 6.4. Classification of 2-$(v, 5, \lambda)$ designs

v	k	λ	b	r	N	References	No
21	5	1	21	5	1		6
25	5	1	30	6	1	Theorem 2.120	11
41	5	1	82	10			45
11	5	2	11	5	1	[396], [574], HD	7
15	5	2	21	7	0	Theorem 2.54	16
21	5	2	42	10			46
10	5	4	18	9	21	[206], [375]	33
11	5	4	22	10	4,393	[407] (Breach)	47
15	5	4	42	14			102
13	5	5	39	15			124
11	5	6	33	15			125
10	5	8	36	18			195

Table 6.5. Classification of 2-$(v, 6, \lambda)$ designs

v	k	λ	b	r	N	References	No
31	6	1	31	6	1	[386]	12
36	6	1	42	7	0	Theorem 2.54	17
46	6	1	69	9	0	[271]	34
51	6	1	85	10			48
16	6	2	16	6	3	[276]	13
21	6	2	28	8	0	Theorem 2.54	25
31	6	2	62	12			74
16	6	3	24	9	18,920	[407] (Spence)	35
21	6	3	42	12			75
16	6	4	32	12			76
12	6	5	22	11	11,603	[407] (Pietsch)	58
13	6	5	26	12	19,072,802	[303]	77
15	6	5	35	14			104
12	6	10	44	22			319

Table 6.6. Classification of 2-$(v, 7, \lambda)$ designs

v	k	λ	b	r	N	References	No
43	7	1	43	7	0	Theorem 2.47	18
49	7	1	56	8	1	Theorem 2.120	26
85	7	1	170	14			105
22	7	2	22	7	0	Theorem 2.47	19
28	7	2	36	9	8	Theorem 2.54	36
43	7	2	86	14			106
15	7	3	15	7	5	[574], HD	20
21	7	3	30	10	3,809	[545]	49
29	7	3	58	14			107
22	7	4	44	14			108
14	7	6	26	13	15,111,019	[303]	89
15	7	6	30	14			109

Table 6.7. Classification of 2-$(v, 8, \lambda)$ designs

v	k	λ	b	r	N	References	No
57	8	1	57	8	1	[479, 248]	27
64	8	1	72	9	1	Theorem 2.120	37
113	8	1	226	16			155
29	8	2	29	8	0	Theorem 2.47	28
36	8	2	45	10	0	Theorem 2.54	50
57	8	2	114	16			156
22	8	4	33	12			78
16	8	7	30	15			130

Table 6.8. Classification of 2-$(v, 9, \lambda)$ designs

v	k	λ	b	r	N	References	No
73	9	1	73	9	1	[249]	38
81	9	1	90	10	7	Theorem 2.120	51
145	9	1	290	18			203
37	9	2	37	9	4	[515]	39
45	9	2	55	11			59
25	9	3	25	9	78	[151]	40
33	9	3	44	12			79
19	9	4	19	9	6	[242], [45], HD	41
27	9	4	39	13			90
21	9	6	35	15			131
18	9	8	34	17			179

Table 6.9. Classification of 2-$(v, 10, \lambda)$ designs.

v	k	λ	b	r	N	References	No
91	10	1	91	10	4	[346]	52
100	10	1	110	11	0	Theorem 2.120	60
136	10	1	204	15			132
46	10	2	46	10	0	Theorem 2.47	53
55	10	2	66	12	0	Theorem 2.54	80
91	10	2	182	20			254
31	10	3	31	10	151	[545]	54
40	10	3	52	13			91
28	10	5	42	15			134
25	10	6	40	16			160
20	10	9	38	19			224

Table 6.10. Classification of 2-$(v, 11, \lambda)$ designs

v	k	λ	b	r	N	References	No
111	11	1	111	11	0	[351]	61
121	11	1	132	12			81
56	11	2	56	11			62
66	11	2	78	13			92
23	11	5	11	5	1,106	[277, 316], HD	63
33	11	5	48	16			161
22	11	10	42	21			293

6.2 t-Designs

The classification techniques for BIBDs readily extend to t-designs with $t \geq 3$. As for BIBDs, the two basic techniques are point-by-point and block-by-block classification. Compared with BIBDs, however, for $t \geq 3$ the number of blocks of a design grows quite rapidly as the other parameters are increased, and not too many complete classification results are known. In many cases a classification relies extensively on a careful combinatorial study of the designs to restrict the search. An excellent example in this respect is the nonexistence result for 4-$(12, 6, 6)$ designs [417], which relies on properties derived in [338]. Note that a classification of Hadamard 3-designs can be obtained from a classification of Hadamard matrices, which is the topic of Sect. 8.2.

6.2.1 Classification Point by Point

Pursuing an analogy with the BIBD case, t-(v, k, λ) designs with $t \geq 3$ can be constructed point by point using the same basic setting as in the BIBD case. That is, incidence matrix representation the designs are constructed in one row at a time.

For $t \geq 3$, the essential difference compared with the BIBD case lies in the problem of finding rows that augment a given partial incidence matrix. Nevertheless, the row generation problem remains an instance of DIOPHANTINE, which we proceed to derive.

By Theorem 2.38, every t-(v, k, λ) design satisfies the property that every ℓ-subset of points, $\ell = 0, 1, \ldots, t$, is incident to exactly b_ℓ blocks, where

$$b_\ell = \lambda \binom{v-\ell}{t-\ell} \Big/ \binom{k-\ell}{t-\ell}.$$

In particular, the number of blocks is $b = b_0$.

Looking at a partial $v' \times b$ incidence matrix $\mathbf{N} = (n_{ij})$ of a t-(v, k, λ) design, $v' < v$, the task of finding rows that extend $\mathbf{N}$ corresponds to finding 0-1 vectors $\mathbf{y}$ that are solutions to the following system of equations. For every $\ell = 1, 2, \ldots, \min\{t, v' + 1\}$ and every $(\ell - 1)$-subset $\{i_1, i_2, \ldots, i_{\ell-1}\} \subseteq \{1, 2, \ldots, v'\}$, the system contains the equation

$$\sum_{j=1}^{b} \Big(\prod_{u=1}^{\ell-1} n_{i_u,j}\Big) y_j = b_\ell, \tag{6.17}$$

where it is assumed that $\prod_{u=1}^{\ell-1} n_{i_u,j} = 1$ for $\ell = 1$. In addition to these equations, the row generation problem has the constraint forcing $y_j = 0$ whenever $\sum_{i=1}^{v'} n_{ij} = k$. We assume that variables forced to 0 are removed from the equation system.

The system (6.17) is linear in the variables y_j, so we have an instance of DIOPHANTINE. Furthermore, for $t = 2$ the system reduces to the system with $\mathbf{N}\mathbf{y} = \lambda\mathbf{1}$ and $\mathbf{1}^T\mathbf{y} = r\mathbf{1}$ encountered in the BIBD case.

Example 6.31. For the parameters 3-(8, 4, 1), a partial 3×14 incidence matrix is

$$\mathbf{N} = \begin{bmatrix} 11111110000000 \\ 11100001111000 \\ 10011001100110 \end{bmatrix}.$$

The resulting system of equations consists of seven equations in three groups based on the value of ℓ.

For $\ell = 3$ we obtain the equations

$$\begin{aligned} y_1 + y_2 + y_3 &&&= 1, \\ y_1 &+ y_4 + y_5 &&= 1, \\ y_1 &&+ y_8 + y_9 &= 1. \end{aligned}$$

For $\ell = 2$ we obtain the equations

$$\begin{aligned} y_1 + y_2 + y_3 + y_4 + y_5 + y_6 + y_7 &&&= 3, \\ y_1 + y_2 + y_3 && + y_8 + y_9 + y_{10} + y_{11} &= 3, \\ y_1 \quad + y_4 + y_5 && + y_8 + y_9 \quad + y_{12} + y_{13} &= 3. \end{aligned}$$

For $\ell = 1$ we obtain the equation

$$y_1 + y_2 + \cdots + y_{14} = 7.$$

One 0-1 solution of this system of equations is

$$\mathbf{y} = (0, 1, 0, 1, 0, 1, 0, 1, 0, 1, 0, 1, 0, 1)^T.$$

If the $\ell = 3$ equations are removed, then we have a row generation problem for a partial 2-(8, 4, 3) design.

Observe that the equations with $\ell < t$ are useful in restricting the number of partial incidence matrices that need to be considered, but redundant in the sense that all $v \times b$ matrices obtained are incidence matrices of t-(v, k, λ) designs even if only equations with $\ell = t$ are employed. In general, the more equations we have, the more constrained the search becomes, but at the cost of keeping track of the equations during the backtrack search.

A basic technique from Sect. 6.1.1 that also applies to the case $t \geq 3$ is that variables associated with identical columns of $\mathbf{N}$ can be merged into a single variable, thus reducing the number of variables in the system. Also the order-based constraints (6.6), (6.7), and (6.11) can be used in row generation, provided that orderly generation based on lexicographic order is used for isomorph rejection.

Possibilities for isomorph rejection include orderly generation similar to the BIBD approach in Sect. 6.1.1 (cf. [153, 154]), or generation by canonical augmentation as employed in [417].

We would like to emphasize, however, that the general-purpose techniques outlined here are not in general sufficient to yield practical classification algorithms for t-designs with larger parameters. Typically a detailed combinatorial study of the target designs is required to restrict the number of partial incidence matrices that need to be traversed.

The nonexistence result for 4-$(12, 6, 6)$ designs [417] is perhaps the most advanced application of a point-by-point classification approach for t-designs to date. The nonexistence is established by three independent techniques, two of which rely on point-by-point classification. All of the techniques depend on a detailed combinatorial study [338] of the parameters 4-$(12, 6, 6)$ and related simple 5-$(12, 6, 3)$ designs. The first technique establishes nonexistence by direct point-by-point construction for the parameters 4-$(12, 6, 6)$. The second technique relies on a point-by-point classification of the simple 5-$(12, 6, 3)$ designs. In both cases the properties derived in [338] are employed extensively to prune the search. (For an illustration of the magnitude of an unrestricted search, observe that for the parameters 4-$(12, 6, 6)$ and 5-$(12, 6, 3)$, the number of blocks is 198 and 396, respectively.)

Studies of t-designs where a point-by-point approach is employed include [153, 154, 278, 417]. With the exception of [417], point-by-point classification is applied only in the case $t \leq 3$ in these studies.

6.2.2 Classification Block by Block

The block-by-block classification approach for BIBDs in Sect. 6.1.3 extends to $t \geq 3$ by requiring that the matrix $\mathbf{A} = (a_{TK})$ be defined for t-subsets $T \subseteq V$ instead of 2-subsets. In accordance with Sect. 6.1.3, we assume in this section that designs are represented as set systems.

A natural family of seeds for a block-by-block approach are the derived designs of a t-design.

Definition 6.32. *Let $\mathcal{X} = (V, \mathcal{B})$ be a t-(v, k, λ) design and let $D \subseteq V$ with $1 \leq |D| = \ell < t$. The* derived design *of $\mathcal{X}$ with respect to D has $V \setminus D$ as the point set, and the blocks consist of all the blocks in $\mathcal{X}$ that intersect D, with the points in D deleted.*

It is immediate that a derived design is a $(t-\ell)$-$(v-\ell, k-\ell, \lambda)$ design, but in general not all designs with these parameters appear as derived designs of a t-(v, k, λ) design. We remark that the term derived design in this context is somewhat unfortunate due to the conflict with Definition 2.46, but the term is too well-established to be changed.

Example 6.33. Figure 6.7 shows an incidence matrix of a 3-$(10, 4, 1)$ design and a derived 2-$(9, 3, 1)$ design in the upper left submatrix.

111100000000	111111110000000000
100011100000	100001101111100000
100000011100	011100001011011000
010010010010	110010000110010110
010001001001	001001010101010101
001010000101	000111001000110011
001000101010	010000111100001011
000101000110	001010100010101101
000100110001	100100010001101110
111111111111	000000000000000000

Fig. 6.7. A 3-$(10, 4, 1)$ design with one 2-$(9, 3, 1)$ derived design individualized

An approach for block-by-block classification of t-(v, k, λ) designs based on derived designs is now immediate. First, classify up to isomorphism all designs with parameters $(t-1)$-$(v-1, k-1, \lambda)$. For each such design, introduce a new point and add it to all blocks of the design. The result is a seed whose possible extensions to a t-(v, k, λ) design correspond to the solutions of an instance of DIOPHANTINE, as discussed in Sect. 6.1.3. Isomorphs must be rejected among the generated designs, which can be accomplished either by using recorded representatives or generation by canonical augmentation. In both cases the techniques in Sect. 6.1.4 apply.

Example 6.34. Steiner quadruple systems can be classified via their derived Steiner triple systems. Seeds of this type have been used in the classification of SQS(14) [423] and SQS(16) [306]. In this case the extension problem is an instance of EXACT COVERS, where the task is to cover 3-subsets of points using 4-subsets of points.

For a given admissible v, generation by canonical augmentation can be employed in a classification of the SQS(v) as follows. Each SQS(v) design $\mathcal{X}$ is associated with a canonical derived STS($v-1$) design $m(\mathcal{X})$ by selecting one point from the design in canonical form, cf. Sect. 6.1.4. Invariants for derived STS$(v-1)$ can be used to obtain a point invariant for SQS(v). For $v = 14, 16$ the multiset consisting of the number of occurrences of each point in Pasch configurations is an invariant that distinguishes between all isomorphism classes of STS$(v-1)$.

The isomorphism classes of the derived designs are used as a point invariant in performing the test $m(\mathcal{X}) \cong_{\mathrm{Aut}(\mathcal{X})} p(\mathcal{X})$ and in expediting the evaluation of a canonical labeling of $\mathcal{X}$. Complete isomorph rejection among generated designs $\mathcal{X}$ satisfying $m(\mathcal{X}) \cong_{\mathrm{Aut}(\mathcal{X})} p(\mathcal{X})$ is achieved by accepting $\mathcal{X}$ only if it is the lexicographic minimum of its $\mathrm{Aut}(p(\mathcal{X}))$-orbit. In this connection it is advantageous to select $m(\mathcal{X})$ so that it has the minimum automorphism group order among all derived STS$(v-1)$ occurring in $\mathcal{X}$.

Example 6.35. For $t \geq 3$, one may carry out a classification recursively via a sequence of derived designs and extensions; the following sequence is utilized in [417]:

$$1\text{-}(8,2,3) \;\rightarrow\; 2\text{-}(9,3,3) \;\rightarrow\; 3\text{-}(10,4,3) \;\rightarrow\; 4\text{-}(11,5,3) \;\rightarrow\; 5\text{-}(12,6,3).$$

In classifying 3-$(8,4,\lambda)$ designs, it is useful to observe that such designs are always resolvable (Theorem 6.41). Then an extension theorem of Alltop [5] applied to each design in a set of isomorphism class representatives of 2-$(7,3,\lambda)$ designs yields a set of 3-$(8,4,\lambda)$ designs with at least one representative from every isomorphism class, and a classification is obtained after rejecting isomorphs. From a computational perspective this is a very efficient extension strategy because no searching is required; however, its applicability is limited.

Among combinatorial arguments that can be employed in the classification of t-designs, those related to the block intersection structure can be especially useful [154, 177, 314, 384].

6.2.3 Results

A list of classification results for t-designs with $t \geq 3$ have been collected into Table 6.11. Hadamard 3-designs are indicated by HD; the reference(s) listed for these occasionally address some of the other families in Theorem 2.127. We do not include parameters for which nonexistence follows from Corollary 2.39. In addition to some of the references given in Sect. 6.1.6, the following works contain (tables of) classification results for t-designs: [169, 170].

Denny and Mathon [154] classified 3-$(8,4,\lambda)$ designs up to $\lambda = 15$. Some parameter sets of simple 3-designs have been considered in, for example, [417].

Research Problem 6.36. Classify t-designs with $t \geq 3$ to extend the list in Table 6.11.

Table 6.11. Classification of t-(v, k, λ) designs

t	v	k	λ	b	r	N	References
3	8	4	1	14	7	1	[28], HD
3	8	4	2	28	14	4	[433], [224]
3	8	4	3	42	21	10	[433], [224]
3	8	4	4	56	28	31	[278]
3	8	4	5	70	35	82	[153]
3	8	4	6	84	42	240	[153]
3	8	4	7	98	49	650	[153]
3	8	4	8	112	56	1,803	[153]
3	8	4	9	126	63	4,763	[153]
3	10	4	1	30	12	1	[28]
3	10	4	2	60	24	132	[153]
3	10	5	3	36	18	7	[206], [66]
3	11	5	2	33	15	0	[153]
3	11	5	4	66	30	1,749	[154]
3	12	6	2	22	11	1	[574], HD
3	14	4	1	91	26	4	[423]
3	16	4	1	140	35	1,054,163	[306]
3	16	8	3	30	15	5	[574], HD
3	17	5	1	68	20	1	[612]
3	20	10	4	38	19	3	[242], HD
3	22	6	1	77	21	1	[612]
3	24	12	5	46	23	130	[277, 316], HD
3	26	6	1	130	30	1	[106]
3	28	14	6	54	27	7,570	[317, 318], HD
4	11	5	1	66	30	1	[28]
4	11	5	2	132	60	12	[153]
4	12	6	4	132	66	11	[154]
4	12	6	6	198	99	0	[417]
4	15	5	1	273	91	0	[423]
4	18	6	1	204	68	0	[612]
4	23	7	1	253	77	1	[612]
5	12	6	1	132	66	1	[28]
5	24	8	1	759	253	1	[612]

6.3 Resolutions of Designs

Resolutions of designs can be classified either via the underlying (resolvable) design or by constructing the resolutions directly.

6.3.1 Classification via the Underlying Design

If a classification of the (resolvable) designs with the relevant parameters is available, then it can be employed to arrive at a classification of resolutions. Since the isomorphism class of the underlying design is an invariant of a resolution, isomorph rejection needs to be performed only among the resolutions of each design. By Theorem 3.82, isomorphism computations on resolutions can be carried out in the framework of OE codes.

One approach to constructing the resolutions of a given design with parameters t, v, k, λ, b, r is to first construct all possible parallel classes of blocks, and then to find all sets of r parallel classes such that no two parallel classes in the set use the same block. This approach can be implemented by using Exact Covers in two stages [458]. In the first stage, we find the parallel classes by covering the points with blocks. In the second stage, we cover the blocks with parallel classes.

A number of issues should be taken into account with the aforementioned approach. First, repeated blocks can induce a lot of unwanted redundancy into the search space. This difficulty can be alleviated by replacing every repeated block with multiplicity $m > 1$ by a single block in the first stage, and requiring that this block is covered m times in the second stage. Second, the underlying design can contain many parallel classes, which can make the second stage search laborious. Third, the automorphism group of the underlying design induces redundancy into the search space. As an illustration, consider generating all resolutions of the unique 2-$(2n, 2, 1)$ design, which is equivalent to generating all 1-factorizations of K_{2n}; see Sect. 8.1.1.

Alternatively, the construction of resolutions can be formulated as a single exact cover problem. Let $p_1, p_2, \ldots, p_v$ be the points of the design, and let $B_1, B_2, \ldots, B_b$ be the blocks (represented as sets of points). Furthermore, let $P_1, P_2, \ldots, P_r$ be a set of distinct labels for the putative parallel classes. The task is now to find an exact cover for all sets of the form $\{P_\ell, p_i\}$ with $1 \le \ell \le r$ and $1 \le i \le v$, and all sets $\{B_j\}$ with $1 \le j \le b$, using $(k+2)$-sets of the form $\{P_\ell\} \cup \{B_j\} \cup \{p_i : p_i \in B_j\}$ with $1 \le \ell \le r$ and $1 \le j \le b$. Each $(k+2)$-set covers the subsets that occur in it.

Also this formulation as an exact cover problem contains unnecessary redundancy induced by the labels for the parallel classes and potential repeated blocks. The former type of redundancy can be easily eliminated by selecting a point, say p_1, and placing the blocks incident with p_1 in distinct parallel classes before starting the search. The latter type of redundancy can be eliminated by replacing each block with multiplicity $m > 1$ by a single block, and requiring that this block be covered m times.

If necessary, redundancy induced by the automorphism group of the design can be reduced by completing one or more parallel classes (the number depending on the automorphism group), rejecting isomorphs, and only then starting the exact cover search. Such seed structures can also be used to perform isomorph rejection on the generated resolutions using generation by canonical augmentation; cf. Sect. 6.1.4.

6.3.2 Direct Classification

Analogously to the methods for classifying designs, there are essentially two approaches for classifying resolutions of designs directly. Either one proceeds point by point – that is, codeword by codeword in the framework of OE codes – or parallel class by parallel class – that is, coordinate by coordinate in the framework of codes. Approaches presented elsewhere in this book are then applicable: the codeword-by-codeword approach discussed in Sect. 7.1.2, and the coordinate-by-coordinate approach discussed in the context of covering codes in Sect. 7.2. If $t \geq 3$, then even further restrictions can be imposed on the OE codes.

Combinatorial properties of the resolutions can be used to restrict the search space. Morales and Velarde [431, 432] use the intersection properties of blocks in different parallel classes to restrict the search. More precisely, let $Q = \{B_1, B_2, \ldots, B_q\}$ and $Q' = \{B'_1, B'_2, \ldots, B'_q\}$ be two different parallel classes in a resolution of a 2-(qk, k, λ) design. The (Q, Q') *parallel class intersection matrix* is the $q \times q$ matrix $\mathbf{A}(Q, Q')$ with entries defined by $a_{ij} = |B_i \cap B'_j|$ for all $1 \leq i, j \leq q$.

Example 6.37. We follow [432]. A resolution of a 2-$(10, 5, 16)$ design has $r = 36$ parallel classes, each consisting of two blocks. Thus, for an arbitrary fixed parallel class Q there are 35 parallel class intersection matrices of the form $\mathbf{A}(Q, Q')$. Clearly, each such matrix must satisfy $\sum_i a_{ij} = 5$ and $\sum_j a_{ij} = 5$ for all $1 \leq i, j \leq 2$. Thus, by relabeling the blocks in the parallel classes if necessary, each of the 35 matrices associated with Q can be assumed to be one of

$$\begin{bmatrix} 5 & 0 \\ 0 & 5 \end{bmatrix}, \quad \begin{bmatrix} 4 & 1 \\ 1 & 4 \end{bmatrix}, \quad \begin{bmatrix} 3 & 2 \\ 2 & 3 \end{bmatrix}.$$

Let n_{50}, n_{41}, and n_{32} be the number of matrices of each type associated with Q. Since there are 35 matrices in total,

$$n_{50} + n_{41} + n_{32} = 35. \tag{6.18}$$

Let B be a block in Q. Counting the occurrences of the $\binom{5}{2} = 10$ pairs of distinct points from B in blocks of the other parallel classes, it follows that

$$10n_{50} + 6n_{41} + 4n_{32} = 15 \times 10. \tag{6.19}$$

The equations (6.18) and (6.19) have two solutions in nonnegative integers: $n_{50} = 0$, $n_{41} = 5$, $n_{32} = 30$; and $n_{50} = 1$, $n_{41} = 2$, $n_{32} = 32$. This observation can then be employed to restrict the search space in a parallel class by parallel class backtrack search.

Example 6.38. Following [353], we consider affine resolvable 2-(27, 9, 4) designs; recall Definition 2.61 and Bose's condition in Theorem 2.63. For 2-(27, 9, 4) designs, the intersection parameter is $\mu = 3$ and the number of parallel classes is $r = 13$. Thus, any 13 blocks incident with a common point belong to different parallel classes and have pairwise exactly three points in common. Removing the common point, these blocks define (the dual of) a 2-(13, 4, 2) design. Starting from the 2,461 nonisomorphic 2-(13, 4, 2) designs, the affine resolvable 2-(27, 9, 4) designs can be classified by adding blocks of size 9 that intersect the initial 13 blocks in either 0 or 3 points.

In certain specific cases such as those considered in Theorems 2.62, 2.63, 2.120, and 2.127, it is possible to arrive at a classification of resolutions directly based on a classification of related objects.

6.3.3 Results

We here tabulate results only for resolutions of BIBDs; such results are presented in Tables 6.12 to 6.15. The parameters for which RBIBDs exist motivate an ordering different from that in the tables for BIBDs, and we now present one table for each value of

$$q = \frac{v}{k} = \frac{b}{r}.$$

Within a table, the rows are ordered lexicographically by k and λ.

The entries of the tables contain the five parameters of the underlying design, the number of nonisomorphic resolvable designs (Nd), the number of nonisomorphic resolutions (N), a reference to the result, and its entry number in the table of [407]. For some parameters, only the number of resolutions is known and not the number of resolvable designs.

As for BIBDs, the first open case in a series of parameters is indicated; here the N/Nd values are omitted. In any case, following [407], we stop at $r = 41$. Some results for $r > 41$ can be found in [298]. The following references contain surveys and additional results related to the classification of RBIBDs: [36, 244, 404, 405, 406, 407].

Resolvable designs that do not exist because of Bose's condition (Theorem 2.63)

$$b \geq v + r - 1$$

are indicated by giving a reference to that theorem.

The following two theorems for the case $v = 2k$ can be obtained with combinatorial arguments; the latter is from [404].

Theorem 6.39. *There is a resolvable* 2-$(6,3,\lambda)$ *design only if* λ *is divisible by* 4. *Moreover, such a design is unique.*

Proof. A resolvable 2-$(6,3,\lambda)$ design is a resolvable 3-$(6,3,\lambda'=\lambda/4)$ design by Theorem 2.59. Hence, a 2-$(6,3,\lambda)$ design can exist only if λ is divisible by 4. Existence and uniqueness follow directly from the fact that $t=k$ in a 3-$(6,3,\lambda')$ design whereby the value of λ' tells how many times each 3-subset occurs as a block. □

Theorem 6.40. *There is no resolvable* 2-$(4t+2,2t+1,2t(2m-1))$ *design for* $m,t \geq 1$.

Proof. By Theorem 2.59, a resolvable 2-$(4t+2,2t+1,2t(2m-1))$ design is a resolvable 3-$(4t+2,2t+1,\lambda)$ design with $\lambda=(2m-1)(2t-1)/2$, which is not an integer. □

For $v=2k$, one may utilize Theorem 2.59 and check whether the 3-(v,k,λ) designs from Table 6.11 are resolvable or not; such entries are indicated with a reference to Theorem 2.59. For example, it turns out that five out of the seven 3-$(10,5,3)$ designs are resolvable. For $v=8$, $k=4$ the following theorem is applicable.

Theorem 6.41. *Any* 3-$(8,4,\lambda)$ *design is resolvable.*

Proof. Partition the set of points into two sets, S and T, of four points each. For a given 3-$(8,4,\lambda)$ design, we denote by a_i the number of blocks that intersect S in i points. Since $|S|=|T|$, the sums of all intersection numbers with respect to S and T must coincide,

$$a_1+2a_2+3a_3+4a_4=4a_0+3a_1+2a_2+a_3,$$

which simplifies to

$$2a_3+4a_4=4a_0+2a_1. \tag{6.20}$$

On the other hand, if we count how many triples within S and T are contained in the blocks, we get

$$a_3+4a_4=4a_0+a_1. \tag{6.21}$$

Combining (6.20) and (6.21) yields $a_1=a_3$ and $a_0=a_4$, where the latter equality implies resolvability. □

Note that Nd equals N in Table 6.12 by the comment at the end of Sect. 2.2.6. Also here we indicate designs mentioned in Theorem 2.127 by HD, and recognize results for related families as well. We know from Theorem 2.62 that affine planes are resolvable and have a unique resolution, so we get the count directly from the corresponding entry in Tables 6.2 through 6.10; in these cases we refer to the aforementioned theorem.

Table 6.12. Classification of 2-(v, k, λ) RBIBDs with $q = 2$

v	k	λ	b	r	Nd	N	References	No
6	3	2	10	5	0	0	Theorem 6.39	4
6	3	4	20	10	1	1	Theorem 6.39	43
6	3	6	30	15	0	0	Theorem 6.39	118
6	3	8	40	20	1	1	Theorem 6.39	236
6	3	10	50	25	0	0	Theorem 6.39	409
6	3	12	60	30	1	1	Theorem 6.39	596
6	3	14	70	35	0	0	Theorem 6.39	816
6	3	16	80	40	1	1	Theorem 6.39	1078
8	4	3	14	7	1	1	Theorem 2.59, HD	15
8	4	6	28	14	4	4	Theorem 2.59	101
8	4	9	42	21	10	10	Theorem 2.59	278
8	4	12	56	28	31	31	Theorem 2.59	524
8	4	15	70	35	82	82	Theorem 2.59	819
10	5	4	18	9	0	0	Theorem 6.40	33
10	5	8	36	18	5	5	Theorem 2.59	195
10	5	12	54	27	0	0	Theorem 6.40	480
10	5	16	72	36	27,121,734	27,121,734	[432]	891
12	6	5	22	11	1	1	[574], HD	58
12	6	10	44	22	545	545	[H]	319
12	6	15	66	33				743
14	7	6	26	13	0	0	Theorem 6.40	89
14	7	12	52	26	1,363,486	1,363,486	[301]	451
14	7	18	78	39	0	0	Theorem 6.40	1038
16	8	7	30	15	5	5	[574], HD	130
16	8	14	60	30				618
18	9	8	34	17	0	0	Theorem 6.40	179
18	9	16	68	34				791
20	10	9	38	19	3	3	[242], HD	224
20	10	18	76	38				1007
22	11	10	42	21	0	0	Theorem 6.40	293
24	12	11	46	23	130	130	[277, 316], HD	346
26	13	12	50	25	0	0	Theorem 6.40	424
28	14	13	54	27	7,570	7,570	[317, 318], HD	499
30	15	14	58	29	0	0	Theorem 6.40	579
32	16	15	62	31			HD	668

Table 6.13. Classification of 2-(v, k, λ) RBIBDs with $q = 3$

v	k	λ	b	r	Nd	N	References	No
9	3	1	12	4	1	1	Theorem 2.62	2
9	3	2	24	8	9	9	[403], [434]	21
9	3	3	36	12	395	426	[402, 463]	66
9	3	4	48	16	119,985	149,041	[463]	145
9	3	5	60	20		203,047,732	[463]	235
9	3	6	72	24				356
12	4	3	33	11	5	5	[431]	56
12	4	6	66	22				316
15	5	2	21	7	0	0	No BIBD exists	16
15	5	4	42	14	0	0	[302]	102
15	5	6	63	21				281
18	6	5	51	17				176
21	7	3	30	10	0	0	Theorem 2.63	49
21	7	6	60	20				250
24	8	7	69	23				343
27	9	4	39	13	68	68	[353]	90
27	9	8	78	26				453
30	10	9	87	29				576
33	11	5	48	16	0	0	Theorem 2.63	161
33	11	10	96	32				704

Table 6.14. Classification of 2-(v, k, λ) RBIBDs with $q = 4$

v	k	λ	b	r	Nd	N	References	No
12	3	2	44	11	62,929	74,700	[452]	55
12	3	4	88	22				314
16	4	1	20	5	1	1	Theorem 2.62	5
16	4	2	40	10	325,062	339,592	[303]	44
16	4	3	60	15				119
20	5	4	76	19				221
24	6	5	92	23				341
28	7	2	36	9	0	0	Theorem 2.63	36
28	7	4	72	18				201
32	8	7	124	31				664

Table 6.15. Classification of 2-(v, k, λ) RBIBDs with $q = 5$

v	k	λ	b	r	Nd	N	References	No
15	3	1	35	7	4	7	[435], [128]	14
15	3	2	70	14				99
20	4	3	95	19				220
25	5	1	30	6	1	1	Theorem 2.62	11
25	5	2	60	12				72
30	6	5	145	29				573

Table 6.16. Classification of some affine 2-(v, k, λ) RBIBDs

v	k	λ	b	r	Nd	N	References	No
36	6	1	42	7	0	0	No BIBD exists	17
49	7	1	56	8	1	1	Theorem 2.62	26
64	8	1	72	9	1	1	Theorem 2.62	37
81	9	1	90	10	7	7	Theorem 2.62	51
100	10	1	110	11	0	0	No BIBD exists	60
121	11	1	132	12				81

6.4 Designs with Additional Properties

Often one is interested in designs that, for example, either lack or possess a particular configuration or subdesign. Other additional properties of interest include resolvability (Sect. 6.3) and prescribed automorphisms (Chap. 9).

In many cases it is not computationally feasible to generate all nonisomorphic designs with given parameters – due to their large number – and filter out those designs that do not have the desired property. Consequently, there is a demand for tailored classification algorithms for designs that have a specified property.

Informally, we can speak of local and global properties of a design. A *local* property is such that its presence can be ascertained by looking only at a (small) subset of points and blocks. The property of having a subdesign is a good example of a local property. In contrast, a *global* property cannot be determined without looking at the design as a whole. For example, resolvability and the absence of a subdesign are global properties.

A classification algorithm tailored for a local property typically proceeds by first determining up to isomorphism all the substructures that cause the property to hold. Then, these substructures are used as seeds in classifying the designs. If the substructure inducing the property is small, it may be necessary to enlarge the seeds somewhat.

Here are some concrete examples in the context of Steiner triple systems.

Example 6.42. An incidence matrix of an STS(19) with an STS(9) subsystem is depicted in Fig. 6.8. Such an STS(19) consists of an STS(9) and a GDD(19, 3) of group type $9^1 1^{10}$ [563]. The latter design is equivalent up to isomorphism to a 1-factorization of the complete graph K_{10}; see Sect. 8.1.1 for further details.

111100000000	111110000000000000000000000000000000000000000
100011100000	000001111100000000000000000000000000000000000
100000011100	000000000011111000000000000000000000000000000
010010010010	000000000000000111110000000000000000000000000
010001001001	000000000000000000001111100000000000000000000
001010000101	000000000000000000000000011111000000000000000
001000101010	000000000000000000000000000000111110000000000
000101000110	000000000000000000000000000000000001111100000
000100110001	000000000000000000000000000000000000000011111
000000000000	100001000010000100001000010000100001000010000
000000000000	100000100001000010000100001000010000100001000
000000000000	010001000001000001000010000100001000010000100
000000000000	010000100010000000100001000010000100001000010
000000000000	001000010000100100000100000010000010000100100
000000000000	001000001000010010001000000001000100010000001
000000000000	000100000100100001000000110000010000001000001
000000000000	000100001000001000010010001000100000000100010
000000000000	000010000100010000010001000100000011000001000
000000000000	000010010000001000100000100001001000100010000

Fig. 6.8. An STS(19) with an STS(9) subsystem

There are essentially two ways the previous structural observation can be exploited to classify the STS(19) with an STS(9) subsystem. Namely, either we can use the 396 nonisomorphic 1-factorizations of K_{10} (Sect. 8.1.1) as seeds, or we can use the STS(9) as a seed. Out of these, the former approach is more practical and is employed in [563]. Indeed, in this way each seed has exactly $9!/432 = 840$ extensions, which correspond to the number of distinct STS(9) on a fixed point set. If the latter approach is used, then additional isomorph rejection is required – for example, in the form of a larger seed.

Example 6.43. To illustrate the use of larger seeds, we discuss a classification of the STS(19) admitting an STS(7) subsystem [307]. See Fig. 6.9 for an example of such a design.

1110000	111111000000000000	00000000000000000000000000000000
1001100	000000111111000000	00000000000000000000000000000000
1000011	000000000000111111	00000000000000000000000000000000
0101010	000000000000000000	11111100000000000000000000000000
0100101	000000000000000000	00000011111100000000000000000000
0011001	000000000000000000	00000000000011111100000000000000
0010110	000000000000000000	00000000000000000011111100000000
0000000	100000100000100000	10000010000010000010000011000000
0000000	100000010000010000	01000001000001000001000000110000
0000000	010000100000010000	00100000100000100000100000001100
0000000	010000010000100000	00010000010000010000010000000011
0000000	001000001000001000	10000000010000001000001000101000
0000000	001000000100000100	00100000001000000100010010010000
0000000	000100001000000100	00001000000101000000100001000010
0000000	000100000100001000	00000101000010000000000100000101
0000000	000010000010000010	01000000001000001010000000000110
0000000	000010000001000001	00010000000100100000000110100000
0000000	000001000010000001	00001010000000000101000000001001
0000000	000001000001000010	00000100100000010000001001010000

Fig. 6.9. An STS(19) with an STS(7) subsystem

Clearly, using only an STS(7) as a seed leaves too much redundancy in the search space, so a larger seed is required. It turns out that we can use the seeds from Example 6.20, now restricted to seeds that contain an STS(7). Namely, a seed in Example 6.20 consists of all the blocks in an STS(19) that have nonempty intersection with a given block. Because the blocks of an STS(7) intersect pairwise, it follows that every STS(19) that contains an STS(7) also contains a set system isomorphic to a seed that contains an STS(7). Out of the 14,648 seeds in Example 6.20, only 157 contain an STS(7). The approach described in Sect. 6.1.4 can be used to reject isomorphs among the generated designs, provided that we modify the canonical parent function m so that it always produces a set system isomorphic to a seed that contains an STS(7).

Compared with local properties, global properties typically require more effort in tailoring a classification algorithm. Techniques must be developed for detecting partial solutions that cannot be extended to a design for which the global property holds. For example, designs that lack a particular configuration of blocks can only be generated (block by block) from partial solutions that lack the configuration. Similarly, resolvable designs can be generated so that every partial solution admits a resolution. Yet another example is generating designs that admit a group of automorphisms H (acting on the points). In this case every design satisfying the property is a union of H-orbits of blocks, so we may restrict the (block by block) partial solutions to unions of H-orbits of blocks; this is discussed in detail in Chap. 9.

Here are some examples of global properties.

Example 6.44. An STS is said to be *anti-Pasch* if it does not contain a Pasch configuration. To classify the anti-Pasch STS(19), we may restrict the seeds in Example 6.20 to those that do not contain a Pasch configuration. There are 1,678 such seeds. In extending the seeds block by block, we require every partial solution to be anti-Pasch; see Example 6.47 for one possibility on how to incorporate an anti-Pasch test into block by block backtrack.

Research Problem 6.45. Classify the anti-Pasch STS(21).

It is not hard to come up with global properties that hardly constrain the search at all.

Example 6.46. Among the STS(19), 10,997,902,498 of the 11,084,874,829 nonisomorphic designs are subsystem-free, that is, contain neither an STS(7) nor an STS(9) [307].

On the other hand, certain properties are very restrictive.

Example 6.47. The *cycle structure* of an STS(v) is the set of all cycle graphs (Example 3.115) defined by pairs of distinct points. An STS(v) is *perfect* if the cycle structure consists of cycles of length $v-3$ only.

A perfect STS(v) is known only for [221]

$$v = 7,\ 9,\ 25,\ 33,\ 79,\ 139,\ 367,\ 811,\ 1531,\ 25771,\ 50923,\ 61339,\ 69991.$$

On the other hand, nonexistence of a perfect STS(v) can be established for $v = 13, 15$ by a straightforward investigation of the classified systems. For order $v = 19$ nonexistence can be established by looking at the 2,591 nonisomorphic anti-Pasch STS(19). Indeed, observe that every Pasch configuration in an STS(v) defines three 4-cycles in the cycle graphs, and conversely a 4-cycle occurs in a cycle graph only if the STS contains a Pasch configuration; thus for $v \neq 7$ a perfect STS(v) is necessarily anti-Pasch.

Nonexistence of a perfect STS(21) can be established by tailoring the block-by-block classification algorithm in Sect. 6.1.3 to keep track of the emerging cycles in the structure. For $v = 21$ there are 219,104 nonisomorphic seeds analogous to the seeds in Example 6.20, 7,116 of which have the property that all cycles in the (partial) cycle structure have length 18. During seed extension a partial STS(21) is pruned whenever a cycle of length less than 18 appears in a cycle graph. Details can be found in [300].

7

Classification of Codes

In this chapter, classification of several main classes of codes are considered: unrestricted error-correcting and covering codes in Sects. 7.1 and 7.2, respectively, and error-correcting linear codes in Sect. 7.3. Various subclasses of these codes are also discussed, including equidistant codes and constant weight codes.

The intersection of the combinatorial objects of this chapter and those of the previous chapter, Chap. 6, is obviously nonempty since, as we have seen, BIBDs correspond to certain optimal binary constant weight error-correcting codes and RBIBDs correspond to certain optimal q-ary error-correcting codes. Algorithms already considered will not be repeated here, but references to appropriate earlier parts will be used whenever necessary.

7.1 Error-Correcting Codes

Given the parameters of an unrestricted error-correcting code – that is, n, M, d, and q – the problem of finding and classifying $(n, M, d)_q$ codes can be transformed into the problem of finding cliques of size M in a certain graph $G = (V, E)$. The vertex set of the graph G is the set of all words in the space, $V = Z_q^n$. The edge set consists of the pairs of words in Z_q^n that are at distance greater than or equal to d from each other,

$$E = \{\{\mathbf{x}, \mathbf{y}\} : \mathbf{x}, \mathbf{y} \in Z_q^n,\ d_H(\mathbf{x}, \mathbf{y}) \geq d\}.$$

If $M = A_q(n, d)$, then the codes are optimal and correspond to the maximum cliques in the graph G.

A straightforward approach for classifying error-correcting codes would be to find all cliques of size M in the aforementioned graph G and then carry out isomorph rejection among the codes found. By the orbit-stabilizer theorem (Theorem 3.20), the number of codes equivalent to a given code C is

$$\frac{|\mathrm{Aut}(Z_q^n, d_H)|}{|\mathrm{Aut}(C)|} = \frac{n!(q!)^n}{|\mathrm{Aut}(C)|},$$

which shows that this approach is not feasible (one often encounters codes with $|\mathrm{Aut}(C)| = 1$). It is therefore obvious that one should utilize the large automorphism group of the space to carry out isomorph rejection among partial codes. In the sequel, we discuss the classification problem in coding-theoretic terms. Whenever clique search has to be carried out, it is understood that this is done in the graph constructed in the aforementioned way. If a set of codewords has been fixed, we must also take into account that the new codewords are at distance at least d from those words. A graph obtained in this manner is called a *compatibility graph.*

One important implementational issue should still be mentioned. Namely, the ordering of vertices is crucial when Algorithm 5.1 from Sect. 5.1 is used to search for cliques. Experience shows that good performance is obtained when the vertices are ordered so that the corresponding codewords are in – increasing or decreasing – lexicographic order.

Until the late 1990s, only sporadic results had been published on classification of unrestricted error-correcting codes. Earlier results had mainly concerned constant weight error-correcting codes (packing designs), within both the coding theory and the design theory community. The idea of constructing codes via certain subcodes pertains to much of this work, and that approach will be the first one considered here. There are two other basic approaches for classifying error-correcting codes: codeword by codeword and coordinate by coordinate. The former of these two basic methods is discussed in this section, whereas the latter is presented for covering codes in Sect. 7.2 (mentioning the minor modifications needed for the error-correcting case). The current section ends with the special case of constant weight codes and a survey of published classification results.

The choice between a classification via subcodes and the codeword-by-codeword approach – very little is yet known about the coordinate-by-coordinate approach – should be based on the parameters of the code; the former method performs better for short length n, large cardinality M, and small minimum distance d, whereas the latter is better for large n and d and small M. If in doubt, preliminary experiments should be carried out to indicate the most efficient approach for the desired code parameters.

7.1.1 Classification via Subcodes

Consider an $(n, M, d)_2$ binary code C. This code can be shortened in a given coordinate by removing the coordinate and taking the codewords with a given value i in the removed coordinate. We denote such shortened codes by C_i, $i = 0, 1$; the parameters of C_i are $(n-1, M_i, d)_2$ with $M_0 + M_1 = M$. Hence

$$\max\{M_0, M_1\} \geq \lceil M/2 \rceil.$$

Observe that both $0C_0 \cup 1C_1$ (we use the notation $iC = \{(i, \mathbf{x}) : \mathbf{x} \in C\}$) and $1C_0 \cup 0C_1$ are equivalent to C; permutations of the coordinate values give equivalent codes. Therefore, we may without loss of generality assume that $M_0 \geq \lceil M/2 \rceil$.

Reversing this procedure by *lengthening* codes to classify the $(n, M, d)_2$ codes, we first classify the $(n-1, M', d)_2$ codes for

$$M/2 \leq M' \leq A_2(n-1, d); \qquad (7.1)$$

obviously the size of an optimal code with the given parameters gives an upper bound on M'. Thereafter, for each such $(n-1, M', d)_2$ code C_0, we construct the compatibility graph $G = (V, E)$ in which we search for the possible values of C_1. Since all words in C_1 must be at distance greater than or equal to $d-1$ from the words in C_0, all words in Z_q^{n-1} that fulfill this property make up the vertex set V. Edges are inserted as in the basic approach, that is, between vertices whose corresponding words are at distance greater than or equal to d from each other. In the graph G, we search for all cliques of size $M - M'$ and finally carry out isomorph rejection.

Isomorph rejection can be carried out via recorded objects among the codes constructed from all possible $(n-1, M', d)_2$ codes with M' fulfilling (7.1). Certificates of codes are most conveniently obtained via a transformation to graphs, see Sect. 3.3.2. This approach is appropriate whenever the number of subcodes is limited, but to solve the hardest instances with an abundance of subcodes, generation by canonical augmentation is to be preferred. We will return to this issue later.

The described approach is recursive: in order to classify the $(n-1, M', d)_2$ codes, one should carry out the aforementioned procedure starting from all $(n-2, M'', d)_2$ codes with $M'/2 \leq M'' \leq A_2(n-2, d)$, and so on. The recursive break-up can be stopped at a length where optimal codes have few words and the necessary codes are obtainable in some other manner (such as the method in Sect. 7.1.2 or perhaps even by hand calculation). This process is illustrated in the following example.

Example 7.1. Consider the proof of $A_2(10, 3) = 72$ and a classification of optimal such codes [459]. To prove nonexistence of $(10, 73, 3)_2$ codes, codes with various parameters (n, M, d) – omitting the subscript 2 – are linked to each other as shown in Fig. 7.1.

An arrow from one parameter to another indicates what codes should be included in the set of subcodes from which a lengthened code is constructed. The recursive classification starts from the two $(4, 2, 3)_2$ codes $\{0000, 0111\}$ and $\{0000, 1111\}$. In the scheme, it is assumed that we know the values of $A_2(n, 3)$ for $n \leq 9$; if this is not the case, the scheme should be widened by introducing one more entry on each level.

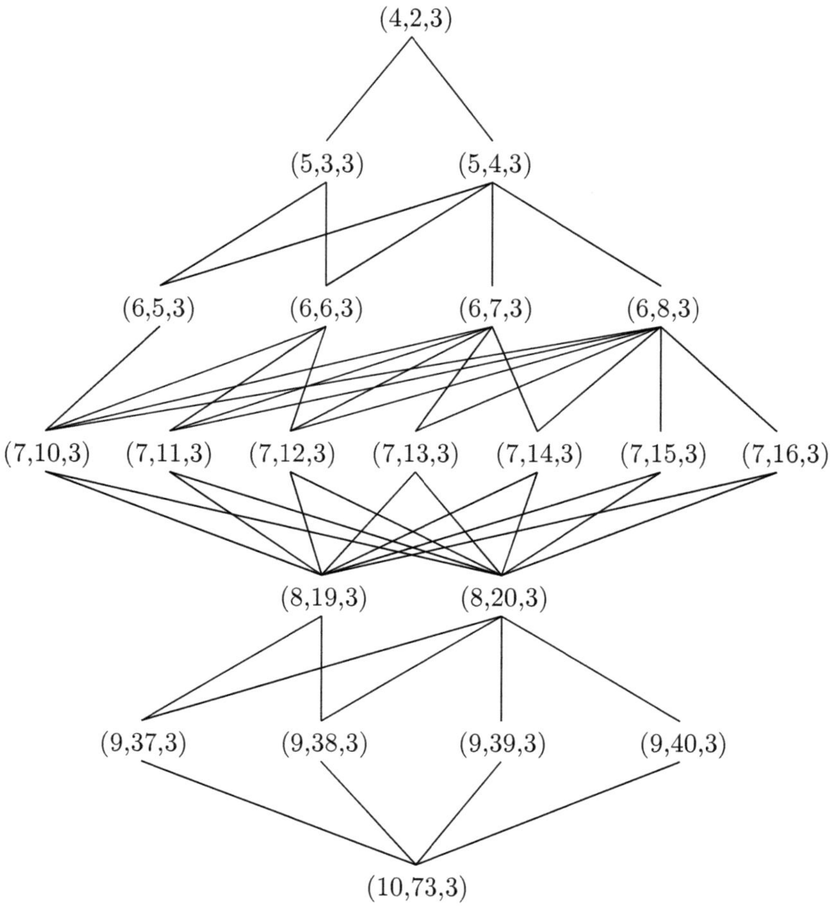

Fig. 7.1. Subcode graph

Example 7.2. In traversing the subcode graph in Fig. 7.1, one step is that of classifying the $(5,4,3)_2$ codes from the $(4,2,3)_2$ codes $\{0000, 0111\}$ and $\{0000, 1111\}$. For lengthening these codes, we get the compatibility graphs in Fig. 7.2. All words in the graph have length 5 and a 1 in the first coordinate to indicate that we search for codewords of C_1. There are obviously three cliques of size 2 in each graph. When the six codes obtained are checked for equivalence – do this by hand calculation! – it turns out that they are all pairwise equivalent, so the $(5,4,3)_2$ code is unique.

In the framework of canonical augmentation, one may test whether the new coordinate is in the canonically first coordinate orbit. See Sects. 3.3.2 and 4.2.3. Implementation of this approach could turn out useful in extending published classification results.

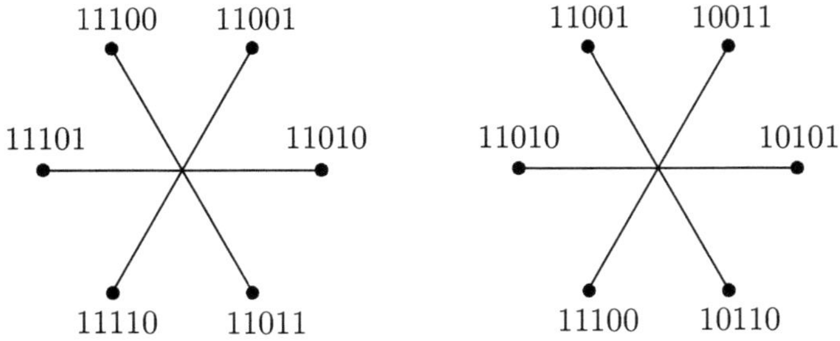

Fig. 7.2. The compatibility graphs for $\{0000, 0111\}$ and $\{0000, 1111\}$ with $d = 3$

Research Problem 7.3. Implement an algorithm based on generation by canonical augmentation for classifying error-correcting codes via subcodes. Use this algorithm to extend known classification results, such as those in [450, 451].

For q-ary codes, we may proceed in an analogous way, and classify $(n, M, d)_q$ codes via $(n-1, M', d)_q$ codes for $M/q \leq M' \leq A_q(n-1, d)$; cf. [370, 371]. However, for all but the smallest values of q this approach is not efficient because the relative size of a subcode decreases with increasing value of q. One may then proceed via mixed codes instead, cf. [450, 451]. For example, $(n, M, d)_3$ codes may be classified via codes with one binary coordinate, $n-1$ ternary coordinates, minimum distance at least d, and M' codewords with $M' \geq 2M/3$.

7.1.2 Classification Codeword by Codeword

We will now present an algorithm for classifying error-correcting codes codeword by codeword that is based on orderly generation. If we modify the algorithm slightly by adding the restriction that the code be equidistant, we get an approach for classifying RBIBDs, linking the contents to Sect. 6.3.2.

The basic problem in this setting is that we are given an $(n, M, d)_q$ code C, and we must find all words $\mathbf{y} = (y_1, y_2, \ldots, y_n) \in Z_q^n$ such that $C \cup \{\mathbf{y}\}$ is an $(n, M+1, d)_q$ code.

When q and n are small, such as for the codes in the previous section, then one can simply exhaustively search all the q^n words in Z_q^n and check whether the required minimum distance is met.

For larger parameters, however, a preferred approach is to use a tailored backtrack algorithm – for example, one that proceeds one coordinate at a time – for generating words that meet the required minimum distance. An alternative is to adopt a reductionist approach and view the word generation problem as an instance of DIOPHANTINE. For each coordinate $j = 1, 2, \ldots, n$, we have 0-1 variables $x_{j,0}, x_{j,1}, \ldots, x_{j,q-1}$ constrained by

$$x_{j,0} + x_{j,1} + \cdots + x_{j,q-1} = 1. \tag{7.2}$$

In other words, for every coordinate $j = 1, 2, \ldots, n$, exactly one $x_{j,v}$ has value 1, and the other variables have value 0, corresponding to the situation $y_j = v$. To achieve the required minimum distance, we have the inequality

$$\sum_{j=1}^{n} \sum_{v \neq c_j} x_{j,v} \geq d \tag{7.3}$$

for every codeword $\mathbf{c} = (c_1, c_2, \ldots, c_n) \in C$. To obtain an instance of DIOPHANTINE, each inequality can be transformed into an equality by introducing a slack variable. If equidistance is required, then the inequalities (7.3) should be replaced with equalities.

Example 7.4. Consider the $(4, 2, 2)_2$ code $C = \{0000, 0011\}$, which we want to augment to get $(4, 3, 2)_2$ codes. From (7.2) and (7.3) we obtain the following system of equalities and inequalities over 0-1 variables:

$$\begin{aligned}
x_{1,0} + x_{1,1} \qquad\qquad\qquad\qquad\qquad\qquad &= 1,\\
x_{2,0} + x_{2,1} \qquad\qquad\qquad\qquad &= 1,\\
x_{3,0} + x_{3,1} \qquad\qquad &= 1,\\
x_{4,0} + x_{4,1} &= 1,\\
x_{1,1} + x_{2,1} + x_{3,1} + x_{4,1} &\geq 2,\\
x_{1,1} + x_{2,1} + x_{3,0} + x_{4,0} &\geq 2.
\end{aligned}$$

To obtain an instance of DIOPHANTINE, we introduce the slack variables $z_1, z_2 \in \{0, 1, 2\}$ for the two inequalities:

$$\begin{aligned}
x_{1,0} + x_{1,1} \qquad\qquad\qquad\qquad\qquad\qquad &= 1,\\
x_{2,0} + x_{2,1} \qquad\qquad\qquad\qquad &= 1,\\
x_{3,0} + x_{3,1} \qquad\qquad &= 1,\\
x_{4,0} + x_{4,1} &= 1,\\
x_{1,1} + x_{2,1} + x_{3,1} + x_{4,1} - z_1 &= 2,\\
x_{1,1} + x_{2,1} + x_{3,0} + x_{4,0} - z_2 &= 2.
\end{aligned} \tag{7.4}$$

It is possible to simplify a system by combining the sets of 0-1 variables associated with identical coordinates in C into one set of variables, with an appropriate increase in the variable bounds. However, after such a transformation there is a one-to-many rather than a one-to-one correspondence between the solutions and the words that augment C.

Example 7.5. Coordinates 1, 2 and coordinates 3, 4 are identical in the code C in Example 7.4. After combining variables in (7.4), we obtain the system

$$
\begin{aligned}
x_{12,0} + x_{12,1} \qquad\qquad\qquad\qquad\qquad &= 2,\\
x_{34,0} + x_{34,1} \qquad\quad &= 2,\\
x_{12,1} \qquad\qquad + x_{34,1} - z_1 \qquad &= 2,\\
x_{12,1} + x_{34,0} \qquad\qquad\qquad - z_2 &= 2,
\end{aligned}
$$

where all variables assume values in $\{0,1,2\}$. This system has four solutions, which we list together with the sets of corresponding words:

$x_{12,0}$	$x_{12,1}$	$x_{34,0}$	$x_{34,1}$	z_1	z_2	
1	1	1	1	0	0	$\{0101, 0110, 1001, 1010\}$
0	2	2	0	0	2	$\{1100\}$
0	2	1	1	1	1	$\{1101, 1110\}$
0	2	0	2	2	0	$\{1111\}$.

We proceed to discuss isomorph rejection. The following orderly generation technique was developed in [302] in the context of OE codes. It will be convenient to work with matrices with entries in Z_q. A *matrix representation* of an $(n, M, d)_q$ code is an $M \times n$ matrix $\mathbf{C} = (c_{ij})$ such that the rows are the codewords of the code. Code equivalence (Definition 2.100) is captured in the matrix representation by viewing two matrices as *equivalent* if one is obtained from the other by permuting the rows, the columns, and the values Z_q independently in each column. In the language of group actions, let E be the group $S_m \times (S_q \wr S_n)$, and let $\mathbf{C} = (c_{ij})$ be an $m \times n$ matrix with entries in Z_q. Let $e = (g, (k, h)) \in E$ act on $\mathbf{C}$ by $e * \mathbf{C} = \mathbf{D}$, where $d_{ij} = k_j(c_{g^{-1}(i),h^{-1}(j)})$ for all $1 \le i \le m$ and $1 \le j \le n$. Two matrices are now equivalent if they are in the same orbit of this action of E.

To employ orderly generation, we require an appropriate lexicographic order for $m \times n$ matrices with entries in Z_q. In analogy with BIBDs in Sect. 6.1.1, we associate with an $m \times n$ matrix $\mathbf{C}$ the mn-tuple $w(\mathbf{C})$ formed by concatenating the rows of $\mathbf{C}$, and order matrices lexicographically based on lexicographic order on tuples, $0 < 1 < \cdots < q-1$. A matrix is *canonical* if it is the lexicographic minimum of its equivalence class. (Note that requiring minimality is the opposite of the situation with BIBDs in Sect. 6.1.1, where maximality was required – for codes it is aesthetically more appealing to require minimality.)

The following theorem enabling orderly generation has essentially the same proof as Theorem 6.1.

Theorem 7.6. *Let* $\mathbf{C}$ *be a canonical* $M \times n$ *matrix with entries in* Z_q*, and let* $1 \le m \le M$*. Then the* $m \times n$ *matrix obtained by restricting* $\mathbf{C}$ *to the first* m *rows is canonical.*

Thus, starting with the unique canonical $1 \times n$ matrix – which contains only 0s – all canonical $M \times n$ matrices of $(n, M, d)_q$ codes can be generated in a row-by-row manner where every encountered matrix that is not canonical is rejected.

Example 7.7. To classify the $(5,3,3)_2$ codes, we start from the all-zero word of length 5. There are three canonical matrices of $(5,2,3)_2$ codes:

$$\begin{matrix}00000 & 00000 & 00000\\ 00111 & 01111 & 11111\end{matrix}$$

When one tries to augment these matrices, only one matrix is canonical, so the following $(5,3,3)_2$ code is unique:

$$\begin{matrix}00000\\ 00111\\ 11011\end{matrix}$$

We proceed to make some observations on canonical matrices. These observations will assist in developing an algorithm for testing the canonicity and in restricting generation of augmenting codewords when the aim is to obtain a canonical matrix.

First, in analogy with Theorem 6.3, the rows and columns of a canonical matrix must appear in increasing lexicographic order – sorting yields a lexicographically smaller equivalent matrix if this is not the case. For example, this implies that only the lexicographic minimum word in a solution set in Example 7.5 can lead to a canonical matrix.

Second, because we also allow permutation of the values Z_q in every column, we obtain a minimality requirement for the columns of a canonical matrix. Let $\mathbf{C}$ be an $m \times n$ matrix with entries in Z_q. We say that a column $1 \le j \le n$ of $\mathbf{C}$ is *minimal* (with respect to permutation of the values Z_q) if every occurrence of a value $u \in Z_q \setminus \{0\}$ in column j is preceded by an occurrence of $u-1$ in an earlier row in column j. Equivalently, column j is minimal if $\{0,1,\ldots,c_{ij}-1\} \subseteq \{c_{1j},c_{2j},\ldots,c_{i-1,j}\}$ for all $i=1,2,\ldots,m$. If a column of $\mathbf{C}$ is not minimal, then it can be transformed into a minimal column – that is, *minimized* – by permuting the values Z_q so that any new value that is discovered in the sequence $c_{1j},c_{2j},\ldots,c_{mj}$ is mapped to the minimum value in Z_q that has not been used so far.

Example 7.8. For $q=4$, a column j with $(c_{1j},c_{2j},\ldots,c_{6j})=(1,0,1,3,0,2)$ is obviously not minimal. The first entry in the sequence is 1, so we set $k_j(1)$ equal to the minimum unused value 0. The next entry is 0. The value 0 has been used, so $k_j(0)=1$ is the minimum unused value. The third entry is 1, which we have already encountered. The fourth entry is 3, so we set $k_j(3)=2$, and so forth. Eventually we obtain the minimized column $(0,1,0,2,1,3)$ via the value permutation $k_j(1)=0$, $k_j(0)=1$, $k_j(3)=2$, $k_j(2)=3$.

In a canonical matrix all columns must be minimal – otherwise a lexicographically smaller equivalent matrix could be obtained by minimizing a column that is not minimal.

Let us now develop these observations into an algorithm for testing the canonicity. It turns out that Algorithm 6.1 adapts to the present context with minor modifications. To test the canonicity of an $m \times n$ matrix $\mathbf{C}$ with entries in Z_q, it suffices to consider only permutations of rows in a backtrack setting instead of considering the entire group $E = S_m \times (S_q \wr S_n)$. Indeed, once a permutation of the rows has been fixed, the minimum equivalent matrix subject to this ordering of the rows is obtained by first minimizing every column and then sorting the columns to increasing lexicographic order; we write $\overrightarrow{\mathbf{C}}$ for a matrix obtained from $\mathbf{C}$ in this manner.

Example 7.9. We show an example matrix $\mathbf{C}$, the matrix obtained by minimizing the columns of $\mathbf{C}$, and the matrix $\overrightarrow{\mathbf{C}}$:

$$\begin{bmatrix} 2220123 \\ 0111111 \\ 1103333 \\ 3321032 \end{bmatrix}, \quad \begin{bmatrix} 0000000 \\ 1111011 \\ 2122120 \\ 3201222 \end{bmatrix}, \quad \begin{bmatrix} 0000000 \\ 0111111 \\ 1012222 \\ 2220123 \end{bmatrix}.$$

Algorithm 6.1 now applies almost verbatim to test the minimality of a matrix $\mathbf{C}$ in the present setting. Indeed, the only required changes are that $\overrightarrow{\mathbf{C}}$ must incorporate column minimization, and the lexicographic order relations in lines 1, 4, and 34 must be reversed to reflect the fact that we are searching for lexicographically smaller counterexamples, not lexicographically greater as in the BIBD setting. A pseudocode implementation of this variant of the algorithm, which however lacks the automorphism pruning in Algorithm 6.1, can be found in [302].

The number of inequivalent $(n, m, d)_q$ codes for increasing values of m follows the trend of an exploding growth at some point followed by a sharp drop when approaching $m = A_q(n, d)$. A considerable speed-up may therefore be achieved by abandoning the described approach at some level and using clique searching to complete the codes. In constructing a compatibility graph, it should be remembered that all words have to be lexicographically greater than the codewords of the fixed part of a code.

For equidistant codes, we have the additional requirement that $d_H(\mathbf{c}, \mathbf{c}') = d$ for all codewords $\mathbf{c}, \mathbf{c}' \in C$. For equidistant codes corresponding to RBIBDs (Theorem 3.82), we also know that the code is equireplicate. These observations can be used to restrict the search. In particular, for an equireplicate code, the canonical matrix representation of a code must have the first column

$$[0\,0\,\cdots\,0\,1\,1\,\cdots\,1\,\cdots\,q-1\;q-1\,\cdots\,q-1]^T,$$

where each value occurs M/q times. Forcing this property during row by row construction has the advantage of restricting the $(n, m, d)_q$ codes for $m < M$ that need to considered beyond plain isomorph rejection, which is analogous to the requirement $y_{j_0} = 1$ discussed for BIBDs in Sect. 6.1.1.

Little experimental work has been done in this area; in particular, it would be interesting to see how generation by canonical augmentation compares with the outlined orderly generation algorithm. Apparently, there are so far no results in the literature on using generation by canonical augmentation in this context. Yet again, effective invariants are of great importance in developing such algorithms. In the current case, the invariants could be built from distance distributions between codewords and value distributions in coordinates (for equidistant codes, invariants of the former type are useless).

Research Problem 7.10. Classify error-correcting codes by canonical augmentation in a codeword-by-codeword manner and make comparisons with orderly generation.

7.1.3 Constant Weight Codes

The two main methods for classifying constant weight codes are the same as for unrestricted error-correcting codes, namely codeword by codeword and through subcodes. It turns out that the codeword-by-codeword approach is directly related to the main BIBD classification algorithm in Sect. 6.1.1.

Only the binary case, $q = 2$, is considered here. In constructing binary constant weight codes codeword by codeword, we want to find vectors of length n and weight w, whose minimum pairwise distance is at least d, that is, the inner product (over nonnegative integers) between two vectors is at most $w-d/2$. When comparing these parameters with a point-by-point classification of BIBDs – see Sect. 6.1.1 – one realizes that the only differences are that we here have an upper bound on the inner product between vectors, and the number of 1s in the positions is not restricted. Therefore the approach in Sect. 6.1.1 applies with only minor modifications. In [475], this approach is considered in the framework of packing designs. As usual, clique finding can be applied to complete a code.

In a classification via subcodes, there are two possibilities in lengthening. To get a code with length n, constant weight w, minimum distance at least d, and cardinality M, one may start from codes with length $n-1$, minimum distance at least d, and constant weight w or $w-1$. In the former case the old codewords get a 0 in the new coordinate, in the latter case they get a 1, and vice versa for the new codewords. Lower bounds on the sizes of the codes that one needs to consider for the respective cases are

$$(n-w)M/n, \quad wM/n. \tag{7.5}$$

These bounds are related to the first Johnson bound for constant weight codes [284].

As with unrestricted codes, the codes can be found through cliques in a compatibility graph after which isomorph rejection is carried out. A transformation of codes to graphs that is analogous to the transformation for BIBDs presented in Sect. 3.3.2 can be utilized in isomorph rejection.

The strategy of classifying constant weight codes via subcodes is used in [72] with details slightly differing from those presented here.

7.1.4 Results

Results on classification of error-correcting codes of various types are scattered throughout the literature. We will mostly consider classification of codes that are optimal, that is, attain $A_q(n,d)$, but suboptimal codes are often interesting as well.

Binary codes with minimum distance at least 1 are just subsets of words of Z_2^n, and do not seem very useful. However, we will give two examples of applications of such codes.

Example 7.11. Subsets of words of Z_2^n can be viewed as 0/1-*polytopes* – polytopes in the n-dimensional space where the coordinates of the vertices take values 0 and 1 – which are of great importance in combinatorial optimization. One often wants the 0/1-polytopes to be full-dimensional, which in coding-theoretic terms means that there must not be a coordinate with only 0s or only 1s. The term 0/1-*equivalence* of polytopes corresponds to the concept of code equivalence.

For a survey of 0/1-polytopes, see [623]. These objects have been classified in dimensions up to 5, see [2]. It would be interesting to see how fast canonical augmentation (Research Problem 7.10) could produce inequivalent objects in dimension 6; in [623] it is mentioned that such a classification is not within reach, but this is perhaps not the case. Preliminary experiments by the authors of this book reveal that such objects can be produced at a rate of over 10^5 per second using a 1-GHz PC.

Research Problem 7.12. Study generation by canonical augmentation for the classification of 6-dimensional 0/1-polytopes. In particular, study the impact of different invariants.

Example 7.13. There is a one-to-one correspondence between Boolean functions of n variables, $f(x_1, x_2, \ldots, x_n)$, and subsets of words of Z_2^n: given a Boolean function, the n-tuples of values for which the function evaluates to 1 form a code, and vice versa. Slepian [539] studied the problem of counting Boolean functions, and Golomb [214] made the first attempts to classify these. Note, however, that in these two studies, the definition of equivalence is slightly more general than ours; in coding-theoretic terms, they also consider C and $Z_2^n \setminus C$ to be equivalent. This slightly modified definition is rather motivated for Boolean functions, since if 0 and 1 are interchangeable in the input, this should hold for the output as well.

We will next look at some general observations and then summarize the published results for specific parameters. The following theorem plays a central role for binary codes.

Theorem 7.14. *For all positive integers d, $A_2(n+1, 2d) = A_2(n, 2d-1)$.*

Proof. If a parity check bit is added to a binary code, the resulting code has only even-weight codewords and even minimum distance, so $A_2(n+1, 2d) \geq A_2(n, 2d-1)$. Since the minimum distance can decrease by at most 1 when a code is punctured by deleting one of the coordinates, $A_2(n, 2d-1) \geq A_2(n+1, 2d)$. □

In determining the size of optimal binary error-correcting codes, one may therefore restrict to either odd or even minimum distances. For the classification problem, codes of one type can be obtained from the other in the following ways.

Binary codes with odd minimum distance $2d-1$ are classified by taking all inequivalent codes with minimum distance $2d$, puncturing these in all possible ways, and rejecting equivalent codes. To go in the other direction, starting with inequivalent codes with odd minimum distance $2d-1$, the procedure is slightly more involved; see [459] and the proof of [378, Theorem 5].

To find an extension from a binary code with minimum distance $2d-1$ to one with minimum distance $2d$ we need to partition the codewords into two sets so that no two codewords in the same set are at distance $2d-1$ from each other. This can be done by finding the proper 2-colorings (if any) of the graph G that has one vertex for each codeword and an edge connecting two vertices exactly when the corresponding codewords have distance $2d-1$. The color of a vertex gives the value of the new coordinate.

A proper 2-coloring can be found by breadth-first search in G. Every connected graph that admits a proper 2-coloring has exactly two proper 2-colorings. The extensions for one of the connected components of G may be fixed up to equivalence, so if G has c connected components, then there are 2^{c-1} extensions to be carried out. The value of c is small for optimal codes – it is 1 or 2 for the optimal codes considered in [459]. Isomorph rejection has to be carried out among the extended codes.

Example 7.15. For $d = 1$, all words in the space are included to get an optimal code, so $A_2(n, 1) = 2^n$ and we have a unique code. To classify the codes attaining $A_2(n+1, 2) = 2^n$ (Theorem 7.14), we construct a graph by connecting vertices whose codewords differ in exactly one coordinate. The graph we get is the n-cube, which is connected, so the optimal binary codes with minimum distance 2 are unique.

The sizes of optimal binary error-correcting codes of length at most 15 are known; see Table 7.1. The values for $d = 1, 2$ follow from Example 7.15 and are not tabulated. By Theorem 7.14, we may restrict to odd values of d.

Codes with one codeword are obviously unique. Codes consisting of two codewords with a given pairwise distance are unique, so the number of inequivalent $(n, 2, d)_2$ codes with $n \geq d$ is $n-d+1$. The following two theorems show that there are no optimal binary codes with three codewords.

Theorem 7.16. *For* $n \geq 1$, $A_2(3n-1, 2n) = A_2(3n-2, 2n-1) = 2$.

Table 7.1. The values of $A_2(n, d)$ for $n \leq 15$ and odd $3 \leq d \leq 11$

$n\backslash d$	3	5	7	9	11
1	1	1	1	1	1
2	1	1	1	1	1
3	2	1	1	1	1
4	2	1	1	1	1
5	4	2	1	1	1
6	8	2	1	1	1
7	16	2	2	1	1
8	20	4	2	1	1
9	40	6	2	2	1
10	72	12	2	2	1
11	144	24	4	2	2
12	256	32	4	2	2
13	512	64	8	2	2
14	1024	128	16	4	2
15	2048	256	32	4	2

Proof. According to the Plotkin bound (Corollary 2.83), $A_2(3n-1, 2n) \leq 2$. A $(3n-1, 2, 2n)_2$ code exists when $3n - 1 \geq 2n$, that is, $n \geq 1$. □

Theorem 7.17. *For* $n \geq 1$, $A_2(3n, 2n) = A_2(3n-1, 2n-1) = 4$. *There are unique* $(3n, 4, 2n)_2$ *and* $(3n-1, 4, 2n-1)_2$ *codes.*

Proof. According to the Plotkin bound (Corollary 2.83), $A_2(3n, 2n) \leq 4$, and a $(3n, 4, 2n)_2$ code attains that bound. Therefore, $d_H(\mathbf{x}, \mathbf{y}) = 2n$ for all pairs $\mathbf{x}, \mathbf{y}$ of codewords of such a code (which also must be equireplicate). Up to equivalence, this code is

$$\begin{array}{ccc}
00\cdots0 & 00\cdots0 & 00\cdots0 \\
00\cdots0 & 11\cdots1 & 11\cdots1 \\
11\cdots1 & 00\cdots0 & 11\cdots1 \\
\underbrace{11\cdots1}_{n} & \underbrace{11\cdots1}_{n} & \underbrace{00\cdots0}_{n}
\end{array}$$

The automorphism group of this code acts transitively on the coordinates, so the $(3n-1, 4, 2n-1)_2$ code obtained by puncturing is also unique. □

In Table 7.2 we survey classification results for the entries in Table 7.1 that are not covered by Theorems 7.16 and 7.17 and the preceding discussion. Now we need to consider both odd and even distances, so for each entry we give both the number of inequivalent codes (N) and the number of inequivalent extended codes (Ne).

For the references, we have used the conventions described in Sect. 6.1.6. For several of the cases, only the even-weight or the odd-weight result appears

Table 7.2. Classification of binary error-correcting codes

n	d	$A_2(n,d)$	N	Ne	References
6	3	8	1	1	2-(8, 4, 3) RBIBD
7	3	16	1	1	[620]
8	3	20	5	3	[24]
9	3	40	1	1	[378]
10	3	72	562	96	[459]
11	3	144	7398	1041	[459]
12	3	256			
13	3	512			
14	3	1024			
15	3	2048			
9	5	6	1	1	2-(6, 3, 4) RBIBD
10	5	12	1	1	2-(12, 6, 5) RBIBD
11	5	24	1	1	[H]
12	5	32	2	1	[208]
13	5	64	1	1	[208]
14	5	128	1	1	[542]
15	5	256	1	1	[542]
12	7	4	9	6	[H]
13	7	8	6	4	2-(8, 4, 6) RBIBD
14	7	16	10	5	2-(16, 8, 7) RBIBD
15	7	32	5	5	[H]
15	9	4	9	6	[H]

in the literature; however, as we have discussed earlier, obtaining the missing entry in these cases by puncturing or extending is rather straightforward. Recall that the reference [H] indicates that the result was obtained in the current work. For entries corresponding to RBIBDs – by Theorem 3.82 – the parameters of the design is given and the value of Ne equals that of N in Table 6.12. If the values of N and Ne are missing, then the instance is still open. Observe that not all of the classification results in the literature are computer-aided.

Some of the results in Table 7.2 have later been verified in [293, 459]. Classification results for many suboptimal codes with $d = 3, 4$ can be found in [459].

The unique $(15, 256, 5)_2$ code – alternatively, its extension – is the celebrated *Nordstrom-Robinson code* [444]. Outside the range of Table 7.1, it was shown by Snover [542] that the following codes are unique: $(23, 4096, 7)_2$, $(23, 2048, 8)_2$, $(24, 4096, 8)_2$; a simpler proof of this result can be found in [147]. These are the *binary Golay code* [209] and two related codes. Classification results for several other parameters outside the range of Table 7.1 can be found in Table 6.12, where results for RBIBDs with $b = 2r$, $v = 2k$ are summarized.

Codes with the same parameters as binary Hamming codes have been extensively studied. There is a unique such code of length 7 – see Table 7.2 – but in general it has been proved that for admissible lengths n there are at least

$$2^{2^{(n+1)/2-\log(n+1)}} 2^{2^{(n-3)/4}}$$

labeled perfect binary one-error-correcting codes [392]; divide this expression by $2^n n!$ to get a lower bound on the number of inequivalent such codes. Similar bounds exist also for nonbinary perfect codes [179].

For q-ary codes with $q > 2$, much less is known than in the binary case. Most of the known results concern codes corresponding to RBIBDs; such classification results can be found in Tables 6.13 to 6.15 and are not repeated here.

Binary optimal codes with $d = 2$ are unique by Example 7.15, but this does not hold for general alphabet sizes q. It is not difficult to see that $(n, q^{n-1}, 2)_q$ optimal codes correspond to orthogonal arrays of size q^{n-1} and strength $n-1$, so for length $n = 3$, these objects correspond to Latin squares by Theorem 2.112. The problem of counting all codes with $d = 2$, not just the inequivalent ones, is studied in [186].

Some classification results for ternary codes are as follows: in [591] it is stated that there is a unique $(6, 4, 5)_3$ code; the uniqueness of $(5, 18, 3)_3$ and $(6, 38, 3)_3$ codes, and $(5, 6, 4)_3$ and $(6, 18, 4)_3$ codes are shown in [451] and [450], respectively. The uniqueness of the $(4, 9, 3)_3$ Hamming code is proved in [292] but was stated without proof even earlier [570]. Delsarte and Goethals [147] proved that the *ternary Golay code* [209] (constructed independently by Virtakallio, see [250] and [109, Sect. 15.3]) and some related unrestricted codes are unique; these codes have parameters $(11, 729, 5)_3$, $(11, 243, 6)_3$, and $(12, 729, 6)_3$. In [370] it is shown that there are ten $(10, 14, 7)_3$ codes, and in [371] that there are 2,703 $(13, 6, 10)_3$ codes and 6,151 $(14, 13, 10)_3$ codes. Classification results for suboptimal ternary codes with $d = 3, 4$ can be found in [450, 451].

The described methods have in several places been used to prove nonexistence of codes and improve upper bounds on $A_q(n, d)$. We list two instances for which the size of an optimal code is not known, and which seem to be tractable.

Research Problem 7.18. Determine the values of $A_2(16, 7) = A_2(17, 8)$ and $A_4(9, 7)$. From [35] and [53], we know that $36 \leq A_2(16, 7) \leq 37$ and $18 \leq A_4(9, 7) \leq 20$, respectively.

The known classification results on constant weight codes almost exclusively concern codes corresponding to Steiner systems and BIBDs (with, respectively, the columns and the rows of the incidence matrix being codewords; for the connection between BIBDs and constant weight codes, see the text following Corollary 2.89).

It is perhaps surprising that constant weight covering codes, which will be discussed subsequently, have been more thoroughly studied from the classification point of view.

Petrenjuk [475] discusses classification of packing designs by orderly generation in a block-by-block manner, and shows that there is a unique optimal packing design with parameters $(v, k, t) = (11, 5, 3)$ and 11 blocks. A tailored algorithm for classifying packing designs with very few blocks is also discussed in [475]; in particular, a scheme for finding all nonisomorphic packing designs with three blocks is presented, and the number of such designs is tabulated for small v, k, and t. Nonexistence results for constant weight codes are reported in [72].

Research Problem 7.19. Carry out a classification of constant weight codes that exhaustively covers parameters up to the computational limit.

Classification methods have been developed for several types of codes with more specific properties, including mixed codes [450, 451], nonbinary constant weight codes [465, 566], and packing designs with index $\lambda > 1$ [558], but there are still many types of codes for which such a study is justified.

Research Problem 7.20. Develop and implement classification algorithms for various types of error-correcting codes – such as asymmetric and unidirectional codes [183], just to mention two examples – and use these to improve bounds in the literature.

7.2 Covering Codes

There is a fundamental difference between error-correcting codes (which can be viewed as packings) and covering codes. Whereas the packing criterion concerns pairs of codewords, the covering criterion concerns all words in the space and the distances between these and the codewords. Obviously, different approaches are therefore needed in classification of covering codes. (But, we will in fact see that the main approaches for these types of codes can in some sense be linked to each other.)

Various approaches for computer-aided classification of covering codes have been tried along the years. Kamps and Van Lint [294] and Stanton and Kalbfleisch [554] were the first to utilize computers to classify covering codes (in [294] expressly to obtain a nonexistence proof). In these two seminal works, combinatorial arguments are intertwined with computational results. Even though the methods have been improved since those days, and several new ideas have seen the light of day, some basic ideas permeate all these approaches.

Consider the Hamming space (Z_q^n, d_H) and a group $G \leq \mathrm{Aut}(Z_q^n, d_H)$. The action of G partitions the space into orbits:

$$Z_q^n = W_1 \cup W_2 \cup \cdots \cup W_m. \tag{7.6}$$

If $G = \mathrm{Aut}(Z_q^n, d_H)$, then we get only one orbit, and if $G = \{1\}$, then the space is partitioned into singleton sets. The following lemma is central in the further development of the classification methods.

Lemma 7.21. *For $1 \le i, j \le m$ and $\mathbf{a} \in W_i$, the distance distribution $(D_{ij0}, D_{ij1}, \ldots, D_{ijn})$, where $D_{ijk} = |\{\mathbf{b} \in W_j : d_H(\mathbf{a}, \mathbf{b}) = k\}|$ does not depend on the choice of the word* $\mathbf{a}$.

Proof. Since $\mathrm{Aut}(Z_q^n, d_H)$ consists of isometries, so does its subgroup G. Hence $d_H(\mathbf{a}, \mathbf{b}) = d_H(g(\mathbf{a}), g(\mathbf{b}))$ for any words $\mathbf{a}, \mathbf{b} \in Z_q^n$ and $g \in G$, so the distance distributions with respect to $\mathbf{a}$ and $g(\mathbf{a})$ coincide. □

To classify covering codes with prescribed parameters, $(n, M)_q R$, we consider the space Z_q^n, choose the group G, and focus on the basic property that must hold: All words in Z_q^n must be covered, that is, must be at distance at most R from some codeword. The next theorem now follows by direct counting.

Theorem 7.22. *For all $1 \le j \le m$,*

$$\sum_{i=1}^{m} \sum_{k=0}^{R} D_{ijk} |W_i| \ge |W_j|.$$

In the next subsections, various choices of these parameters and the methods they lead to are considered. The general approach is actually closely connected to the homomorphism principle. Isomorph rejection is discussed separately.

7.2.1 Some Basic Approaches

When $G = \mathrm{Aut}(Z_q^n, d_H)$, Theorem 7.22 corresponds to the Hamming bound, Theorem 2.78. At the other extreme, when $G = \{1\}$, Theorem 7.22 coincides with the definition of a code with covering radius at most R. It is tempting to assume that a computer search based directly on the inequalities obtained from this basic definition does not lead very far. However, even if such a method is not competitive with the most advanced methods, some interesting results have still been achieved.

Stanton and Bate [552] were the first to consider a search problem based on the basic definition of a covering code – but they restrict their consideration to the subclass of covering designs, that is, constant weight covering codes – and apply a branch-and-bound algorithm similar to Algorithm 5.4 to solve instances of SET COVERS. They use the method only for proving nonexistence, which is the case also in [461], where the method is applied to general (in fact, mixed) covering codes. Complete classification results are obtained in

[468], where much effort is put on isomorph rejection of partial solutions. This approach is particularly appealing for perfect codes, for which we solve EXACT COVERS instead of SET COVERS.

Having discussed the cases $G = \mathrm{Aut}(Z_q^n, d_H)$ and $G = \{1\}$, we are left with a large number of other possible choices of G:

$$\{1\} \leq G \leq \mathrm{Aut}(Z_q^n, d_H).$$

The choice of G and its interaction with the other parts of the algorithm are crucial for the overall performance; a few possible approaches for choosing G that have been discussed in the literature will now be discussed.

Let $G \cong S_n$ be the subgroup of $\mathrm{Aut}(Z_q^n, d_H)$ that permutes the n coordinates arbitrarily but keeps the coordinate values fixed; this choice leads us to the approach taken by Stanton and Kalbfleisch in [554]. For simplicity, we consider binary codes, $q = 2$. Clearly, the space is then partitioned by the action of G into one set for each Hamming weight; we let y_i be the number of codewords of weight i.

By applying Theorem 7.22, the following set of $n + 1$ inequalities is obtained:

$$\begin{array}{rrrl}
y_0 + & y_1 & & \geq \binom{n}{0}, \\
ny_0 + & y_1 + 2y_2 & & \geq \binom{n}{1}, \\
& (n-1)y_1 + \;\; y_2 + 3y_3 & & \geq \binom{n}{2}, \\
& \ddots & & \vdots \\
& & 2y_{n-2} + y_{n-1} + ny_n & \geq \binom{n}{n-1}, \\
& & y_{n-1} + \;\; y_n & \geq \binom{n}{n}.
\end{array} \tag{7.7}$$

To classify covering codes with M codewords, we search for the (nonnegative integer) solutions to this set of inequalities with the additional constraint

$$\sum_{i=0}^{n} y_i = M.$$

The specific structure of the inequalities (7.7) makes it possible to solve these instances of DIOPHANTINE with a fairly straightforward backtrack algorithm; we refer the interested reader to [291].

Note that solving the set of inequalities (7.7) with a prescribed code size M does not yet finish the classification (unless it turns out that there is no solution). Further computer search has to be carried out (to be discussed in Sect. 7.2.2) or combinatorial arguments have to be applied [554] to fix the codewords.

Example 7.23. To classify the $(4,4)_2 1$ covering codes, we choose G to permute the four coordinates arbitrarily and get

$$\begin{aligned}
y_0 + \ \ y_1 \qquad\qquad\qquad\qquad\quad &\geq 1,\\
4y_0 + \ \ y_1 + 2y_2 \qquad\qquad\qquad &\geq 4,\\
3y_1 + \ \ y_2 + 3y_3 \qquad\quad &\geq 6,\\
2y_2 + \ \ y_3 + 4y_4 &\geq 4,\\
y_3 + \ \ y_4 &\geq 1,
\end{aligned}$$

which together with $y_0+y_1+y_2+y_3+y_4 = 4$ has the solutions $(y_0, y_1, y_2, y_3) =$ $(0, 1, 2, 1, 0)$, $(0, 2, 1, 0, 1)$, $(1, 2, 0, 0, 1)$, $(1, 1, 0, 1, 1)$, $(1, 0, 1, 2, 0)$, $(1, 0, 0, 2, 1)$. The reader is encouraged to complete the classification; there are, up to equivalence, two optimal $(4, 4)_2 1$ covering codes: $\{0000, 0001, 1110, 1111\}$ and $\{0000, 0011, 1101, 1110\}$.

The observant reader has realized that there is some symmetry in (7.7) which one obviously should not ignore. Another useful remark is that any code has an equivalent code that contains the all-zero word, so we could assume that $y_0 = 1$. However, we will not waste time tuning a suboptimal method, but instead we turn to a more efficient approach.

7.2.2 Stepwise Refinement of Hamming Spaces

The ideas behind the next choice of G can actually be traced as far back as to the 1960s and results by Kamps and Van Lint [294] and Stanton and Kalbfleisch [555]. The full strength of it, however, was achieved through improvements that were obtained very much later by Blass and Litsyn [51] and Östergård and Blass [460]. Some related ideas have also been presented by Ma [385].

Actually, in this case there is not a single choice of G, but we have a series of groups,

$$\{1\} = G_0 \leq G_1 \leq \cdots \leq G_n = \mathrm{Aut}(Z_q^n, d_H),$$

where $G_i \cong \mathrm{Aut}(Z_q^i, d_H)$ is the subgroup of $\mathrm{Aut}(Z_q^n, d_H)$ that fixes the $n-i$ first coordinates and the values in these coordinates. The orbits of words under the actions of these groups are in fact just stepwise refinements of the space with respect to the values of the coordinates. For example, the orbits under G_{n-1} partition the space Z_q^n into q sets consisting of the words starting with $0, 1, \ldots, q-1$. We denote the number of codewords in the set of words starting with i by y_i, and generalize this notation to any number of specified coordinates.

Having prescribed the size of the code, M, Theorem 2.78 should be checked in the very first step. Now we are ready to apply Theorem 7.22, and we do this for each pair of groups $G_{i-1} \leq G_i$, $1 \leq i \leq n$, starting from $i = n$. For $i = n$ we have the additional equality

$$M = \sum_{j=0}^{q-1} y_j,$$

for $i = n - 1$ we have (for $k = 0, 1, \ldots, q - 1$)

$$y_k = \sum_{j=0}^{q-1} y_{kj},$$

and so on. The whole procedure is demonstrated by the following example.

Example 7.24. As in Example 7.23, we want to classify the $(4, 4)_2 1$ codes, whose existence is not ruled out by the Hamming bound. Initially, we get the instance

$$\begin{aligned} 4y_0 + y_1 &\geq 8, \\ y_0 + 4y_1 &\geq 8, \\ y_0 + y_1 &= 4, \end{aligned}$$

whose only solution is $y_0 = y_1 = 2$. In the next step, we get

$$\begin{aligned} 3y_{00} + y_{01} + y_{10} \phantom{+ y_{11}} &\geq 4, \\ y_{00} + 3y_{01} + y_{11} &\geq 4, \\ y_{00} + 3y_{10} + y_{11} &\geq 4, \\ y_{01} + y_{10} + 3y_{11} &\geq 4, \\ y_{00} + y_{01} &= 2, \\ y_{10} + y_{11} &= 2, \end{aligned}$$

which has the following solutions for $(y_{00}, y_{01}, y_{10}, y_{11})$: $(2, 0, 0, 2)$, $(0, 2, 2, 0)$, $(1, 1, 1, 1)$. At this point, we abandon the example. In the last two stages, one gets, respectively, 8 and 16 inequalities, and isomorph rejection should be included to keep the number of (intermediate and final) solutions small.

To solve the instances of DIOPHANTINE, one may use any of the algorithms discussed in Sect. 5.4.

In comparing this method with the method for classifying error-correcting codes discussed in Sect. 7.1.1, one observes that these are in some sense dual to each other. For error-correcting codes, the method is bottom-up, where small codes are classified and used in the classification of larger and larger codes. For covering codes, the approach is top-down, and possible codeword distributions in the space are considered, dividing the space into smaller and smaller subspaces.

A coordinate-by-coordinate approach for equidistant, equireplicate error-correcting codes is considered in [305]. The tools of this section are not utilized in [305], where pruning is based on distances between partial codewords (and the partial codes are all the time kept equireplicate).

For error-correcting codes with odd minimum distance d, the spheres of radius $R = (d-1)/2$ must be nonoverlapping, so the only modification needed to apply the approach in this section is a replacement of $\geq$ by $\leq$ in Theorem 7.22.

Research Problem 7.25. Implement a coordinate-by-coordinate algorithm for classifying error-correcting codes, and compare it with the approaches presented in Sect. 7.1 and the one from [305] for various sets of parameters. Perhaps it is also possible to develop an algorithm applicable to even d, for example, by using ideas from [71, p. 156]. It should also be tested whether Delsarte's linear programming bounds [145], [388, Chap. 17] lead to an essentially more effective pruning strategy.

7.2.3 Further Improvements

Theorem 7.22, the core of the methods for classifying covering codes that have been discussed here, is derived from the covering property that such codes must have. The covering property can also be expressed as follows: for a given code C and all $\mathbf{x} \in Z_q^n$,

$$\mathcal{A}_0(\mathbf{x}) + \mathcal{A}_1(\mathbf{x}) + \cdots + \mathcal{A}_R(\mathbf{x}) \geq 1, \tag{7.8}$$

where

$$\mathcal{A}_i(\mathbf{x}) = |B_i(\mathbf{x}) \cap C|;$$

the definition of a Hamming sphere, $B_i(\mathbf{x})$, is given in (2.12). If, for all $\mathbf{x} \in Z_q^n$, a covering code must satisfy

$$\lambda_0\mathcal{A}_0(\mathbf{x}) + \lambda_1\mathcal{A}_1(\mathbf{x}) + \cdots + \lambda_l\mathcal{A}_l(\mathbf{x}) \geq \beta,$$

where l is an integer and λ_i $(0 \leq i \leq l)$ and β are rational numbers, then this set of inequalities is denoted by

$$(\lambda_0, \lambda_1, \ldots, \lambda_l)\beta. \tag{7.9}$$

Using this notation, the inequality (7.8) can be expressed as

$$\underbrace{(1, 1, \ldots, 1)}_{R+1}\, 1. \tag{7.10}$$

Obtaining new inequalities that can be used efficiently in the search for covering codes is a highly nontrivial task. Fortunately, for our purposes, we need not go into these constructions, but we may just use inequalities produced by others. In general, the inequalities have been obtained by combinatorial methods or by combining known inequalities. The latter approach has been automatized by Habsieger and Plagne [240]. We here present one important inequality for binary codes obtained by Johnson [285] and rediscovered by Van Wee [600]:

$$\underbrace{\left(\left\lceil\frac{n+1}{R+1}\right\rceil, \left\lceil\frac{n+1}{R+1}\right\rceil, \ldots, \left\lceil\frac{n+1}{R+1}\right\rceil, 1, 1\right)}_{R+2} \left\lceil\frac{n+1}{R+1}\right\rceil.$$

Other important inequalities include those by Zhang [621] and Zhang and Lo [622]. To make use of these inequalities in the algorithms, one should replace the bound given by Theorem 7.22 by the following generalization.

Theorem 7.26. *For all* $1 \le j \le m$,

$$\sum_{i=1}^{m} \sum_{k=0}^{R} \lambda_k D_{ijk} |W_i| \ge \beta |W_j|.$$

The strength of these inequalities is highly dependent on the parameters of the code. For example, for binary codes with covering radius 1, the known inequalities are much stronger for even lengths than for odd lengths. In any case, it is obvious that for binary Hamming codes, which have odd lengths, there cannot be any stronger inequalities than (7.10).

7.2.4 Isomorph Rejection

As with previously discussed structures, isomorph rejection is required for two reasons: partial structures are rejected to speed up the search, and complete structures are rejected to get the final result. Obviously, if no complete structures are found and nonexistence is thereby established, that part of isomorph rejection is not needed. We will here look at how isomorph rejection can be used together with the previously discussed approaches.

In using the method in the beginning of Sect. 7.2.1 where instances of SET COVERS are solved, one may carry out isomorph rejection via recorded objects until a prescribed level of the search tree, and also at the leaves of the tree (that is, for complete codes). See [468] for details. It seems hard to implement canonical augmentation in a nice way, since the choice of the next word to cover in Algorithm 5.4 heavily affects the order in which codewords are added to the code.

For the approach in Sect. 7.2.2, generation by canonical augmentation is possible and even desirable. One may then encode a code into a graph as described in Sect. 3.3.2 and check whether the new coordinate is in the canonically first coordinate orbit The following details, presented in [305], complete the algorithm.

In the extension step, C' is obtained from C by adding one coordinate. The length of C' is n and $|C| = |C'| = M$. The automorphism group $\mathrm{Aut}(C)$ acts on $C = \{\mathbf{c}_1, \mathbf{c}_2, \ldots, \mathbf{c}_M\}$ by permuting the codewords among themselves (this group can be saved from the previous extension step). Let A be the corresponding permutation group that acts on the indices $\{1, 2, \ldots, M\}$ of the words.

An extension of C into C' can be seen as an M-tuple $(c_{1n}, c_{2n}, \ldots, c_{Mn})$ of symbols such that $c_{in} \in Z_q$ extends the word $\mathbf{c}_i$ for $1 \le i \le M$. The group $S_q \times A$ acts on the set of M-tuples of symbols by permuting the symbols and the entries. Finally, we assume a total (for example, lexicographic) order on the M-tuples.

In summary, a code is accepted if the new coordinate is in the canonically first coordinate orbit and if the extension is minimal in its orbit under $S_q \times A$. If A is large, then some care is needed in implementing the latter test; in [305] this is done by utilizing orderly generation that simultaneously checks that the distance criteria are fulfilled. The former test is carried out in [305] only after an extension is completed; see the original paper for details.

Example 7.27. Ignoring the distance criterion and focusing on the minimality test of the extensions, consider the following three-word code of length 5:

00000
00111
11011

The reader is encouraged to verify that $A = \langle(1\ 3)(2)\rangle$ so $|A| = 2$. The extensions that are minimal under the action of $S_2 \times A$ are $(0,0,0)$, $(0,0,1)$, and $(0,1,0)$.

7.2.5 Constant Weight Covering Codes

As mentioned in Sect. 7.2.1, Stanton and Bate [552] used an algorithm for solving SET COVERS to prove nonexistence of certain constant weight covering codes. A coordinate-by-coordinate approach does not seem to be efficient for covering codes with the additional property of having constant weight. On the other hand, as we will now see, constant weight covering codes may be classified via subcodes, an approach that does not seem suitable for general covering codes.

The following method of classifying constant weight covering codes via subcodes is due to Applegate, Rains, and Sloane [10]. We restrict the discussion to binary codes that are covering designs in the restricted sense as defined in Sect. 2.3.1 and continue the discussion in the framework of designs.

For any point p of a covering design with b blocks and parameters v, k, and t, consider the blocks containing p and delete p. These blocks form a covering design with parameters $v-1$, $k-1$, and $t-1$. Moreover, by a straightforward counting argument there must be such a covering design with b' blocks where

$$C(v-1, k-1, t-1) \le b' \le kb/v. \tag{7.11}$$

Compare the bounds of (7.11) with the bounds of (7.1) and (7.5). We now have a framework for classifying covering designs in a recursive manner. For a complete classification with given parameters and cardinality, all inequivalent *minimal* covering designs – analogous to minimal covers as defined in Sect. 5.3 – given by (7.11) are used as starting points and an algorithm for SET

COVERS is used to find the remaining blocks (which do not contain the new point).

Additional conditions for pruning any SET COVERS algorithm applied to instances of the problem under consideration are obtained by observing that each point must occur in at least $C(v-1,k-1,t-1)$ blocks, each pair of distinct points in at least $C(v-2,k-2,t-2)$ blocks, and so on; cf. [395, Eq. (5.6)].

Research Problem 7.28. Study classification of constant weight covering codes in the general sense as defined in Sect. 2.3.1.

7.2.6 Results

We will now look at some general results and summarize the published results on the classification of optimal covering codes for specific parameters. As for perfect codes, the results overlap those of Sect. 7.1.4. In the binary case, all values of $K_2(n,R)$ are known for $n \le 9$ and are shown in Table 7.3.

Table 7.3. The values of $K_2(n,R)$ for $n \le 9$ and $R \le 4$

$n \backslash R$	1	2	3	4
1	1	1	1	1
2	2	1	1	1
3	2	2	1	1
4	4	2	2	1
5	7	2	2	2
6	12	4	2	2
7	16	7	2	2
8	32	12	4	2
9	62	16	7	2

As with error-correcting codes, covering codes with very small cardinality can be classified using combinatorial techniques. Results of this type are summarized in the next two theorems, which are from [468]; the second theorem is part of [468, Theorem 9]. When $n \le R$, the optimal one-word code is obviously unique.

Theorem 7.29. *For* $R+1 \le n \le 2R+1$, $K_2(n,R) = 2$ *and there are* $2R-n+2$ *inequivalent optimal codes.*

Theorem 7.30. *For* $R \ge 1$, $K_2(2R+2,R) = 4$ *and there are* $\lfloor (\frac{R}{2}+1)^2 \rfloor$ *inequivalent optimal codes.*

The entries in Table 7.3 that are not covered by these combinatorial results are listed in Table 7.4. The number of inequivalent codes is given in the column N, and that value is omitted for open cases. One result obtained in this work is denoted by [H]. It might be possible to settle the number of inequivalent codes attaining $K_2(2R+3, R) = 7$, $R \geq 1$, perhaps by utilizing techniques from [310, 311].

Research Problem 7.31. Determine the number of inequivalent codes attaining $K_2(2R+3, R) = 7$.

Table 7.4. Classification of binary covering codes

n	R	$K_2(n, R)$	N	References
5	1	7	1	[554]
6	1	12	2	[468]
7	1	16	1	[620]
8	1	32	10	[468]
9	1	62		
7	2	7	3	[468, 32]
8	2	12	277	[468]
9	2	16	4	[32]
9	3	7	8	[H]

As the following theorem shows, the number of inequivalent optimal codes has no upper limit. The *partition number* $P(m)$ is the number of multisets of positive integers whose sum is m.

Theorem 7.32. *There are $P(q)$ optimal codes attaining $K_q(2, 1) = q$.*

Proof. A $(2, q)_q 1$ code must have all values of Z_q in (at least) one coordinate, say the first one. The values of the second coordinate may then be arbitrarily chosen. Two such codes are equivalent exactly when they have the same integer partition corresponding to the number of times the values of Z_q occur. □

For optimal ternary codes we know that there are three $(2, 3)_3 1$ codes [449] (and Theorem 7.32), a unique $(3, 5)_3 1$ code [449], a unique $(4, 9)_3 1$ (Hamming) code [292], 17 $(5, 27)_3 1$ codes [469], and a unique $(5, 8)_3 2$ code [32]. For optimal quaternary codes we know that there are a unique $(4, 24)_4 1$ code [464] and eight $(4, 7)_4 2$ codes [310].

Research Problem 7.33. It is known [466] that $65 \leq K_3(6, 1) \leq 73$, and it is conjectured that $K_3(6, 1) = 73$. Settling this case – by proving nonexistence of $(6, 72)_3 1$ codes – would be a major achievement in classifying covering codes.

Research Problem 7.34. Classify optimal binary/ternary mixed covering codes. Only sporadic results have been obtained for this problem. In [329] it is reported that there are two inequivalent codes with 4 binary coordinates, 1 ternary coordinate, cardinality 8, and covering radius 1. A classification algorithm proceeding coordinate by coordinate – analogous to the algorithm for q-ary codes that we have seen here – is discussed in [32], but is there used only for nonexistence proofs.

Classification results for covering designs are listed in Tables 7.5 and 7.6. The columns of Tables 7.5 and 7.6 contain the parameters of the design, v, k, and t, the size of a minimal covering, M, the number of nonisomorphic optimal coverings, N, and references to the classification results following the conventions outlined in Sect. 6.1.6. Steiner systems are optimal covering designs; since these have been tabulated in Chap. 6, they are not included here. The cases $v = k + 1$ and $v = k + 2$ are settled by the following two theorems and are therefore also omitted. The result in Theorem 7.36 was obtained by Turán [585].

Theorem 7.35. *For any* $t \leq v - 1$, *we have* $C(v, v - 1, t) = t + 1$. *There is a unique covering design attaining* $C(v, v - 1, t)$.

Proof. For any set of t blocks, consider the set S (of size less than or equal to t) consisting of the points that do not occur in the respective blocks. If necessary, add arbitrary points to S so that $|S| = t$. Obviously, S is not covered by any of the blocks, so $C(v, v - 1, t) \geq t + 1$. On the other hand, any set of $t + 1$ distinct blocks covers all t-sets, so $C(v, v - 1, t) = t + 1$. Since all such sets of blocks are isomorphic, the optimal covering is unique. □

Theorem 7.36. *For any* $t \leq v - 2$, *we have* $C(v, v - 2, t) = s(s - 1)(v - t - 1)/2 + sr$, *where* $s = \lfloor v/(v - t - 1) \rfloor$ *and* $r = v - (v - t - 1)s$. *There is a unique covering design attaining* $C(v, v - 2, t)$.

Many of the results in Tables 7.5 and 7.6 have been verified in [10]; see also [30, 64, 84, 550, 551, 502].

Research Problem 7.37. Carry out a systematic classification of covering designs: verify old results, fill gaps around entries in Tables 7.5 and 7.6, and extend these tables.

There are several classification results in the literature that concern variants of the discussed covering codes, for example, with respect to the index λ. Some classification results for multiple constant weight covering codes are listed in [429, Sect. 11]. It is shown in [460] that the approach used in Sect. 7.2.2 classifies certain multiple covering codes as a by-product. Classification of codes with both the minimum distance and the covering radius prescribed is discussed in [464].

Table 7.5. Classification of covering designs with $t \le 3$

v	k	t	M	N	References
6	3	2	6	1	[10]
8	3	2	11	5	[10]
10	3	2	17	1	[10]
7	4	2	5	4	[565]
9	4	2	8	17	[551, 428], [30]
10	4	2	9	4	[551, 428]
15	4	2	19	4	[3]
22	4	2	39	1	[553]
8	5	2	4	1	[565]
9	5	2	5	1	[550]
10	5	2	6	2	[550]
11	5	2	7	2	[550]
7	4	3	12	4	[10]
9	4	3	25	77	[10]
8	5	3	8	3	[565]
11	5	3	20	1	[502]
9	6	3	7	4	[565]
12	6	3	15	68	[216]

Table 7.6. Classification of covering designs with $t \ge 4$

v	k	t	M	N	References
8	5	4	20	6	[10]
9	5	4	30	3	[84]
10	5	4	51	40	[10], [395]
9	6	4	12	1	[565]
10	6	4	20	1	[30]
10	7	4	10	2	[565]
9	6	5	30	2	[10]
10	6	5	50	1	[533]
10	7	5	20	5	[64]
10	7	6	45	20	[10]
11	7	6	84	3	[10]
11	8	6	29	1	[64]
11	8	7	63	40	[10]
12	8	7	126	3	[10]
12	9	7	40	16	[64]
12	9	8	84	4	[10]
13	9	8	185	1	[10]
13	10	8	52	1	[64]
14	10	9	259	1	[10]

7.3 Linear Codes

Our discussion of linear codes will be restricted to error-correcting codes, that is, codes with minimum distance greater than or equal to d, for some prescribed d. Before we can discuss classification of linear codes, we must first make precise what is meant by equivalence of linear codes.

7.3.1 Equivalence of Linear Codes

Recall from Sect. 3.2.1 that equivalence transformations of unrestricted codes correspond to isometries of the Hamming space (Z_q^n, d_H). This notion of equivalence is too general for linear codes because such an equivalence transformation may destroy linearity. An equivalence transformation for linear codes is by convention *linearity-preserving* in the sense that it transforms linear codes onto linear codes.

Definition 7.38. *Two linear codes $C, C' \in \mathbb{F}_q^n$ are* equivalent *if there exists a linearity-preserving isometry $\psi \in \mathrm{Aut}(\mathbb{F}_q^n, d_H)$ such that $\psi(C) = C'$.*

We proceed to characterize the subgroup of linearity-preserving isometries in the isometry group $\mathrm{Aut}(\mathbb{F}_q^n, d_H)$. Recall that the multiplicative group of $\mathbb{F}_q$ is denoted by $\mathbb{F}_q^*$. Moreover, the group of field automorphisms of $\mathbb{F}_q$ is denoted by $\mathrm{Aut}(\mathbb{F}_q)$. A group element $(z, h) \in \mathbb{F}_q^* \wr S_n$ is a pair consisting of a permutation $h \in S_n$ and an n-tuple $z = (z_1, z_2, \ldots, z_n)$, where $z_i \in \mathbb{F}_q^*$ for all $1 \le i \le n$; cf. Sect. 3.1.4. A field automorphism $\alpha \in \mathrm{Aut}(\mathbb{F}_q)$ acts on z by $\alpha(z) = (\alpha(z_1), \alpha(z_2), \ldots, \alpha(z_n))$.

Define a group homomorphism $\theta : \mathrm{Aut}(\mathbb{F}_q) \to \mathrm{Aut}(\mathbb{F}_q^* \wr S_n)$ by setting $\theta_\alpha((z, h)) = (\alpha(z), h)$ for all $\alpha \in \mathrm{Aut}(\mathbb{F}_q)$ and $(z, h) \in \mathbb{F}_q^* \wr S_n$. Let the semidirect product $(\mathbb{F}_q^* \wr S_n) \rtimes_\theta \mathrm{Aut}(\mathbb{F}_q)$ act on $\mathbb{F}_q^n$ by

$$\big((z, h, \alpha)\mathbf{x}\big)_i = z_i \alpha(x_{h^{-1}(i)}) \tag{7.12}$$

for all $\alpha \in \mathrm{Aut}(\mathbb{F}_q)$, $(z, h) \in \mathbb{F}_q^* \wr S_n$, $\mathbf{x} \in \mathbb{F}_q^n$, and $1 \le i \le n$. The reader is encouraged to check that this indeed defines a group action. The action (7.12) can be described as follows. First, all coordinate values in a word are permuted using α. Then, the coordinates are permuted so that coordinate i becomes coordinate $h(i)$ for all $1 \le i \le n$. Finally, the value in coordinate i is multiplied by z_i for all $1 \le i \le n$. Each of these individual transformations maps linear subspaces onto linear subspaces, so the compound transformation (7.12) is also linearity-preserving.

The proof of the following theorem is from [448].

Theorem 7.39. *For $n \ge 3$ the linearity-preserving subgroup of $\mathrm{Aut}(\mathbb{F}_q^n, d_H)$ is isomorphic to the semidirect product $(\mathbb{F}_q^* \wr S_n) \rtimes_\theta \mathrm{Aut}(\mathbb{F}_q)$.*

Proof. It is easily checked that the kernel of the group homomorphism given by (7.12) is trivial, so it suffices to show that every linearity-preserving isometry $\psi \in \mathrm{Aut}(\mathbb{F}_q^n, d_H)$ can be obtained as $\psi(\mathbf{x}) = (z, h, \alpha)\mathbf{x}$ for some $(z, h, \alpha) \in (\mathbb{F}_q^* \wr S_n) \rtimes_\theta \mathrm{Aut}(\mathbb{F}_q)$.

Because ψ is linearity-preserving, it must fix the subspace consisting only of the all-zero word. Because ψ is an isometry, by Theorem 3.54 we may assume – by composing ψ with an appropriate isometry induced by a $(z, h, \alpha) \in (\mathbb{F}_q^* \wr S_n) \rtimes_\theta \mathrm{Aut}(\mathbb{F}_q)$ if necessary – that ψ fixes all the coordinates and the all-one word. Thus, for all $\mathbf{x} \in \mathbb{F}_q^n$ we have $\psi(\mathbf{x}) = (\alpha_1(x_1), \alpha_2(x_2), \ldots, \alpha_n(x_n))$, where $\alpha_i \in \mathrm{Sym}(\mathbb{F}_q)$, $\alpha_i(0) = 0$, $\alpha_i(1) = 1$ for all $1 \leq i \leq n$. Consider the subspace $C = \{c(1, 1, \ldots, 1) : c \in \mathbb{F}_q\}$. Since $(1, 1, \ldots, 1) \in \psi(C)$ and ψ preserves the dimension of a subspace, we have $\psi(C) = C$. Thus, $\alpha_1 = \alpha_2 = \cdots = \alpha_n = \alpha$. To complete the proof, we show that $\alpha \in \mathrm{Aut}(\mathbb{F}_q)$.

Let $a, b \in \mathbb{F}_q$ and consider the subspace $C' = \{c(1, 0, a) + d(0, 1, b) : c, d \in \mathbb{F}_q\}$ – for spaces with more than three coordinates, the remaining coordinates are set to 0. Since $(1, 1, a + b) \in C'$, we have $(1, 1, \alpha(a + b)) \in \psi(C')$. On the other hand, $(1, 0, \alpha(a)) \in \psi(C')$ and $(0, 1, \alpha(b)) \in \psi(C')$, so $(1, 1, \alpha(a) + \alpha(b)) \in \psi(C')$ since $\psi(C')$ is a subspace. It follows that $\alpha(a+b) = \alpha(a)+\alpha(b)$ for all $a, b \in \mathbb{F}_q$.

Letting $d = 1$, we get $\psi((c, 1, ac + b)) = (\alpha(c), 1, \alpha(ac) + \alpha(b)) \in \psi(C')$ for all $c \in \mathbb{F}_q$. On the other hand, $\alpha(c)(1, 0, \alpha(a)) + (0, 1, \alpha(b)) = (\alpha(c), 1, \alpha(a)\alpha(c) + \alpha(b)) \in \psi(C')$, so $\alpha(ac) = \alpha(a)\alpha(c)$ for all $a, c \in \mathbb{F}_q$. Thus, $\alpha \in \mathrm{Aut}(\mathbb{F}_q)$. □

The following lemma gives an important special case of Theorem 7.39.

Lemma 7.40. *For a parity check matrix of a linear code C over $\mathbb{F}_q$, and any $a \in \mathbb{F}_q^*$, if a column vector $\mathbf{y}$ of the parity check matrix is replaced by $a\mathbf{y}$ to get a new code C', then C' is equivalent to C.*

Proof. Multiply the values in the coordinate corresponding to the altered column by a to get C from C' – an equivalent code is then obtained by Theorem 7.39. □

We are now ready to discuss classification of linear codes. Since $[n, k, d]_q$ linear codes make up a subset of unrestricted $(n, q^k, d)_q$ codes, one possibility of getting at this classification would be through a classification of the latter codes. However, since such an approach is applicable only for trivially small parameters, it is not discuss further.

As in the unrestricted case, one possibility of classifying linear error-correcting codes is through subcodes.

7.3.2 Constructing Linear Codes via Subcodes

An $[n, k, d]_q$ linear code C with $k \geq 1$ has at least one coordinate in which not all codewords have a 0. If we shorten C in such a coordinate, and let C'

consist of the codewords which had a 0 in the deleted coordinate, then C' is obviously a linear $[n-1, k-1, d]_q$ code.

If we look back at what we did with unrestricted codes in Sect. 7.1.1, it is obvious that the arguments above are leading us towards an analogous approach here. Lengthening a linear code, however, is easier than lengthening an unrestricted code, and the clique search part may be omitted. To describe the lengthening step – and the whole approach – it is convenient to define a linear code through a parity check matrix. Moreover, by Theorem 2.95, to test whether the code with a given parity check matrix has minimum distance greater than or equal to d, it suffices to check whether every $d-1$ columns of the matrix are linearly independent. Generator matrices may here be considered instead of parity check matrices – and some authors indeed do so.

All $[n, k, d]_q$ codes can be obtained by, for each $[n-1, k-1, d]_q$ code, adding a new column in all possible ways to its parity check matrix, checking the minimum distance of the new code, and finally carrying out isomorph rejection. This recursive procedure may be started from the unique $[n-k, 0, d]_q$ code with parity check matrix $[\mathbf{I}_{n-k}]$; see the discussion preceding Theorem 2.93. Details of this procedure will be discussed when we put all pieces together in Sect. 7.3.5.

Actually, we may start the construction procedure from $[n-k+1, 1, d]_q$ codes rather than $[n-k, 0, d]_q$ codes.

Lemma 7.41. *The number of inequivalent* $[n, 1, d]_q$ *codes is* $n-d+1$. *The parity check matrices of these codes are, up to equivalence,* $\mathbf{H} = [\mathbf{I}\ \mathbf{A}]$, *where* $\mathbf{A} = [0 \ \cdots \ 0 \ \underbrace{1 \ \cdots \ 1}_{t}]^T$ *and* $t \geq d-1$.

Proof. An $[n, 1, d]_q$ code is generated by a single word of weight at least d. All nonzero coordinates of this word may be transformed into 1s by multiplying the coordinate values with their multiplicative inverses (such transformations lead to equivalent codes, see Sect. 7.3.1). The codes obtained are, up to permutation of the coordinates, the codes defined by the parity check matrix stated in the theorem. □

If one is expressly constructing codes with minimum distance exactly d, one may start the search from the $[n-k+1, 1, d]_q$ code with $d-1$ 1s in the last column of the parity check matrix.

One may also construct linear codes via their residual codes. The *residual code* of C with respect to a codeword $\mathbf{c}$ is the code obtained by deleting all coordinates of C where $\mathbf{c}$ has a nonzero entry. The following result is from [164].

Lemma 7.42. *Suppose* C *is an* $[n, k, d]_q$ *code and suppose* $\mathbf{c} \in C$ *has weight* w, *where* $d > w(q-1)/q$. *Then the residual code of* C *with respect to* $\mathbf{c}$ *is an* $[n-w, k-1, d']_q$ *code with* $d' \geq d - w + \lceil w/q \rceil$.

At this point we will present two methods for deciding equivalence of linear codes. Both are important, since one is practical for codes with small dimension and the other for codes with large dimension.

7.3.3 Isomorph Rejection using Words of Given Weights

Leon [365] developed an algorithm for computing the automorphism group of a linear code; this algorithm can with minor modifications be utilized to test equivalence of codes. Since the algorithm is fairly advanced – the pseudocode description contains approximately 200 lines – we do not discuss it here, but refer the interested reader to [365]. Instead, we will once again transform the object under consideration into a graph (so that, for example, *nauty* can be used). The exposition builds on that of [455].

Definition 7.38 and Theorem 7.39 give the framework for studying equivalence of linear codes. Notably, the isometries in Theorem 7.39 all preserve the weight of a codeword to which they are applied.

Theorem 7.43. *Consider two linear codes $C, C' \in \mathbb{F}_q^n$. For a given set $S \subseteq \{0, 1, \ldots, n\}$, let $W = \{c \in C : \mathrm{wt}(c) \in S\}$ and $W' = \{c \in C' : \mathrm{wt}(c) \in S\}$. If the words in W and W' generate C and C', respectively, then for any linearity-preserving $\psi \in \mathrm{Aut}(\mathbb{F}_q^n, d_H)$ we have $\psi(W) = W'$ if and only if $\psi(C) = C'$.*

Proof. By Theorem 7.39, ψ preserves the weight of a word. Thus, $\psi(C) = C'$ implies $\psi(W) = W'$. Conversely, because the words of $W \subseteq C$ generate C, any word $\mathbf{c} \in C$ can be written as a sum $\mathbf{c} = \mathbf{w}_1 + \mathbf{w}_2 + \cdots + \mathbf{w}_m$ for some m – note that we need not consider a weighted sum since $\mathrm{wt}(\mathbf{w}) = \mathrm{wt}(a\mathbf{w})$ for $a \in \mathbb{F}_q^*$, so if $\mathbf{w} \in W$, then $a\mathbf{w} \in W$. Consequently, $\psi(\mathbf{c}) = \psi(\mathbf{w}_1 + \mathbf{w}_2 + \cdots + \mathbf{w}_m) = \psi(\mathbf{w}_1) + \psi(\mathbf{w}_2) + \cdots + \psi(\mathbf{w}_m) \in C'$ since $\psi(\mathbf{w}_i) \in W'$. Hence $\psi(C) \subseteq C'$. Similarly, from $\psi^{-1}(W') = W$ we obtain $\psi^{-1}(C') \subseteq C$, so $\psi(C) = C'$. □

To compute the automorphism group and a certificate for a linear code C, first find the minimum value of w such that $\{\mathbf{c} \in C : \mathrm{wt}(\mathbf{c}) \leq w\}$ generates C, and call this set W (so far, but no longer, this follows the approach in [365]). Almost always it suffices to take the codewords of minimum nonzero weight. For small codes one might as well take all codewords, but it is important that one consistently uses the same algorithm for the choice of weights, as this is a part of the certificate.

For a moment, let us ignore the isometries with a nontrivial field automorphism component in Theorem 7.39; we are then left with what is called *monomial* transformations. We now transform the words in W to a graph in the following way [455]. Index the coordinates by $1, 2, \ldots, n$. For a codeword set $C = \{\mathbf{c}(1), \mathbf{c}(2), \ldots, \mathbf{c}(M)\}$, we construct a vertex-colored graph with vertex set $C \cup (\{1, 2, \ldots, n\} \times \mathbb{F}_q^*)$ and edge set $\{\{\mathbf{c}(i), (j, y)\} : c(i)_j = y \neq 0\} \cup \{((j, y), (j, z)) : 1 \leq j \leq n,\ y, z \in \mathbb{F}_q^*,\ z = ay\}$, where a is a primitive root of $\mathbb{F}_q$, that is, a generator of its multiplicative group. Note that some of the edges are presented as ordered pairs to indicate that they are directed. For binary codes there are no such edges and the transformation coincides with that for binary constant weight codes.

Example 7.44. By taking the four nonzero codewords of the unique $[2, 1, 2]_5$ code, we get the graph in Fig. 7.3.

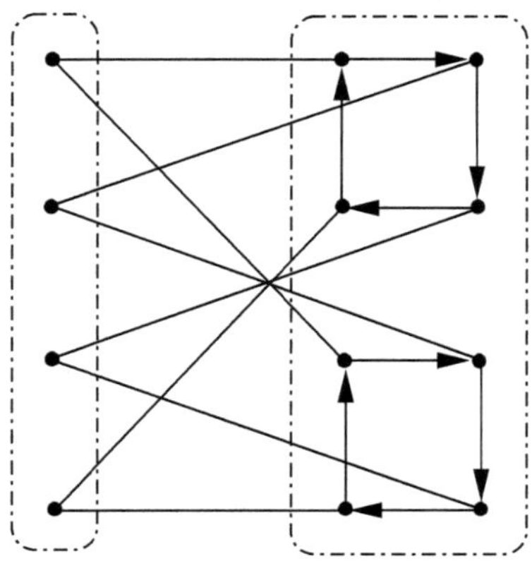

Fig. 7.3. Transforming a linear code to a graph

Fields of order p^a, where p is a prime and $a > 1$ have nontrivial automorphisms. The smallest such fields are $\mathbb{F}_4$, $\mathbb{F}_8$, $\mathbb{F}_9$, $\mathbb{F}_{16}$, $\mathbb{F}_{25}$,.... There are several possibilities of taking the field automorphisms into account.

Since the number of field automorphisms is very small for small fields and they act globally, one may simply construct $|\mathrm{Aut}(\mathbb{F}_q)|$ codes (one for each automorphism, let it act globally on the original code), compute the certificates, and let, say, the lexicographically smallest certificate be the certificate of the code [455]. The number of coinciding certificates among these must be taken into account in calculating the order of the automorphism group of the code.

7.3.4 Isomorph Rejection in Projective Geometries

We will here bring forward and utilize a well-known (cf. [272]) connection between linear codes and sets of points in projective geometries.

The one-dimensional subspaces of $\mathbb{F}_q^r$ can be viewed as the points of an incidence structure $(P, \mathcal{L}, I)$, where the blocks – or *lines* – are the two-dimensional subspaces, and incidence is given by containment; that is, $(p, L) \in I$ if and only if $p \subseteq L$. This incidence structure is called the *projective geometry* $\mathrm{PG}(r-1, q)$ of order q and *projective* dimension $r - 1$. The two-dimensional projective geometries were encountered already in Chap. 2. See [78] for a handbook of projective and other geometries; a brief survey can be found in [43].

Let us now consider linear codes in the framework of projective geometries. Throughout this section we assume that the minimum distance of a linear code is at least 3. An $[n,k,d]_q$ code is uniquely determined by a parity check matrix $\mathbf{H}$ of size $r \times n$, where $r = n - k$. By our assumption $d \geq 3$, the columns of a parity check matrix span distinct one-dimensional subspaces. Furthermore, by Lemma 7.40 any column $\mathbf{h}$ of $\mathbf{H}$ can be replaced by $a\mathbf{h}$ with $a \in \mathbb{F}_q^*$ to get an equivalent code. Thus, we can determine C up to equivalence from the n-set of one-dimensional subspaces of $\mathbb{F}_q^r$ spanned by the columns of $\mathbf{H}$ – simply take a nonzero vector from each subspace to get a parity check matrix of a code equivalent to C. In projective geometry terms, the preceding is equivalent to saying that we can determine C up to equivalence from an n-set of points of $\mathrm{PG}(r-1,q)$. However, a given equivalence class of linear codes in general admits multiple such representations as an n-set of points. To perform isomorph rejection on linear codes using this representation, we require a characterization of the distinct n-sets of points that define equivalent codes. We proceed to show that two n-sets of points represent equivalent linear codes precisely when there exists an automorphism – the term *collineation* is preferred by geometers – of $\mathrm{PG}(r-1,q)$ relating the two point sets.

The structure of the collineation group of $\mathrm{PG}(r-1,q)$ is as follows. It will be convenient to regard a collineation as a permutation of the points only. First, observe that every field automorphism $\alpha \in \mathrm{Aut}(\mathbb{F}_q)$ defines a collineation by acting on $\mathbf{x} = (x_1, x_2, \ldots, x_r) \in \mathbb{F}_q^r$ by

$$\alpha * \mathbf{x} = (\alpha(x_1), \alpha(x_2), \ldots, \alpha(x_r)).$$

Similarly, every invertible $r \times r$ matrix $\mathbf{L}$ over $\mathbb{F}_q$ defines a collineation by acting on $\mathbf{x} \in \mathbb{F}_q^r$ by $\mathbf{L} * \mathbf{x} = \mathbf{L}\mathbf{x}$. Here $\mathbf{L}$ and $a\mathbf{L}$ define the same collineation for all $a \in \mathbb{F}_q^*$. The group $\mathrm{P\Gamma L}_r(q)$ is the permutation group on the points of $\mathrm{PG}(r-1,q)$ generated by these collineations.

Theorem 7.45 (Fundamental theorem of projective geometry). *The collineation group of* $\mathrm{PG}(r-1,q)$ *is the group* $\mathrm{P\Gamma L}_r(q)$.

Three references for a proof of this result are listed in [388, p. 700]. See also [509, Theorem 9.43].

Now, two linear codes $C, C' \subseteq \mathbb{F}_q^n$ with $r \times n$ parity check matrices $\mathbf{H}, \mathbf{H}'$, respectively, are equivalent if and only if there exists an invertible $r \times r$ matrix $\mathbf{L}$ and a linearity-preserving $\psi \in \mathrm{Aut}(\mathbb{F}_q^n, d_H)$ such that $\mathbf{H}' = \mathbf{L}\psi(\mathbf{H})$, where ψ acts on each row of $\mathbf{H}$. Recalling Theorem 7.39, if we view the matrices $\mathbf{H}'$ and $\mathbf{H}$ as defining n-sets of points in $\mathrm{PG}(r-1,q)$, the monomial transformation part of ψ obviously fixes the n-set of points defined by $\mathbf{H}$. What remains is the field automorphism component α and the matrix $\mathbf{L}$, and this is exactly the notion of equivalence for two n-sets of points under the induced action of $\mathrm{P\Gamma L}_r(q)$. Therefore we have that $\mathbf{H}$ and $\mathbf{H}'$ determine equivalent linear codes if and only if the n-sets of points defined by their columns are in the same orbit of the action of $\mathrm{P\Gamma L}_r(q)$.

The approach of classifying linear codes via sets of points in projective geometries is presented by Davies and Royle [141], who use available algebra software to classify small binary and ternary linear codes, calculating orbits of point sets under the action of $\mathrm{P\Gamma L}_r(q)$.

To classify sets of points in $\mathrm{PG}(r-1,q)$, we may either employ the projective geometry directly, or consider a more compact incidence structure which we now describe. This idea is stated without proof in [510]; the proof of the following theorem was communicated to us by G. F. Royle.

A *hyperplane* in $\mathbb{F}_q^r$ is a subspace of dimension $r-1$. Analogously to a projective geometry, we can define the point-hyperplane incidence structure with incidence defined by the subspace relation. The advantage of this incidence structure is that the number of hyperplanes is equal to the number of points, which is in general smaller than the number of lines in a projective geometry.

Again we consider the automorphism group on the points only.

Theorem 7.46. *The automorphism group of the point-hyperplane incidence structure is the group* $\mathrm{P\Gamma L}_r(q)$.

Proof. It is obvious that $\mathrm{P\Gamma L}_r(q)$ is a group of automorphisms. To show that no other automorphisms exist, we prove that the lines of $\mathrm{PG}(r-1,q)$ can be uniquely determined from the hyperplanes. The result then follows by Theorem 7.45.

Any subspace of $\mathbb{F}_q^r$ is clearly uniquely determined by the set of 1-dimensional subspaces that it contains. From the point-hyperplane incidence structure we know all the $(r-1)$-dimensional subspaces. Given all the k-dimensional subspaces, we can obtain all the $(k-1)$-dimensional subspaces as pairwise intersections of k-dimensional subspaces, where the cardinality of an intersection reveals its dimension. Thus, proceeding one dimension at a time, we obtain the 2-dimensional subspaces, which constitute the lines of $\mathrm{PG}(r-1,q)$. □

To study a set of points in $\mathrm{PG}(r-1,q)$, we construct the incidence graph of the point-hyperplane incidence structure, and introduce a new color for the distinguished set of points. (Alternatively, an extra vertex may be added that is connected by vertices to the subset of points under consideration.) We then have a graph of order $2(q^r-1)/(q-1)$ which can, for example, be handled by *nauty*. In practice, it turns out that such instances can be very hard due to their regularity and symmetry, whereby careful attention has to be paid to invariants and to pruning with automorphisms.

Example 7.47. The geometry $\mathrm{PG}(2,2)$ is a projective plane of order 2, which we have already seen in Example 2.34 and Fig. 2.9, and which we know as the Fano plane. Hyperplanes of geometries with projective dimension 2 are lines, so a point-hyperplane incidence matrix is then a point-line incidence matrix of $\mathrm{PG}(2,2)$, which is isomorphic to the incidence matrix

$$\begin{bmatrix} 1\,1\,1\,0\,0\,0\,0 \\ 1\,0\,0\,1\,1\,0\,0 \\ 1\,0\,0\,0\,0\,1\,1 \\ 0\,1\,0\,1\,0\,1\,0 \\ 0\,1\,0\,0\,1\,0\,1 \\ 0\,0\,1\,1\,0\,0\,1 \\ 0\,0\,1\,0\,1\,1\,0 \end{bmatrix}.$$

7.3.5 Implementation Issues

We now have all ingredients needed for classification via recorded objects (Sect. 4.2.1). If we, for given parameters, start from the $n+1-d$ parity check matrices given by Lemma 7.41, process these in lexicographic order (with respect to the values of the columns), and go through the candidates for a new column in lexicographic order, then canonical representatives are generated in this classification and it is sufficient to test only candidate columns that are lexicographically greater than the previous columns (that is, greater than the last column). For nonbinary codes, Lemma 7.40 is also useful; it implies that we need only consider columns with a 1 in the most significant nonzero position.

Example 7.48. We classify the $[6,3,3]_2$ codes, and start from the $[4,1,3]_2$ codes given by Lemma 7.41:

$$\begin{bmatrix} 1\,0\,0\,0 \\ 0\,1\,0\,1 \\ 0\,0\,1\,1 \end{bmatrix}, \quad \begin{bmatrix} 1\,0\,0\,1 \\ 0\,1\,0\,1 \\ 0\,0\,1\,1 \end{bmatrix}.$$

There are no columns lexicographically greater than the last column of the second matrix. Three columns can be adjoined to the first matrix: $[1\ 0\ 1]^T$, $[1\ 1\ 0]^T$, and $[1\ 1\ 1]^T$ ($[1\ 0\ 0]^T$ leads to a code with minimum distance 2). It turns out that the three possible matrices are equivalent – this is doable by hand, check it – so there is a unique $[5,2,3]_2$ code,

$$\begin{bmatrix} 1\,0\,0\,0\,1 \\ 0\,1\,0\,1\,0 \\ 0\,0\,1\,1\,1 \end{bmatrix}.$$

In the same manner, it turns out that there is a unique $[6,3,3]_2$ code,

$$\begin{bmatrix} 1\,0\,0\,0\,1\,1 \\ 0\,1\,0\,1\,0\,1 \\ 0\,0\,1\,1\,1\,0 \end{bmatrix}.$$

In fact, we can proceed one step further to get to the unique $[7,4,3]_2$ Hamming code (but no computer search is needed to obtain that result, since there is only one possible choice – disregarding the order of the columns – of seven distinct nonzero binary columns of length 3).

Let us redo this example in projective geometry.

Example 7.49. With $r = 3$ and $q = 2$, we study sets of points in PG(2, 2), the Fano plane. For clarity, no combinatorial arguments are used in this example; we merely search for sets of points of the Fano plane (see Example 7.47). Because the collineation group of the Fano plane acts 2-transitively both on the points and on the lines, it follows that there are unique sets – displayed as rows of a matrix – of one and two points: [1 1 1 0 0 0 0] and

$$\begin{bmatrix} 1\,1\,1\,0\,0\,0\,0 \\ 1\,0\,0\,1\,1\,0\,0 \end{bmatrix}.$$

For three points, it turns out that there are two possibilities:

$$\begin{bmatrix} 1\,1\,1\,0\,0\,0\,0 \\ 1\,0\,0\,1\,1\,0\,0 \\ 1\,0\,0\,0\,0\,1\,1 \end{bmatrix}, \quad \begin{bmatrix} 1\,1\,1\,0\,0\,0\,0 \\ 1\,0\,0\,1\,1\,0\,0 \\ 0\,1\,0\,1\,0\,1\,0 \end{bmatrix}.$$

The discrepancy compared with Example 7.48 is due to the required form of the parity check matrix with the identity matrix in the first three columns in that example, where we thereby need not consider a matrix with (in projective geometry terms) three collinear points:

$$\begin{bmatrix} 1\,0\,1 \\ 0\,1\,1 \\ 0\,0\,0 \end{bmatrix}.$$

The classification process may be continued by adding further points of the Fano plane and rejecting isomorphs. The fact that the collineation group of the Fano plane is 2-transitive implies that the sets of five and six points are unique. Obviously, all seven points in the Fano plane gives the unique $[7, 4, 3]_2$ (Hamming) code.

Linear codes differ from unrestricted codes in the sense that it is possible to *count* some important classes of such codes without having to construct the corresponding codes. These classes include, for example, linear codes with minimum distance at least 3. Obviously, this is not easy for arbitrary d; if that was the case, the problem of determining the maximal d for which $[n, k, d]_q$ codes exist would also be easy. But that would then solve one of the main open problems in coding theory!

Slepian [540] developed formulae for counting binary linear codes and Fripertinger and Kerber [194, 195, 196, 197] studied the problem of counting linear codes with $q > 2$ (considering, in those papers, only monomial transformations for nonprime fields). In classifying certain linear codes, one is therefore in the rare but fortunate situation of knowing – or at least being able to calculate – the number of inequivalent objects one should get. It is even more straightforward to count inequivalent self-dual linear codes since the number of such codes is given by a simple formula [497].

It seems difficult to come up with a method for finding a proper definition of a canonical parity check (or generator) matrix, which could be used for orderly generation of linear codes (already Slepian [540] noticed this). Isomorph rejection via recorded objects can be used for the smallest cases, but in the general case one would certainly need to utilize generation by canonical augmentation (Sect. 4.2.3).

In the projective geometry framework, generation by canonical augmentation has been used, for example, in [510]. In the framework of codes – where we classify $[n, k, d]_q$ codes through a sequence of classifications of shorter codes with co-dimension $r = n-k$ – the property $\mathcal{P}$ that the co-dimension should be r is not hereditary, meaning that not all codes obtained from a code with the property $\mathcal{P}$ by shortening have the property $\mathcal{P}$. However, we can guarantee that there is at least one shortened code with the property $\mathcal{P}$, so generation by canonical augmentation is applicable. The following approach is presented in [61].

Shortening an $[n, k, d]_q$ linear code by deleting one coordinate and keeping the codewords with a 0 in that coordinate gives an $[n-1, k', d]_q$ code with $k' = k$ if the original code has only 0s in the coordinate to be deleted, and $k' = k - 1$ otherwise. Therefore, in the parent test of generation by canonical augmentation – after adding one coordinate via a new column in the parity check matrix – one should first check which coordinates are all-zero. In the test itself, only coordinates that are not all-zero should be considered. In implementing the test, the ideas in Sect. 7.3.3 can be applied. For fields with nontrivial automorphisms, like $\mathbb{F}_4$, if one uses the idea of producing one graph for each automorphism, a code passes the parent test if at least one of the $|\mathrm{Aut}(\mathbb{F}_q)|$ instances passes the test. Finally, observe that care should be taken when choosing candidates for the next column of the matrix; one cannot in general assume that the added columns should be in lexicographic order.

Research Problem 7.50. Study and implement generation by canonical augmentation for classification of linear codes and use this approach to attack, for example, Research Problem 7.52.

Studies where linear codes have been constructed and classified via their residuals include [59, 61, 62].

Other techniques that have been proposed and used for classifying linear codes include utilization of split weight enumerators [280] and various group theoretic methods [181, 197].

7.3.6 Results

For linear codes, a wide variety of manual and computer-aided classification results have been published, and a complete coverage of all those results are out of the scope of this book. In particular, compared with unrestricted codes, nonexistence results leading to an optimality proof are occasionally orders of magnitude easier than a classification of optimal codes – this is the reason why comparatively many manual optimality proofs have been published for linear codes.

We here give references to the main results for the most important classes of linear codes, and restrict the tabulated results to binary codes with minimum distance 3 for small parameters.

In the binary case, the following observation can be used for odd values of d. Since an $[n, k, 2d-1]_2$ code can be extended to an $[n+1, k, 2d]_2$ code – Theorem 7.14 also holds for linear codes, since extension by adding a parity check bit preserves linearity – we can classify the former codes by first classifying the latter, followed by a puncturing of these in all $n+1$ possible ways together with a final isomorph rejection. If we know the coordinate orbits under the automorphism group of an $[n+1, k, 2d]_2$ code, then we need only puncture the code in one coordinate of each orbit and no additional isomorph rejection is required. As a special case, as Simonis [536] points out, if the automorphism group of a unique $[n+1, k, 2d]_2$ code acts transitively on the coordinates, then there is also a unique $[n, k, 2d-1]_2$ code.

Many classification result for optimal linear codes, that is, codes attaining $d_{\max}(n, k)$, are fairly easily obtainable and well-known, and it was not until the late 1990s that there was a growing interest in publishing classification results generally and not only for the most important code parameters. We will now list some of the main references and their results (some of which are overlapping).

The results in [280] include classification of (optimal) binary codes with the following parameters: $[14, 5, 6]_2$, $[15, 6, 6]_2$, $[16, 7, 6]_2$, $[17, 8, 6]_2$, $[18, 9, 6]_2$, $[20, 4, 10]_2$, $[21, 5, 10]_2$, $[17, 5, 8]_2$, $[18, 6, 8]_2$, $[19, 7, 8]_2$, $[20, 8, 8]_2$, $[21, 9, 8]_2$, $[22, 10, 8]_2$, $[23, 11, 8]_2$, $[24, 12, 8]_2$, $[20, 7, 8]_2$, $[21, 8, 8]_2$,$[22, 9, 8]_2$, $[23, 6, 10]_2$, $[24, 7, 10]_2$, and for some sets of parameters with the length n in the range $25 \le n \le 30$. In [534], $[16, 7, 6]_2$ codes are classified. In [535], $[18, 9, 6]_2$ codes are classified. In [332], $[15, 7, 5]_2$ and $[16, 7, 6]_2$ codes are classified. In [165], $[18, 6, 8]_2$, $[19, 7, 8]_2$, $[20, 8, 8]_2$, $[21, 9, 8]_2$, $[22, 10, 8]_2$, and $[23, 11, 8]_2$ codes are classified. In [482], the uniqueness of the $[23, 12, 7]_2$ Golay code is proved implying that the related $[24, 12, 8]_2$ code is also unique (and the ternary Golay code is proven unique in the same place); note, however, that we have reported the stronger result in Sect. 7.1.4 that the unrestricted codes with these parameters are unique. In [297], $[14, 6, 5]_2$, $[23, 7, 9]_2$, and $[24, 7, 10]_2$ codes are classified. In [573], $[21, 5, 10]_2$ codes are classified, and the following theorem is proved.

Theorem 7.51. *There are unique* $[2^k - 2^u, k, 2^{k-1} - 2^{u-1}]_2$, $1 \le u \le k-1$, $[2^k - 2^{k-2} - 3, k, 2^{k-1} - 2^{k-3} - 2]_2$, $k \ge 6$, *and* $[2^{k-1} + k, k, 2^{k-2} + 2]_2$, $k \ge 3$, $k \ne 5$, *binary linear codes.*

Classification results for infinite families of binary codes can also be found in [261]. Classification results for (long) binary codes with $k \le 7$ were obtained in [60]. Many results are also known for ternary codes, cf. [181], and codes with larger values of q.

Classification results for binary codes with minimum distance at least 3 and small parameters are given in Table 7.7. The number of inequivalent codes were actually determined already by Slepian [540]. Arnold [11] classified these codes up to length 7, Betten [38] up to length 12, and Östergård [455] up to length 14. The entries for $k = 1$ follow from Lemma 7.41.

Research Problem 7.52. Extend the classification of binary linear codes with minimum distance at least 3 outside the range of Table 7.7. See also Research Problem 7.50.

Table 7.7. The number of inequivalent $[n, k, 3]_2$ codes for $n \leq 14$

$n \backslash k$	1	2	3	4	5	6	7	8	9	10
3	1									
4	2									
5	3	1								
6	4	4	1							
7	5	8	5	1						
8	6	14	15	6						
9	7	22	38	29	5					
10	8	32	80	105	46	4				
11	9	44	151	312	273	64	3			
12	10	59	266	821	1,285	700	89	2		
13	11	76	440	1,948	5,098	5,632	1,794	112	1	
14	12	96	695	4,288	17,934	37,191	26,792	4,579	128	1

Ternary and quaternary linear codes with minimum distance at least 3 have been classified in [455].

There are many types of linear codes that have received particular interest because of their applications. We list references to some of the most studied such codes.

Optimal rate-1/2 codes have been classified in [37, 232], [235], and [234] for binary, ternary, and quaternary codes, respectively, and in [233] for $q = 5, 7$.

Classification of self-dual linear codes has been thoroughly studied; see [497] for a survey that includes many classification results for special types of self-dual codes. The main results for general self-dual (and some related self-orthogonal) codes can be found in [46, 47, 48, 133, 135, 483, 485], [134, 390, 482, 486], [134], [253, 369], and [252, 488] for $q = 2$ and $n \leq 34$, $q = 3$ and $n \leq 20$, $q = 4$ (with Hermitian inner product) and $n \leq 16$, $q = 5$ and $n \leq 16$, and $q = 7$ and $n \leq 12$, respectively. Classification results for self-orthogonal codes can be found in [59, 61, 63].

Certain, so-called extremal, self-dual codes are particularly interesting. The uniqueness of a $[48, 24, 12]_2$ self-dual doubly-even codes has been proved by Houghten et al. [269, 270] in a sophisticated and extensive computer search. The next open case for extremal self-dual doubly-even codes has parameters $[72, 36, 16]_2$.

Research Problem 7.53. Conjecturally, a $[72, 36, 16]_2$ self-dual doubly-even code does not exist. Prove this.

The discussion of linear codes has been restricted to error-correcting codes. We finally point out that little is known about classification of linear *covering* codes [142, 313].

Research Problem 7.54. Study classification of linear covering codes.

8

Classification of Related Structures

In this chapter, we study classification of combinatorial objects that are closely related to codes and designs. The objects considered can if fact in most cases be defined as certain codes or designs; algorithms and ideas from Chaps. 6 and 7 are then directly applicable.

Most of this chapter is devoted to objects that can be formulated as triple systems, including Latin squares and 1-factorizations of complete graphs, which are considered in Sect. 8.1. Section 8.2 is devoted to Hadamard matrices and Sect. 8.3 to orthogonal arrays.

8.1 Triple Systems

The classification framework described in Sect. 6.1.3 for Steiner triple systems can be used essentially without change to classify certain other combinatorial objects that admit representation as set systems. In this section we study some examples of such objects. For the framework to be directly applicable, the set system representation must have the following three properties:

The construction problem for an object can be formulated as an exact cover problem, in which the blocks are used to cover certain subsets of points. (8.1)

No two blocks intersect the same set of blocks. (8.2)

The set system is strongly reconstructible from its line graph. (8.3)

Property (8.1) guarantees that the exact cover algorithm can be used for exhaustive generation. Properties (8.2) and (8.3) guarantee that generation by canonical augmentation can be applied in a manner analogous to that for Steiner triple systems in Sect. 6.1.3, defining a seed as a subsystem consisting of a block and all blocks that intersect that block. Note that (8.3) is required only when line graphs are used for isomorph rejection. An invariant based on

Pasch configurations can be used if we have block size 3, which in fact holds for all objects in this section.

We here consider two types of objects defined in Sect. 2.4: 1-factorizations of the complete graph K_{2n} and uniform 3-GDDs; the latter include Latin squares. For each of the objects considered, we will derive a representation as a set system and establish that (8.1) to (8.3) hold.

Obviously, if only (8.1) holds for a particular type of object, we may still classify the objects with an algorithm for Exact Covers by using other techniques for (intermediate and final) isomorph rejection. One such classification problem is that of finding starters in Abelian groups of odd order. See [161] for a definition of starters; orderly generation of such objects is considered in [327].

Research Problem 8.1. Classify starters in a framework utilizing Exact Covers, and compare the algorithm with that in [327].

8.1.1 One-Factorizations of Complete Graphs

We first derive a set system representation for 1-factorizations of K_{2n}. Let $n \geq 1$ and let $\mathcal{F}$ be a 1-factorization of K_{2n} with vertex set $\{1, 2, \ldots, 2n\}$. Denote the 1-factors by $F_1, F_2, \ldots, F_{2n-1}$.

We represent $\mathcal{F}$ using a set system of the following form. There is one point – labeled $1, 2, \ldots, 2n$ – for each vertex of K_{2n}, and one point – labeled $f_1, f_2, \ldots, f_{2n-1}$ – for each 1-factor in $\mathcal{F}$. Each block of the set system is associated with an edge of K_{2n}. More precisely, for each edge $\{i, j\}$, the set system contains a block $\{f_k, i, j\}$, where k is determined from $\{i, j\} \in E(F_k)$. It is straightforward to check that a 1-factorization of K_{2n} is strongly reconstructible from this representation.

The exhaustive generation of 1-factorizations of K_{2n} can be formulated as an exact cover problem in which the task is to cover exactly once all

$$\binom{2n}{2} + 2n(2n-1) = 3n(2n-1)$$

2-subsets of the forms

$$\begin{aligned} &\{i, j\}, \quad 1 \leq i < j \leq 2n, \\ &\{f_k, i\}, \quad 1 \leq k \leq 2n-1,\ 1 \leq i \leq 2n \end{aligned} \tag{8.4}$$

using 3-subsets of the form

$$\{f_k, i, j\}, \quad 1 \leq i < j \leq 2n,\ 1 \leq k \leq 2n-1. \tag{8.5}$$

The solutions of this problem are precisely the set system representations of 1-factorizations of K_{2n}. This establishes (8.1).

It follows directly from (8.4) and (8.5) that a set system representation of a 1-factorization of K_{2n} is a GDD$(4n-1, 3)$ with group type $(2n-1)^1 1^{2n}$.

Conversely, every GDD of this type is a set system representation of a 1-factorization of K_{2n}.

A seed is the subsystem consisting of a block $\{f_k, i, j\}$ and all blocks that intersect this block. There are n blocks that contain f_k, $2n-1$ blocks that contain i, and $2n-1$ blocks that contain j. Thus, a seed consists of

$$1 + (n-1) + 2(2n-2) = 5n - 4$$

blocks. The block that induces a seed is easily seen to be unique for $n \geq 3$. Thus, (8.2) holds for $n \geq 3$. An example of a seed for $2n = 12$ is given in Example 8.2.

Example 8.2. One seed for finding 1-factorizations of K_{12}, with the points f_i corresponding to the 1-factors in the first 11 rows, is displayed below.

111111	0000000000	0000000000
000000	1000000000	1000000000
000000	0100000000	0100000000
000000	0010000000	0010000000
000000	0001000000	0001000000
000000	0000100000	0000100000
000000	0000010000	0000010000
000000	0000001000	0000001000
000000	0000000100	0000000100
000000	0000000010	0000000010
000000	0000000001	0000000001
100000	1111111111	0000000000
100000	0000000000	1111111111
010000	1000000000	0100000000
010000	0100000000	1000000000
001000	0010000000	0001000000
001000	0001000000	0010000000
000100	0000100000	0000010000
000100	0000010000	0000100000
000010	0000001000	0000000100
000010	0000000100	0000001000
000001	0000000010	0000000001
000001	0000000001	0000000010

We will now consider the line graphs of (set system representations of) 1-factorizations of K_{2n}. Property (8.3) for a line graph of a 1-factorization can be proved by a slight modification of the proof of Corollary 3.93. This result can also be found in [305, Theorem 1].

Theorem 8.3. *For $2n \geq 8$, a set system of a 1-factorization of K_{2n} is strongly reconstructible from its line graph.*

Proof. Every block of the set system has 3 points, every point corresponding to a 1-factor occurs in n blocks, and every point corresponding to a vertex occurs in $2n-1$ blocks. By Theorem 3.92 the points corresponding to vertices can be reconstructed from the cliques of size $2n-1$ when $2n-1 > 3^2-3+1 = 7$. This can be strengthened by the fact [155, 156] that if a clique of size 7 does not correspond to one common point for the blocks, then it corresponds to the Fano plane. It is straightforward to check that a Fano plane is not possible here.

Consequently, for $2n-1 \geq 7$ the maximum cliques of size $2n-1$ in the line graph correspond to the vertices of K_{2n}. Excluding edges occurring in the subgraphs of the line graph induced by these cliques, the remaining edges define $2n-1$ pairwise disjoint cliques of size n, which correspond to the 1-factors in the 1-factorization. To reconstruct the blocks, each vertex of the line graph occurs in precisely three of the aforementioned cliques, which reveals the point incidences of the corresponding block.

Any isomorphism between line graphs must map maximum cliques onto maximum cliques, which induces a unique isomorphism between the underlying set systems. This establishes strong reconstructibility. □

A set of six blocks that are pairwise intersecting and do not all have one single point in common is possible:

$$\{f_1,1,2\},\ \{f_1,3,4\},\ \{f_2,1,3\},\ \{f_2,2,4\},\ \{f_3,1,4\},\ \{f_3,2,3\}.$$

Such a set in fact corresponds to a 1-factorization of K_4.

A Pasch configuration in our incidence structure representation of a 1-factorization corresponds to a 4-cycle with the edges from two 1-factors. For example, the blocks (columns) number 1, 2, 7, and 17 in Example 8.2 form a Pasch configuration. Such incidence structure representations that are anti-Pasch thus correspond to 1-factorizations in which no pair of 1-factors contains a 4-cycle.

The approach of considering 1-factorizations of complete graphs as set systems is not the only possible one; however, the results of [305] seem to indicate that it is faster than previously used approaches. We will here briefly mention a few other possible methods, viewing 1-factorizations as certain RBIBDs.

A 1-factorization of K_{2n} corresponds to a resolution of a 2-$(2n,2,1)$ design. Classification of resolutions of designs are considered in Sect. 6.3 and we may apply those methods here as well. To sum up, the methods can roughly be divided into those proceeding point by point and those proceeding parallel class by parallel class (the underlying design is here the complete graph, so the methods of 6.3.1 are not useful). Equivalently, in the framework of equidistant n-ary codes, one may construct the codes codeword by codeword or coordinate by coordinate.

For 1-factorizations of complete graphs, the approach that proceeds parallel class by parallel class has been the method of choice in published studies, including the seminal classification of 1-factorizations of K_{12} utilizing orderly generation [162].

We conclude our discussion of these objects by giving the numbers of nonisomorphic 1-factorizations of K_{2n} for $2n \leq 12$ in Table 8.1. The nonisomorphic 1-factorizations of K_6 and K_8 are displayed in Figs. 2.6 and 2.7, respectively, and those of K_{10} are listed in [6]. See also [522] for some related results. Classification of 1-factorizations of regular graphs of degree smaller than $2n-1$ (K_{2n} is $(2n-1)$-regular) has been considered in [305, 507, 522].

Table 8.1. Classification of 1-factorizations of complete graphs K_{2n}

$2n$	N	References
2	1	Trivial
4	1	Trivial
6	1	Trivial
8	6	[157]
10	396	[202]
12	526,915,620	[162]

8.1.2 Group Divisible Designs with Block Size 3

The subsequent discussion of classification of group divisible designs with block size 3 is restricted to 3-GDDs that are *uniform*, that is, whose group type is of the form t^u. Exhaustive generation of GDDs of this type can clearly be formulated as an exact cover problem, in which the task is to cover exactly once all 2-subsets of points from different groups using 3-subsets of points such that no two points in a 3-subset are in the same group. This establishes Property (8.1).

By a double counting argument (cf. Theorem 2.38) one obtains that in a 3-GDD of type t^u, each point occurs in $r = t(u-1)/2$ blocks and the number of blocks is $b = t^2u(u-1)/6$.

Since each 2-subset of points occurs in at most one block, the number of blocks that intersect a block of a uniform 3-GDD is $1 + 3(r-1) = 3r - 2$. Once again, such a set will form a seed for the classification. We will now see that a block that induces a seed is unique for $r \geq 4$. Any two blocks that intersect in a point p can intersect at most 4 common blocks that do not contain p. However, for the blocks to induce the same seed, they must intersect $2(r-1)$ common blocks that do not contain p. Thus, Property (8.2) holds for $2(r-1) > 4$, that is, $r \geq 4$. We next analyze the line graphs of uniform 3-GDDs.

Theorem 8.4. *For* $t(u-1)/2 \geq 8$, *a* 3-GDD *is strongly reconstructible from its line graph.*

Proof. The proof is analogous to the proof of Theorem 8.3. Now every point occurs in at exactly $t(u-1)/2$ blocks, and the points can be reconstructed from the line graphs whenever $t(u-1)/2 > 7$ (by Theorem 3.92). The group partition can be reconstructed using the observation that two maximum cliques in the line graph are disjoint if and only if the corresponding points are in the same group. Finally, strong reconstructibility follows from the fact that any isomorphism between line graphs must map maximum cliques onto maximum cliques, which induces a unique isomorphism between the underlying set systems. □

The assertion of Theorem 8.4 can be strengthened to $t(u-1)/2 \geq 7$ when $u < 7$, since a 3-GDD can contain a Fano plane only if $u \geq 7$.

In the following we discuss one particular instance of 3-GDDs, namely, Latin squares. Several other types of 3-GDDs can be considered in this setting, generalized Steiner (triple) systems [178] just to mention an example.

Research Problem 8.5. Classify generalized Steiner triple systems.

8.1.3 Latin Squares

The line graph of a Latin square – or a 3-GDD of type 3^n, or a $\mathrm{TD}(3,n)$ – is an $(n^2, 3(n-1), n, 6)$ strongly regular graph. A short case-by-case analysis shows that the maximum cardinality of a set of pairwise intersecting blocks in a $\mathrm{TD}(3,n)$ that do not all have one single point in common is 4, which occurs precisely when the four blocks form a Pasch configuration.

A Pasch configuration is a $\mathrm{TD}(3,2)$, that is, a Latin square of order 2. Thus, an anti-Pasch $\mathrm{TD}(3,n)$ corresponds to a Latin square with no subsquares of order 2. A Latin square with no subsquares of order 2 is called an N_2 square [117].

The approach used in Sect. 6.1.3 and earlier in this chapter for other types of triple systems is applicable here, and we omit the details. Isomorphic triple systems correspond to Latin squares in the same main class (Definition 2.114); the classified main class Latin squares may be used to construct the types, isotopy classes, and the isomorphism classes of Latin squares.

The traditional approach for classifying Latin squares is to construct them one row at a time (and only the use of isomorph rejection has varied). The number of main classes of Latin squares is given in Table 8.2 for squares of side $n \leq 10$ – the horizontal line in the table indicated that the squares of side 9 and 10 are only *counted* in [415]. See also [73] for orderly generation of Latin squares (but note that the numerical results are in error; cf. [328]). In fact, it appears that a classification of Latin squares row by row might be faster than a classification based on seeds and Exact Covers. One reason for this is the well-known fact [117, Theorem 1.38] that a Latin rectangle can always be completed to a Latin square, so the search never reaches a dead end.

Research Problem 8.6. Compare the aforementioned two approaches for classifying Latin squares (using different strategies for isomorph rejection).

Most of the earliest classification results, especially those obtained in the 19th century, considered the number of reduced Latin squares rather than the number of main classes. We have ignored those references except for the results for $n \leq 5$; there are so few such main classes (and reduced squares) that Euler [180] could easily have obtained the classification result in any form. The references for $n = 6$ consider the number of types. Unfortunately, erroneous classification results for Latin squares occur in the literature; see [415] for a historical account.

Table 8.2. Classification of Latin squares

n	N	References
1	1	Trivial
2	1	Trivial
3	1	Trivial
4	2	[180]
5	2	[180]
6	12	[473, p. 81] (Clausen), [569]
7	147	[446, 514]
8	283,657	[328]
9	19,270,853,541	[415]
10	34,817,397,894,749,939	[415]

Research Problem 8.7. Classify the Latin squares of side 9. Use this classification to check, for example, [117, Conjectures 1.18 and 1.22].

8.2 Hadamard Matrices

By Theorem 2.127 there are several ways of getting at a classification of Hadamard matrices. For example, using the algorithms presented in Chap. 6, one may classify the 2-$(4n-1, 2n-1, n-1)$ designs and obtain the Hadamard matrices of order $4n$ via these. However, as Example 3.62 shows, the number of nonisomorphic designs with the aforementioned parameters may be much larger than the number of inequivalent Hadamard matrices; an Hadamard matrix may lead to up to $4n \cdot 4n = 16n^2$ inequivalent 2-$(4n-1, 2n-1, n-1)$ designs (cf. [547]). With respect to the number of designs expected, classification via 3-$(4n, 2n, n-1)$ designs is therefore to be preferred; this is in fact the approach in the original classification of Hadamard matrices of order 24 ([277], later corrected in [316]). We will now briefly look at classification within the framework of Hadamard matrices.

Hadamard matrices may be classified in a row-by-row manner by orderly generation. For (partial) Hadamard matrices we allow row and column permutations as well as row and column negations (Definition 2.125) – a matrix is in canonical form if the tuple formed by concatenating its rows is the lexicographic maximum of its equivalence class. Algorithm 6.1 from Sect. 6.1.2 with extensions discussed in Sect. 7.1.2 can be used to test maximality in the case when no row negations occur. To incorporate row negations, one possibility is to employ the extended Algorithm 6.1 as a subroutine. Namely, given an input matrix, we select one column and its sign in all possible ways as the first column, normalize the first column by row negations to consist of only 1s, and then employ the extended Algorithm 6.1 as a subroutine in searching for counterexamples to the maximality of the input matrix; that is, the counterexample candidate being constructed is always compared against the input matrix rather than the input to the subroutine.

Every new row that augments a partial Hadamard matrix must have inner product 0 with the existing rows. It turns out that sets of one, two, and three rows of an Hadamard matrix are unique up to equivalence. Any set of three rows – and columns as well, transposed – of an Hadamard matrix of order $4n$ is equivalent to

$$\left[\begin{array}{ccc|ccc|ccc|ccc} 1\,1 & & 1 & 1\,1 & & 1 & 1\,1 & & 1 & 1\,1 & & 1 \\ 1\,1 & \cdots & 1 & 1\,1 & \cdots & 1 & -\,- & \cdots & - & -\,- & \cdots & - \\ 1\,1 & & 1 & -\,- & & - & 1\,1 & & 1 & -\,- & & - \end{array}\right], \qquad (8.6)$$

where the submatrices are of size $3 \times n$; cf. the proof of Theorem 2.123. This is the fundamental reason why we get 3-designs from Hadamard matrices (Theorem 2.127), and also indicates that finding (easily computable) invariants for Hadamard matrices is a nontrivial task.

Spence [547] remarks that canonicity testing in orderly generation is very time-consuming already when the partial matrix contains a rather small number of rows. Consequently, beyond a certain number of rows, canonicity testing cannot be performed for each partial matrix.

Research Problem 8.8. Develop an algorithm for classifying (parts of) Hadamard matrices in a row-by-row manner via generation by canonical augmentation. Compare this algorithm with the outlined orderly algorithm.

Completing a partial matrix, disregarding isomorph rejection, is an instance of CLIQUES. A $4n$-tuple is admissible and corresponds to a vertex in the particular graph exactly when its inner product with the rows in the partial matrix is 0. Two admissible rows are compatible, and the corresponding vertices are adjacent, exactly when their inner product is 0. One should also observe that the first three columns of an Hadamard matrix in canonical form are the transpose of (8.6); this supports intermediate stopping points, with isomorph rejection, at n (which, however, can be reached by the initial row-by-row classification part for parameters one might consider), $2n$, and perhaps even $3n$ rows. See [547, 548] for more details.

Another way of attacking this classification problem is to focus on certain submatrices of the Hadamard matrices and dichotomize the search space by separately classifying matrices that contain and do not contain such submatrices. For a fourth row extending (8.6), we denote the number of 1s under the respective submatrices by a, b, c, and d. Then

$$\begin{aligned} a + {} & b + {} & c + {} & d = 2n, \\ a + {} & b + (n-c) + {} & (n-d) = 2n, \\ a + (n-b) + {} & c + {} & (n-d) = 2n, \end{aligned} \tag{8.7}$$

whose solution is $a = d$, $b = c = n - a$. Moreover, since the new row may be complemented up to equivalence, we may assume that $0 \le a \le \lfloor n/2 \rfloor$. The extremal case, $a = 0$, leads to the matrix

$$\left[\begin{array}{ccc|ccc|ccc|ccc} 1\ 1 & & 1 & 1\ 1 & & 1 & 1\ 1 & & 1 & 1\ 1 & & 1 \\ 1\ 1 & \cdots & 1 & 1\ 1 & \cdots & 1 & -\ - & \cdots & - & -\ - & \cdots & - \\ 1\ 1 & & 1 & -\ - & & - & 1\ 1 & & 1 & -\ - & & - \\ -\ - & & - & 1\ 1 & & 1 & 1\ 1 & & 1 & -\ - & & - \end{array}\right], \tag{8.8}$$

which, however, can be excluded for certain parameters as the next theorem shows.

Theorem 8.9. *An Hadamard matrix of order* $8k+4$ *cannot have a* $4 \times (8k+4)$ *submatrix equivalent to* (8.8).

Proof. For a row extending (8.8), we denote the number of 1s under the respective submatrices by a, b, c, and d. Then

$$\begin{aligned} a + b + c + d &= 4k+2, \\ a + b + (n-c) + (n-d) &= 4k+2, \\ a + (n-b) + c + (n-d) &= 4k+2, \\ (n-a) + b + c + (n-d) &= 4k+2, \end{aligned}$$

whose solution is $a = b = c = d = (4k+2)/4$, which is not an integer. □

A set of four rows with $a = 1$ in (8.7) is called a *Hall set*; such submatrices play a central role in the classification of Hadamard matrices of order 28 in [317, 318]. By straightforward counting arguments one can prove that an Hadamard matrix of order $8k + 4$ with a Hall set is equivalent to a matrix of the form

$$\begin{bmatrix} \mathbf{A}_{11} & \mathbf{B}_{12} & \mathbf{B}_{13} & \mathbf{B}_{14} & \mathbf{D}_1^T \\ \mathbf{B}_{21} & \mathbf{A}_{22} & \mathbf{B}_{23} & \mathbf{B}_{24} & \mathbf{D}_2^T \\ \mathbf{B}_{31} & \mathbf{B}_{32} & \mathbf{A}_{33} & \mathbf{B}_{34} & \mathbf{D}_3^T \\ \mathbf{B}_{41} & \mathbf{B}_{42} & \mathbf{B}_{43} & -\mathbf{A}_{44} & \mathbf{D}_4^T \\ \mathbf{D}_1 & \mathbf{D}_2 & \mathbf{D}_3 & \mathbf{D}_4 & \mathbf{E} \end{bmatrix}, \tag{8.9}$$

where the matrices $\mathbf{A}_{ii}$ of size $2k \times 2k$ have row and column sum -2, the matrices $\mathbf{B}_{ij}$ of size $2k \times 2k$ have row and column sum 0, $[\mathbf{D}_1\ \mathbf{D}_2\ \mathbf{D}_3\ \mathbf{D}_4]$ coincides with (8.8), and

$$\mathbf{E} = \begin{bmatrix} 1 & 1 & 1 & 1 \\ 1 & 1 & - & - \\ 1 & - & 1 & - \\ 1 & - & - & 1 \end{bmatrix}.$$

See [318] and its references for details on utilizing the structure (8.9) in classifying such Hadamard matrices. A corollary of (8.9) is that the transpose of this matrix also has a Hall set.

The known classification results for Hadamard matrices are summarized in Table 8.3.

Uniqueness for the cases $n = 8$ and $n = 12$ follows from Example 3.81 and the uniqueness of the 2-$(7, 3, 1)$ and 2-$(11, 5, 2)$ designs, respectively, which was known long before this problem was studied within the framework of Hadamard matrices. The classification results have been verified for $n = 16$ in [241] and for $n = 24$ and $n = 28$ in [547].

Table 8.3. Classification of Hadamard matrices

n	N	References
1	1	Trivial
2	1	Trivial
4	1	Trivial
8	1	Trivial, [28]
12	1	[396], [574]
16	5	[574]
20	3	[242]
24	60	[277, 316]
28	487	[317, 318]

8.3 Orthogonal Arrays

Similarly to the situation for Hadamard matrices, we have seen in Chap. 2 that various classes of orthogonal arrays can be obtained – and therefore classified – via other objects. Consequently, by using Theorems 2.112, 2.120, 2.121, and 2.127, many results presented in this book for BIBDs, Latin squares, and Hadamard matrices lead directly (or with minor computational effort) to classification results for orthogonal arrays. In all the theorems mentioned, the strength of the orthogonal array obtained is $t = 2$. We have also seen in

Theorem 2.106 that orthogonal arrays with $\lambda = 1$ and arbitrary strength t correspond to certain error-correcting codes.

In existence problems for t-designs, $t = 2$ is the basic case, with the case $t = 1$ being trivial and the cases $t \geq 3$ being more special and involved. For orthogonal arrays, referring to the parameter t as well, the situation is similar. In the subsequent discussion of classification of orthogonal arrays, we therefore focus on orthogonal arrays with strength $t = 2$.

The obvious way of proceeding in a classification of general orthogonal arrays $\mathrm{OA}_\lambda(2, n, q)$ is a row-by-row manner, where each row has length λq^2. Since an $\mathrm{OA}_\lambda(2, n, q)$ is an $\mathrm{OA}_{\lambda q}(1, n, q)$, every row contains each value exactly λq times. The first two rows are obviously unique:

$$\left[\begin{array}{c|c|c|c} 0\,0 \ldots 0 & 0\,0 \ldots 0 & \ldots & q-1 \;\; q-1 \;\ldots\; q-1 \\ 0\,0 \ldots 0 & 1\,1 \ldots 1 & \ldots & q-1 \;\; q-1 \;\ldots\; q-1 \end{array}\right]$$

with each pair of values occurring λ times. The procedure of extending an orthogonal array – note that all objects in the search tree are orthogonal arrays with strength $t = 2$ – consists of the tasks of finding rows leading to orthogonal arrays and carrying out isomorph rejection. The total number of λq^2-tuples containing each value exactly λq times is

$$\prod_{i=0}^{q-1} \binom{\lambda q(q-i)}{\lambda q},$$

which grows rapidly with growing q. The computational problem of finding admissible tuples is a recurrent topic of this book; the details are not repeated here. For the isomorph rejection part, viewing the columns as codewords, see Chap. 7.

To classify the $\mathrm{OA}_\lambda(2, n, q)$ with given parameters, one may start from a classification of the $\mathrm{OA}_\lambda(2, n', q)$ for some $n' < n$, and utilize the approach of finding cliques of size $n - n'$. Here a λq^2-tuple is admissible if it forms an $\mathrm{OA}_\lambda(2, 2, q)$ together with each row of the fixed, smaller array. Two admissible λq^2-tuples are compatible if they form an $\mathrm{OA}_\lambda(2, 2, q)$. We leave the discussion here; since this classification problem has not been thoroughly studied there is still room for many variations and improvements. A more detailed consideration would without doubt be superseded by forthcoming studies.

To get at a classification of $\mathrm{OA}_\lambda(t, n, q)$, we may utilize a classification of $\mathrm{OA}_\lambda(t-1, n-1, q)$; cf. Theorem 2.109. This idea is mentioned in [618], which discusses classification of orthogonal arrays in the framework of solving linear systems of equations.

Research Problem 8.10. Study classification algorithms for orthogonal arrays. Develop algorithms for the cases $t = 1$, $t = 2$, as well as $t \geq 3$.

No systematic classification of general orthogonal arrays has been carried out. We summarize some of the known results. By Theorem 2.112, the data in Table 8.2 apply to $\mathrm{OA}(3,n)$ as well and need not be repeated. Classification results for $\mathrm{OA}_n(4n-1,2)$ can be gathered from Table 6.12 because the equivalence classes of $\mathrm{OA}_n(4n-1,2)$ coincide by Theorem 2.127 with those of $(4n-1,4n,2n)_2$ OE codes, and therefore, by Theorem 3.82, with the isomorphism classes of resolutions of 2-$(4n,2n,2n-1)$ designs. These results are shown in Table 8.4. The entries for $n=4$ and $n=5$ are discussed in [614] and [615], respectively. Alternatively, one could start from the classified Hadamard matrices, viewed in Table 8.3, and compute their row automorphism orbits. It turns out that every Hadamard matrix of order at most 20 has an automorphism group that acts transitively on the rows (and thereby on the columns, as well) [257, p. 165].

Table 8.4. The number of $\mathrm{OA}_n(4n-1,2)$

n	$4n-1$	N
1	3	1
2	7	1
3	11	1
4	15	5
5	19	3
6	23	130
7	27	7,570

We will now look at parameters for which existence of orthogonal arrays follows by Theorems 2.63, 2.66, and 2.121. The above mentioned orthogonal arrays related to Hadamard matrices form one of the families via Theorem 2.66; the parameters of the other family are

$$\mathrm{OA}_{q^n}\left(\frac{q^{n+2}-1}{q-1},q\right), \tag{8.10}$$

where q is a prime power and n is nonnegative.

For $n=0$ in (8.10) we get $\mathrm{OA}(q+1,q)$; the number of nonisomorphic $\mathrm{OA}(n+1,n)$ coincides with the number of inequivalent $(n+1,n^2,n)_n$ OE codes, and therefore, by Theorem 3.82, with the number of nonisomorphic 2-$(n^2,n,1)$ designs (each admitting a unique resolution by Theorem 2.62). Entries for such designs are tabulated in Tables 6.2 through 6.10 (alternatively, Tables 6.13 through 6.16, cf. Theorem 2.62): the number of nonisomorphic $\mathrm{OA}(q+1,q)$ for $q=3,4,\ldots,10$ is 1, 1, 1, 0, 1, 1, 7, and 0, respectively.

For $q=2$ in (8.10) we get $\mathrm{OA}_{2^n}(2^{n+2}-1,2)$; these parameters are already taken care of in Table 8.4. The only remaining case within the scope of the tables in Chap. 6 is $q=3$, $n=1$; by Table 6.13 there are 68 $\mathrm{OA}_3(13,3)$, classified by Lam and Tonchev [353].

The connection between orthogonal arrays and error-correcting codes is in fact much stronger than what we have seen so far (MDS codes, etc.). The principal theorem is [257, Theorem 4.9], which is due to Delsarte [146].

Theorem 8.11. *A code with dual distance* $d^{\perp} \geq d'$ *and the parameters* $(n, \lambda q^{d'-1}, d)_q$ *can be transformed into an* $\mathrm{OA}_{\lambda}(d'-1, n, q)$ *and vice versa.*

Note that the minimum distance d of the code in Theorem 8.11 is irrelevant. The dual distance of a linear code is simply the minimum distance of its dual code, cf. Definition 2.98. For an unrestricted code, the dual distance is obtained from the MacWilliams transform of the weight distribution of the code [257, p. 69]. Hence, specific coding-theoretic methods are of central importance in utilizing Theorem 8.11. For an exhaustive discussion of this topic, we refer the reader to [257, Chaps. 4 and 5]. Three examples of unique orthogonal arrays are listed in [257, p. 109]: $\mathrm{OA}_8(5, 16, 2)$, $\mathrm{OA}_{32}(7, 24, 2)$, and $\mathrm{OA}_3(5, 12, 3)$. The corresponding codes are the Nordstrom-Robinson code, the binary Golay code, and the ternary Golay code, respectively; we know from Sect. 7.1.4 that these are unique.

Other classification results for orthogonal arrays include those in [198, 256, 438, 617, 619].

The fact that the classification results in the literature are rather scattered inspires the restatement (slightly reformulated) of the following open problem from [257, Research Problem 5.13]. Known existence results are summarized in [257, Chap. 12], which is a useful reference in pursuing a further study.

Research Problem 8.12. Collect exhaustive tables of classification results for orthogonal arrays with small parameters, in particular for strength $t = 2$, and settle missing entries.

9

Prescribing Automorphism Groups

So far we have considered the problem of classifying some general families of combinatorial objects. In practice, however, one is often interested in a subfamily of these with certain additional properties. There are two possibilities for carrying out such a classification, which we have already encountered in discussing classification algorithms for resolutions of designs in Sect. 6.3: either all objects are classified and these are searched for the particular properties, or the additional properties are made inherent in the classification. The former approach can often be implemented easily using existing computer programs, but the latter approach is computationally more efficient.

Many of the properties that one might be interested in are of a very specific nature and should be considered on a case-by-case basis. In the present book it is neither possible to list all possible such properties nor to include a complete survey of related classification results; some examples in the context of Steiner triple systems were discussed in Sect. 6.4. However, there is one type of property that is inherent in all types of combinatorial objects, namely the automorphism group. Prescribing of automorphism groups has been extensively studied and used particularly in searching for objects whose existence is unknown. If an object is found, the search is commonly stopped, but one may obviously continue and find all objects in an exhaustive manner; of course, if an existence question is answered in the negative, we have a classification result as well.

Some general results on classifying combinatorial objects with prescribed automorphism groups are presented in Sect. 9.1. Most of the published results on this topic – including the celebrated Kramer–Mesner method – concern designs and are considered in Sect. 9.2. Codes are treated in this setting in Sect. 9.3, and other objects in Sect. 9.4.

9.1 Preliminaries

Throughout this chapter we assume the following setup. Our interest is to classify objects in an implicitly defined finite set Ω, where isomorphism is defined by an action of a group G on Ω (Definition 3.16). The *automorphism group* of an object $X \in \Omega$ is the group $\mathrm{Aut}(X) = \mathrm{Iso}(X, X) \leq G$. An object $X \in \Omega$ admits a group $H \leq G$ as a *group of automorphisms* if $H \leq \mathrm{Aut}(X)$.

The focus of classification with prescribed automorphisms may be very wide considering, for example, all objects with a nontrivial automorphism group, or very narrow, fixing a specific group order $|H|$ and possibly even the group H up to conjugacy in G. Moreover, we might require that H be the automorphism group $\mathrm{Aut}(X)$ instead of merely a group of automorphisms. We do not here contemplate on the various motivations behind different choices of H and $|H|$.

It is usually desirable to break up problems of this type into subproblems where the prescribed group H is fixed. So if one only prescribes the group order, then the first task is to find all applicable groups up to conjugacy in G. Combinatorial arguments may often be applied to reject some of the groups immediately. Since such arguments are generally problem-specific, they have to be applied on a case-by-case basis and we will not be able to give any general guidelines.

What comes to finding the groups, one may use group-theoretic results or consult the literature – for example, generators for the transitive permutation groups of degree up to 15 are listed in [132], and the possible permutation representations of large groups on a relatively small number of points are generally known. One may obviously implement an independent classification of possible groups up to conjugacy; the need for such an approach is accentuated if one is considering objects where the isomorphism-inducing group G is not a symmetric group S_v. For example, this issue is discussed in [385] for ternary codes and cyclic subgroups. Since group-theoretic classification problems are out of the scope of this work, we omit further discussions and refer the reader to [33, 132].

One recurrent special case that is simple enough to solve by hand calculation is the situation where $G = S_v$ and we want to produce up to conjugacy in S_v the subgroups $H \leq S_v$ of prime order p. Since every nonidentity element of a prime-order group generates the group, it suffices to produce up to conjugacy all permutations $h \in S_v$ with order p. The *cycle type* $a_1^{n_1} a_2^{n_2} \cdots a_t^{n_t}$ of a permutation $h \in S_v$ gives for each cycle length a_i the number n_i of such cycles in the cycle decomposition of h, where $v = \sum_i n_i a_i$. Observing that for every $g \in S_v$ and k-cycle $(x_1\ x_2\ \cdots\ x_k) \in S_v$ we have

$$g \cdot (x_1\ x_2\ \cdots\ x_k) \cdot g^{-1} = (g(x_1)\ \ g(x_2)\ \ \cdots\ \ g(x_k)),$$

it follows that two permutations are conjugate in S_v if and only if they have the same cycle type. Furthermore, a permutation $h \in S_v$ has order p if and only if its cycle type is of the form $1^f p^m$, where $v = f + pm$, $f \geq 0$, and

$m \geq 1$. Integral solutions f, m to this system determine the conjugacy classes of prime order subgroups in S_v.

Prime order groups can be employed also in cases where the group(s) of interest are larger. In this connection the following basic result from group theory is useful [509, Corollary 4.15].

Theorem 9.1. *Let H be a finite group and let p be a prime. If p^k divides $|H|$, then H contains a subgroup of order p^k.*

In particular, if we can solve the classification problem for all applicable subgroups of order p, then we obtain also all objects with automorphism groups of order divisible by p in the process. If all applicable primes p are considered, we obtain all objects with a nontrivial automorphism group.

Example 9.2. To find designs on 10 points with an automorphism group of order 12, one may apply Theorem 9.1 to deduce that the group must have subgroups of orders 2 and 3. The classification is generally easier for larger groups, so we consider groups of order 3. Up to conjugacy in S_{10}, three groups of order 3 must be considered, namely those with nonidentity elements of type $1^7 3^1$, $1^4 3^2$, and $1^1 3^3$ generated by the permutations

$$(1\ 2\ 3)(4)(5)(6)(7)(8)(9)(10),$$
$$(1\ 2\ 3)(4\ 5\ 6)(7)(8)(9)(10),$$
$$(1\ 2\ 3)(4\ 5\ 6)(7\ 8\ 9)(10),$$

respectively. A classification of all designs with these automorphisms is then filtered for designs with automorphism group order 12. To find all designs on 10 points with a nontrivial automorphism group, one can apply analogous reasoning for all primes at most 10; that is, 2, 3, 5, and 7. (In many cases it is possible to eliminate some primes and/or groups using combinatorial arguments. For example, in many cases a prime order automorphism cannot have too many fixed points.)

Once a group H has been fixed, the main task is to generate the objects admitting H as a group of automorphisms. As a rule of thumb, the larger the prescribed group H, the easier the generation problem for objects admitting H is to solve. Indeed, because H fixes an object X with $H \leq \mathrm{Aut}(X)$, it follows that H moves the elementary objects that make up X – say, the blocks of a design or the codewords of a code – among themselves. Thus, every X with $H \leq \mathrm{Aut}(X)$ consists of H-orbits of elementary objects, and to generate such objects it suffices to consider orbits instead of individual elementary objects. The larger the group, the larger and fewer orbits there are to consider.

Example 9.3. To find designs on 10 points and block size 3 admitting the group $H = \langle(1\ 2\ 3\ 4\ 5)(6\ 7\ 8\ 9\ 10)\rangle$, it suffices to consider the 24 H-orbits of 3-subsets in constructing the designs instead of the $\binom{10}{3} = 120$ individual 3-subsets. Figure 9.1 shows a 2-$(10, 3, 2)$ design admitting H as a group of automorphisms. The point and block orbits of H are separated by horizontal and vertical lines, respectively.

10011	10010	10000	10000	10000	10000
11001	01001	01000	01000	01000	01000
11100	10100	00100	00100	00100	00100
01110	01010	00010	00010	00010	00010
00111	00101	00001	00001	00001	00001
00000	00010	11000	00101	01001	00110
00000	00001	01100	10010	10100	00011
00000	10000	00110	01001	01010	10001
00000	01000	00011	10100	00101	11000
00000	00100	10001	01010	10010	01100

Fig. 9.1. A 2-$(10,3,2)$ design admitting a group of order 5

Apart from the reduction in problem size, classification with prescribed automorphisms retains the same basic characteristics as the original classification problem. In particular, the reduced problem may still have symmetry that results in redundancy requiring isomorph rejection techniques. Moreover, isomorphs must still be rejected among the generated objects. Again much depends on the prescribed group H.

If H is a small prime order group, say $|H| = 2$, then the reduction in problem size may not be very large and the reduced problem may still have significant symmetry left. Typically the normalizer

$$N_G(H) = \{g \in G : gHg^{-1} = H\}$$

captures the symmetry in the reduced problem. Indeed, elements of $N_G(H)$ always map H-orbits onto H-orbits whereas the elements of G in general do not. However, we emphasize that final isomorph rejection for complete objects must be performed with respect to the group G. One useful exception to the previous remark is presented in the following theorem.

Theorem 9.4. *Let* $X, Y \in \Omega$. *If* $\mathrm{Aut}(X) = \mathrm{Aut}(Y) = H$, *then* $\mathrm{Iso}(X,Y) \subseteq N_G(H)$.

Proof. The claim is obvious if $\mathrm{Iso}(X,Y)$ is empty. Let $g \in \mathrm{Iso}(X,Y)$ be arbitrary; in other words, $gX = Y$. For an arbitrary $a \in \mathrm{Aut}(X)$, we get that $gag^{-1}Y = gag^{-1}gX = gaX = gX = Y$; and for an arbitrary a such that $gag^{-1}Y \in \mathrm{Aut}(Y)$, we get that $aX = ag^{-1}Y = g^{-1}(gag^{-1}Y) = g^{-1}Y = X$. We have thus shown that $a \in \mathrm{Aut}(X)$ if and only if $gag^{-1} \in \mathrm{Aut}(Y)$. Hence, $gHg^{-1} = H$, implying $g \in N_G(H)$. □

Consequently, isomorph rejection with respect to $N_G(H)$ suffices if we know that $\mathrm{Aut}(X) = H$ for all objects X considered.

On the other hand, if H is large, then typically a big reduction in problem size is achieved and the remaining symmetry is negligible. However, large groups are typically employed for large parameters, which results in challenges in constructing the reduced problem and rejecting isomorphs among the generated objects. One basic observation that is useful for large groups is that if H is a maximal subgroup of the isomorphism-inducing group G with $N_G(H) = H$, then by Theorem 9.4 all distinct objects admitting H are nonisomorphic (unless $\mathrm{Aut}(X) = G$, which is usually impossible). In particular, for $G = S_v$ we have $N_{S_v}(H) = H$ unless the maximal subgroup H is the alternating group A_v, eliminating the need for isomorph rejection in classifying designs with such prescribed groups. See [239, 357, 358, 520, 521] for further techniques for classifying designs with large prescribed groups.

If several prescribed groups are employed, then one has to account for the possibility that isomorphic objects are encountered from different groups in the case when $\mathrm{Aut}(X)$ contains up to conjugacy more than one prescribed group. In many cases it is possible to carry out isomorph rejection via recorded objects. A less memory-intensive strategy is obtained if only the objects with $|\mathrm{Aut}(X)| > |H|$ are treated in this manner, and isomorph rejection with respect to the normalizer $N_G(H)$ is carried out for objects with $\mathrm{Aut}(X) = H$; a very fast implementation of this strategy in the context of Latin squares can be found in [415]. Yet another technique is generation by canonical augmentation such that the canonical parent $m(X)$ associated with each generated object X involves a subgroup in $\mathrm{Aut}(X)$ conjugate to one of the prescribed groups. To apply the framework in Sect. 4.2.3 in this context, it is useful to observe that objects with an individualized group of automorphisms can be viewed as ordered pairs (H, Z), where Z is a combinatorial object and $H \leq \mathrm{Aut}(Z)$ is a group. A group element $g \in G$ acts on such a pair by $g*(H, Z) = (gHg^{-1}, gZ)$. An example of this approach is discussed in Sect. 9.2.3.

The rest of this chapter is concerned with implementing these ideas in practice in the context of designs and codes.

9.2 Designs

The most common methods for classifying designs with prescribed automorphism groups are the Kramer–Mesner method and a method based on tactical decompositions, which are the main topics of this section. As an example we also discuss how the block-by-block approach considered in Sect. 6.1.3 can be modified when required symmetries have been prescribed; actually, this approach turns out to be closely related to the Kramer–Mesner method. As far as we know, little or no work has been carried out on using the approach in Sect. 6.1.1 in the prescribed automorphism group setting.

Research Problem 9.5. Study orderly generation (and generation by canonical augmentation) for producing incidence matrices point by point in classifying designs with prescribed automorphisms. Carry out a comparison with other approaches.

Analytical rather than computational results on designs and their automorphism groups can be found in any design theory textbook as well as in the survey articles [91, 168].

9.2.1 The Kramer–Mesner Method

The general approach that has become known as the *Kramer–Mesner method* is based on results for simple designs presented by Alltop [4], which where further developed by Kramer and Mesner [335].

Recall the general parameters of a design, t-(v, k, λ). In this section it is convenient to view a design as a set system $(V, \mathcal{B})$ over a fixed point set V. Isomorphism is considered under the action of $S_v = \mathrm{Sym}(V)$ on the points in the blocks.

A prescribed group $H \leq S_v$ partitions the set of all candidate blocks (that is, the set of all k-subsets of V) into orbits $\mathcal{K}_1, \mathcal{K}_2, \dots, \mathcal{K}_n$. In constructing designs admitting H as a group of automorphisms, we may now study orbits of k-subsets instead of individual k-subsets. Namely, if a k-subset is in the final design, then all other k-subsets in the same orbit are also in the design, because the design admits H.

Also the t-subsets of V are partitioned into H-orbits: $\mathcal{T}_1, \mathcal{T}_2, \dots, \mathcal{T}_m$.

Example 9.6. We consider the classification of 2-$(7, 3, 1)$ designs with respect to the prescribed group $H = \langle (1\ 2\ 3)(4\ 5\ 6)(7) \rangle$. This group partitions the 2-subsets and the 3-subsets of $V = \{1, 2, 3, 4, 5, 6, 7\}$ into the following orbits:

$$
\begin{aligned}
\mathcal{T}_1 &= \{\{1,2\},\{2,3\},\{1,3\}\}, & \mathcal{K}_1 &= \{\{1,2,3\}\},\\
\mathcal{T}_2 &= \{\{1,4\},\{2,5\},\{3,6\}\}, & \mathcal{K}_2 &= \{\{1,2,4\},\{2,3,5\},\{1,3,6\}\},\\
\mathcal{T}_3 &= \{\{1,5\},\{2,6\},\{3,4\}\}, & \mathcal{K}_3 &= \{\{1,2,5\},\{2,3,6\},\{1,3,4\}\},\\
\mathcal{T}_4 &= \{\{1,6\},\{2,4\},\{3,5\}\}, & \mathcal{K}_4 &= \{\{1,2,6\},\{2,3,4\},\{1,3,5\}\},\\
\mathcal{T}_5 &= \{\{1,7\},\{2,7\},\{3,7\}\}, & \mathcal{K}_5 &= \{\{1,2,7\},\{2,3,7\},\{1,3,7\}\},\\
\mathcal{T}_6 &= \{\{4,5\},\{5,6\},\{4,6\}\}, & \mathcal{K}_6 &= \{\{1,4,5\},\{2,5,6\},\{3,4,6\}\},\\
\mathcal{T}_7 &= \{\{4,7\},\{5,7\},\{6,7\}\}, & \mathcal{K}_7 &= \{\{1,4,6\},\{2,4,5\},\{3,5,6\}\},\\
& & \mathcal{K}_8 &= \{\{1,4,7\},\{2,5,7\},\{3,6,7\}\},\\
& & \mathcal{K}_9 &= \{\{1,5,6\},\{2,4,6\},\{3,4,5\}\},\\
& & \mathcal{K}_{10} &= \{\{1,5,7\},\{2,6,7\},\{3,4,7\}\},\\
& & \mathcal{K}_{11} &= \{\{1,6,7\},\{2,4,7\},\{3,5,7\}\},\\
& & \mathcal{K}_{12} &= \{\{4,5,6\}\},\\
& & \mathcal{K}_{13} &= \{\{4,5,7\},\{5,6,7\},\{4,6,7\}\}.
\end{aligned}
$$

Theorem 9.7. *For given i and j, every t-subset in $\mathcal{T}_i$ occurs the same number of times in k-subsets of $\mathcal{K}_j$, and every k-subset in $\mathcal{K}_j$ contains the same number of t-subsets in $\mathcal{T}_i$.*

Proof. Let $T \in \mathcal{T}_i$ occur exactly in the k-subsets $K_1, K_2, \dots, K_s \in \mathcal{K}_j$. Then any other t-subset $T' \in \mathcal{T}_i$ can be obtained as $T' = h(T)$ for some $h \in H$, and T' occurs in the k-subsets $h(K_1), h(K_2), \dots, h(K_s) \in \mathcal{K}_j$. The first part of the theorem now follows from the fact that h induces bijections on the t-subsets and on the k-subsets. The proof of the second part is analogous. □

Since the number of times a t-subset in $\mathcal{T}_i$ occurs in the k-subsets in $\mathcal{K}_j$ only depends on the values of i and j, we denote this number by a_{ij}. If we form an $m \times n$ matrix $\mathbf{A}_{tk} = (a_{ij})$, called a *Kramer–Mesner matrix*, then to find the collections of k-subset orbits that form a t-(v, k, λ) design, we have to find the nonnegative integer solutions $\mathbf{x}$ of the system

$$\mathbf{A}_{tk}\mathbf{x} = \lambda\mathbf{1},$$

where the solution vector $\mathbf{x}$ has length n and the all-λ vector has length m.

Example 9.8. In the setting of Example 9.6, we obtain the following 7×13 Kramer–Mesner matrix, where row i and column j correspond to orbits $\mathcal{T}_i$ and $\mathcal{K}_j$, respectively:

$$\mathbf{A}_{tk} = \begin{bmatrix} 1&1&1&1&1&0&0&0&0&0&0&0&0 \\ 0&1&1&0&0&1&1&1&0&0&0&0&0 \\ 0&0&1&1&0&1&0&0&1&1&0&0&0 \\ 0&1&0&1&0&0&1&0&1&0&1&0&0 \\ 0&0&0&0&2&0&0&1&0&1&1&0&0 \\ 0&0&0&0&0&1&1&0&1&0&0&1&1 \\ 0&0&0&0&0&0&0&1&0&1&1&0&2 \end{bmatrix}.$$

There are six solutions to $\mathbf{A}_{tk}\mathbf{x} = \mathbf{1}$, for example,

$$\mathbf{x} = [1\;0\;0\;0\;0\;1\;0\;0\;0\;0\;1\;0\;0]^T,$$

which corresponds to the design $(V, \mathcal{B})$, where

$$\mathcal{B} = \mathcal{K}_1 \cup \mathcal{K}_6 \cup \mathcal{K}_{11} = \{\{1,2,3\},\{1,4,5\},\{2,5,6\},\{3,4,6\},\\ \{1,6,7\},\{2,4,7\},\{3,5,7\}\}.$$

All solutions obviously lead to the unique 2-$(7, 3, 1)$ design, the Fano plane.

Looking at the Kramer–Mesner method from a computational perspective, the following two subproblems have to be solved.

1. Determine the matrix $\mathbf{A}_{tk}$.
2. Find (up to isomorphism) the solutions of $\mathbf{A}_{tk}\mathbf{x} = \lambda\mathbf{1}$.

The difficulty of the subproblems in general depends on the size of the parameters t-(v, k, λ) and the prescribed group H. The first subproblem is easy if the t-subset and k-subset orbits can be explicitly constructed, and becomes more difficult for large groups and large parameters. Approaches for solving the first subproblem are described in [72, p. 1354] and [356, 521]; the latter references consider group-theoretic methods, which are essential when H is large. See also the survey in [205].

The difficulty of the second subproblem depends on the prescribed group H and its normalizer $N_{S_v}(H)$, where the latter induces symmetry into the system $\mathbf{A}_{tk}\mathbf{x} = \lambda\mathbf{1}$. In essence, the larger the group H, the smaller and hence easier to solve the equation system becomes. The role of the normalizer $N_{S_v}(H)$ is best illustrated with an example.

Example 9.9. In Example 9.6 the normalizer is

$$N_{S_7}(H) = \langle (2\ 3)(5\ 6), (1\ 5\ 3\ 6\ 2\ 4) \rangle, \quad |N_{S_7}(H)| = 36.$$

Each normalizer element permutes the t-subset and k-subset orbits among themselves. For example, the generator (1 5 3 6 2 4) induces the following permutations of orbits in Example 9.6:

$$(\mathcal{T}_1\ \mathcal{T}_6)(\mathcal{T}_2\ \mathcal{T}_3\ \mathcal{T}_4)(\mathcal{T}_5\ \mathcal{T}_7),$$
$$(\mathcal{K}_1\ \mathcal{K}_{12})(\mathcal{K}_2\ \mathcal{K}_6\ \mathcal{K}_4\ \mathcal{K}_7\ \mathcal{K}_3\ \mathcal{K}_9)(\mathcal{K}_5\ \mathcal{K}_{13})(\mathcal{K}_8\ \mathcal{K}_{10}\ \mathcal{K}_{11}).$$

In terms of Example 9.8, these two permutations can be viewed as permuting the columns and rows of the matrix $\mathbf{A}_{tk}$, respectively, in such a way that the matrix is transformed onto itself.

In general, the larger the index $|N_{S_v}(H)|/|H|$, the more symmetry the matrix $\mathbf{A}_{tk}$ has. Because the right-hand side of the system $\mathbf{A}_{tk}\mathbf{x} = \lambda\mathbf{1}$ is constant, each symmetry transformation of $\mathbf{A}_{tk}$ is also a symmetry of the equation system. Thus, whenever the normalizer is large, isomorph rejection techniques must typically be employed to eliminate redundancy from the search when solving the system. One approach is to locate an appropriate set of partial solutions occurring in every design admitting H, classify these partial solutions up to $N_{S_v}(H)$-isomorphism, and then solve the remaining system

$$\mathbf{A}_{tk}\mathbf{y} = \lambda\mathbf{1} - \mathbf{A}_{tk}\mathbf{s}$$

for each classified partial solution $\mathbf{s}$. An example of this approach is discussed in Sect. 9.2.3.

The problem of solving Diophantine linear systems of equations has been encountered several times earlier in this text, and algorithms and references can be found in Sect. 5.4. If we require that the designs be simple, then we search for 0-1 solution vectors $\mathbf{x}$, which (often significantly) speeds up the search.

A useful tool for constructing t-designs with prescribed automorphism groups is DISCRETA [41], developed at Universität Bayreuth and publicly available.

Isomorph rejection for the constructed designs can in many cases be carried out via recorded objects. More sophisticated techniques, such as generation by canonical augmentation, are required if insufficient storage space is available to store all the designs. An example is discussed in Sect. 9.2.3.

For large parameters and large groups, group-theoretic techniques can be used to carry out isomorph rejection for designs that are too large to handle with conventional algorithms but are known to admit a select group of automorphisms (such as a projective or affine linear group) [358]; see also [239, 357, 521].

For classification results obtained using the Kramer–Mesner method, we refer the reader to the extensive corpus of published case studies, including [39, 40, 177, 239, 314, 339, 340, 341, 357, 384, 520, 521, 598]. Many of these studies focus on simple designs.

9.2.2 Tactical Decompositions

If the number of candidate blocks, $\binom{v}{k}$, is fairly large and the order of the group H is fairly small, then there is a prohibitively large number of orbits of k-subsets, and it is not reasonable to try to construct the $\mathbf{A}_{tk}$-matrix. Tactical decompositions form one possible way of circumventing this obstacle – this idea dates back to work by Dembowski [148].

To discuss tactical decompositions, it is convenient to view a design as an incidence structure $(P, \mathcal{B}, I)$ over a fixed point set P and a fixed block set $\mathcal{B}$. Isomorphism is considered under the action of $\mathrm{Sym}(P) \times \mathrm{Sym}(\mathcal{B})$ on the incidence relation $I \subseteq P \times \mathcal{B}$. Accordingly, a prescribed group $H \leq \mathrm{Sym}(P) \times \mathrm{Sym}(\mathcal{B})$ acts explicitly both on the points and on the blocks.

A *decomposition* of an incidence structure $(P, \mathcal{B}, I)$ is a partition of the blocks and the points: $\mathcal{B} = \mathcal{B}_1 \cup \mathcal{B}_2 \cup \cdots \cup \mathcal{B}_n$ and $P = P_1 \cup P_2 \cup \cdots \cup P_m$. For $1 \leq i \leq m$ and $p \in P_i$ fixed, let

$$s_{ij} = |\{B \in \mathcal{B}_j : (p, B) \in I\}|,$$

and for $1 \leq j \leq n$ and $B \in \mathcal{B}_j$ fixed, let

$$t_{ij} = |\{p \in P_i : (p, B) \in I\}|.$$

Definition 9.10. *A decomposition is* tactical *if the numbers s_{ij} and t_{ij} are independent of the choices of $p \in P_i$ and $B \in \mathcal{B}_j$.*

Associated with a tactical decomposition are the two $m \times n$ matrices $\mathbf{S} = (s_{ij})$ and $\mathbf{T} = (t_{ij})$.

Let H be a group of automorphisms of a given incidence structure $\mathcal{X}$. The *H-decomposition* of $\mathcal{X}$ is the decomposition of the blocks and the points into H-orbits. The following theorem resembles Theorem 9.7.

Theorem 9.11. *Let $\mathcal{X}$ be an incidence structure and let $H \leq \mathrm{Aut}(\mathcal{X})$. Then, the H-decomposition of $\mathcal{X}$ is tactical.*

Proof. Consider an arbitrary pair i, j and select $p, p' \in P_i$ arbitrarily. Let $\{B_1, B_2, \ldots, B_{s_{ij}}\} = \{B \in \mathcal{B}_j : (p, B) \in I\}$ and $\{B'_1, B'_2, \ldots, B'_{s'_{ij}}\} = \{B \in \mathcal{B}_j : (p', B) \in I\}$. Because P_i is an H-orbit, there exists an $h \in H$ with $p' = h(p)$. Because an automorphism by definition maps flags onto flags, $p' = h(p)$ is incident with $h(B_1), h(B_2), \ldots, h(B_{s_{ij}}) \in \mathcal{B}_j$, and thus $s_{ij} \leq s'_{ij}$. Similarly, $p = h^{-1}(p')$ gives $s'_{ij} \leq s_{ij}$ and hence $s_{ij} = s'_{ij}$. The proof for t_{ij} is analogous with the roles of the points and blocks interchanged. □

If H is trivial, then we get a *discrete* decomposition where the blocks and points are partitioned into singletons (sets with one element); the matrices $\mathbf{S}$ and $\mathbf{T}$ associated with such a tactical decomposition coincide with an incidence matrix. If H is a nontrivial group – in most studies where this approach is used, $|H| = p$ is a prime so that the orbits are of size 1 or p – the classification consists of two steps:

1. Find all matrices $\mathbf{T} = (t_{ij})$ associated with a tactical decomposition.
2. For each matrix $\mathbf{T}$, refine the matrix in all possible ways to an incidence matrix of a design. Each refinement step replaces one entry t_{ij} with a H-invariant 0-1 matrix of size $|P_i| \times |\mathcal{B}_j|$ and constant column sum t_{ij}.

It is customary in this context to call a matrix $\mathbf{T}$ a tactical decomposition, although in many cases also matrices that admit no refinement to an incidence matrix are considered in the first step.

For BIBDs, the candidate matrices $\mathbf{T} = (t_{ij})$ satisfy the following system of equations for $1 \le i_1 \le i_2 \le m$ (corresponding to occurrences of points and pairs of distinct points):

$$\sum_{j=1}^{n} t_{i_1 j} |\mathcal{B}_j| = r|P_i|, \tag{9.1}$$

$$\sum_{j=1}^{n} t_{i_1 j} t_{i_2 j} |\mathcal{B}_j| = \lambda |P_{i_1}||P_{i_2}| + \delta_{i_1 i_2}(r - \lambda)|P_{i_1}|, \tag{9.2}$$

where the Kronecker delta $\delta_{i_1 i_2}$ is 1 if $i_1 = i_2$ and 0 otherwise (this takes care of the two situations where two points are from the same orbit and from different orbits in the decomposition). We also know the block sizes: for $1 \le j \le n$,

$$\sum_{i=1}^{m} t_{ij} = k, \tag{9.3}$$

but it turns out that (9.3) is redundant given (9.1) and (9.2); this can be shown with a proof analogous to that of [458, Theorem 3.3]. Note however, that although (9.3) is redundant for complete $m \times n$ matrices, these additional equations may be useful in pruning partial matrices that cannot be extended to a full matrix. For t-designs with $t \ge 3$, equations may be stated in an analogous way.

From a computational perspective the construction of a matrix $\mathbf{T}$ amounts to solving a quadratic system of equations, for example, using row-by-row backtrack search (cf. Sect. 6.1.1). The normalizer $N_{\mathrm{Sym}(P)\times\mathrm{Sym}(\mathcal{B})}(H)$ induces symmetry to the equation system by acting on the orbits $P_1, P_2, \ldots, P_m$ and $\mathcal{B}_1, \mathcal{B}_2, \ldots, \mathcal{B}_n$. Isomorph rejection on partial solutions is typically required to eliminate redundant work in solving the system.

Example 9.12. We yet again consider the classification of 2-$(7,3,1)$ designs. Let $P = \{1,2,3,4,5,6,7\}$ and $\mathcal{B} = \{B_1, B_2, B_3, B_4, B_5, B_6, B_7\}$. Let $H = \langle (1\ 2\ 3)(4\ 5\ 6)(7)(B_1\ B_2\ B_3)(B_4\ B_5\ B_6)(B_7) \rangle$ be the prescribed group of automorphisms. Accordingly, there are 3 orbits on the points and 3 orbits on the blocks:

$$P_1 = \{1,2,3\}, \qquad \mathcal{B}_1 = \{B_1, B_2, B_3\},$$
$$P_2 = \{4,5,6\}, \qquad \mathcal{B}_2 = \{B_4, B_5, B_6\},$$
$$P_3 = \{7\}, \qquad \mathcal{B}_3 = \{B_7\}.$$

From (9.1) and (9.2), we obtain the equation system

$$\begin{aligned}
3t_{11} + \quad 3t_{12} + \quad t_{13} &= 9,\\
3t_{21} + \quad 3t_{22} + \quad t_{23} &= 9,\\
3t_{31} + \quad 3t_{32} + \quad t_{33} &= 3,\\
3t_{11}t_{21} + 3t_{12}t_{22} + t_{13}t_{23} &= 9,\\
3t_{11}t_{31} + 3t_{12}t_{32} + t_{13}t_{33} &= 3,\\
3t_{21}t_{31} + 3t_{22}t_{32} + t_{23}t_{33} &= 3,\\
3t_{11}^2 + \quad 3t_{12}^2 + \quad t_{13}^2 &= 15,\\
3t_{21}^2 + \quad 3t_{22}^2 + \quad t_{23}^2 &= 15,\\
3t_{31}^2 + \quad 3t_{32}^2 + \quad t_{33}^2 &= 3,
\end{aligned}$$

where $t_{ij} \in \{0,1,2,3\}$ if $i \neq 3$ and $j \neq 3$, and $t_{ij} \in \{0,1\}$ otherwise. This system has the solution

$$\mathbf{T} = \begin{bmatrix} 2\,1\,0 \\ 1\,1\,3 \\ 0\,1\,0 \end{bmatrix},$$

which is unique up to $N_{\mathrm{Sym}(P)\times\mathrm{Sym}(\mathcal{B})}(H)$-isomorphism. In this case the normalizer acts on T by transposing either the first and second row or the first and second column, or both.

Having constructed all matrices $\mathbf{T}$ that satisfy (9.1) and (9.2), we still have to refine the matrices to find any associated designs. This can be accomplished, for example, by refining one row of the matrix at a time. Partial solutions may be pruned by observing that the inner product of any two distinct complete rows must be λ. Again isomorph rejection on partial solutions may be required.

Example 9.13. Starting with the top left corner of the matrix $\mathbf{T}$ in Example 9.12, there are three possibilities to replace the 1,1 entry with an H-invariant 0-1 submatrix, namely

$$\begin{bmatrix} 0\,1\,1 \\ 1\,0\,1 \\ 1\,1\,0 \end{bmatrix}, \quad \begin{bmatrix} 1\,0\,1 \\ 1\,1\,0 \\ 0\,1\,1 \end{bmatrix}, \quad \begin{bmatrix} 1\,1\,0 \\ 0\,1\,1 \\ 1\,0\,1 \end{bmatrix}.$$

Likewise there are three possible choices for the 1,2 entry, and so forth. One possible complete refinement of the matrix $\mathbf{T}$ is

$$\left[\begin{array}{c|c|c} 0\,1\,1 & 1\,0\,0 & 0 \\ 1\,0\,1 & 0\,1\,0 & 0 \\ 1\,1\,0 & 0\,0\,1 & 0 \\ \hline 1\,0\,0 & 1\,0\,0 & 1 \\ 0\,1\,0 & 0\,1\,0 & 1 \\ 0\,0\,1 & 0\,0\,1 & 1 \\ \hline 0\,0\,0 & 1\,1\,1 & 0 \end{array}\right].$$

In total there are 27 possible refinements, each leading to the unique Fano plane.

Classification results obtained by this approach include [98, 103, 140, 259, 260, 334, 336, 399, 472, 544, 546, 582]; in a large number of other studies this method has been used to obtain existence rather than classification results.

9.2.3 Example: STSs with Nontrivial Groups

In this section we will look at a concrete example of classification of designs with nontrivial automorphism groups, with the focus on Steiner triple systems, in particular on STS(21). The discussion is based on [299] with many of the technical details omitted.

Following the general treatment in Sect. 9.1, the first task is to determine the possible prime order groups of automorphisms an STS(21) can admit – recall that by Theorem 9.1 any nontrivial group admits a subgroup of prime order, and therefore it suffices to consider all such groups in generating all designs with a nontrivial automorphism group.

With the eventual aim of applying the Kramer–Mesner framework, we will view an STS(21) as a set system $(V, \mathcal{B})$ over the fixed point set $V = \{1, 2, \dots, 21\}$, where isomorphism is given by the action of $S_{21} = \mathrm{Sym}(V)$. Accordingly, determining the prime order subgroups $H \leq S_{21}$ is equivalent to determining up to conjugacy in S_{21} the possible prime order automorphisms. Each such automorphism has cycle type $1^f p^m$, where p is prime and $f + pm = 21$.

It is typical that some of the possible prime order groups can be eliminated by combinatorial arguments based on properties of the designs. The following results allow us to restrict the prime order groups in the case of Steiner triple systems (cf. [121]).

Theorem 9.14. *The set of points fixed by any automorphism of a Steiner triple system induces a subsystem.*

Proof. Let h be an automorphism of an STS, and let F be the set of all points fixed by h. We prove that the blocks contained in F form an STS(f), $f = |F|$ by showing that there exists no block $\{x, y, z\}$ with $F \cap \{x, y, z\} = \{x, y\}$. Indeed, for distinct $x, y \in F$, the automorphism h maps the block $\{x, y, z\}$ onto a block $\{h(x), h(y), h(z)\} = \{x, y, h(z)\}$, so $h(z) = z$ and also $z \in F$. □

Theorem 9.15. *If an* STS(v) *contains a subsystem of order w, then either $v \geq 2w + 1$ or $v = w$.*

Proof. The claim is trivial when $v = w$, so suppose $0 < w < v$. Let the points of the subsystem be $\{x_1, \dots, x_w\}$ and let the other points of the STS(v) be $\{y_1, \dots, y_{v-w}\}$. Consider the blocks that contain y_1. Since no pair of the form $\{x_i, x_j\}$, $i \neq j$, occurs in these blocks, there exist w blocks of the form $\{x_i, y_1, y_\ell\}$. Since there are exactly $v - w - 1$ pairs of the form $\{y_1, y_\ell\}$, $\ell \neq 1$, we have $v - w - 1 \geq w$. □

In the case of STS(21), the possible numbers for fixed points are thus $f \in \{0, 1, 3, 7, 9\}$. We obtain the following prime factorizations for each possible number of moved points:

$$\begin{aligned}21 - 0 &= 21 = 7 \times 3,\\ 21 - 1 &= 20 = 5 \times 2^2,\\ 21 - 3 &= 18 = 3^2 \times 2,\\ 21 - 7 &= 14 = 7 \times 2,\\ 21 - 9 &= 12 = 3 \times 2^2.\end{aligned}$$

Up to conjugacy in S_{21}, this gives ten candidates for automorphisms of prime order:

$$7^3,\quad 3^7,\quad 1^1 5^4,\quad 1^1 2^{10},\quad 1^3 3^6,\quad 1^3 2^9,\quad 1^7 7^2,\quad 1^7 2^7,\quad 1^9 3^4,\quad 1^9 2^6.$$

Frequently it is possible to eliminate some candidates with further combinatorial arguments. In this case it is possible to eliminate the type $1^1 2^{10}$; the following observation is implied by the results in [167, 506, 571].

Lemma 9.16. *A permutation with cycle type* $1^1 2^{10}$ *cannot occur as an automorphism of an* STS(21).

Proof. To reach a contradiction, consider an STS(21) admitting an automorphism of type $1^1 2^{10}$, let 0 be the fixed point, and let $\{\{-i, i\} : 1 \leq i \leq 10\}$ be the pairs of points fixed by such an automorphism. Thus, $\{x, y, z\}$ is a block if and only if $\{-x, -y, -z\}$ is a block. In particular, $\{0, -i, i\}$ is a block for all $1 \leq i \leq 10$. The remaining 60 blocks must cover the remaining 90 pairs of points with opposite signs. Thus, each of the 30 remaining orbits of blocks covers either 0 or 4 of the remaining pairs of points with opposite signs. This is a contradiction because 4 does not divide 90. □

We are left with nine prime order groups in S_{21}; each of these groups actually occurs as a group of automorphisms of an STS(21). Table 9.1 shows the problem sizes for each of the nine prime order groups. The original classification problem with no prescribed automorphisms has $\binom{21}{2} = 210$ 2-subsets of points to be covered and $\binom{21}{3} = 1{,}330$ candidate 3-subsets to cover these.

Looking at the normalizer orders in Table 9.1, it is clear that at least for groups with order 2 and 3 isomorph rejection is required to eliminate normalizer-induced symmetry from the Kramer–Mesner search. Thus, there are essentially three subproblems that need to be solved to obtain a classification of the STS(21) with a nontrivial automorphism group.

1. Constructing the STSs using the Kramer–Mesner method for each H.
2. Isomorph rejection with respect to $N_{S_{21}}(H)$ on partial solutions to the Kramer–Mesner system.
3. Isomorph rejection for constructed STSs.

Table 9.1. Problem sizes for eligible prime order subgroups

H	$\lvert N_{S_{21}}(H)\rvert$	2-subset orbits	3-subset orbits
7^3	12,348	30	190
3^7	22,044,960	70	448
$1^1 5^4$	60,000	42	266
$1^3 3^6$	6,298,560	72	448
$1^3 2^9$	1,114,767,360	111	679
$1^7 7^2$	2,963,520	48	220
$1^7 2^7$	3,251,404,800	119	707
$1^9 3^4$	1,410,877,440	94	502
$1^9 2^6$	16,721,510,400	126	734

We adopt a solution approach analogous to block-by-block BIBD construction based on seeds in Sect. 6.1.3; that is, essentially we apply the Kramer–Mesner method with initial isomorph rejection based on seeds. The problem of completing seeds is solved using an algorithm for EXACT COVERS – the task is to cover the remaining 2-subset orbits with 3-subset orbits.

Obviously the main challenge lies in isomorph rejection, mostly because we have to work with the prescribed groups and the associated normalizers, but also because there are too many isomorphism classes of STSs to store in memory. Fortunately, generation by canonical augmentation is a versatile technique that can be adapted to the prescribed group setting.

The basic algorithmic setup consists of three layers of backtrack search. Let $H \leq S_{21}$ be one of the nine eligible prime order groups – all nine groups are to be considered to achieve exhaustive generation. We proceed to outline the three backtrack layers.

The first layer classifies up to isomorphism a set of seeds to remove $N_{S_{21}}(H)$-induced symmetry from the Kramer–Mesner system. Seeds are triples of the form $(H, T, \mathcal{S})$, where $T \subseteq V$ is a nonempty set of points and $\mathcal{S}$ is a union of H-orbits of 3-subsets of V. The intuition here is that $\mathcal{S}$ is the set of blocks that we get if we consider an STS(21) admitting H and constructed from this seed, take the set of blocks that contain at least one point from T, and form the union of the H-orbits of these blocks. In particular, every point in T occurs in $r = 10$ 3-subsets in $\mathcal{S}$. Isomorphism of seeds is induced from the action of S_{21} defined for all $g \in S_{21}$ by $g * (H, T, \mathcal{S}) = (gHg^{-1}, gT, g\mathcal{S})$. In particular, for fixed H, the seeds $(H, T_1, \mathcal{S}_1)$ and $(H, T_2, \mathcal{S}_2)$ are isomorphic if and only if there exists a $g \in N_{S_{21}}(H)$ with $gT_1 = T_2$ and $g\mathcal{S}_1 = \mathcal{S}_2$.

Example 9.17. For group type $1^3 3^6$ one example of a seed is as follows. Let $H = \langle (4\ 5\ 6)(7\ 8\ 9) \cdots (19\ 20\ 21) \rangle$, $T = \{1, 2, 3\}$, and let $\mathcal{S}$ be given by the incidence matrix in Fig. 9.2. The point and block orbits are separated by horizontal and vertical lines, respectively. Note that all points in T are fixed by H. Furthermore, each point in T occurs in $r = 10$ blocks.

1	111	111	111	000	000	000	000	000	000
1	000	000	000	111	111	111	000	000	000
1	000	000	000	000	000	000	111	111	111
0	100	000	000	100	000	000	100	000	000
0	010	000	000	010	000	000	010	000	000
0	001	000	000	001	000	000	001	000	000
0	100	000	000	001	000	000	010	000	000
0	010	000	000	100	000	000	001	000	000
0	001	000	000	010	000	000	100	000	000
0	000	100	000	000	100	000	000	100	000
0	000	010	000	000	010	000	000	010	000
0	000	001	000	000	001	000	000	001	000
0	000	100	000	000	001	000	000	010	000
0	000	010	000	000	100	000	000	001	000
0	000	001	000	000	010	000	000	100	000
0	000	000	100	000	000	100	000	000	100
0	000	000	010	000	000	010	000	000	010
0	000	000	001	000	000	001	000	000	001
0	000	000	100	000	000	001	000	000	010
0	000	000	010	000	000	100	000	000	001
0	000	000	001	000	000	010	000	000	100

Fig. 9.2. A seed for group type $1^3 3^6$

Example 9.18. For group type $1^1 5^4$ one example of a seed is as follows, let $H = \langle (2\ 3\ 4\ 5\ 6)(7\ 8\ 9\ 10\ 11) \cdots (17\ 18\ 19\ 20\ 21) \rangle$, $T = \{1, 2\}$, and let $\mathcal{S}$ be given by the incidence matrix in Fig. 9.3. The point and block orbits are separated by horizontal and vertical lines, respectively. Note that T contains one point fixed and one point moved by H. Furthermore, each point in T occurs in $r = 10$ blocks.

Given a seed $(H, T, \mathcal{S})$ as input, the second layer extends $(H, T, \mathcal{S})$ in all possible ways into an STS(21), $\mathcal{X} = (V, \mathcal{B})$, such that $\mathcal{S} \subseteq \mathcal{B}$ and $H \leq \mathrm{Aut}(\mathcal{X})$. This is an instance of Exact Covers on H-orbits of 2- and 3-subsets of V, and can be solved using techniques from Sect. 5.2.

The third layer uses generation by canonical augmentation to reject isomorphs. An input to the layer consists of a generated STS(21), $\mathcal{X}$, and the associated parent seed $p(\mathcal{X}) = (H, T, \mathcal{S})$. To apply generation by canonical augmentation, we must associate a canonical parent $m(\mathcal{X})$ with every STS(21) having a nontrivial automorphism group.

Let κ be a canonical labeling map for the action of S_{21} on STS(21) over V. Let $\mathcal{X}$ be the STS(21) given as input. First, let $\hat{\mathcal{X}} = \kappa(\mathcal{X})\mathcal{X}$. Identify a group $\hat{H}$ conjugate to an eligible prime order group among the subgroups of $\mathrm{Aut}(\hat{\mathcal{X}})$ – such a group must clearly exist based on Theorem 9.1 and our analysis of the prime order automorphisms of STS(21). For example, take the group generated by the minimum prime order element in $\mathrm{Aut}(\hat{\mathcal{X}})$ with respect

11111	11111	00000	00000	00000	00000	00000	00000	00000
10000	00000	10001	10000	10000	10000	10000	10000	10010
01000	00000	11000	01000	01000	01000	01000	01000	01001
00100	00000	01100	00100	00100	00100	00100	00100	10100
00010	00000	00110	00010	00010	00010	00010	00010	01010
00001	00000	00011	00001	00001	00001	00001	00001	00101
00000	10000	00000	00000	01000	00001	00100	10010	00000
00000	01000	00000	00000	00100	10000	00010	01001	00000
00000	00100	00000	00000	00010	01000	00001	10100	00000
00000	00010	00000	00000	00001	00100	10000	01010	00000
00000	00001	00000	00000	10000	00010	01000	00101	00000
10000	00000	00000	00001	01000	00100	00010	00000	00000
01000	00000	00000	10000	00100	00010	00001	00000	00000
00100	00000	00000	01000	00010	00001	10000	00000	00000
00010	00000	00000	00100	00001	10000	01000	00000	00000
00001	00000	00000	00010	10000	01000	00100	00000	00000
00000	10000	10000	00010	00000	00000	00000	00000	00100
00000	01000	01000	00001	00000	00000	00000	00000	00010
00000	00100	00100	10000	00000	00000	00000	00000	00001
00000	00010	00010	01000	00000	00000	00000	00000	10000
00000	00001	00001	00100	00000	00000	00000	00000	01000

Fig. 9.3. A seed for group type $1^1 5^4$

to lexicographic order. Then, identify a nonempty set $\hat{T} \subseteq V$ based on $\hat{\mathcal{X}}$ and $\hat{H}$. Let $\hat{\mathcal{S}}$ be the union of all $\hat{H}$-orbits of blocks in $\hat{\mathcal{X}}$ that contain a block that has nonempty intersection with $\hat{T}$. Finally, put $m(\mathcal{X}) = \kappa(\mathcal{X})^{-1}(\hat{H}, \hat{T}, \hat{\mathcal{S}})$.

Because κ is a canonical labeling map and $\hat{H}$ and $\hat{\mathcal{S}}$ are selected from the canonical representative $\hat{\mathcal{X}}$, it is immediate that (4.12) holds. Also it is straightforward to check that (4.15) holds for a natural search tree associated with the search; cf. Sect. 6.1.4. Also (4.16) holds if we assume that $\hat{T}$ is always selected in such a manner that $(\hat{H}, \hat{T}, \hat{\mathcal{S}})$ is isomorphic to a seed classified in the first layer. Essentially this amounts to correlating for every eligible prime order group the seed classification algorithm and the algorithm for selecting $\hat{T}$. With this assumption, Algorithm 4.7 applies for isomorph rejection.

Example 9.19. For $\hat{H}$ of type $1^3 3^6$ we let $\hat{T}$ consist of the 3 points fixed by $\hat{H}$. Accordingly, the seed classification algorithm classifies all such seeds $(H, T, \mathcal{S})$ for H of type $1^3 3^6$ by completing one point p fixed by H at a time so that p occurs in $r = 10$ blocks in $\mathcal{S}$. Example 9.17 gives one example of such a seed.

Example 9.20. For $\hat{H}$ of type $1^1 5^4$ we let $\hat{T}$ consist of the point fixed and the minimum point moved by $\hat{H}$. The seed classification algorithm proceeds by first completing the fixed point and then considering all possible ways to complete a moved point. Example 9.18 gives one example of such a seed.

This completes the outline of the backtrack layers. Obviously the algorithm is analogous to the one described in Sect. 6.1.3, only now we have the prescribed group H and the point set T inducing $\mathcal{S}$ complicating the implementation. Also, the entire algorithm operates on H-orbits of 3-subsets instead of individual 3-subsets.

The main implementation challenge is that isomorph rejection must be performed with respect to the normalizer $N_{S_{21}}(H)$ in classifying the seeds. Also, the implementation of Algorithm 4.7 and the test (4.13) require checking whether $m(\mathcal{X})$ and $p(\mathcal{X})$ are in the same orbit of $\mathrm{Aut}(\mathcal{X})$. Furthermore, computation of the automorphism group $\mathrm{Aut}(H, T, \mathcal{S}) = \{g \in N_{S_{21}}(H) : gT = T,\ g\mathcal{S} = \mathcal{S}\}$ is required if automorphisms of the parent are employed to reject isomorphic siblings in Algorithm 4.7. A more detailed discussion of the algorithm implementation can be found in [299].

There are 62,336,617 nonisomorphic STS(21) with a nontrivial automorphism group; Table 9.2 partitions these into classes based on the order of the automorphism group and the types of prime order groups admitted. In a row of the table there is one letter for each such class.

9.2.4 Some Results

Numerous published studies consider classification of designs with specific parameters admitting a specific group of automorphisms. The present section attempts to collect some of these results together roughly based on the type of design being considered.

In the case of BIBDs, perhaps the two most studied families are square designs and Steiner systems $S(2, k, v)$. Studies focusing on classification of square designs with prescribed automorphisms include [98, 99, 100, 101, 102, 103, 104, 105, 140, 259, 394, 472, 476, 544, 546, 578, 582]. For some of the preceding studies also a complete classification of the BIBDs in question has been obtained, see [545]. A special case of square designs are the Hadamard designs, which are discussed in connection with Hadamard matrices in Sect. 9.4.

Two extensively studied families among Steiner systems are Steiner triple systems $S(2, 3, v)$ and related Steiner quadruple systems $S(3, 4, v)$. Of special interest in these families are the systems admitting the *cyclic* automorphism $(1\ 2\ \cdots\ v)$ on points and, more generally, the systems where the automorphism group acts transitively on the points. Other types of automorphisms often considered include a *k-rotational* automorphism consisting of one fixed point and k cycles of equal length, and a *k-cyclic* automorphism with no fixed points and k cycles of equal length.

Classification results for cyclic $S(2, 3, v)$ can be found in [127] for $v \leq 33$, in [124] for $v \leq 45$. Classification results for $S(2, 3, v)$ with rotational automorphisms appear in [478]. Point-transitive $S(2, 3, v)$ are classified in [579] for $v = 25$ and in [120] for $v = 27$. A classification of all $S(2, 3, 21)$ with a nontrivial automorphism has been obtained in [304]; see Sect. 9.2.3.

Table 9.2. The STS(21) with a nontrivial automorphism group

$\|\mathrm{Aut}(\mathcal{D})\|$	3^7	7^3	$1^1 5^4$	$1^3 2^9$	$1^3 3^6$	$1^7 2^7$	$1^7 7^2$	$1^9 2^6$	$1^9 3^4$	N
2				*a*		*b*		*c*		60,588,267
3	*a*				*b*				*c*	1,732,131
4				*ab*		*a*		*abc*		11,467
5			*a*							1,772
6	*abc*			*ade*	*dfg*	*bfh*		*cgi*	*ehi*	2,379
7		*a*								66
8				*ab*		*a*		*abc*		222
9	*ab*				*abc*				*a*	109
12	*ab*			*acd*	*ce*	*cd*		*abcde*	*d*	85
14		*a*				*a*				14
16				*a*		*a*		*a*		12
18	*abc*			*ad*	*abcdef*	*bce*		*f*	*bf*	33
21	*a*	*ab*			*bc*		*c*			10
24	*abc*			*abde*	*def*	*ad*		*abcdef*		19
27	*a*				*a*					3
36	*a*			*b*	*ab*	*b*		*ab*		5
42	*a*	*ab*		*a*	*bc*	*bc*	*c*			7
48				*a*	*a*	*a*		*a*		2
54	*a*			*a*	*a*				*a*	1
72	*ab*			*ab*	*ab*	*a*		*ab*		5
108				*a*	*a*	*a*		*a*	*a*	1
126	*ab*	*ab*		*a*	*ab*	*b*				2
144	*a*			*a*	*a*	*a*		*a*		1
294		*a*			*a*	*a*	*a*			1
504	*a*	*a*			*a*			*a*		1
882	*a*	*a*			*a*	*a*	*a*			1
1,008	*a*	*a*		*a*	*a*	*a*		*a*		1

Classification results for cyclic $S(3,4,v)$ appear in [236] for $v = 8, 14, 16$, in [28] for $v = 10$, in [477] for $v = 20$, and in [193] for $v = 22$ (the classification reported in [158] is erroneous).

Other studied families of Steiner systems include the $S(2,4,v)$ family, where a complete classification of the systems admitting a nontrivial automorphism group has been obtained in [334] for $v = 25$ (see also [549] for the complete classification) and in [336] for $v = 28$. In [127] the cyclic $S(2,4,v)$ and $S(2,5,v)$ are classified for $v \leq 52$ and $v \leq 65$, respectively.

Other studies focusing on classification of Steiner systems with prescribed automorphisms include [337] for the family $S(2,5,41)$.

Among other types of BIBDs, in [581] the cyclic 2-$(13,5,5)$ designs are classified and in [191] the dicyclic 2-$(v,4,2)$ designs are classified for $v \leq 22$. In [160] a family of quasi-derived 2-$(28,12,11)$ designs with prescribed automorphisms is classified.

Studies focusing on classification of t-designs with $t \geq 3$, $\lambda > 1$, and prescribed automorphisms include [39, 40, 176, 177, 239, 314, 339, 340, 341, 357, 384, 520, 521, 598].

9.3 Codes

In classifying codes with prescribed automorphisms, we consider the two main types of codes, error-correcting and covering codes, as well as the particular case of linear codes. In particular, a Kramer–Mesner-type approach turns out to be useful for these objects.

9.3.1 Covering Codes

In the context of unrestricted codes, equivalence is given by the product action (3.9) of $G = S_q \wr S_n$ on Z_q^n. A prescribed group $H \leq G$ partitions this space into orbits of words: $Z_q^n = W_1 \cup W_2 \cup \cdots \cup W_s$, and we may study sets of orbits instead of sets of words along the ideas of [385, 467]. The following theorem is in the same flavor as Theorem 9.7.

Theorem 9.21. *For given i and j and an arbitrary $\mathbf{x} \in W_i$, $\min\{d_H(\mathbf{x}, \mathbf{y}) : \mathbf{y} \in W_j\}$ is independent of the choice of $\mathbf{x}$.*

Proof. Let $\mathbf{x}, \mathbf{x}' \in W_i$ be arbitrary. By definition of an orbit, there exists an $h \in H$ with $h(\mathbf{x}) = \mathbf{x}'$. Because h defines an isometry of Z_q^n, we have $d_H(\mathbf{x}, \mathbf{y}) = d_H(\mathbf{x}', h(\mathbf{y}))$ for all $\mathbf{y} \in W_j$. In particular, if $\mathbf{y}' \in W_j$ has the smallest distance to $\mathbf{x}$, then $h(\mathbf{y}') \in W_j$ has the same distance to $\mathbf{x}'$. □

For a given covering radius R, we may form an $s \times s$ matrix $\mathbf{A} = (a_{ij})$ similar to the Kramer–Mesner matrix for designs by letting $a_{ij} = 1$ if and only if $d_H(W_i, W_j) = \min\{d_H(\mathbf{x}, \mathbf{y}) : \mathbf{x} \in W_i, \mathbf{y} \in W_j\} \leq R$, and $a_{ij} = 0$ otherwise. By Theorem 9.21, either $\mathbf{x} \in W_i$ or $\mathbf{y} \in W_j$ can be fixed in calculating $d_H(W_i, W_j)$. To find codes admitting H and having covering radius R, we have to find solutions of the system

$$\mathbf{A}\mathbf{x} \geq \mathbf{1}, \tag{9.4}$$

where the 0-1 column vector $\mathbf{x}$ and the all-one vector have length s. Since we are generally interested only in the covering codes with minimum cardinality, we wish to minimize $\mathbf{c}\mathbf{x}$ over all solutions of (9.4), where $\mathbf{c} = [|W_1| \ |W_2| \ \cdots \ |W_s|]$. Consequently, we have instances of WEIGHTED SET COVERS.

Example 9.22. With $q = 2$, $n = 3$, $R = 1$, and H generated by the coordinate permutation (1 2 3), we obtain the orbits

$$\begin{aligned} W_1 &= \{000\}, \\ W_2 &= \{001, 010, 100\}, \\ W_3 &= \{011, 101, 110\}, \\ W_4 &= \{111\}, \end{aligned}$$

the cost vector $\mathbf{c} = [1\ 3\ 3\ 1]$, and

$$\mathbf{A} = \begin{bmatrix} 1\,1\,0\,0 \\ 1\,1\,1\,0 \\ 0\,1\,1\,1 \\ 0\,0\,1\,1 \end{bmatrix}.$$

This instance has a unique solution with 2 codewords corresponding to the vector $\mathbf{x} = [1\ 0\ 0\ 1]^T$.

To classify the covering codes of minimum cardinality with a prescribed automorphism group, we have to find all solutions to the aforementioned optimization problem and reject isomorphs. Little has been done on these problems; the problem of rejecting isomorphs can be solved with the methods used for designs in Sect. 9.2 with slight modifications as we are now facing subgroups of the wreath product $S_q \wr S_n$ rather than subgroups of the symmetric group S_v.

9.3.2 Error-Correcting Codes

For error-correcting codes with d odd, we could with minor modifications proceed in the same manner as for covering codes in Sect. 9.3.1. However, this would just be an intermediate step leading to a clique problem that we will now discuss.

For error-correcting codes the basic requirement is that the distance between two distinct codewords must be at least d. Since a code admitting H is a union of H-orbits $W_1, W_2, \ldots, W_s$ of words, this basic requirement transforms into two criteria for an error-correcting code admitting H.

1. For any orbit W_i occurring in the code, the distance between any two distinct words must be at least d.
2. For any two orbits W_i, W_j occurring in the code, $d_H(W_i, W_j) \geq d$ must hold.

These criteria yield the following approach for constructing error-correcting codes admitting H. First, we disregard all orbits that do not satisfy the first criterion. The remaining orbits are called *admissible*. From the admissible orbits, we create a compatibility graph as follows (cf. Sect. 7.1): there is one vertex for each admissible orbit, and two vertices are adjacent if and only if the second criterion is fulfilled for the corresponding orbits. This approach is analogous to that in Sect. 7.1 but now we want to find the maximum-weight cliques in a weighted graph, where the weight of a vertex is the number of words in the corresponding admissible orbit. The extent of isomorph rejection required in the process again depends on the prescribed group H and its normalizer.

Error-correcting (unrestricted and constant-weight) codes with prescribed automorphisms have so far been studied mainly to obtain existence proofs, see, for example, [72, 290, 447]. For the problem of rejecting isomorphs among codes corresponding to maximum-weight cliques, see the comment at the end of Sect. 9.3.1.

One particular type of prescribed group for codes, leading to the so-called matrix method, deserves a separate treatment.

9.3.3 The Matrix Method

If the coordinate values belong to a field $\mathbb{F}_q$ and the prescribed group H consists only of value permutations of the type

$$\mathbf{x} \mapsto \mathbf{x} + \alpha\mathbf{g}, \quad \mathbf{x} \in \mathbb{F}_q^n \tag{9.5}$$

for a set of vectors $\mathbf{g}$ and all $\alpha \in \mathbb{F}_q$, such codes are more conveniently considered in a different framework. Equation (9.5) implies that the orbits under the group H are the cosets of a linear code (generated by the vectors $\mathbf{g}$), so all orbits will have length q^k for some k, the dimension of the linear code.

As we have seen in Sect. 2.3.3, the codewords in a coset of a linear code can be detected by calculating syndromes: if $\mathbf{H}$ is an $(n-k) \times n = r \times n$ parity check matrix of the linear code, then $\mathbf{H}\mathbf{x}^T = \mathbf{H}\mathbf{y}^T$ if and only if $\mathbf{x}$ and $\mathbf{y}$ are in the same coset. For clarity, we assume in the sequel that $\mathbf{H}$ has full rank (but essentially the same result is obtained even if this assumption is relaxed).

Codes of type (9.5) can be represented as a set of syndromes S and the matrix $\mathbf{H}$, which gives the method its name, the *matrix method.* Moreover, from S and $\mathbf{H}$, it is fairly straightforward to directly calculate the minimum distance and the covering radius of the code, as we will now see.

Definition 9.23. *We say that $S \subseteq \mathbb{F}_q^r$ R-*covers *$\mathbb{F}_q^r$ using $\mathbf{H}$ if every column vector in $\mathbb{F}_q^r$ can be represented as a sum of one column vector of S and a $\mathbb{F}_q$-linear combination of at most R columns of $\mathbf{H}$.*

Note that for $\mathbf{H} = \mathbf{I}$, we get the definition of an unrestricted covering code with covering radius at most R. If $|S| = 1$ (and S contains, without loss of generality, the all-zero vector), on the other hand, we have a linear code with covering radius R as shown by Theorem 2.96.

The following theorem, presented independently by Van Lint, Jr. [377] and Carnielli [94] is a slight generalization of a result by Blokhuis and Lam [52], which in turn was inspired by an idea of Kamps and Van Lint [295].

Theorem 9.24. *If S R-covers $\mathbb{F}_q^r$ using an $r \times n$ full rank matrix* $\mathbf{H}$, *then* $K_q(n,R) \le |S|q^k$, *which is attained by the code* $\{\mathbf{w} \in \mathbb{F}_q^n : \mathbf{H}\mathbf{w}^T \in S\}$.

Proof. Consider an arbitrary word $\mathbf{x} \in \mathbb{F}_q^n$. Then, since S R-covers $\mathbb{F}_q^r$ using $\mathbf{H}$, we get that $\mathbf{H}\mathbf{x}^T \in \mathbb{F}_q^r$ can be written as $\mathbf{H}\mathbf{y}^T + \mathbf{s}^T$ where $\mathrm{wt}(\mathbf{y}) \le R$ and $\mathbf{s} \in S$. But $\mathbf{H}\mathbf{x}^T = \mathbf{H}\mathbf{y}^T + \mathbf{s}^T$ implies that $\mathbf{H}(\mathbf{x}^T - \mathbf{y}^T) \in S$, so the word $\mathbf{x}$ is within distance R from a word in $\{\mathbf{w} \in \mathbb{F}_q^n : \mathbf{H}\mathbf{w}^T \in S\}$.

Since $\mathbf{H}$ has full rank, the number of solutions to $\mathbf{H}\mathbf{w}^T \in S$, and the size of the code obtained is $|S|q^{n-r} = |S|q^k$. □

To classify covering codes of this type, we get an instance of SET COVERS: for each vector $\mathbf{s} \in \mathbb{F}_q^r$, there is a set consisting of the elements that can be represented as a sum of $\mathbf{s}^T$ and an $\mathbb{F}_q$-linear combination of at most R columns of $\mathbf{H}$; we now want to cover all words in $\mathbb{F}_q^r$ using such sets.

For error-correcting codes, we have a similar theorem [462], but for these it is still more convenient to use the clique approach. The criterion for the distance between words in the same orbit means that the linear code with parity check matrix $\mathbf{H}$ must have minimum distance at least d. As all orbits have the same length, we get an instance of the maximum clique problem instead of the maximum-weight clique problem.

The problem of finding all possible groups of the aforementioned type is in itself a code classification problem, namely that of classifying linear codes with prescribed parameters (Sect. 7.3).

9.3.4 Linear Codes

In the previous discussion we saw that a classification of linear $[n,k]_q$ codes could be seen as a classification of codes with a particular prescribed group of order q^k acting regularly on the codewords.

For linear codes, we may prescribe even further automorphisms. This approach quickly leads to intricate algebraic problems, which are outside the scope of this book. We will therefore just give a few references, where the interested reader can find further information on this topic. Some general theory on automorphism groups of linear codes can be found in [92].

Cyclic codes [388, Chaps. 7 and 8] make up the most studied class of linear codes with prescribed automorphisms; these are codes of length n admitting the coordinate permutation $(1\ 2\ \cdots\ n)$ as an automorphism. Classification of cyclic linear codes in general is studied in [166]; indeed, the automorphisms of such codes are so restrictive that covering radius and minimum distance checks can be carried out after the classification. In fact, this is reasonable from the point of view that calculating these parameters from a generator or parity check matrix representation of the code is believed to be computationally intractable in general; see Sect. 11.2.

Cyclic self-dual linear codes are considered in [541]; see also [166]. More general automorphism groups are prescribed by Braun, Kohnert, and Wassermann [65] (with the main focus on existence).

9.4 Other Objects

We conclude this chapter by giving links to the main results on classifying Latin squares, 1-factorizations of complete graphs, Hadamard matrices, and resolutions of BIBDs with prescribed automorphism groups.

Latin squares and 1-factorizations have been studied in this context in [415] and [524] ([523] also contains related results), respectively. Within the framework of triple systems, discussed in Sect. 8.1, Latin squares and 1-factorizations of complete graphs with prescribed automorphism group may be classified following the approach in Sect. 9.2.3. It would be interesting to compare the performance of these algorithms.

Research Problem 9.25. Compare the classification algorithms in [415] and [524], respectively, with an approach analogous to that in Sect. 9.2.3 using the triple system representation given in Sect. 8.1.

In [415] the limits for what can currently be done with respect to classifying Latin squares with a nontrivial isomorphism group have been reached by using state-of-the-art algorithms and an extensive amount of computing time. For 1-factorizations of complete graphs, however, it should be possible to settle the next open instance.

Research Problem 9.26. Classify the 1-factorizations of K_{14} with nontrivial automorphism groups.

As pointed out in [161, Theorem 45.11], a 1-factorization of the complete graph K_{2n} with a 1-rotational automorphism (acting on the vertices) is equivalent to a starter $\mathbb{Z}_{2n-1}$, so Research Problem 8.1 is relevant for objects of this type as well.

Classification of Hadamard matrices with prescribed automorphism groups is inevitably connected to classification of the associated Hadamard designs. Studies where either Hadamard designs or matrices are considered in this setting include [260, 315, 362, 575, 576, 577, 583]. However, in studying those references, one should keep in mind that the work on classifying Hadamard matrices and the associated designs has been completed up to order 28; see Sect. 8.2.

For resolutions of BIBDs, Kirkman triple systems form one family that has attracted classification interest in the setting of prescribing automorphism groups. In [110] a complete classification of KTS(21) with a nontrivial automorphism is obtained (however, we remark that a preliminary study based on a classification of the STS(21) with a nontrivial automorphism group has indicated a possible error in the classification in [110]). In [120] the point-transitive KTS(27) are classified. Other results include a classification of the 1-rotational KTS(33) in [79] and for $v \in \{27, 33, 39\}$ a classification of the KTS(v) admitting an automorphism with 3 fixed points and 3 cycles of length $(v-3)/3$ in [119].

10

Validity of Computational Results

In the Introduction, problems related to combinatorial objects were divided into those of *existence*, *counting*, and *classification*, all of which can be attacked with computational methods. The existence problem – whenever answered in the positive by explicit construction – differs from the others as the constructed object almost without exception can be presented in the published work and checked by hand calculation or by easy computations. Results related to the other types of problems, however, are of intrinsically different nature, and lead us to the somewhat controversial issue of validity of computational results.

What makes this issue controversial is that we are dealing with exact, mathematical questions, to which answers are presented that cannot be checked by a human being alone as in classical mathematical proofs. We are therefore facing the philosophical problem of the epistemic status of such results, a multifaceted problem involving the disciplines of computing, mathematics, and philosophy.

Extensive mathematical (hand) calculations – such as finding the decimals of π – were carried out already before the era of computers, so this paradigm as well as the related polemic are much older than many conceive. It was, however, the celebrated result by Appel and Haken [9] in 1976 that every planar graph is four-colorable that brought the debate into the open. Since then an ever increasing number of important results – also regarding nondiscrete mathematical problems – have been obtained by computational methods; the main contemporary question is therefore *how* computational results should be obtained, presented, and evaluated rather than whether they should be accepted at all.

For a discussion from a philosophical point of view, the reader is referred to [586, 587]. We will focus on what Rall [498] calls *mathematical computations*: computations in which assertions about the validity and reliability of results are addressed in addition to the results themselves. Our discussion is along the lines of [343].

In Sect. 10.1 we discuss the main types of errors that can occur in a computer-aided classification, and give an overview of general strategies for reducing the error probability (which, unfortunately, always will be nonzero). Consistency checking based on the basic principle of double counting is considered in detail in Sects. 10.2 and 10.3, and some related general observations are presented in Sect. 10.4.

10.1 Errors and Remedies

There are three main classes of errors related to computer-aided results:

methodological errors Errors in the algorithm or in the approach in general.

software errors Errors in implementing and running the program. This includes errors in input data and programming errors (for example, in the written program, in the library programs used, or in the operating system).

hardware errors Errors occurring because the computer is not acting as it is supposed to with the programs and the data. One example is the random changing of bits in a computer memory.

The possibilities of detecting these errors vary. The results should be documented so that methodological errors can be detected by studying the published work. One further possibility of making it possible for others to partially check the results (without having to write a program of one's own and redoing the complete study) is to make the written program public. Software errors in the implemented program are under the user's control, but errors in the operating system or hardware are not. The latter are particularly harmful, as they often occur at random and are then not reproducible. Obviously, to add up, even if all conceivable countermeasures are taken, errors cannot be completely eliminated. Lam [343] suggests four general approaches to increase confidence in computational results. From that list – which we reproduce slightly edited – we omit the case of proving results by hand calculation; producing, say, a list of over a million designs is not possible by hand calculation.

N-version programming. The method of executing two or more programs that have been developed independently based on a given problem specification is known as N-version programming [16]. This method may be used in the original study, or someone may publish a verification of earlier results. Both hardware and software errors can be detected by executing the programs in different environments (as for the operating system, libraries, computer architecture, etc.).

Solving a problem by using implementations of different algorithms is closely related to N-version programming, but does not fall strictly under this definition. In the current taxonomy, the following group of methods is perhaps more appropriate for that approach.

Consistency checking. Consistency checking means that some relationships among the results – preferably involving the whole computation – are checked. For classification algorithms, most methods for consistency checking are based on *double counting*, also known as *two-way counting*: the objects are counted in two different ways. The basic principles of such methods are considered in Sect. 10.2 (using the orbit-stabilizer theorem) and in Sect. 10.3 (using a method due to Lam and Thiel [347] and others) along with illustrative examples.

Using well-tested programs. Not only should one use well-tested programs written by others, but the programs of one's own should also be well tested. Tuning the programs for particular cases should be avoided whenever possible, to minimize the risk of introducing errors. Optimally, a general program should be used to as large an extent as possible. Many errors in published studies have occurred because a minor subcase was solved by hand calculation (with details omitted) or with another (special) program.

One should be particularly careful with nonexistence proofs; it goes without saying that if the correct result of a classification instance is that no such objects exist, the program would arrive at that result in spite of a wide variety of errors. To be able to compare the results of subsequent studies with such a result, it is advisable that details regarding partial objects encountered in the search be published.

We conclude this discussion by bringing forward the fact that there are some similarities between computational results and empirical data in other sciences.

10.2 Double Counting Using the Orbit-Stabilizer Theorem

The results in this section are based on the orbit-stabilizer theorem (Theorem 3.20). Let Ω be the set of all labeled objects and let G be the group whose action induces isomorphism on Ω. Denote by N_i the number of nonisomorphic objects whose automorphism group has order i. Then, by the orbit-stabilizer theorem, the total number $N = |\Omega|$ of labeled objects is

$$N = |G| \sum_i \frac{N_i}{i}. \tag{10.1}$$

Equation (10.1) is called a *mass formula* and can clearly be used as one part of a double counting argument whenever the value of N also can be obtained in an alternative way.

For some types of objects the total number of labeled objects can be obtained analytically. Examples of such objects considered in this book are self-dual linear codes and Latin squares. For the number of self-dual codes there are rather simple formulae [497]; unfortunately, the known formulae for the number of Latin squares [203, 440, 531] appear to be of little practical importance.

The fastest known method for counting the total number of Latin squares is not purely analytic but involve some classification [420, 421]; this method is still orders of magnitude faster than a complete classification of Latin squares. Due to the close relationship between Latin squares and 1-factorizations of complete graphs, apparent from the discussion in Chap. 8, a similar counting approach is applicable to the latter objects as well [162].

In the majority of cases that we have encountered earlier in this book there is no obvious way of counting the total number of labeled objects that is essentially different from and faster than carrying out a complete classification up to isomorphism. Fortunately, a computer-aided classification produces in many cases as a by-product data that can be used for double counting. We illustrate this with two examples.

Example 10.1. In the classification of Steiner triple systems in Sect. 6.1.3, the designs are obtained from a set of nonisomorphic seeds, $\mathcal{S}_i$. Now let E_i denote the total number of completions of $\mathcal{S}_i$ to a Steiner triple system, a number that is easily recorded during the search.

By the orbit-stabilizer theorem, the total number of labeled Steiner triple systems of order v with one seed individualized is

$$N' = \sum_i \frac{v! \cdot E_i}{|\mathrm{Aut}(\mathcal{S}_i)|}. \tag{10.2}$$

In each labeled Steiner triple system there are b ways to individualize a seed, so

$$N' = bN,$$

where N and N' are from (10.1) and (10.2), respectively.

The classification of Steiner triple systems of order 19 in [305] was checked for consistency in this manner.

Example 10.2. Consider a classification of (binary, but a generalization is straightforward) error-correcting codes via subcodes, discussed in Sect. 7.1.1. Let E_i denote the total number of ways to lengthen an $(n-1, M', d)_2$ code C_i to an $(n, M, d)_2$ code. By the orbit-stabilizer theorem, the total number of $(n, M, d)_2$ codes of type $0C \cup 1C'$ where C is equivalent to the code C_i is then

$$\frac{2^{n-1}(n-1)! \cdot E_i}{|\mathrm{Aut}(C_i)|}. \tag{10.3}$$

To obtain the total number of all codes, one should sum this expression over all i *and* take the expression *twice* whenever $|C_i| > M/2$. Namely, assuming (7.1), we do not otherwise count the codes with more 1s than 0s in the first coordinate.

As a concrete example, we lengthen the two inequivalent $(4,2,3)_2$ codes, $C_1 = \{0000, 0111\}$ and $C_2 = \{0000, 1111\}$ codes to get the unique $(5,3,3)_2$ code, see Example 7.2 and Fig. 7.1.

By Fig. 7.2 both instances have 6 cliques of size 1, so $E_1 = E_2 = 6$. Since $|\mathrm{Aut}(C_1)| = 12$ and $|\mathrm{Aut}(C_2)| = 48$, we get by (10.3) and the related observations a total count of

$$N' = 2 \times 2^4 \times 4! \times \left(\frac{6}{12} + \frac{6}{48}\right) = 480.$$

The order of the automorphism group of the unique $(5,3,3)_2$ code is 8, so (10.1) gives

$$N = \frac{2^5 \times 5!}{8} = 480,$$

and the two counts agree.

As these examples show, this method is widely applicable, and it deserves much more attention than it has received so far in the literature. One more study where the method has been used could be mentioned: classification of designs with prescribed automorphism groups (as described in Sect. 9.2.3) [299].

If all the other values in (10.1) are known except N_1, then N_1 and thereby $\sum_i N_i$ can be determined. This means that is we know the value of N, then a classification of the objects with a nontrivial automorphism group makes it possible to count the total number of nonisomorphic objects. This can be seen as one motivation for classifying objects with nontrivial automorphism groups.

10.3 Double Counting by Identifying Subobjects

The methods in Sect. 10.2 have negligible impact on the overall performance of the classification algorithms. At the cost of efficiency, however, more advanced consistency checking can be built into a classification algorithm. In this section a framework for consistency checking based on ideas presented by Lam and Thiel [347] is considered.

The general idea is that it is often possible to determine in two different ways how many times a certain object should occur in a search tree. Counting can take place both for the objects that we want to classify and for partial objects. For the sake of clarity we do not discuss the ideas in the general search tree model of Chap. 4, but make use of the following somewhat simplified setting, where the notion of a subobject is clear and intuitive.

In many cases it is possible to view the objects considered by a search as subsets of a finite set Π, where isomorphism is induced by the action of a group G on Π. More precisely, $\mathrm{Iso}(X,Y) = \{g \in G : gX = Y\}$ for all $X, Y \subseteq \Pi$. For example, codes are subsets of Z_q^n with isomorphism induced by the action of $\mathrm{Aut}(Z_q^n, d_H)$, and simple designs are subsets of $\binom{V}{k}$ with isomorphism induced by the action of $\mathrm{Sym}(V)$.

A search procedure can now be viewed as constructing the objects that we want to classify by extending a given subset $X \subseteq \Pi$ with new elements – say, by inserting new words into a code or blocks into a (partial) design. Accordingly, for $X, Y \subseteq \Pi$ we say that X is a *subobject* of Y if $X \subseteq Y$.

Let P and Q be isomorphism invariant properties of objects. For example, think of P as the property that an object is isomorphic to a node at level ℓ of a search tree, and similarly for Q and level $\ell + 1$.

To arrive at a double counting approach, consider any object $X \subseteq \Pi$ with property P. Let $N(X)$ be the number of objects $Y \subseteq \Pi$ with property Q that have X as a subobject. By the orbit-stabilizer theorem,

$$\frac{|G| \cdot N(X)}{|\mathrm{Aut}(X)|} \tag{10.4}$$

counts the number of pairs (Z, W) of objects such that $Z \cong X$, $Z \subseteq W$, and W has property Q. To arrive at a second count, let $S(Y, X)$ be the number of subobjects of Y isomorphic to X. The second count is now

$$\sum_{Y} \frac{|G| \cdot S(Y, X)}{|\mathrm{Aut}(Y)|}, \tag{10.5}$$

where the sum considers exactly one object Y from every isomorphism class of objects with property Q.

Note that by summing (10.4) and (10.5) over all isomorphism classes of objects with property P, we arrive at the results of Sect. 10.2.

Equations (10.4) and (10.5) can be combined into [151]

$$N(X) = |\mathrm{Aut}(X)| \sum_{Y} \frac{S(Y, X)}{|\mathrm{Aut}(Y)|}. \tag{10.6}$$

Application of (10.6) is illustrated by the following example.

Example 10.3. Consider Example 7.7, where the three $(5, 2, 3)_2$ codes

```
00000  00000  00000
00111  01111  11111
```

with automorphism groups of order 24, 48, and 240, respectively, are augmented to get the unique $(5, 3, 3)_2$ code

```
00000
00111
11011
```

whose automorphism group has order 8.

The number of ways to augment the three two-word codes to a $(5,3,3)_2$ code is 6, 6, and 0, respectively.

If, respectively, the first, second, or the third codeword of the $(5,3,3)_2$ code is removed, we get the first, second, and the first two-word code as a subcode. Now we have all information needed for applying (10.6), once for each of the subcodes:

$$
\begin{aligned}
6 &= 24 \times \frac{2}{8}, \\
6 &= 48 \times \frac{1}{8}, \\
0 &= 240 \times \frac{0}{8}.
\end{aligned}
$$

Thus, equality holds in (10.6) in all three cases.

The two counts (10.4) and (10.5) can now be used to perform consistency checks at a number of levels of sensitivity. The basic idea is that we look at objects with property P or Q encountered in a search. Assuming that isomorph rejection is employed, each object encountered is either rejected or accepted as the unique representative of its isomorphism class.

For each accepted object X with property P, we keep track of all the objects Y encountered that have property Q and X as a subobject. From these objects we can typically determine $N(X)$ – in many cases $N(X)$ is simply the number of distinct Y encountered. Once $N(X)$ is available, the count (10.4) for the isomorphism class of X can be evaluated.

On the other hand, the second count (10.5) is accumulated across different isomorphism classes. Whenever an object Y with property Q is accepted, we determine all the subobjects $X \subseteq Y$ with property P. For each isomorphism class of subobjects, we increment the count (10.5) for the isomorphism class by $|G|/|\mathrm{Aut}(Y)| \cdot S(Y,X)$.

When the search terminates, the two counts (10.4) and (10.5) should agree for every isomorphism class of objects with property P.

A straightforward approach to implement this counting technique is to associate two counters to every certificate of an isomorphism class of objects with property P. Whenever a counter for an isomorphism class needs to be accumulated, the correct counter is found by computing the certificate of an object in the isomorphism class. This approach clearly requires that we have enough memory available to store the certificates of all isomorphism classes. Furthermore, computing a certificate can be expensive.

To obtain a consistency check with less intensive resource requirements, we can trade sensitivity to errors for performance. A simple strategy is to replace a certificate with an isomorphism invariant that is faster to evaluate and that assumes less values in the relevant isomorphism classes, resulting in a smaller memory requirement. An extreme example is an invariant that has the same value for all isomorphism classes – in this case the consistency check reduces to checking whether (10.4) and (10.5) are equal when summed over all isomorphism classes of objects with property P.

A slightly different approach to trading sensitivity for a less heavy memory requirement can be obtained as follows. A set of counters, each associated with a certificate (or invariant value), can be viewed as a multiset $\mathcal{F}$ of certificates, where the multiplicity of each certificate is equal to the value of the counter. Now, instead of representing the multiset $\mathcal{F}$ explicitly, we maintain only a hash value $H(\mathcal{F})$. Accordingly, instead of comparing two sets of counters individually, a consistency check is carried out by testing hash values for equality. To accumulate a hash value, the hash function H should be equipped with a commutative operation $+_H$ such that

$$H(\mathcal{F}_1 \cup \mathcal{F}_2) = H(\mathcal{F}_1) +_H H(\mathcal{F}_2),$$

where $\cup$ is multiset union. In particular, we may take $\mathcal{F}_2$ to be (the certificate of) a single object. The term *hash accumulator* is occasionally used to stress the incremental nature of $H(\mathcal{F})$. Functions of this type have been studied, in particular, in checking data integrity [108] (but some of the cryptographic properties of the proposed schemes are redundant in the current application); for their use in a consistency checking context, see [299]. The basic theory of hash functions is treated in [322, Sect. 6.4].

It is also possible to increase the sensitivity of the consistency check by associating counters to other isomorphism classes of objects than those with property P. For example, we can also associate counters to the isomorphism classes with property Q. Similarly, we can associate a pair of counters to each isomorphism class of pairs (Z, W) of the form counted by (10.4) and (10.5). These are by no means the only possibilities. In general, the double counting check should be tailored to suit the generation and isomorph rejection techniques employed. Preferably the counters should depend on all the objects encountered. For example, associating counters to the pairs (Z, W) can be seen as counting the augmentations – cf. Sect. 4.2.3 – that occur in a search, which yields a consistency check suited for generation by canonical augmentation; cf. [299].

Only sporadic results have been reported on the use of consistency checking methods of these types. Two notable exceptions are [151] and [346]. One would like to see consistency checks based on double counting to be more widely adopted, even if the additional design and implementation effort may be substantial. Not only do such checks increase the degree of confidence in the computed result, but their adoption also forces one to develop a detailed understanding as to exactly what is encountered in the search, which may in itself reveal subtle (methodological) errors that would otherwise go unnoticed.

10.4 Some Final Observations

The basic method of isomorph rejection via recorded objects, considered in Sect. 4.2.1, is self-correcting in the sense that if an object is missed on some level of the search tree, objects on deeper levels that also contain other subobjects are still obtainable. This property is lost in the more advanced methods discussed in Chap. 4, but is regained (albeit error-detecting rather than error-correcting) in the double counting methods.

If the set of objects that can be constructed from a certain subobject cannot be obtained in any other manner, then a loss of the subobject cannot be detected by a double counting argument. Moreover, this reveals that there is a trade-off with respect to the number of times a certain object occurs in a search tree: if it occurs many times, this is good for consistency checks (via double counting) but bad for efficiency reasons, and vice versa.

The problem of verifying nonexistence results, where double counting is not applicable in the final step, was briefly mentioned earlier. In such situations it is particularly important that a putative object would have been obtained in many different ways. For example, in their proof of the nonexistence of projective planes of order 10 – to be considered in Chap. 12 – Lam, Thiel, and Swiercz [351] argue that such a plane, if it exists, could be constructed as the extension of 24,675 subobjects of a certain kind.

11

Computational Complexity

Judging by practical experience, solving classification problems for combinatorial objects such as codes and designs is a computationally demanding task. A question of a different nature is what can be said from the perspective of computational complexity theory, where it is common to regard a computational problem tractable only if it admits a polynomially resource-bounded algorithm, whereas problems for which no such algorithm exists are regarded as intractable.

This chapter examines the computational complexity of problems related to classification of codes and designs as studied in this book. Surveys focusing on more general computational complexity aspects of codes and designs include [27, 593] and [125], respectively.

Section 11.1 contains a review of some of the central concepts and tools in complexity theory. The complexity of decision problems associated with exhaustive generation of codes and designs is examined in Sect. 11.2. The complexity of isomorphism problems is studied in Sect. 11.3. In particular, the aim is to illustrate the known complexity connections of isomorphism problems for codes and designs with the extensively studied graph isomorphism problem. The chapter is concluded in Sect. 11.4 by a discussion of the complexity of classification problems. Here apparently not a great deal is known, certainly not in the case of codes and designs – we will argue that short of a major breakthrough in complexity theory, it is likely that the only decisive progress on the existence of efficient classification algorithms for the restricted types of classification problems studied in this book has to come from the discovery of such algorithms. A brief discussion on selected efficient listing and classification algorithms is provided.

11.1 Preliminaries

Our objective in this section is to give a review of the central complexity-theoretic concepts and tools used in what follows. Standard texts on complexity theory where a more extensive treatment can be found include [201, 471]. More advanced or broader treatises include [262, 361].

Computational complexity theory studies computational problems with the ultimate aim of characterizing the computational resources – for example, time and storage space – required to solve a problem. Intuitively, a *problem* consists of a collection of finite *problem instances*. Each problem instance has a *solution* that one wishes to compute.

Problem 11.1. (CLIQUE) Given a graph G and a nonnegative integer k, decide whether G contains a clique of size k.

An instance of CLIQUE is a pair (G, k). The solution to (G, k) is "yes" if G contains a k-clique and "no" otherwise.

To accommodate algorithms and computation into this intuitive setting, we must first represent both the problem instances and the solutions in a manner that is suitable for computation. In practice it is customary to use (binary) strings, where the intuitive analog is a sequence of bits stored in computer memory.

Let Σ be a finite alphabet of symbols, say $\Sigma = \{0, 1\}$, and denote by Σ^* the set of all finite strings over Σ. Some basic operations on strings are as follows. For $x \in \Sigma^*$ we write $|x|$ for the length of x. For $x, y \in \Sigma^*$ we write xy for the concatenation of x and y. For $x \in \Sigma^*$ and a nonnegative integer k we write x^k for the string obtained by concatenating k copies of x. We assume that Σ^* is lexicographically ordered first by length and then by lexicographic order on strings of equal length.

A formally precise way to define a computational problem is now as follows.

Definition 11.2. *A* problem π *associates to every* instance $x \in \Sigma^*$ *a nonempty set* $\pi(x) \subseteq \Sigma^*$ *of* solutions.

A large part of complexity theory deals with problems admitting a simple "yes" or "no" answer, such as CLIQUE.

Definition 11.3. *A* decision problem *is a problem in which every instance has a unique solution that is either* 1 *("yes") or* 0 *("no").*

To study CLIQUE in the precise setting of Definition 11.3, we must agree on a string representation for the instances (G, k) and the solutions. For the solutions we obviously set 1 ("yes") and 0 ("no"), but for the instances there are more degrees of freedom. One possibility is as follows. A graph G has the natural string representation consisting of the $|V(G)|^2$ entries of an adjacency matrix, listed row by row. Similarly, a nonnegative integer k is naturally represented in its binary form as a string of binary digits, which we denote by $\langle k \rangle$ For example, $\langle 0 \rangle = 0$, $\langle 1 \rangle = 1$, $\langle 2 \rangle = 10$, $\langle 3 \rangle = 11$, and so forth. For two strings $x, y \in \Sigma^*$, the ordered pair (x, y) is represented by the string $1^{|\langle |x| \rangle|}0\langle |x| \rangle xy$, from which x and y can be uniquely determined.

Example 11.4. Consider the graph G in Fig. 2.1 and $k = 3$. The adjacency matrix of G is given in Example 2.6, from which we obtain the string

$$010000101001010101001010000100011000$$

of length $6^2 = 36$ and $\langle 36 \rangle = 10100$. Thus, with the previous assumptions, the ordered pair (G, k) is represented as the string

$$11111010100010000101001010101001010000100011000 11.$$

Looking at Fig. 2.1, the associated solution is 1.

In a strictly precise setting it must be observed that not all strings $x \in \Sigma^*$ encode a proper pair (G, k), but such strings are also required to have an associated solution by Definition 11.3. We assume (arbitrarily) that such strings have solution 0, and make this assumption also in the context of other decision problems and encodings.

Thus, with a bit of effort we have arrived at a precise string representation for CLIQUE that conforms to Definition 11.3. However, in practice our interest will be on very coarse properties such as polynomial time solvability, whereby certain general assumptions about the string representation suffice, and problem definitions can be given in a more relaxed form such as Problem 11.1. We return to the representation issue in what follows.

Informally an *algorithm* is a step-by-step procedure for solving a problem. In a formally precise setting we require a model of computation with respect to which algorithms are specified. As our model of computation we adopt the deterministic random access machine (RAM) [471, Sect. 2.6]; an extensive treatment of RAMs and other models of computation can be found in [174].

Definition 11.5. *An* algorithm *is a RAM program with the associated input/output semantics.*

A detailed discussion of a RAM and RAM programs is beyond the present scope. A sufficient intuition is that a RAM is a rough model of a contemporary computer CPU, and a RAM program constitutes a program written in the machine language of the CPU. The main deviation from practice in the model is that a RAM has an infinite amount of memory (registers) at its disposal.

For a string $x \in \Sigma^*$ and an algorithm A, we write $A(x)$ for the output of algorithm A on input x; in so writing we also assume that A halts on input x.

Definition 11.6. *An algorithm A* solves *a problem π if $A(x) \in \pi(x)$ for all $x \in \Sigma^*$.*

In solving an instance $x \in \Sigma^*$ of a problem π, the algorithm A consumes computational resources that are defined with respect to the model of computation. Here we are interested in the required *time* – measured by the number of instructions executed by the RAM – and in the required *space* – measured by the maximum length of all register contents when concatenated into one binary string, where the maximum is taken over all configurations of the RAM during the computation. Again a practical intuition suffices – the running time of an algorithm is the number of CPU instructions executed and the space usage is the maximum number of memory bits required by the algorithm.

The resource usage of an algorithm is studied as a function of the instance size $|x|$.

Definition 11.7. *A* polynomial-time algorithm *is an algorithm for which there exists a polynomial p such that for every input $x \in \Sigma^*$ the algorithm halts within time $p(|x|)$.*

Definition 11.8. *A* polynomial-space algorithm *is an algorithm for which there exists a polynomial p such that for every input $x \in \Sigma^*$ the space required by the algorithm is at most $p(|x|)$.*

Before proceeding further we now briefly return to the issue of string representations. As long as we are only interested in very coarse properties, such as whether a problem is solvable by a polynomial-time algorithm or not, the precise details of the string representation are irrelevant. In the context of polynomial-time computability we only have to specify the string representation so that it is determined up to polynomial variance in length. For example, in this case it becomes irrelevant whether a graph is represented by the adjacency matrix or as a list of edges – a polynomial-time algorithm exists in one representation if and only if it exists in the other. What we do have to worry about, however, are superpolynomial differences in length. A concrete example in the context of this book is that a linear code can be represented succinctly via a generator (or parity check) matrix, which admits a natural string representation that can be exponentially shorter than a string explicitly listing all the codewords in the code.

These observations withstanding, we assume that an explicit string representation (as opposed to a succinct representation) is always used unless indicated otherwise. In more precise terms, graphs are represented by listing an adjacency matrix, incidence structures by listing an incidence matrix, and unrestricted codes by explicitly listing all the codewords – a q-ary word of length n is represented using $n\lceil \log_2 q \rceil$ bits. A nonnegative integer m is represented in binary as $\langle m \rangle$; in some cases the *unary* representation 1^m is used and this is explicitly indicated. Ordered pairs of strings are formed using the pairing construction $(x, y) \mapsto 1^{|\langle |x| \rangle|}0\langle |x| \rangle xy$. Sets and tuples of strings are formed by applying the pairing construction repeatedly.

Having properly set up the basics, we now proceed to define the central concepts and tools employed.

In general it is very difficult to characterize the computational resources that are required to solve a problem. To combat this obstacle, complexity theory uses a somewhat indirect set of tools in dividing problems into different types with respect to computational difficulty. Central concepts in this respect are complexity classes, reductions, and complete problems in a complexity class.

A *complexity class* is a collection of problems possessing certain common properties.

Definition 11.9. *The complexity class* **P** *consists of all decision problems solvable by polynomial-time algorithms.*

The class **P** is often viewed as capturing, or at least containing, the decision problems that can be efficiently solved. (If one allows randomized algorithms with a small failure probability, then a potentially larger class than **P** can be viewed to represent the efficiently solvable problems, but this is beyond the current scope; see [471].) Decision problems that are provably not in **P** are known, but for a vast number of practically relevant decision problems the question whether the problems are in **P** remains open. In the absence of decisive results, indirect suggestive evidence has been accumulated. A central tool in this respect is a *reduction* relating two problems. Intuitively, a reduction makes precise the notion that one problem is "at least as hard to solve" as another problem.

To define reducibility between problems we require the notion of an algorithm A with an *oracle* for a problem π. The semantics of an oracle algorithm are otherwise identical to an algorithm without an oracle, except that an oracle algorithm is allowed to submit a *query* $z \in \Sigma^*$ to the oracle for π. Once a query is submitted, the oracle (magically) determines in one computation step a string $w \in \pi(z)$ which it places into a sequence of RAM registers, each containing one alphabet symbol. The intuition is that the algorithm A has been reinforced with a subroutine that solves instances of π in one time step.

Definition 11.10. *A problem π_1 is* polynomial-time Turing reducible *to a problem π_2 if there exists a polynomial-time algorithm A that solves π_1 using an oracle for π_2.*

The intuition is that (up to polynomial time overhead) the problem π_2 is "at least as hard" as problem π_1 because π_1 can be solved using a π_2 oracle.

Between decision problems a restricted notion of reducibility is appropriate.

Definition 11.11. *Let π_1 and π_2 be decision problems. A* polynomial-time transformation *from π_1 to π_2 is a polynomial-time algorithm A such that $\pi_1(x) = \pi_2(A(x))$ holds for all $x \in \Sigma^*$.*

In other words, a polynomial-time transformation takes an instance $x \in \Sigma^*$ of π_1 as input, and outputs an instance $A(x) \in \Sigma^*$ of π_2 so that x has the same solution (either 0 or 1) as $A(x)$. Note that A essentially defines a polynomial-time oracle algorithm that solves π_1 by asking one query from a π_2 oracle and halting with the answer as output. Thus, the existence of a polynomial-time transformation implies polynomial-time Turing reducibility.

In what follows we assume that whenever only decision problems are considered, reducibility refers to the existence of a polynomial-time transformation; otherwise reducibility refers to polynomial-time Turing reducibility.

Reducibility is transitive: if π_1 reduces to π_2 and π_2 reduces to π_3, then π_1 reduces to π_3.

Definition 11.12. *Two problems π_1, π_2 are* polynomial-time equivalent *if π_1 reduces to π_2 and vice versa.*

Definition 11.13. *Let $\mathscr{C}$ be a complexity class. A problem π is* $\mathscr{C}$-hard *if every problem $\pi_1 \in \mathscr{C}$ reduces to π. A problem π is* $\mathscr{C}$-complete *if $\pi \in \mathscr{C}$ and π is $\mathscr{C}$-hard.*

Intuitively, if a complexity class $\mathscr{C}$ has a complete problem π, then π is "one of the hardest problems in $\mathscr{C}$" because if we can solve π, then we can solve all problems in $\mathscr{C}$ up to the overhead caused by reduction.

To define more complexity classes we require some preliminaries. A relation $R \subseteq \Sigma^* \times \Sigma^*$ is *polynomially balanced* if there exists a polynomial p such that for all $(x, y) \in R$ it holds that $|y| \leq p(|x|)$. For a complexity class $\mathscr{C}$ consisting of decision problems, the relation R is *$\mathscr{C}$-decidable* if the decision problem asking whether $(x, y) \in R$ is in $\mathscr{C}$.

Definition 11.14. *The complexity class* **NP** *consists of all decision problems π for which there exists a polynomially balanced,* **P***-decidable relation $R \subseteq \Sigma^* \times \Sigma^*$ such that*

$$\{x \in \Sigma^* : \pi(x) = \{1\}\} = \{x \in \Sigma^* : \exists y \in \Sigma^* \; (x, y) \in R\}.$$

Intuitively, **NP** consists of all decision problems with the property that every "yes"-instance x has a polynomially balanced *witness* y such that the pair (x, y) can be *verified* in polynomial time.

Example 11.15. Clique is in **NP** because a witness for a "yes"-instance (G, k) is provided by a k-clique $C \subseteq V(G)$, where an associated verification algorithm checks in polynomial time for a given (G, k) and $C \subseteq V(G)$ that C defines a k-clique in G.

Clearly, $\mathbf{P} \subseteq \mathbf{NP}$. Whether $\mathbf{P} \neq \mathbf{NP}$ is a long-standing open question in complexity theory.

A classical result in complexity theory is that **NP**-complete problems exist in abundance – an extensive list can be found in [201]. For example, Clique is **NP**-complete, along with numerous other problems of considerable practical importance, including the decision versions of Set Covers, Exact Covers, and Diophantine from Chap. 5, where in the decision version the task is to decide whether a solution exists or not. **NP**-complete problems are fundamental to the $\mathbf{P} \neq \mathbf{NP}$ question because an immediate consequence of the definitions is the following result.

Theorem 11.16. **P** = **NP** *if and only if any one* **NP***-complete problem admits a polynomial-time algorithm.*

One basic **NP**-complete problem is 3-SATISFIABILITY (3SAT), defined as follows. Let $\{v_1, v_2, \ldots, v_k\}$ be a set of variables. Each variable can assume one of two truth values: 0 (false) or 1 (true). A *truth assignment* is a map $T : \{v_1, v_2, \ldots, v_k\} \to \{0, 1\}$ associating each variable with a truth value. Associated with every variable v_i there are two *literals*, the positive literal, denoted by v_i, and the negative literal, denoted by $\bar{v}_i$. A (K-)*clause* is a set of (K) distinct literals. A truth assignment T *satisfies* a positive (respectively, negative) literal if $T(v_i) = 1$ (respectively, $T(v_i) = 0$).

Problem 11.17. (3-SATISFIABILITY) Given a set of 3-clauses, decide whether there exists a truth assignment that satisfies at least one literal in every clause.

Analogously to **NP**, the complexity class **coNP** consists of those decision problems whose "no"-instances have polynomially balanced, polynomial-time verifiable witnesses. An intriguing open question in complexity theory is whether **NP** $\neq$ **coNP**. For example, given the unary parameters v, k, λ, b, r of a BIBD, it is easy to verify in polynomial time a witness consisting of a BIBD with the required parameters. However, when a design with the given parameters does not exist, it is not at all clear whether it is possible to give a polynomially balanced, polynomial-time verifiable witness of this fact.

Analogously to **P**, **NP**, and **coNP** it is possible to define the following natural hierarchy of complexity classes that is believed to extend beyond **NP** and **coNP**. The base level of the hierarchy consists of the complexity class $\mathbf{\Delta}_0 = \mathbf{\Sigma}_0 = \mathbf{\Pi}_0 = \mathbf{P}$. For $i = 1, 2, \ldots$, the levels of the hierarchy are inductively defined as follows.

Definition 11.18. *The class* $\mathbf{\Delta}_i$ *consists of all decision problems* π *for which there exists a problem* $\pi_1 \in \mathbf{\Sigma}_{i-1}$ *and a polynomial-time algorithm* A *such that* A *solves* π *using* π_1 *as an oracle.*

Definition 11.19. *The class* $\mathbf{\Sigma}_i$ *consists of all decision problems* π *for which there exists a polynomially-balanced,* $\mathbf{\Delta}_i$*-decidable relation* $R \subseteq \Sigma^* \times \Sigma^*$ *such that*

$$\{x \in \Sigma^* : \pi(x) = \{1\}\} = \{x \in \Sigma^* : \exists y \in \Sigma^* \; (x, y) \in R\}.$$

Definition 11.20. *The class* $\mathbf{\Pi}_i$ *is defined similarly to* $\mathbf{\Sigma}_i$*, except that* $\pi(x) = \{1\}$ *is replaced by* $\pi(x) = \{0\}$ *in the definition.*

Intuitively, $\mathbf{\Delta}_i$ is the analog of **P** reinforced with an oracle that is capable of solving problems in the previous level of the hierarchy. Similarly, $\mathbf{\Sigma}_i$ is the analog of **NP** and $\mathbf{\Pi}_i$ is the analog of **coNP**.

The complexity classes $\mathbf{\Delta}_i$, $\mathbf{\Sigma}_i$, $\mathbf{\Pi}_i$ for $i = 1, 2, \ldots$ define the *polynomial-time hierarchy*. Note that $\mathbf{\Delta}_1 = \mathbf{P}$, $\mathbf{\Sigma}_1 = \mathbf{NP}$, and $\mathbf{\Pi}_1 = \mathbf{coNP}$. Furthermore, $\mathbf{\Sigma}_{i-1} \cup \mathbf{\Pi}_{i-1} \subseteq \mathbf{\Delta}_i$ and $\mathbf{\Delta}_i \subseteq \mathbf{\Sigma}_i, \mathbf{\Pi}_i$ for all $i = 1, 2, \ldots$. It is generally believed that this hierarchy is proper, that is, the classes for $i = 1, 2, \ldots$ are all distinct. Complete problems are known for all classes in the polynomial-time hierarchy.

11.2 Completion Problems

A basic type of decision problem associated with exhaustive generation of combinatorial objects such as codes and designs is that we are given a partial object with some parts undefined, and are asked whether it is possible to complete the undefined part in such a way that an object with the required properties is obtained. A problem of this type is called a *completion problem*; alternative names include *extension problem* and *embedding problem*.

It is possible to prove that various completion problems for codes and designs are **NP**-complete. We will first look at completion problems for designs, where our aim is to show that both point-by-point and block-by-block completion of partial designs is **NP**-complete. The tools developed will also allow us to establish the **NP**-completeness of completion problems for various types of codes and other objects.

Following [111], we start with a well-known **NP**-complete problem [267].

Problem 11.21. (EDGE-COLORING A 3-REGULAR GRAPH) Given a 3-regular graph G, decide whether it admits a proper edge coloring with 3 colors.

Observe that an equivalent problem is to decide whether a 3-regular graph has a 1-factorization.

In what follows we first transform an arbitrary instance G of EDGE-COLORING A 3-REGULAR GRAPH into an instance H where the complement graph has a known 1-factorization. This will then allow us to construct a partial Steiner triple system that admits a completion if and only if the original graph G admits a proper edge 3-coloring.

Following [111], it is convenient to use the language of Latin squares. A Latin square L of side m is *symmetric* if $L(i,j) = L(j,i)$ for all $i,j \in Z_m$. The square has *constant diagonal* if $L(i,i) = L(j,j)$ for all $i,j \in Z_m$.

A 1-factorization of the complete graph K_m can be represented as a symmetric Latin square L of side m with constant diagonal, where any two edges, $\{u,v\}$ and $\{x,y\}$, occur in the same 1-factor if and only if $L(u,v) = L(x,y)$.

Example 11.22. A 1-factorization of K_6 represented as a symmetric Latin square of side 6 with constant diagonal.

5	0	1	2	3	4
0	5	2	3	4	1
1	2	5	4	0	3
2	3	4	5	1	0
3	4	0	1	5	2
4	1	3	0	2	5

Analogously, a 1-factorization of the complement of a 3-regular graph H of order n can be represented as a partial symmetric Latin square L with constant diagonal and side n, where the entries corresponding to 1s in an adjacency matrix of H are undefined, and all other entries contain one of the symbols $\{3, 4, \ldots, n-1\}$. A completion of L into a symmetric Latin square with constant diagonal corresponds to a proper edge coloring of H with 3 colors.

An arbitrary 3-regular graph G does not necessarily admit a 1-factorization for its complement. However, G can be embedded as a connected component into a 3-regular graph H whose complement always admits a 1-factorization. Furthermore, it is possible to choose H in such a way that its components other than G always admit a 1-factorization, implying that H admits a 1-factorization if and only if G does.

We develop the embedding tools using the language of Latin squares.

Definition 11.23. *Let G be an r-regular graph of order n and let $m \geq s \geq n$. A* Latin background $\mathrm{LB}(G; m, s)$ *for G is an $s \times s$ symmetric array with entries either undefined or chosen from Z_m so that the following five properties hold:*

1. *every element in Z_m appears at most once in every row and column,*
2. *every diagonal entry contains the element $m-1$,*
3. *in the first n rows, every entry is either undefined or contains an element from $\{r, r+1, \ldots, m-1\}$,*
4. *in the last $s-n$ rows, every entry is defined and contains an element from $\{0, 1, \ldots, m-1\}$,*
5. *the pattern of undefined entries is precisely the pattern of 1s in an adjacency matrix of G.*

Observe that given an $\mathrm{LB}(G; m, m)$ for a 3-regular graph G, we obtain the adjacency matrix of a graph H with the aforementioned properties by first replacing the undefined entries and the entries containing $\{0, 1, \ldots, r-1\}$ with 1s, and then replacing the other entries with 0s.

Example 11.24. An $\mathrm{LB}(K_4; 8, 8)$.

7				3	4	5	6
	7			4	5	6	3
		7		5	6	3	4
			7	6	3	4	5
3	4	5	6	7	0	1	2
4	5	6	3	0	7	2	1
5	6	3	4	1	2	7	0
6	3	4	5	2	1	0	7

The following two lemmata and their proofs are essentially from [111]. The first lemma provides a base case for constructing an $\mathrm{LB}(G; m, m)$.

Lemma 11.25. *For any regular graph G of order n, there is an* $\mathrm{LB}(G; m, n)$ *for every $m \geq 2n$.*

Proof. Let $\mathbf{A} = (a_{ij})$ be an adjacency matrix of G, where $i, j \in Z_n$. For all $i, j \in Z_n$ the background L has the entry $L(i, j)$ undefined if $a_{ij} = 1$; otherwise $L(i, j) = m - 1 - ((i - j) \bmod n)$. □

The second lemma requires some preliminaries. First, we require a classic result on systems of distinct representatives from [393]. Let $S_0, S_1, \ldots, S_{s-1}$ be a collection of finite sets. A set $\{a_0, a_1, \ldots, a_{s-1}\}$ is called a *system of distinct representatives* for $S_0, S_1, \ldots, S_{s-1}$ if $a_i \in S_i$ for all $i \in Z_s$.

Theorem 11.26. *Let $S_0, S_1, \ldots, S_{s-1}$ be a collection of finite sets such that every $\ell = 1, 2, \ldots, s$ sets contain at least ℓ distinct elements in their union. Furthermore, let $e_0, e_1, \ldots, e_{k-1}$ be elements such that every e_i occurs in at least t of the sets S_j, and each set S_j contains at most t of the elements e_i. Then, $S_0, S_1, \ldots, S_{s-1}$ has a system of distinct representatives containing $e_0, e_1, \ldots, e_{k-1}$.*

We require some further definitions for a concise statement of the lemma. Let L be an $\mathrm{LB}(G; m, s)$ for an r-regular graph G of order n. Denote by $P(u)$ the number of occurrences of $u \in Z_m$ in L. Define $N(u)$ for all $u \in Z_m$ by

$$N(u) = \begin{cases} P(u) + n & \text{if } u \in \{0, 1, \ldots, r-1\}, \\ P(u) & \text{if } u \notin \{0, 1, \ldots, r-1\}. \end{cases}$$

We say that L is *admissible* if $N(u) \geq 2s - m$ for all $u \in Z_m$. Note that the backgrounds constructed in Lemma 11.25 are trivially admissible.

Lemma 11.27. *Let G be an r-regular graph of even order n. An admissible $\mathrm{LB}(G; m, s)$ with $s < m$ and m even can be extended to an admissible $\mathrm{LB}(G; m, s+1)$.*

Proof. Let L be an admissible $\mathrm{LB}(G; m, s)$ with $s < m$ and m even. For $i \in Z_s$, let T_i be the set of elements appearing in the ith row of L, and define S_i by

$$S_i = \begin{cases} \{r, r+1, \ldots, m-1\} \setminus T_i & \text{if } i < n, \\ \{0, 1, \ldots, m-1\} \setminus T_i & \text{if } i \geq n. \end{cases}$$

We have $|S_i| = m - s$ for all $i \in Z_s$. Indeed, for $i < n$ we have $|T_i| = s - r$ so $|S_i| = (m - r) - (s - r) = m - s$; for $i \geq n$ we have $|T_i| = s$ so $|S_i| = m - s$.

We next determine in how many sets S_i an element $u \in Z_m$ occurs. Consider first the case $u < r$. Clearly, u occurs in neither S_i nor T_i for $i < n$. Thus, u occurs in $P(u)$ sets T_i for $n \leq i < s$, and thereby in $(s - n) - P(u) = (s - n) - (N(u) - n) = s - N(u)$ sets S_i. Similarly, in the case $u \geq r$ we have that u occurs in $s - P(u) = s - N(u)$ sets S_i. By admissibility we thus have that each $u \in Z_m$ occurs in $s - N(u) \leq m - s$ of the sets S_i.

An element $u \in Z_m$ that occurs in exactly $m - s$ sets S_i is called *critical*; then $N(u) = 2s - m$. Note that $N(u)$ is even for all $u \neq m - 1$ since $P(u)$ is even (follows from the fact that L is symmetric and has $m - 1$ on the diagonal) and n is even. Consequently, as m is even, $N(u) \geq 2s - m + 2$ holds for any noncritical $u \neq m - 1$. Note that $m - 1$ cannot be critical because $N(m-1) = P(m-1) = s$ and $s < m$.

Taking ℓ of the sets S_i there is a total of $\ell(m-s)$ occurrences of elements, so the union of these sets contains at least $\ell(m-s)/(m-s)=\ell$ elements as every element in Z_m appears in at most $m-s$ of the sets. Theorem 11.26 invoked with $t=m-s$ and the elements e_i being the critical elements shows that the collection $S_0,S_1,\ldots,S_{s-1}$ admits a system of distinct representatives $\{a_0,a_1,\ldots,a_{s-1}\}$ containing all the critical elements.

Setting $L(s,s)=m-1$ and $L(i,s)=L(s,i)=a_i$ for all $i\in Z_s$ extends L into an $\mathrm{LB}(G;m,s+1)$. To verify admissibility, observe that we had $N(u)\geq 2(s+1)-m$ for all noncritical elements $u\neq m-1$ before extension, and two occurrences of each critical element were added during extension. □

Lemmata 11.25 and 11.27 give the following embedding result [111].

Theorem 11.28. *For any regular graph G of even order n and any even $m\geq 2n$ there exists an $\mathrm{LB}(G;m,m)$. Moreover, such an $\mathrm{LB}(G;m,m)$ can be computed from G in time polynomial in m.*

Proof. Apply Lemma 11.25 to obtain an $\mathrm{LB}(G;m,n)$, and then extend it step by step into an $\mathrm{LB}(G;m,m)$ using Lemma 11.27. A system of distinct representatives guaranteed by Theorem 11.26 can be found in time polynomial in m using a maximum matching algorithm for bipartite graphs (see, for example, [331]): first find an arbitrary system of distinct representatives and then transform it into a system containing all critical elements using the lemma in [393, p. 399]. □

The following result is an almost immediate corollary [111].

Theorem 11.29. *It is an* **NP**-*complete problem to decide whether a given set of 1-factors of K_m can be completed into a 1-factorization of K_m.*

Proof. The problem is in **NP** because a 1-factorization of K_m extending the given factors can be verified in time polynomial in m. To establish completeness, we exhibit a polynomial-time transformation from EDGE-COLORING A 3-REGULAR GRAPH. Let G be an arbitrary 3-regular graph of order n. Put $m=2n$, and construct in time polynomial in n an $\mathrm{LB}(G;m,m)$, L. Construct a set of 1-factors of K_m as follows. The vertex set of each factor is Z_m. For each $u\in\{3,4,\ldots,m-1\}$, define a 1-factor consisting of the edges $\{\{i,j\}: L(i,j)=u,\ 0\leq i<j\leq m-1\}$. These 1-factors can be completed to a 1-factorization of K_m if and only if G admits a proper edge coloring with 3 colors. □

Let us proceed to completion problems for designs. The following two problems are motivated by Chap. 6.

Problem 11.30. (BIBD COMPLETION BLOCK BY BLOCK) Given the parameters v,k,λ,b,r in unary and a 0-1 matrix of size $v\times b'$ with $b'\leq b$, decide whether the matrix can be completed to an incidence matrix of a BIBD with the given parameters.

Problem 11.31. (BIBD Completion Point by Point) Given the parameters v, k, λ, b, r in unary and a 0-1 matrix of size $v' \times b$ with $v' \leq v$, decide whether the matrix can be completed to an incidence matrix of a BIBD with the given parameters.

The following result is contained in the proof of [111, Theorem 3.1].

Theorem 11.32. BIBD Completion Block by Block *is* **NP**-*complete.*

Proof. The problem is in **NP** because an incidence matrix of a BIBD can be verified in time polynomial in the parameters.

To establish completeness, we exhibit a polynomial-time transformation from Edge-Coloring a 3-Regular Graph. Let G be an arbitrary 3-regular graph of order n. We construct a partial incidence matrix from G as follows. Choose the smallest even m such that $2n \leq m \leq 2n+5$ and $m-1$ is an admissible order of a Steiner triple system. Using Theorem 11.28, let L be an $\mathrm{LB}(G; m, m)$. Note that L can be constructed in time polynomial in n.

From L, construct a partial symmetric Latin square L' of side $m-1$ first by setting $L'(i,j) = L(i,j)$ for all distinct $i, j \in Z_{m-1}$ and $L'(i,i) = L(m-1,i)$ for all $i \in Z_{m-1}$. Then, rearrange the diagonal elements by permuting the rows and columns simultaneously so that $L'(i,i) = i$ for all $i \in Z_{m-1}$. Observe that L' can be completed to a symmetric Latin square if and only if L can be completed to a symmetric Latin square.

Put $v = 2m-1$, $k = 3$, $\lambda = 1$, and $b' = v(v-1)/6 - 3n/2$. Let $\{x_0, x_1, \ldots, x_{m-2}, y_0, y_1, \ldots, y_{m-2}, z\}$ be a set of v points. First, include the blocks of an $\mathrm{STS}(m-1)$ on the points $\{x_0, x_1, \ldots, x_{m-2}\}$. Observe that an $\mathrm{STS}(m-1)$ can be constructed in time polynomial in n using, say, Examples 2.49 and 2.50. Next, include the block $\{x_i, y_i, z\}$ for each $i \in Z_{m-1}$. Finally, for all $0 \leq i < j \leq m-2$ such that $L'(i,j)$ is defined, include the block $\{y_i, y_j, x_{L'(i,j)}\}$.

Construct a $v \times b'$ incidence matrix defined by these points and blocks. This incidence matrix can be completed by addition of blocks to an incidence matrix of an $\mathrm{STS}(v)$ if and only if G admits a proper edge coloring with 3 colors. Moreover, the incidence matrix can be computed in time polynomial in n. □

The previous theorem can be extended to show that it is an **NP**-complete problem to decide whether a given partial $\mathrm{STS}(v)$ can be embedded in an $\mathrm{STS}(w)$ for any $w \leq 2v-1$ [111].

Example 11.33. Figure 11.1 illustrates a partial STS(15) resulting from the construction in the proof of Theorem 11.32 when applied to the $\mathrm{LB}(K_4, 8, 8)$ in Example 11.24. The partial symmetric Latin square L' used in the construction is also displayed.

We can apply analogous embedding techniques to demonstrate **NP**-completeness also in the case of point by point completion.

0	2	1	6	3	4	5
2	1	0	5	6	3	4
1	0	2	4	5	6	3
6	5	4	3			
3	6	5		4		
4	3	6			5	
5	4	3				6

x_0	1110000	1000000	000000100000000	
x_1	1001100	0100000	010000000000000	
x_2	1000011	0010000	100000000000000	
x_3	0101010	0001000	000100000100001	
x_4	0100101	0000100	000010000011000	
x_5	0011001	0000010	000001010000100	
x_6	0010110	0000001	001000001000010	
y_0	0000000	1000000	111111000000000	
y_1	0000000	0100000	100000111110000	
y_2	0000000	0010000	010000100001111	
y_3	0000000	0001000	001000010001000	
y_4	0000000	0000100	000100001000100	
y_5	0000000	0000010	000010000100010	
y_6	0000000	0000001	000001000010001	
z	0000000	1111111	000000000000000	

Fig. 11.1. An illustration of Theorem 11.32

Theorem 11.34. BIBD COMPLETION POINT BY POINT *is* **NP**-*complete.*

Proof. Again the problem is evidently in **NP**. To establish completeness, we exhibit a polynomial-time transformation from EDGE-COLORING A 3-REGULAR GRAPH. Let G be an arbitrary 3-regular graph of order n. We construct a partial incidence matrix from G as follows. Choose the smallest even m such that $2n \leq m \leq 2n + 5$ and $m - 1$ is an admissible order of a Steiner triple system.

Put $v = 2m - 1$, $k = 3$, $\lambda = 1$, and $v' = v - 3$. Let the set of v points be $\{x_0, x_1, \ldots, x_{m-1}, y_0, y_1, \ldots, y_{m-2}\}$. Construct an STS$(m-1)$ on the points $\{y_0, y_1, \ldots, y_{m-2}\}$. By relabeling the points if necessary, we can assume that $\{y_0, y_1, y_2\}$ is a block of the STS$(m-1)$. Delete the points y_0, y_1, y_2 from all the blocks, thereby creating one empty block and $3(m/2 - 2)$ blocks of size 2. Using Theorem 11.28, let L be an LB$(G; m, m)$. For every $0 \leq i < j \leq m - 1$, introduce the block $\{x_i, x_j, y_{L(i,j)}\}$ if $L(i, j)$ is defined; otherwise introduce the block $\{x_i, x_j\}$.

Construct a $v' \times b$ incidence matrix defined by these blocks on the points $\{x_0, x_1, \ldots, x_{m-1}, y_3, y_4, \ldots, y_{m-2}\}$; note that points y_0, y_1, y_2 are not included. This incidence matrix can be completed by addition of points to an incidence matrix of an STS(v) if and only if G admits a proper edge coloring with 3 colors. Moreover, the partial incidence matrix can be computed in time polynomial in n. □

Example 11.35. Figure 11.2 illustrates a partial STS(15) resulting from the construction in the proof of Theorem 11.34 when applied to the LB($K_4, 8, 8$) in Example 11.24.

y_0				
y_1				
y_2				
y_3	0101010	000000	1000000100100100	000000
y_4	0100101	000000	0100100000010010	000000
y_5	0011001	000000	0010010010000001	000000
y_6	0010110	000000	0001001001001000	000000
x_0	0000000	111000	1111000000000000	000000
x_1	0000000	100110	0000111100000000	000000
x_2	0000000	010101	0000000011110000	000000
x_3	0000000	001011	0000000000001111	000000
x_4	0000000	000000	1000100010001000	111000
x_5	0000000	000000	0100010001000100	100110
x_6	0000000	000000	0010001000100010	010101
x_7	0000000	000000	0001000100010001	001011

Fig. 11.2. An illustration of Theorem 11.34

Turning now to codes, Theorem 11.34 together with the second Johnson bound (Theorem 2.88) implies that it is an **NP**-complete problem to decide whether a given $(n, M, d)_2$ code of constant weight w can be augmented to an $(n, M + 3, d)_2$ code of constant weight w by adding new codewords. For illustration, note that the rows in Fig. 11.2 define the codewords of a constant weight code.

For unrestricted error-correcting codes there are two basic problems associated with exhaustive generation: the problem of extending a given code by adding new coordinates so that given distance constraints are met, and the problem of augmenting a given code with new codewords so that given distance constraints are met.

One example of an **NP**-completeness result is as follows. Observing that a 1-factorization of K_{2u} corresponds to a resolution of the unique 2-$(2u, 2, 1)$ design, and hence, to a $(2u-1, 2u, 2u-2)_u$ OE code by Theorem 3.82, Theorem 11.29 implies that it is an **NP**-complete problem to decide whether a given code can be extended to an OE code by adding 3 new coordinates.

Research Problem 11.36. Study the complexity of augmenting a given code to an OE code by adding new codewords. One may conjecture that the problem is **NP**-complete for OE codes corresponding to Kirkman triple systems; that is, for $(3u+1, 6u+3, 3u)_{2u+1}$ codes.

In terms of lengthening a code, the central problem is to find, for a given code $C \subseteq Z_q^n$, words whose distance to all the codewords in C is at least a given value. The complexity of this problem has been studied in the language of covering codes.

Problem 11.37. (COVERING RADIUS LOWER BOUND) Given a code $C \subseteq Z_q^n$ and a nonnegative integer ℓ, decide whether there exists a word $\mathbf{x} \in Z_q^n$ such that $d_H(\mathbf{x}, \mathbf{c}) \geq \ell$ for all $\mathbf{c} \in C$.

As the problem name suggests, an equivalent formulation is to decide whether $R(C) \geq \ell$.

Following [190], we prove that COVERING RADIUS LOWER BOUND is **NP**-complete for $q = 2$. Call a word $\mathbf{x} = (x_1, x_2, \ldots, x_{2n}) \in Z_2^{2n}$ *doubled* if $x_{2i-1} = x_{2i}$ for all $i = 1, 2, \ldots, n$. In the proof we require a subcode that guarantees that the complement of a word meeting the distance constraint is doubled. To this end, denote by Y_1 the set of the following four words of length $2n$:

$$\begin{array}{l} 10\,|\,01010101\cdots0101, \\ 01\,|\,01010101\cdots0101, \\ 10\,|\,10101010\cdots1010, \\ 01\,|\,10101010\cdots1010. \end{array}$$

Observe that if a word $\mathbf{x} \in Z_q^{2n}$ satisfies $d_H(\mathbf{x}, \mathbf{y}) \leq n$ for all $\mathbf{y} \in Y_1$, then $x_1 = x_2$. To extend this property to each successive coordinate pair, denote by Y_j the code obtained from Y_1 by the circular right shift of the words by $2j-2$ coordinates, and let $Y = Y_1 \cup Y_2 \cup \cdots \cup Y_n$.

Lemma 11.38. *A word* $\mathbf{x} \in Z_2^{2n}$ *is doubled if* $d_H(\mathbf{x}, \mathbf{y}) \leq n$ *for all* $\mathbf{y} \in Y$.

The proof of the next theorem follows [190].

Theorem 11.39. COVERING RADIUS LOWER BOUND *is* **NP**-*complete.*

Proof. The problem is in **NP** because the distance bound for a given word $\mathbf{x}$ can be verified in time polynomial in n and $|C|$. To establish completeness, we exhibit a polynomial-time transformation from 3-SATISFIABILITY. Let $\mathcal{C} = \{C_1, C_2, \ldots, C_m\}$ be a set of m clauses of size 3 over a set of k variables $\{v_1, v_2, \ldots, v_k\}$. Without loss of generality we can remove any clause(s) containing a pair $\{v_i, \bar{v}_i\}$ of complementary literals. Similarly, we can assume $k \leq 3m$. We construct from $\mathcal{C}$ a binary code $C \subseteq Z_2^{2n}$ of length $2n = 2k+2$ as follows. First, include the words in Y to the code. Then, for every clause C_j include the word $\mathbf{z}(C_j) = (z(C_j)_1, z(C_j)_2, \ldots, z(C_j)_{2n})$ defined for $i = 1, 2, \ldots, n$ by

$$\begin{cases} z(C_j)_{2i-1} = 0, \ z(C_j)_{2i} = 0 & \text{if } i = n \text{ or } v_i \in C_j, \\ z(C_j)_{2i-1} = 1, \ z(C_j)_{2i} = 1 & \text{if } \bar{v}_i \in C_j, \\ z(C_j)_{2i-1} = 0, \ z(C_j)_{2i} = 1 & \text{otherwise.} \end{cases}$$

Finally, put $\ell = n$. This completes the description of the transformation from $\mathcal{C}$ to C, which is obviously computable in time polynomial in m. It remains to show that the clause set $\mathcal{C}$ is satisfiable if and only if there exists a word $\mathbf{x} \in Z_2^{2n}$ with $d_H(\mathbf{x}, \mathbf{c}) \geq n$ for all $\mathbf{c} \in C$.

To establish the "only if" direction, let T be truth assignment that satisfies at least one literal in every clause in $\mathcal{C}$. Define the word $\mathbf{x} \in Z_2^n$ by setting $x_{2i-1} = x_{2i} = T(v_i)$ for $i = 1, 2, \ldots, k$ and $x_{2n-1} = x_{2n} = 1$. Because $\mathbf{x}$ is doubled, we have $d_H(\mathbf{x}, \mathbf{y}) = n$ for all $\mathbf{y} \in Y$. Because T satisfies at least one literal in C_j, we have $d_H(\mathbf{x}, \mathbf{z}(C_j)) \geq 2 + 0 + 0 + k - 3 + 2 = k + 1 = n$ for all $\mathbf{z}(C_j) \in C$. Thus, $d_H(\mathbf{x}, \mathbf{c}) \geq n$ for all $\mathbf{c} \in C$.

To establish the "if" direction, let $\mathbf{x} \in Z_2^{2n}$ satisfy $d_H(\mathbf{x}, \mathbf{c}) \geq n$ for all $\mathbf{c} \in C$. Denote by $\bar{\mathbf{x}}$ be the binary complement of $\mathbf{x}$; that is, the word obtained by transposing the values 0 and 1 in every coordinate of $\mathbf{x}$. Clearly, $d_H(\mathbf{x}, \mathbf{w}) + d_H(\bar{\mathbf{x}}, \mathbf{w}) = 2n$ for every word $\mathbf{w} \in Z_2^{2n}$. Thus, $d_H(\bar{\mathbf{x}}, \mathbf{c}) \leq n$ for all $\mathbf{c} \in C$, which implies by $Y \subseteq C$ and Lemma 11.38 that $\bar{\mathbf{x}}$ is doubled. Because $\bar{\mathbf{x}}$ is doubled, also $\mathbf{x}$ is doubled. Define T for all $i = 1, 2, \ldots, k$ by $T(v_i) = x_{2i}$. Because $\mathbf{x}$ is doubled and $d_H(\mathbf{x}, \mathbf{z}(C_j)) \geq n > 0 + 0 + 0 + k - 3 + 2$, we have that T satisfies at least one literal in every $C_j \in \mathcal{C}$. □

For linear codes given by a generator matrix, COVERING RADIUS LOWER BOUND is $\mathbf{\Sigma}_2$-complete [422], see also [237]. Another example of this increase in complexity caused by a succinct representation is that computing the minimum distance of a linear code given by a generator matrix is **NP**-hard [594], see also [171], whereas for an explicitly given set of codewords the minimum distance is obviously computable in polynomial time.

The following problem is relevant in the classification of linear error-correcting codes.

Problem 11.40. (LINEAR CODE LENGTHENING) Given a generator matrix for an $[n, k, d]_q$ code C, decide whether it is possible to lengthen C to an $[n+1, k+1, d]_q$ code.

With some straightforward effort it can be observed that lengthening is possible if and only if $R(C) \geq d - 1$. Thus, it is likely that LINEAR CODE LENGTHENING is a difficult problem; however, because of the constraint $d(C) \geq d$ in LINEAR CODE LENGTHENING – which the codes constructed in the proofs in [237, 422] apparently do not meet – it is not immediate that LINEAR CODE LENGTHENING is $\mathbf{\Sigma}_2$-complete.

Research Problem 11.41. Study the complexity of COVERING RADIUS LOWER BOUND when the code C has more restricted structure. For example, does the problem remain $\mathbf{\Sigma}_2$-complete in the linear case – or, **NP**-complete in the unrestricted case – if it is required that $d(C) \geq \ell$ or $d(C) \geq \ell - 1$?

We have now seen that many completion problems for combinatorial objects are **NP**-complete or beyond, providing evidence that the completion problems for partial objects are computationally challenging. What should be explicitly pointed out, however, is that essentially all of the hardness results concern only very specific types of partial objects, which may or may not be encountered as partial solutions during exhaustive generation with backtrack search. Consequently, one should be careful when making claims on the hardness of exhaustive generation based on the complexity of completion problems. A case in point occurs with Latin squares. In general, the problem of deciding whether a partial Latin square with some entries undefined can be completed is **NP**-complete [112, 172]. However, it is a classic result that a $k \times m$ Latin rectangle with $k < m$ can always be extended to a complete $m \times m$ Latin square, see, for example, [376, Chap. 17]; for algorithms, see [331]. Thus, if we generate Latin squares row by row, the hard instances in [112, 172] are never encountered. In this respect complexity results for completion problems provide evidence of computational difficulty only for specific exhaustive generation techniques, such as point-by-point or block-by-block generation of BIBDs. Even in these cases it is not always clear whether the difficult partial objects are encountered in practice due to the heuristics and pruning techniques employed.

11.3 Isomorphism Problems

From a theory perspective one of the central open problems in combinatorial computation is whether the isomorphism problem for explicitly given graphs admits a polynomial-time algorithm.

Problem 11.42. (GRAPH ISOMORPHISM) Given two graphs, decide whether they are isomorphic.

The graph isomorphism problem has been extensively studied, surveys include [12, 17, 18, 213, 501]. Extensive bibliographical references can be found also in [266, 324, 601].

One of the reasons behind the central role of graph isomorphism is that most types of explicitly given combinatorial objects can be transformed in polynomial time into a graph for purposes of computing isomorphism [258, 427]. Formally, the isomorphism-respecting properties of such transformations into graphs can be analyzed in the setting of reconstructible functors, see Sect. 3.2. The existence of such transformations is often due to the fact that the acting group inducing isomorphism for the objects in question can be concisely encoded as the automorphism group of a graph, after which the relevant structure in the objects can usually be encoded in a straightforward manner by adding edges and vertices.

For many families of combinatorial objects also a converse polynomial-time transformation from graphs to the objects can be exhibited. In the absence of a polynomial-time algorithm for graph isomorphism, the worst case performance of isomorphism algorithms on such objects therefore remains bad. A decision problem polynomial-time equivalent to GRAPH ISOMORPHISM is called a *graph isomorphism complete* problem, see [56].

Among the many graph isomorphism complete problems are the isomorphism problem for BIBDs [126] and equivalence of unrestricted codes, which we now proceed to prove.

As a preliminary step, we show that the isomorphism problem for colored graphs is graph isomorphism complete. This enables us to transform into colored graphs rather than graphs, which simplifies subsequent proofs because of the colored graph representations developed in Sect. 3.3.2.

Lemma 11.43. *Deciding isomorphism of colored graphs is graph isomorphism complete.*

Proof. A graph can be transformed into a colored graph by introducing a coloring of the vertices where all vertices have the same color. Clearly, any two such colored graphs are isomorphic if and only if the associated graphs are isomorphic.

Conversely, given a colored graph (G, π) with $\pi = (V_1, V_2, \ldots, V_m)$, we transform it into a graph H as follows. Let n be the order of G. Start with the graph G. For each $i = 1, 2, \ldots, m$, add $n + 1 + i$ new vertices into the graph, join these vertices by edges so that they form a cycle of order $n+1+i$, and join all the vertices in V_i by edges into any one vertex of the cycle. The resulting graph is H.

For two colored graphs, (G, π) and (G', π'), any isomorphism $f : V(G) \to V(G')$ obviously extends into an isomorphism of H onto H'. Thus, $(G, \pi) \cong (G', \pi')$ implies $H \cong H'$. The colored graph (G, π) can be reconstructed up to isomorphism from H by processing the longest cycles in decreasing order until all vertices have been encountered either in such a cycle or in the color class identified by the cycle.

The transformations in both directions are obviously computable in time polynomial in the input size. □

Let us now proceed to consider the isomorphism problem for BIBDs.

Problem 11.44. (BIBD ISOMORPHISM) Given two BIBDs, decide whether they are isomorphic.

The graph isomorphism completeness proof for BIBDs that we present is essentially from [126]. A convenient starting point is the following result from [113].

A graph G is *self-complementary* if $G \cong \bar{G}$.

Theorem 11.45. *Deciding isomorphism of regular self-complementary graphs is graph isomorphism complete.*

A self-complementary graph can be seen as the $k = 2$ special case of a graph factorization $\mathcal{F} = \{F_1, F_2, \ldots, F_k\}$ with the requirement that the factors F_i be pairwise isomorphic. Let us call such factorizations *homogeneous*. Following [126], we require a technical result for $k = 3$.

Lemma 11.46. *Deciding isomorphism of homogeneous m-factorizations of the complete graph K_{3m+1} is graph isomorphism complete.*

Proof. We first transform a graph factorization $\mathcal{F} = \{F_1, F_2, \ldots, F_k\}$ with vertex set V into a colored graph (G, π), which is sufficient by Lemma 11.43. Start with an empty graph with vertex set V. Add k new vertices $f_1, f_2, \ldots, f_k$. For each $i = 1, 2, \ldots, k$, add the factor F_i by adding a copy V_i of V and placing the edges of F_i onto V_i. Denote the copy of $v \in V$ in V_i by v_i. Add the edges $\{f_i, v_i\}$ and $\{v, v_i\}$ for all $v \in V$ and $i = 1, 2, \ldots, k$. The result graph is G. To complete the construction, let the coloring of the vertices be $\pi = (V, \{f_1, f_2 \ldots, f_k\}, \cup_{i=1}^k V_i)$. It is immediate that isomorphic graph factorizations are transformed into isomorphic colored graphs. Furthermore, the graph factorization is strongly reconstructible from the colored graph.

Conversely, we transform a graph G into a homogeneous m-factorization $\mathcal{F} = \mathcal{F}(G)$ of K_{3m+1}. By Theorem 11.45 we can assume that G is regular and self-complementary. Let V be the vertex set of G, $|V| = m$. Take 3 disjoint copies V_0, V_1, V_2 of V, and for $v \in V$ denote by v_i the copy of v in V_i. Let x be a vertex not in $V_0 \cup V_1 \cup V_2 \cup V$. For $i = 0, 1, 2$, let addition on subscripts be modulo 3 and define

$$
\begin{aligned}
V(F_i) &= V_0 \cup V_1 \cup V_2 \cup \{x\}, \\
E(F_i) &= \{\{x, v_i\} : v \in V\} \cup \\
&\quad \{\{u_i, v_i\} : \{u, v\} \in E(G)\} \cup \\
&\quad \{\{u_{i+2}, v_{i+2}\} : \{u, v\} \in E(\bar{G})\} \cup \\
&\quad \{\{u_i, v_{i+1}\} : \{u, v\} \in E(\bar{G})\} \cup \\
&\quad \{\{u_{i+1}, v_{i+2}\} : \{u, v\} \in E(G)\} \cup \\
&\quad \{\{v_{i+1}, v_{i+2}\} : v \in V\}.
\end{aligned}
$$

By the cyclic construction, $F_0 \cong F_1 \cong F_2$. Because G is regular and self-complementary, all vertices in G have degree d, where $m = 2d + 1$, based on which it is easy to check that each F_i is m-regular. Furthermore, the edges in F_0, F_1, F_2 partition the edge set of the complete graph on $V_0 \cup V_1 \cup V_2 \cup \{x\}$. Thus, $\mathcal{F} = \{F_0, F_1, F_2\}$ is a homogeneous m-factorization of K_{3m+1}.

For two regular self-complementary graphs, G and G', any isomorphism $f : V \to V'$ of G onto G' defines an isomorphism $g : V_0 \cup V_1 \cup V_2 \cup \{x\} \to V'_0 \cup V'_1 \cup V'_2 \cup \{x'\}$ of $\mathcal{F}(G)$ onto $\mathcal{F}(G')$ where $g(x) = x'$ and $g(v_i) = w'_i$ for all $i = 0, 1, 2$, $v \in V$, and $w' \in V'$ such that $f(v) = w'$. Thus, $G \cong G'$ implies $\mathcal{F}(G) \cong \mathcal{F}(G')$.

To reconstruct G up to isomorphism from $\mathcal{F}(G)$, consider any factor F_i. Observe that x is the only vertex in $V(F_i)$ whose neighborhood induces a d-regular graph (namely, G) in F_i.

The proof is completed by observing that the transformations in both directions are obviously computable in time polynomial in the input size. □

Following [113], we proceed to the main result for BIBDs.

Theorem 11.47. BIBD ISOMORPHISM *is graph isomorphism complete.*

Proof. A colored graph representation for a BIBD is provided by the incidence graph (Definition 3.89).

Conversely, we transform a graph into a BIBD. By Lemma 11.46 it suffices to describe a transformation from a homogeneous m-factorization $\mathcal{F} = \{F_0, F_1, F_2\}$ of K_{3m+1} into a BIBD, $\mathcal{D} = \mathcal{D}(\mathcal{F})$. We assume that $\mathcal{F} = \mathcal{F}(G)$ has been constructed from a graph G as in the proof of Lemma 11.46. In particular, $m \equiv 1 \pmod 2$.

We let $\mathcal{D}$ be a triple system with parameters

$$v = 6m + 1, \quad k = 3, \quad \lambda = m, \quad b = (6m + 1)m^2, \quad r = 3m^2.$$

Let V be the vertex set of the factors in $\mathcal{F}$, and let $W = \{w_{ij} : i \in \{0, 1, 2\},\ j \in \{1, 2, \ldots, m\}\}$ be a set disjoint from V. Let the point set of $\mathcal{D}$ be $V \cup W$. Because $3m \equiv 3 \pmod 6$, there exists an STS($3m$). Construct an STS($3m$) over the point set W using, say, Example 2.49. Include m copies of each block of the STS into $\mathcal{D}$. Then include the block $\{u, v, w_{ij}\}$ for all $i = 0, 1, 2$, $j = 1, 2, \ldots, m$, and $u, v \in V$ for which $\{u, v\} \in E(F_i)$ holds.

Each block is clearly incident to 3 points, so to establish that $\mathcal{D}$ is a BIBD with the stated parameters it remains to check that every pair of distinct points occurs in m blocks. For pairs with both elements in W this is clear. For pairs with both elements in V this follows from the fact that each such pair occurs as an edge in exactly one F_i. For pairs with one element in both V and W, observe that F_i is m-regular.

Let $g : V \to V'$ be an isomorphism of $\mathcal{F}(G)$ onto $\mathcal{F}(G')$. By the construction of g in the proof of Lemma 11.46 we can assume that g takes F_i onto F_i' for all $i = 0, 1, 2$. By using always the same STS for the same value of m, and implementing the transformation $\mathcal{D}$ so that $w_{ij} \mapsto w_{ij}'$ is an isomorphism of STSs used for the same value m, we have that g extends to an isomorphism of $\mathcal{D}(\mathcal{F}(G))$ onto $\mathcal{D}(\mathcal{F}(G'))$ by setting $g(w_{ij}) = w_{ij}'$ for all $i = 0, 1, 2$ and $j = 1, 2, \ldots, m$. Thus, $\mathcal{F}(G) \cong \mathcal{F}(G')$ implies $\mathcal{D}(\mathcal{F}(G)) \cong \mathcal{D}(\mathcal{F}(G'))$.

To reconstruct $\mathcal{F}(G)$ up to isomorphism from $\mathcal{D}(\mathcal{F}(G))$, observe that the only repeated blocks in $\mathcal{D}(\mathcal{F}(G))$ are blocks of the STS, which enables us to identify W. Selecting a $w_{ij} \in W$ and looking at all the blocks containing w_{ij} but not contained in W, we obtain F_i. Letting w_{ij} vary over W, we obtain $\mathcal{F}(G)$. From $\mathcal{F}(G)$ we can reconstruct G by Lemma 11.46.

The transformations are clearly computable in time polynomial in the input size. □

Also the restriction of BIBD ISOMORPHISM to simple BIBDs remains graph isomorphism complete because the repeated blocks in the previous proof can be replaced with a set of disjoint STSs, in which case a pair-closure argument allows us to reconstruct W; see [126].

We now proceed to equivalence problems for codes. Equivalence of constant weight codes is induced by the action (3.9) of $S_q \wr S_n$ restricted to the group in Theorem 3.55; for binary constant weight codes, equivalence coincides with Definition 2.99.

Problem 11.48. (CONSTANT WEIGHT CODE EQUIVALENCE) Given two constant weight codes, decide whether they are equivalent.

Equivalence of unrestricted codes is given by Definition 2.100, or equivalently, induced by the action (3.9) of $S_q \wr S_n$.

Problem 11.49. (UNRESTRICTED CODE EQUIVALENCE) Given two unrestricted codes, decide whether they are equivalent.

Equivalence of linear codes is induced by the action (3.9) of $S_q \wr S_n$ restricted to the group in Theorem 7.39; for binary linear codes, equivalence coincides with Definition 2.99.

Problem 11.50. (LINEAR CODE EQUIVALENCE) Given generator matrices for two linear codes, decide whether the linear codes are equivalent.

In the case of constant weight and unrestricted codes, the graph isomorphism completeness results are almost immediate corollaries of Theorem 11.47.

Theorem 11.51. CONSTANT WEIGHT CODE EQUIVALENCE *is graph isomorphism complete.*

Proof. A colored graph representation for a constant weight code is obtained from the colored graph representation for unrestricted codes in Sect. 3.3.2 by introducing a new color to the vertices that encode the coordinate value 0 in each coordinate.

Conversely, it suffices by Theorem 11.47 to exhibit a transformation from a BIBD to a constant weight code. This is straightforward: take as codewords the rows of an incidence matrix of a BIBD, in which case we obtain from the proof of Theorem 11.47 a constant weight code with parameters $n = b = (6m + 1)m^2$, $M = v = 6m + 1$, $d = 2(r - \lambda) = 6m^2 - 2m$, $w = r = 3m^2$, and $q = 2$. □

Theorem 11.52. UNRESTRICTED CODE EQUIVALENCE *is graph isomorphism complete.*

Proof. A colored graph representation for unrestricted codes appears in Sect. 3.3.2.

Conversely, it suffices to exhibit a transformation from the constant weight codes constructed in the proof of Theorem 11.51. Obviously any equivalent constant weight codes are equivalent as unrestricted codes. Also, the constant weight codes in question have the property that the number of 1s in every coordinate is 3 and the number of 0s is $6m - 2 > 3$. Thus, the constant weight code can be reconstructed after arbitrary permutation of the coordinate values. □

Note that in the last two theorems graph isomorphism completeness holds also in the restricted case when the code is equidistant. Furthermore, the codes in Theorem 11.51 are optimal as constant weight codes by the second Johnson bound (Theorem 2.88).

Linear codes given by a generator matrix form a natural example of objects whose isomorphism problem admits a polynomial-time transformation from graph isomorphism [474] but it is not known whether the problem is graph isomorphism complete. We give here an alternative proof to that in [474] for hardness of LINEAR CODE EQUIVALENCE.

The *cycle matroid* $\mathcal{M}(G)$ of a graph G is the set system with point set $E(G)$ that consists of all subsets of $E(G)$ that do not contain a cycle in G. A graph G is *k-connected* if no deletion of at most $k - 1$ vertices results in a disconnected graph. The following theorem is a standard corollary to Whitney's 2-isomorphism theorem [605], which characterizes the graphs that have isomorphic cycle matroids. A more recent account is given in [470, Sect. 5.3].

Theorem 11.53. *For 3-connected graphs G and G' we have $G \cong G'$ if and only if $\mathcal{M}(G) \cong \mathcal{M}(G')$.*

The *cycle space* $C(G)$ of a graph G is the dual of the binary linear code generated by the rows of an incidence matrix of G.

Lemma 11.54. *For graphs G and G' we have $C(G) \cong C(G')$ if and only if $\mathcal{M}(G) \cong \mathcal{M}(G')$.*

Proof. Let $\mathcal{C}(G)$ be the set system over $E(G)$ that consists of all subsets of $E(G)$ that form a cycle in G. For all sets $I \subseteq E(G)$ we have $I \in \mathcal{M}(G)$ if and only if no set in $\mathcal{C}(G)$ is a subset of I. Thus, $\mathcal{M}(G)$ and $\mathcal{C}(G)$ determine each other.

Associate with every vector $\mathbf{x} \in \mathbb{F}_2^{|E(G)|}$ the set $X = X(\mathbf{x}) \subseteq E(G)$ of edges corresponding to the 1s in $\mathbf{x}$. We have $\mathbf{x} \in C(G)$ if and only if X defines a subgraph of G where all vertices have even degree. Observe that each set $X \subseteq E(G)$ defining an even-degree subgraph of G is a union of edge sets of cycles in G. Moreover, the nonempty minimal such sets X are the edge sets of cycles in G. Thus, $\mathcal{C}(G)$ and $C(G)$ determine each other. □

Theorem 11.55. *There exists a polynomial-time transformation from* GRAPH ISOMORPHISM *to* LINEAR CODE EQUIVALENCE.

Proof. A graph G with $n \geq 3$ vertices can be transformed into a 3-connected graph H in a reconstructible manner by introducing three disjoint copies of K_n into the graph, and connecting each vertex of G by edges to corresponding vertices in each of the complete graphs K_n. Let the generator matrix

$\mathbf{G} = \mathbf{G}(G)$ be an incidence matrix of H. Observing that two linear codes are equivalent if and only if their duals are equivalent, it follows by Lemma 11.54 and Theorem 11.53 that $G \cong G'$ if and only if the binary linear codes generated by $\mathbf{G}(G)$ and $\mathbf{G}(G')$ are equivalent. □

We have now established that the isomorphism problems for most of the central families of codes and designs in this book are graph isomorphism complete.

Research Problem 11.56. Study the complexity of the isomorphism problem for resolutions of BIBDs. In particular, is the problem graph isomorphism complete?

Research Problem 11.57. Study the complexity of isomorphism problems in more restricted settings. For example, does the isomorphism problem for BIBDs remain graph isomorphism complete for $k > 3$ or bounded λ? What about square designs? See [19, 114, 364, 426].

In a general setting, other central problems polynomial-time equivalent to Graph Isomorphism include the following [397].

Problem 11.58. (Graph Automorphism Orbit) Given a graph and two of its vertices, decide whether there is an automorphism mapping one vertex onto the other.

Problem 11.59. (#Graph Automorphisms) Given a graph, compute the order of its automorphism group.

Problem 11.60. (Graph Automorphism Group Generators) Given a graph, output a generator set of size bounded by a polynomial in the number of vertices for the automorphism group of the graph.

It is an open problem whether evaluating a canonical representative map or a canonical labeling map (see Sect. 3.3.6) for graphs is polynomial-time equivalent to Graph Isomorphism. The complexity of canonical labeling of graphs is discussed in [19]. In [50] the complexity relationships between isomorphism, certificates, and canonical representatives are analyzed in the context of a more general equivalence relation.

Although no polynomial-time algorithm is known for Graph Isomorphism, there is strong theoretical evidence that the problem is not **NP**-complete. Indeed, **NP**-completeness of Graph Isomorphism would imply the collapse of the polynomial-time hierarchy [21]; further evidence can be found in [320, 324]. Yet another problem for which a polynomial-time algorithm is not known – but which is generally believed to be easier than Graph Isomorphism – is the following:

Problem 11.61. (Graph Automorphism) Given a graph, decide whether it admits other automorphisms than the trivial automorphism fixing all the vertices.

A contrast to the previous evidence is that the following problem is already **NP**-complete [381].

Problem 11.62. (FIXED-POINT-FREE GRAPH AUTOMORPHISM) Given a graph, decide whether it has an automorphism moving all the vertices.

More recent complexity lower bounds for GRAPH ISOMORPHISM and GRAPH AUTOMORPHISM can be found in [584].

Extensive connections exist between isomorphism computations and computation in permutation groups. Group-theoretic approaches to isomorphism computations are surveyed in [17, 18, 266, 383], see also [19, 182, 382, 436]. A comprehensive treatment of permutation group algorithms is [528]; other recommended references include [82, 187, 188, 266, 296].

Many families of combinatorial objects studied in this book admit tailored isomorphism algorithms that have a superpolynomial upper bound for their worst-case running time, but the bound is considerably better than the current best upper bound for the worst-case running time of general graph isomorphism. Such objects include Steiner triple systems [426] (see also [114, 121, 560]), 1-factorizations of connected graphs and complete multigraphs [114], Hadamard matrices [364], Latin squares [73], and affine and projective planes [426].

Research Problem 11.63. Develop tailored isomorphism algorithms for the aforementioned families of combinatorial objects.

11.4 Classification Problems

From a complexity-theoretic perspective the classification problems studied in this book are largely an unexplored terrain. On one hand, no polynomial-time algorithms (in a sense to be made precise shortly) are known. On the other hand, there is no theoretical evidence suggesting nonexistence thereof either. Accordingly, the treatment in the present section will be largely inconclusive from a complexity-theoretic perspective.

We will show that proving the nonexistence of a polynomial-time algorithm for any explicit-representation classification problem studied in this book is as difficult as proving $\mathbf{P} \neq \mathbf{NP}$. On the other hand, we show that $\mathbf{P} = \mathbf{NP}$ implies that all classification problems in this book admit polynomial-time algorithms. Thus, short of settling the $\mathbf{P} \neq \mathbf{NP}$ question, any decisive theoretical progress on the polynomial-time solvability of classification problems studied in this book has to come from the discovery of polynomial-time algorithms for increasingly challenging families of objects.

We begin by formally defining listing and classification problems. It is convenient to base the definitions on string relations (cf. [282]). For a relation $R \subseteq \Sigma^* \times \Sigma^*$, let $D_R(x) = \{y \in \Sigma^* : (x, y) \in R\}$. In what follows we assume that $D_R(x)$ is finite for all $x \in \Sigma^*$. Intuitively, the strings in $D_R(x)$ are the objects to be listed or, respectively, classified.

Definition 11.64. *Given $x \in \Sigma^*$, the* listing problem λ_R *asks for a string that concatenates in arbitrary order all the strings in $D_R(x)$.*

A classification problem builds on top of this setting by introducing an equivalence relation $E \subseteq \Sigma^* \times \Sigma^*$.

Definition 11.65. *Given $x \in \Sigma^*$, the* classification problem $\lambda_{R,E}$ *asks for a string that concatenates in arbitrary order all the strings in a set consisting of exactly one string from every equivalence class induced by E on $D_R(x)$.*

As examples of classification problems studied in this book, we can consider:

Problem 11.66. (t-DESIGNS) Given the parameters t, v, k, λ, b, r in unary, output exactly one representative from every isomorphism class of t-(v, k, λ) designs.

Problem 11.67. (UNRESTRICTED ERROR-CORRECTING CODES) Given the parameters n, M, d, q in unary, output exactly one representative from every equivalence class of $(n, M, d)_q$ codes.

Problem 11.68. (LINEAR ERROR-CORRECTING CODES) Given the parameters n, k, d, q in unary, output exactly one representative – in generator matrix representation – from every equivalence class of $[n, k, d]_q$ codes.

Listing and classification problems differ from the types of problems discussed earlier in this chapter in two respects when resource usage of algorithms is considered. First, it is not immediately clear what should be regarded as an efficient algorithm in terms of running time. Many natural listing and classification problems produce an output whose size grows exponentially in the input size. Thus, "time polynomial in the input size" as the definition for tractable computation must be altered to accommodate the potentially large size of the output. Second, the space usage of an algorithm requires attention because some algorithms require access to objects output earlier and some do not. In the latter case it is unreasonable to include the entire output of an algorithm in the space requirement. To this end, we assume that a listing algorithm reports each generated object to be output by executing a special instruction whose parameters involve the area of memory (in the RAM model, a sequence of RAM registers) where the object resides; the instruction takes one time step and has otherwise no effect. The execution of the output instruction is called an *output event* in what follows.

A number of notions of efficiency have been proposed for listing and classification algorithms; cf. [173, 210, 211, 283, 511, 568, 592].

polynomial time The running time of the algorithm is bounded by a polynomial in the instance size and the size of the solution. Typically a stronger variant is considered where the running time bound is linear in the number of objects, and polynomial in the instance size.

polynomial delay The delay from the start to the first object output event, and thereafter the delay between any two consecutive object output events, is bounded by a polynomial in the instance size.

constant delay The delay from the start to the first object output event is bounded by a polynomial in the instance size, after which the delay between any two consecutive object output events is bounded by a constant. Constant-delay algorithms are alternatively called *loopless algorithms.* A weaker variant is to require *constant amortized time* (CAT); that is, the time period between the first and the last object output events is bounded by a constant times the number of objects output.

polynomial space The space usage of the algorithm is bounded by a polynomial in the instance size.

All listing and classification problems studied in this book have the property that if one is given an object in an explicit representation – as opposed to a succinct representation such as in LINEAR ERROR-CORRECTING CODES – it is possible to verify in polynomial time that the object meets the required properties. Also the size of each object is typically bounded by a polynomial in the relevant parameters. (However, note that in the case of t-DESIGNS this requires that we consider the number of blocks b as a parameter because by Theorem 2.38 the number of blocks in a t-design is not in general bounded by a polynomial in the parameters t, v, k, λ.) Thus, it is natural to assume that a listing problem λ_R is defined by a polynomially balanced, **P**-decidable relation R.

The existence of efficient listing algorithms in this general setting is connected with the $\mathbf{P} \neq \mathbf{NP}$ question. The essential ideas in the following proof appear in [359] (see also [211, Sect. 5.1]).

Theorem 11.69. $\mathbf{P} = \mathbf{NP}$ *if and only if all listing problems* λ_R *defined by a polynomially balanced,* **P***-decidable relation* R *admit a polynomial-time algorithm.*

Proof. For the "if" direction, let π be any **NP**-complete decision problem, and let R be an associated polynomially balanced, **P**-decidable relation. By assumption, the associated listing problem λ_R admits a polynomial-time listing algorithm, A. Using A we now describe a polynomial-time algorithm A' for π, from which $\mathbf{P} = \mathbf{NP}$ follows. Given an instance $x \in \Sigma^*$ of π, the algorithm A' simulates the operation of A for the maximum number of steps permitted by the polynomial running time bound for A if the output of A is going to be empty. If A has not halted when the simulation time runs out, A' declares x a "yes"-instance. Otherwise A' looks at the output of A and decides accordingly.

For the "only if" direction, let λ_R be an arbitrary listing problem defined by a polynomially balanced, **P**-decidable relation R. Because of our assumptions on R, the following decision problem is in **NP**: given $x \in \Sigma^*$ and $w \in \Sigma^*$, decide whether there exists a $z \in \Sigma^*$ such that $(x, wz) \in R$. Consequently, by the assumption $\mathbf{P} = \mathbf{NP}$ it follows that the decision problem admits a polynomial-time algorithm A.

A polynomial-time listing algorithm for λ_R can now be obtained as follows. Start with a backtrack search that given $x \in \Sigma^*$ generates, one alphabet symbol at a time, all strings $w \in \Sigma^*$ with $|w| \le p(|x|)$, where p is the polynomial guaranteed by the polynomial balance condition on R. For each generated w, the algorithm checks – in polynomial time because of $\mathbf{P}$-decidability of R – whether $(x, w) \in R$, and outputs w if this is the case. Following this, the algorithm checks, by using the algorithm A as a subroutine, for each $a \in \Sigma$ whether wa extends to some waz with $(x, waz) \in R$. If so, the backtrack search proceeds to consider wa; otherwise wa and its extensions are not considered. It is straightforward to check that employing A to prune the backtrack search in this manner results in a polynomial delay algorithm for λ_R. □

Corollary 11.70. $\mathbf{P} = \mathbf{NP}$ *if and only if any problem that asks for a list of all witnesses to an instance of an* $\mathbf{NP}$*-complete decision problem admits a polynomial-time listing algorithm.*

Thus, natural listing problems not admitting a polynomial-time algorithm exist in abundance conditional to $\mathbf{P} \neq \mathbf{NP}$.

If the assumptions on R are removed, then it is possible to define listing problems that provably do not admit a polynomial-time algorithm (see [211, Sect. 5.1]), or even undecidable listing problems that provably admit no algorithm at all. Also classification problems can easily be made arbitrarily difficult if we do not constrain the structure of the equivalence relation E.

The isomorphism problems for every type of object studied in this book, including Linear Code Equivalence, are in $\mathbf{NP}$. Thus, in a formal study it is natural to assume that E is $\mathbf{NP}$-decidable.

Theorem 11.71. $\mathbf{P} = \mathbf{NP}$ *if and only if all classification problems* $\lambda_{R,E}$ *defined by a polynomially balanced,* $\mathbf{P}$*-decidable relation* R *and an* $\mathbf{NP}$*-decidable equivalence relation* E *admit a polynomial-time algorithm.*

Proof. Since a listing problem is a special case of a classification problem where $E = \{(y, y) : y \in \Sigma^*\}$, the "if" direction follows from Theorem 11.69.

For the "only if" direction, let us start with the backtrack algorithm developed in the proof of Theorem 11.69. We will modify the algorithm to list with polynomial delay only canonical representatives of the equivalence classes induced by E on $D_R(x)$. For a given $x \in \Sigma^*$, call a string $y \in \Sigma^*$ *canonical* if it is the lexicographic minimum of its equivalence class induced by E on $D_R(x)$; similarly, call a string $y \in \Sigma^*$ *extendable* if there exists a $w \in \Sigma^*$ such that yw is canonical. We show that given (x, y) both canonicity and extendability can be tested using polynomial-time algorithms, assuming $\mathbf{P} = \mathbf{NP}$. The backtrack algorithm then uses the canonicity algorithm to decide when to output a partial solution, and the extendability algorithm to prune the search.

First, observe that given (x, y), deciding $y \in D_R(x)$ is in $\mathbf{NP}$ because of the assumptions on R. Second, because E is $\mathbf{NP}$-decidable, we have that given

(x, y), deciding whether there exists a $z \in D_R(x)$ lexicographically less than y such that $(y, z) \in E$ is in $\mathbf{\Sigma}_2$. Thus, deciding canonicity is in $\mathbf{\Pi}_2$, which places deciding extendability into $\mathbf{\Sigma}_3$. Because of the assumption $\mathbf{P} = \mathbf{NP}$ we have $\mathbf{P} = \mathbf{\Delta}_i = \mathbf{\Sigma}_i = \mathbf{\Pi}_i$ for all $i = 1, 2, \ldots$, implying that canonicity and extendability admit polynomial time algorithms. □

Consequently, for any classification problem in the assumed setting, a proof of nonexistence of a polynomial-time algorithm immediately gives us a proof that $\mathbf{P} \neq \mathbf{NP}$. Short of settling the $\mathbf{P} \neq \mathbf{NP}$ question, we are thus left with essentially two possible directions to proceed in a complexity-theoretic study of classification problems. One direction is to attempt to obtain indirect evidence of computational difficulty by tools such as reductions and completeness. Unfortunately, this appears tedious at best because of the restricted nature of classification problems for specific families of objects. For illustration, it is difficult to see how the ability to obtain a complete classification of $(n, M, d)_q$ codes with any given parameters in one time step would significantly assist in solving any other computational problem than UNRESTRICTED ERROR-CORRECTING CODES or a strictly related problem. Thus, this direction does not look too promising at least if a direct analysis of problems such as t-DESIGNS, UNRESTRICTED ERROR-CORRECTING CODES,, and LINEAR ERROR-CORRECTING CODES is attempted. Allowing for richer instances – for example, asking for a classification of BIBDs that extend a given partial BIBD – leads to complexity results as in Sect. 11.2, but it is unclear whether these results tell us anything about the complexity of the original classification problem.

The other direction is to attempt to develop provably efficient classification algorithms (in particular, polynomial-time algorithms) for increasingly challenging families of combinatorial objects. Arguably this is a more realistic goal and has more application potential. Nevertheless, complexity theory certainly brings forth the possibility that there are natural classification problems that do not admit an efficient algorithm.

We conclude this chapter by a brief discussion of efficient listing and classification algorithms. In either case we do not claim that an exhaustive survey is provided. Generally recommended references include [211, 342, 516, 608].

First, it should be observed that polynomial-space listing is typically achievable by a straightforward backtrack algorithm (cf. the proof of Theorem 11.69), with no guarantees on the running time. Also polynomial-space classification is in most cases possible by relying on a space-efficient isomorph rejection strategy such as orderly generation or generation by canonical augmentation.

Polynomial-time algorithms or better are known for many listing problems, but general techniques for obtaining such algorithms are more sporadic. We proceed to discuss some techniques.

Backtrack search provides a universal but generally inefficient listing technique. However, in certain cases there exists a polynomial-time algorithm A

for deciding whether a given partial solution can be completed into an object to be listed. With appropriate assumptions on the size of the objects and the structure of the backtrack search, such algorithm A employed as a bounding function transforms the backtrack search into a polynomial-delay listing algorithm. One problem with frequent combinatorial applications – cf. Sect. 11.2 – that can be efficiently solved in this setting is as follows.

Problem 11.72. (SYSTEMS OF DISTINCT REPRESENTATIVES) Given a collection $S_0, S_1, \ldots, S_{s-1}$ of finite sets, list all of its systems of distinct representatives.

Given a partial system of distinct representatives $A = \{a_{i_1}, a_{i_2}, \ldots, a_{i_u}\}$, $a_{i_\ell} \in S_{i_\ell}$, the problem of deciding whether A can be extended to a system of distinct representatives for $S_0, S_1, \ldots, S_{s-1}$ is the same as deciding whether the collection consisting of the sets $S_i \setminus A$ for all $i \notin \{i_1, i_2, \ldots, i_u\}$ has a system of distinct representatives. The existence of a system of distinct representatives can be decided in polynomial time (see [331]); thus it is not difficult to show that the listing problem is solvable with polynomial delay. Tailored listing algorithms for systems of distinct representatives appear in [199, 589, 590].

A well-known consequence of SYSTEMS OF DISTINCT REPRESENTATIVES admitting a polynomial-delay algorithm is that labeled Latin squares of side n can be listed with delay polynomial in n. Indeed, the problem of finding all extensions of a given $k \times n$ Latin rectangle L to a $(k+1) \times n$ Latin rectangle essentially asks for the systems of distinct representatives for the collection $S_0, S_1, \ldots, S_{n-1}$, where $S_j = Z_n \setminus \{L(i,j) : i \in Z_k\}$. Because this collection always admits a system of distinct representatives for $k < n$, a polynomial-delay listing algorithm is obtained.

Another general listing technique is to view the objects in the set to be listed as the vertices of an implicit, connected *configuration graph* G, where two vertices are adjacent if the corresponding objects are obtainable from each other by an easily computable "local" transformation. (Of course this requires that we understand the objects well enough to define such a graph in a manner that is accessible to computation.) Examples can be found in [15, 516, 608]; even labeled Latin squares can be studied in this setting [279].

Because a configuration graph G is by assumption connected, a listing algorithm can be obtained by starting at a vertex, and systematically following the edges to visit all the vertices. Because G is typically a very large graph, the challenge in terms of space-efficiency is how to visit all the vertices without keeping a record of the vertices visited. A space-efficient traversal technique for a configuration graph that is applicable in many contexts is the technique of *reverse search* [15] (see also [442]). Reverse search typically leads to polynomial-time or polynomial-delay listing algorithms.

Constant-delay (respectively, CAT) listing algorithms are known for surprisingly many combinatorial listing problems. Clearly, the defining property for such an algorithm is that the next object can be computed, with the help of appropriate auxiliary data structures, from the current object in constant (amortized) time. Thus, it is necessary that successive objects differ only by a very small amount (on average), which makes such listing algorithms closely connected with configuration graphs – the output of a constant-delay (CAT) listing algorithm can be viewed as tracing a Hamiltonian path in an appropriate configuration graph. Listing with prescribed, typically small, differences between successive objects has been extensively studied under the name *combinatorial Gray codes* [286], see [516, 608] for excellent surveys. To mention one nontrivial example of constant-delay listing, it is possible to list the linear extensions of a given partial order with constant delay [93, 330] (see also [68, 493]). A technique for obtaining CAT listing algorithms is the ECO method [23, 25, 26, 144].

Turning now to classification, the two main challenges in efficient classification are that we must have a means for producing candidate representatives for the equivalence classes, and we must have a means for dealing with the equivalence problem, which may not be at all straightforward in light of Sect. 11.3.

We restrict the consideration to problems where the equivalence relation on each set of objects to be classified can be viewed as being induced by the orbits of a group action – this is clearly the case for the classification problems in this book. Furthermore, without giving formally precise definitions (cf. [212] and [528, Chap. 2]), we assume that the group action is computable in time polynomial in the instance size: this includes group element recognition, taking of products and inverses, and computing the action of a group element on an object.

Apparently with only one exception to be discussed shortly, polynomial-time classification algorithms are known only in cases where polynomial-time algorithms are known for equivalence computations. Polynomial-time classification algorithms can be divided roughly into two types based on the group action inducing equivalence.

First, we have classification problems where the acting group inducing equivalence has order at most polynomial in the instance size, whereby equivalence computations can be achieved by exhaustive search on the acting group, and the classification problem essentially reduces to the underlying listing problem. For problems of this type, the main challenge is to improve upon the obvious approach of listing followed by equivalence testing; for example, by developing CAT or constant-delay algorithms. Classification problems for necklaces and bracelets are excellent examples of problems of this type, see [97, 192, 492, 512, 517, 588].

Second, we have classification problems where the group action inducing equivalence has superpolynomial order, but where both the equivalence classes are well understood and where polynomial-time algorithms are known for equivalence computations. In the context of graphs, classifying the trees of given (unary) order n up to isomorphism provides an excellent example of a problem of this type: not only is a polynomial-time algorithm possible, but also CAT [44, 613] and constant-delay [437] algorithms are known.

The exceptional case where a polynomial-delay, polynomial-space classification algorithm is known despite the unresolved status of the equivalence problem occurs with classifying the graphs of given (unary) order n up to isomorphism [210, 211]. In essence, the classification algorithm relies on the fact that for the majority of graphs equivalence computations are polynomial-time. By interleaving the execution of a polynomial-delay classification algorithm for "easy" graphs with an algorithm for "hard" graphs, an overall polynomial-delay algorithm is obtained because the "hard" graphs can be handled in total time that is bounded by the number of isomorphism classes of "easy" graphs multiplied by a polynomial in n. Both component algorithms essentially rely on generation by canonical augmentation.

This positive result brings forth the possibility that polynomial-time classification could be possible in contexts where the labeled objects are reasonably well understood and – with appropriate tools – equivalence computations are most likely polynomial-time for the majority of equivalence classes. In the context of this book, Latin squares are perhaps the primary such candidate for further study (cf. Research Problem 11.63 and the preceding discussion).

Research Problem 11.73. With the aim of obtaining a polynomial-time classification algorithm, study the classification problem for main classes of Latin squares of given (unary) side n.

Another classification problem whose study could provide fruitful is the problem of listing orbit representatives in a Pólya theory setting; see [212]. Special cases of this problem include the classification problems for necklaces and bracelets mentioned earlier, and the classification problem for graphs of given order.

12

Nonexistence of Projective Planes of Order 10

There is not a more appropriate conclusion of a book on classification methods for designs than a discussion of the nonexistence of projective planes of order 10. Not only is this one of the main achievements on this topic – probably, *the* main – but it also illustrates how a variety of combinatorial results and ideas combined with extensive computations lead to a solution of this problem. Interestingly, codes, the other main type of objects considered in this book, play a central role in the approach.

We start off in Sect. 12.1 by recalling the main parameters of a putative projective plane of order 10. Codes related to designs are introduced in Sect. 12.2, and several basic properties of the code related to a projective plane of order 10 are proved. Different parts of the main search and its final resolution (reported in [351]; see [344] for an excellent and entertaining account of the various stages of this achievement) are discussed in the subsections of Sect. 12.3. Another survey of some parts of the search can be found in [244, Chap. 17], which goes deeper into the coding-theoretical issues.

12.1 Projective Planes of Order 10

By Example 2.36 – or the definitions in Sect. 2.2.5 – a projective plane of order 10 is a 2-$(111, 11, 1)$ design. From (2.3) we then obtain $v = b = 111$ and $k = r = 11$, and, since the design is square, any two rows or columns of an incidence matrix have exactly one common 1. When discussing such designs – in particular, when this is done in a geometric context – the blocks are often called lines, cf. Sect. 2.2.1. In the subsequent discussion we follow our convention of letting rows correspond to points and columns to blocks (which should be remembered when consulting the literature, where these correspondences are sometimes transposed).

By Example 2.36 we know that projective planes exist for orders that are prime powers. For orders $n \equiv 2 \pmod 4$, which cannot be prime powers when $n > 2$, the Bruck–Ryser–Chowla theorem (Theorem 2.47) proves nonexistence whenever n cannot be written as $n = a^2 + b^2$ for two integers a and b. This shows nonexistence for order 6, and leaves order 10 the first open case.

If orderly generation point by point is attempted for the classification of these designs – as described in Sect. 6.1.1 – one soon realizes that the task is hopeless. The first 21 rows of the incidence matrix are unique up to isomorphism, but the number of nonisomorphic structures grows rapidly after that. In fact, having completed 31 rows, we have a classification of the main classes of Latin squares of order 10 (cf. Sect. 8.1.3). See Figure 12.1, where the rows and columns are divided into sets of ten (except for the first row and column). Knowing [415] that there are 34,817,397,894,749,939 main classes of Latin squares of order 10, it is clear that this strategy is impracticable. (But it turns out that if this approach would be taken, one would probably not get even near the 111 rows corresponding to a design. In fact, comparison with the approach in Sect. 8.3 shows that $1 + 10i$ completed rows correspond to an $\mathrm{OA}(i, 10)$, and an $\mathrm{OA}(4, 10)$ is known but no $\mathrm{OA}(5, 10)$, so the height of the search tree is possibly just under 50.)

1	1111111111	0000000000	$\cdots$	0000000000
1	0000000000	1111111111		0000000000
1	0000000000	0000000000		0000000000
$\vdots$	$\vdots$	$\vdots$	$\ddots$	$\vdots$
1	0000000000	0000000000		0000000000
1	0000000000	0000000000		1111111111
0	1000000000	1000000000		1000000000
0	1000000000	0100000000		0100000000
$\vdots$	$\vdots$	$\ddots$	$\cdots$	$\ddots$
0	1000000000	0000000010		0000000010
0	1000000000	0000000001		0000000001
0	0100000000			
0	0100000000			
$\vdots$	$\vdots$	$\mathbf{M}_1$	$\cdots$	$\mathbf{M}_{10}$
0	0100000000			
0	0100000000			

Fig. 12.1. Partial canonical incidence matrix of a projective plane of order 10

An alternative approach for building up a projective plane is discussed in Example 6.21. That approach – which was utilized in the classification of projective planes of orders $n = 8$ and $n = 9$ reported in [249] and [346], respectively – starts from seeds that are obtained from the main classes of Latin squares of order $n - 1$ (rather then n). See Fig. 6.6. However, for a classification of projective planes of order 10, this still means that as many as

19,270,853,541 seeds should be considered. Further ideas are therefore required to make the search feasible. The idea that eventually led to the resolution of this problem is to study the linear codes generated by the blocks (or points) of a design.

Before starting the discussion of the nonexistence proof, we remark that the hope of Lam et al. [351] for an independent verification of this result has at the moment of writing not yet been fulfilled.

12.2 Codes of Designs

Earlier in this text we have discussed various links between codes and designs. Here, yet another connection will be presented. Any set of words in a given space generates a code; these generating words can be taken, for example, from the columns or the rows of an incidence matrix of a design.

Definition 12.1. *Given a field $\mathbb{F}_q$ and an incidence matrix of a design, the* code of the design *is the* linear *code over $\mathbb{F}_q$ generated by the columns or by the rows of the incidence matrix. These two types of codes are called the* code of blocks, *or just* code*; and the* code of points, *or* point code, *respectively.*

We talk about *the* code as isomorphic designs lead to equivalent codes. The *codes of blocks* are sometimes called *block codes*, but by Definition 2.72 that term has another meaning in coding theory. General results on connections between codes and designs can be found in [13, 580].

The search for projective planes of order 10 and a report by Assmus and Mattson [14] have been strong catalysts for further study of codes of designs, and have inspired many later results of utilizing codes in classification and construction of various types of designs [67, 559]. In particular, such an approach has played a central role in the still ongoing work on classifying (conjecturally, proving nonexistence of) 2-$(22, 8, 4)$ designs, see [29, 48, 246, 457] and the survey article [503].

We now turn our discussion to projective planes of order 10 (for an analogous treatment of the much simpler case of projective planes of order 4, see [175]). In the sequel, we consider only binary codes and set $q = 2$. Since we are dealing with parameters of a square design, any of the two types of codes of a design may be considered; the code of blocks is considered in this work. Hence, we consider the binary code generated by the columns of an incidence matrix $\mathbf{N}$ of a putative projective plane of order 10, and we refer to this set of generating words by W and to the entire code by C. The code obtained by extending C with a parity check bit is denoted by $\widehat{C}$. The extended generators, denoted by $\widehat{W}$, all have weight 12 and a 1 in the new coordinate.

Before proving some basic properties of C and $\widehat{C}$ we remark that the term *intersecting* – which is well-defined when considering blocks in the framework of set systems – is used quite broadly in the sequel to express common 1s also among columns, rows, and codewords.

Theorem 12.2. *The code $\widehat{C}$ is self-dual.*

Proof. The 111 words $\mathbf{w} \in \widehat{W} \subseteq \mathbb{F}_2^{112}$ all have weight 12 and intersect pairwise in two coordinates so the inner product of any two (not necessarily distinct) words in W is 0. Hence all generators of $\widehat{C}$ are in $\widehat{C}^{\perp}$, so $\widehat{C} \subseteq \widehat{C}^{\perp}$, that is, $\widehat{C}$ is self-orthogonal. As $\dim(\widehat{C}) + \dim(\widehat{C}^{\perp}) = 112$, in order to prove that $\widehat{C}$ is self-dual, it suffices to prove that $\dim(\widehat{C}) = \dim(\widehat{C}^{\perp}) = 112/2 = 56$. Since $\dim(C) = \dim(\widehat{C})$, we can study the code C.

Consider an incidence matrix $\mathbf{N}$ of a projective plane. Then (for the moment carrying out operations over the integers)

$$(\det(\mathbf{N}))^2 = \det(\mathbf{N}^T\mathbf{N}) = \det(10\mathbf{I} + \mathbf{J}) = 11^2 \times 10^{110}$$

by (2.6). Hence $|\det(\mathbf{N})| = 11 \times 10^{55}$. On the other hand, by elementary row and column operations – interchanging two rows; adding a multiple of a row to another row; and corresponding operations for columns – it is possible to transform the matrix into a diagonal integer matrix with the same absolute value of the determinant. It is not difficult to see that a transformation into a diagonal matrix is possible in this way, but to see that we can get a matrix with integer entries is a little bit more trickier; a proof can be found in [354, Appendix C] and many books on matrix algebra.

From the diagonal matrix we get the absolute value of the determinant directly. Since

$$|\det(\mathbf{N})| = \prod_{i=1}^{111} d_i = 11 \times 5^{55} \times 2^{55},$$

at least $111 - 55 = 56$ of the d_i in the diagonal $d_1, d_2, \ldots, d_{111}$ must be odd. But this means that if we carry out the elementary row and column operations in $\mathbb{F}_2$, we get a diagonal with at least 56 1s, and thereby $\dim(C) \geq 56$. □

The weaker result that $\widehat{C}$ is self-orthogonal is proved in [389]. Actually, this weaker result suffices in many places where Theorem 12.2 is used in the sequel. A proof of Theorem 12.2 is hard to find in the literature since many authors of early papers refer to manuscripts that were never published. Graham and MacWilliams [218] proved the result in the mid-1960s for *Desarguesian planes*, a class of projective planes arising from finite fields. A complete proof can be found in a report by Assmus and Mattson [14], and an alternative proof, different from the one given here, was presented by Carter [96, Propostion 2.1]. Later proofs (with more general results) occur in [244, 354]. These sources also give accounts of the history of this result; in particular, see [354, p. 77].

It can now be shown that $\widehat{C}$ is doubly-even.

Theorem 12.3. *For all* $\mathbf{c} \in \widehat{C}$, $\mathrm{wt}(\mathbf{c}) \equiv 0 \pmod 4$.

Proof. The weight of all words in $\widehat{W}$ is 12 and therefore divisible by 4. To see that all codewords generated by the words in $\widehat{W}$ have weights divisible by 4, the formula

$$\mathrm{wt}(\mathbf{c}+\mathbf{c}') = \mathrm{wt}(\mathbf{c}) + \mathrm{wt}(\mathbf{c}') - 2\sum_{i=1}^{112} c_i c'_i$$

is applied repeatedly. The results follows as $\sum_{i=1}^{112} c_i c'_i \equiv 0 \pmod 2$ by Theorem 12.2. □

Corollary 12.4. *For all* $\mathbf{c} \in C$, $\mathrm{wt}(\mathbf{c}) \equiv 0 \pmod 4$ *or* $\mathrm{wt}(\mathbf{c}) \equiv 3 \pmod 4$.

Theorem 12.5. *For any word* $\mathbf{w} \in W$ *and any word* $\mathbf{c} \in C$, $\mathbf{w} \cdot \mathbf{c} \equiv \mathrm{wt}(\mathbf{c}) \pmod 2$.

Proof. Given $\mathbf{w} \in W$ and $\mathbf{c} \in C$, for the extended words we have by Theorem 12.2 that $\widehat{\mathbf{w}} \cdot \widehat{\mathbf{c}} \equiv 0 \pmod 2$. If $\mathrm{wt}(\mathbf{c})$ is odd we get that $\mathbf{w} \cdot \mathbf{c} \equiv \widehat{\mathbf{w}} \cdot \widehat{\mathbf{c}} + 1 \equiv 1 \pmod 2$, and if $\mathrm{wt}(\mathbf{c})$ is even we get that $\mathbf{w} \cdot \mathbf{c} \equiv \widehat{\mathbf{w}} \cdot \widehat{\mathbf{c}} \equiv 0 \pmod 2$. □

We will now prove that the words in W are of minimum weight in C, and, moreover, that there are no other minimum-weight codewords.

Theorem 12.6. *For all codewords* $\mathbf{c} \in C \setminus \{\mathbf{0}\}$, *we have* $\mathrm{wt}(\mathbf{c}) \geq 11$.

Proof. For any codeword $\mathbf{c} \neq \mathbf{1}$ of odd weight, select a coordinate i with $c_i = 0$. There are 11 words in W that have a 1 in this coordinate, and all these codewords intersect $\mathbf{c}$ by Theorem 12.5, and in different positions. Therefore $\mathrm{wt}(\mathbf{c}) \geq 11$. For a codeword $\mathbf{c} \neq \mathbf{0}$ of even weight, select a coordinate i with $c_i = 1$ and again look at the 11 words in W that have a 1 in this coordinate. By Theorem 12.5 these must intersect $\mathbf{c}$ in at least one more coordinate and we get that $\mathrm{wt}(\mathbf{c}) \geq 12$. □

Theorem 12.7. *The weight* $\mathrm{wt}(\mathbf{c}) = 11$ *exactly when* $\mathbf{c} \in W$.

Proof. Assume that we have a codeword $\mathbf{c} \notin W$ with $\mathrm{wt}(\mathbf{c}) = 11$. Select any two distinct coordinates, i and j, for which $c_i = c_j = 1$. Now there is exactly one word $\mathbf{w} \in W$ that intersects $\mathbf{c}$ in these positions and by Theorem 12.5 $\mathbf{w}$ must intersect $\mathbf{c}$ in at least one more coordinate. Moreover, as $\mathbf{w} \neq \mathbf{c}$, there is a coordinate k for which $w_k = 1$ and $c_k = 0$. Now we focus on the other ten words in W that have a 1 in coordinate k. These must intersect $\mathbf{c}$ in different coordinates, which all differ from the at least three coordinates where $\mathbf{w}$ and $\mathbf{c}$ intersect. Hence $\mathrm{wt}(\mathbf{c}) \geq 13$, a contradiction. □

At this stage we already have a fair amount of information about the weight enumerator of C: $A_0 = 1$, $A_i = 0$ for $1 \leq i \leq 10$, $A_{11} = 111$, and $A_i = 0$ for $i \equiv 1, 2 \pmod 4$. The first four unknown values are A_{12}, A_{15}, A_{16}, and A_{19}. The following result regarding the weight enumerator of C, which is a part of [391, Theorem 1], is of central importance – this is the only place where we need the stronger result of self-duality rather than self-orthogonality in Theorem 12.2.

Theorem 12.8. *For $n \equiv 7 \pmod 8$, let C be an $[n,(n-1)/2,d]_2$ doubly-even self-orthogonal code. Then, the weight enumerator of C has the form*

$$W(x,y) = f_7(x,y)F(f_8(x,y), f_{24}(x,y)) + f_{23}(x,y)G(f_8(x,y), f_{24}(x,y)),$$

where

$$\begin{aligned} f_7(x,y) &= x^7 + 7x^3y^4, \\ f_8(x,y) &= x^8 + 14x^4y^4 + y^8, \\ f_{23}(x,y) &= x^{23} + 506x^{15}y^8 + 1288x^{11}y^{12} + 253x^7y^{16}, \\ f_{24}(x,y) &= x^4y^4(x^4 - y^4)^4, \end{aligned}$$

and $F(f,g)$, $G(f,g)$ are polynomials in f and g.

If we shorten a $[112,56,12]_2$ doubly-even self-dual code by deleting any coordinate and taking the codewords that have a 0 in the deleted coordinate, then we get a $[111,55,12]_2$ doubly-even self-orthogonal code, and Theorem 12.8 can be applied. In other words, we may take the linear code of length 111 generated by the words in W to get codewords with $\mathrm{wt}(\mathbf{c}) \equiv 0,3 \pmod 4$. These two weight classes modulo 4 divide the code into two subcodes of equal size and the doubly-even codewords form a linear subcode. By looking at the degrees of the polynomials, we get that the weight enumerator has the form

$$\begin{aligned} W = {} & \alpha_0 f_7 f_8^{13} + \alpha_1 f_7 f_8^{10} f_{24} + \alpha_2 f_7 f_8^{7} f_{24}^{2} + \alpha_3 f_7 f_8^{4} f_{24}^{3} + \\ & \alpha_4 f_7 f_8 f_{24}^{4} + \alpha_5 f_{23} f_8^{11} + \alpha_6 f_{23} f_8^{8} f_{24} + \alpha_7 f_{23} f_8^{5} f_{24}^{2} + \\ & \alpha_8 f_{23} f_8^{2} f_{24}^{3}. \end{aligned} \tag{12.1}$$

To obtain the complete weight enumerator, we require the nine coefficients α_i. These can be determined if we know the number A_i of codewords of weight i for nine values $i \equiv 0 \pmod 4$. So far we have $A_0 = 1$, $A_3 = A_{108} = 0$ (this directly gives $\alpha_0 = 0$), $A_4 = 0$, $A_7 = A_{104} = 0$, $A_8 = 0$, and $A_{11} = A_{100} = 111$, so by determining three more values of A_i, we can calculate the weight enumerator.

Actually, the weight enumerator is here only a device for demonstrating nonexistence. Indeed, one may view the subsequent proof as a calculation of values of A_i – one obtains $A_i = 0$ by proving that the code of a projective plane cannot have codewords of weight i – until the system of equations given by (12.1) for the unknowns A_i does not have any solutions in nonnegative integers.

In the next section, nonexistence proofs for various weights of codewords are considered. By determining the values of A_{12}, A_{15}, A_{16}, and A_{19}, we would get an overdetermined system of equations, but it turns out that it is not necessary to prove $A_i = 0$ for all these four values. If we look at the weight enumerator when $A_{19} = 0$ and $A_i \neq 0$ for exactly one $i \in \{12, 15, 16\}$, then for the choices $i = 12, 16$ the weight enumerator gets negative coefficients, and for the choice $i = 15$ the weight enumerator gets (many) noninteger coefficients. We can actually get an even stronger result. If we set $A_{15} = A_{19} = 0$ and solve (12.1), we get that

$$A_{12} = \alpha_8 + \frac{91054}{7},$$
$$A_{16} = -\frac{141}{7}\alpha_8 - \frac{13011339}{49},$$

which has no solution that fulfills $A_{12}, A_{16} \geq 0$. This result can be written in a more comprehensive form by writing A_{19} as a function of A_{12} and A_{16} when $A_{15} = 0$, cf. [96, 243]:

$$A_{19} = 141A_{12} + 7A_{16} + 24675.$$

12.3 The Main Search

As we have seen, it is not necessary to eliminate all four cases for codewords of weight 12, 15, 16, and 19, but one may restrict the search to weights 15 and 19. On the other hand, for weight 19 it turns out that we arrive at two subcases that address existence of codewords of weight 12 and 16, respectively. In the current approach, we are therefore bound to consider all four cases in a complete proof.

12.3.1 There are No Codewords of Weight 12

The subcase with codewords of weight 12 is fairly straightforward and was settled by Lam et al. in [352]; see [348] for some initial results. The approach taken here is from [348, 352]; some of the theoretical results are originally due to others, cf. [389].

Theorem 12.9. *A codeword* $\mathbf{c} \in C$ *with* $\mathrm{wt}(\mathbf{c}) = 12$ *intersects the words in* W *in* 0 *or* 2 *coordinates.*

Proof. Take a codeword $\mathbf{c} \in C$ of weight 12, and a word $\mathbf{w} \in W$ that intersects $\mathbf{c}$ – and by Theorem 12.5 does this in an even number s of coordinates. Assume that $s \geq 4$, and consider one of these coordinates. The other ten words in W that intersect $\mathbf{c}$ in this coordinate must intersect $\mathbf{c}$ in at least one more coordinate, leading to $\mathrm{wt}(\mathbf{c}) \geq 4 + 10 = 14$, a contradiction. □

To a geometer, a set of $n + 2$ (here, 12) points of a projective plane of order n no three of which are collinear (occur in the same block) is an *oval*.

Let us now look at the structure of an incidence matrix of a putative projective plane of order 10 whose code of blocks contains a codeword of weight 12. By the result of Theorem 12.9 we know up to isomorphism the part of an incidence matrix leading to the 1s in a weight-12 codeword. We let these be the first 12 rows of the incidence matrix as shown in Fig. 12.2. The first $\binom{12}{2} = 66$ columns contain a column for each possible pair of points, and the rest of the columns contain only 0s.

111111111110000000000		0000000000	0000		0000
100000000001111111111		0000000000	0000		0000
010000000001000000000		0000000000	0000		0000
001000000000100000000		0000000000	0000		0000
000100000000010000000		0000000000	0000		0000
000010000000001000000	…	0000000000	0000	…	0000
000001000000000100000		0000000000	0000		0000
000000100000000010000		1111000000	0000		0000
000000010000000001000		1000111000	0000		0000
000000001000000000100		0100100110	0000		0000
000000000100000000010		0010010101	0000		0000
000000000010000000001		0001001011	0000		0000

Fig. 12.2. Incidence matrix part corresponding to a weight-12 codeword

We will now try to complete this incidence matrix to an incidence matrix of a projective plane by first completing the first 66 blocks in a row-by-row manner. We denote the set of $111 - 12 = 99$ such rows (of length 66) by R.

Lemma 12.10. *Every row in R has six* 1*s.*

Proof. A given row in R and any of the first 12 rows intersect in exactly one column. Every 1 in a row in R contributes to two such intersections, so there must be $12/2 = 6$ 1s in the row. □

The nonzero columns of the first 12 rows can be viewed as an incidence matrix of the complete graph K_{12}, where the rows are vertices and the columns are edges. Obviously, the six columns (edges) corresponding to the 1s of a row in R must not have common 1s, and they therefore define a 1-factor of K_{12}. This observation yields a natural alternative formulation of the completion problem.

Different rows in R correspond to different 1-factors. Two blocks (columns) in a projective plane intersect in exactly one point (row), and for the 1-factors this corresponds to the requirement that every pair of disjoint edges must occur in exactly one 1-factor. Such combinatorial objects are known as hyperfactorizations and have been studied in [57, 86, 268, 288].

Definition 12.11. *A* λ-hyperfactorization *of* K_{2n} *is a collection of* 1*-factors in which each pair of disjoint edges appears in exactly* λ *of the* 1*-factors.*

The value of λ is called the *index* of the hyperfactorization. Cameron [86] places these objects into a larger family.

Definition 12.12. *An* s*-*(t, n) *partition system is a collection of partitions of an* n*-set* X *into* t*-subsets having the property that, given any* s *pairwise disjoint* t*-subsets of* X*, there is a unique partition containing these.*

A 1-hyperfactorization of K_{2n} is hence a 2-$(2, n)$ partition system. A 1-hyperfactorization of K_{2n} is also known as a pg$(n, 2n-3, n-2)$ *partial geometry* [398]. It follows by direct counting arguments that a λ-hyperfactorization of K_{2n} contains $\lambda(2n-1)(2n-3)$ 1-factors, and each edge is in $\lambda(2n-3)$ 1-factors; for $\lambda = 1$, $2n = 12$ (our case), these values are 99 and 9, respectively.

For $\lambda = 1$, hyperfactorizations are known to exist for $2n = 2^a + 2$, see [86], and can be constructed from certain projective planes of order 2^a. This result takes care of the cases $2n = 4, 6, 10, 18, \ldots$. There are unique 1-hyperfactorizations of K_4 and K_6 (trivial, see also [86, 398]). For K_8 and K_{12}, no 1-hyperfactorizations exist (proved in [86, 398] and [352], respectively). The latter result is what we need here; a short discussion of its proof is given below. The minimum index of a hyperfactorization of K_8 and K_{12} is 2 (from [268]) and at most 15 (from [288]), respectively. Observe that the existence of λ-hyperfactorizations and λ'-hyperfactorizations of K_{2n} implies the existence of $(\lambda + \lambda')$-hyperfactorizations of K_{2n}.

As mentioned in Sect. 8.1.1, a 1-factorization of K_{2n} can be viewed as an n-ary equidistant code. To illustrate the 1-factors in a hyperfactorization, we use here the same coding-theoretic formulation. With 1-factors in the columns, we obtain (up to equivalence) the 12×99 matrix in Fig. 12.3. Note that the empty sections are not filled with 0s, but with values from $\{1, 2, 3, 4, 5\}$.

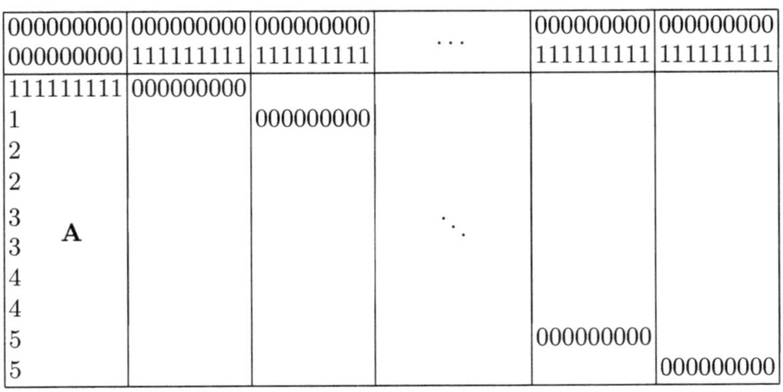

Fig. 12.3. Form of 1-hyperfactorization of K_{12}

As every pair of disjoint edges should be covered exactly once using 1-factors, we have an instance of EXACT COVERS. One obvious approach for the whole search is then to first determine, up to equivalence, all possible 10×9 submatrices indicated by **A** in Fig. 12.3, and use an algorithm for EXACT COVERS for completing the seeds. This is apparently the type of algorithm used in [348, 352], although the discussion in [348] alludes to an algorithm for CLIQUES.

Observe that the inequivalent completions of **A** (viewed as codewords of length 9) are in a bijective correspondence with the nonisomorphic 1-factorizations of K_{10}. We have earlier considered a classification algorithm for 1-factorizations in Sect. 8.1.1; by Table 8.1 there are 396 nonisomorphic 1-factorizations of K_{10}.

By the computer search in [352] there is no 1-hyperfactorization of K_{12}, so we have the following theorem.

Theorem 12.13. *The code of a projective plane of order* 10 *has no codewords of weight* 12.

Research Problem 12.14. Classify the 1-hyperfactorizations of K_{10}; at least two are known [398].

Research Problem 12.15. Whenever hyperfactorizations do not exist, one may try to determine the maximum number of 1-factors such that each pair of disjoint edges is contained in *at most* λ 1-factors. Alternatively, one may try to determine the minimum number of 1-factors such that each pair of disjoint edges is contained in *at least* λ 1-factors. Solve these problems for $2n = 8$ and $2n = 12$ ($\lambda = 1$); the former case is likely tractable, but the latter seems very hard.

12.3.2 There are No Codewords of Weight 15

The first complete proof of nonexistence of codewords of weight 15 was published by MacWilliams et al. [389]. However, as pointed out in [75], the core results of this proof – which are computational – were obtained earlier by Denniston [150] in another context.

We consider the points that correspond to a codeword of weight 15. The subsequent results are due to MacWilliams et al. [389].

Theorem 12.16. *A codeword* $\mathbf{c} \in C$ *with* $\mathrm{wt}(\mathbf{c}) = 15$ *intersects the words in* W *in* 1, 3, *or* 5 *coordinates.*

Proof. By Theorem 12.5, a codeword $\mathbf{c} \in C$ of weight 15 intersects a word $\mathbf{w} \in W$ in an odd number of coordinates. If $\mathbf{c}$ and $\mathbf{w}$ intersect in at least 9 coordinates, then $\mathrm{wt}(\mathbf{c} + \mathbf{w}) \leq 8$, which is impossible. If they intersect in 7 coordinates, then $\mathrm{wt}(\mathbf{c} + \mathbf{w}) = 12$, which is excluded by Theorem 12.13. Alternatively, the latter case follows from the observation that $\mathbf{c}+\mathbf{w}$ intersects $\mathbf{w}$ in $11 - 7 = 4$ coordinates, which is excluded by Theorem 12.9. □

For a codeword $\mathbf{c}$ of weight 15, we denote the number of words $\mathbf{w} \in W$ that intersect $\mathbf{c}$ in exactly i coordinates by b_i. The combinatorial structure in these 15 points is a PBD$(15, \{1,3,5\})$, and we now get a system of equations in b_i – termed B-equations in [29] – by counting the blocks, the occurrences of the points in the blocks, and the pairs that the blocks cover:

$$\begin{aligned} b_1 + b_3 + b_5 &= 111, \\ b_1 + 3b_3 + 5b_5 &= 165, \\ 3b_3 + 10b_5 &= 105. \end{aligned}$$

The unique solution to this system is

$$b_1 = 90, \quad b_3 = 15, \quad b_5 = 6.$$

To classify the PBD$(15, \{1,3,5\})$, we start by considering the blocks of weight 5.

Lemma 12.17. *All six blocks of weight* 5 *in a* PBD$(15, \{1,3,5\})$ *with* 111 *blocks intersect pairwise in exactly one point, and every point occurs in exactly two such blocks.*

Proof. View the blocks as codewords of length 15, weight 5, and minimum distance 8. The result follows from Theorem 2.88 (the second Johnson bound) and the subsequent remarks. □

Lemma 12.18. *In a* PBD$(15, \{1,3,5\})$ *with* 111 *blocks, every point occurs in exactly three blocks of weight* 3.

Proof. Every point occurs together with 8 other points in the blocks of weight 5. It has to occur together with the final $15 - 8 - 1 = 6$ points in the blocks of weight 3, so every point must occur in three such blocks. □

Using these lemmata, the PBD$(15, \{1,3,5\})$ with 111 blocks can be classified even by hand calculation. It turns out that there is a unique set of blocks of weight 5, up to isomorphism; moreover, there is a unique choice for the blocks of weight 3 given the blocks of weight 5. The unique PBD$(15, \{1,3,5\})$ with 111 blocks is presented in Fig. 12.4. In the sequel, we refer to the fifteen points of this structure by $D = \{1, 2, \ldots, 15\}$.

It is possible to make some further observations about how these blocks can be completed to a putative projective plane of order 10. The blocks that contain five points in the partial incidence matrix are pairwise intersecting in this part, so they are pairwise nonintersecting in the rest of the points. These $6 \times 6 = 36$ points are denoted by $E = \{16, 17, \ldots, 51\}$. All blocks of weight 3 in the partial incidence matrix intersect the first six blocks in that part and the rest of their points is therefore among the (60) points $F = \{52, 53, \ldots, 111\}$.

Lemma 12.19. *Every point in* F *occurs in exactly two blocks that intersect* D *in three points.*

110000	111000000000000	111111000000		000000000000
101000	000111000000000	000000111111		000000000000
100100	000000111000000	000000000000		000000000000
100010	000000000111000	000000000000		000000000000
100001	000000000000111	000000000000		000000000000
011000	000000100100100	000000000000		000000000000
010100	000100000010010	000000000000		000000000000
010010	000010010000001	000000000000	···	000000000000
010001	000001001001000	000000000000		000000000000
001100	100000000001001	000000000000		000000000000
001010	010000001000010	000000000000		000000000000
001001	001000010010000	000000000000		000000000000
000110	001001000000100	000000000000		000000000000
000101	010010000100000	000000000000		111111000000
000011	100100100000000	000000000000		000000111111

Fig. 12.4. Incidence matrix part corresponding to a weight-15 codeword

Proof. For a given point $p \in F$, let x and y denote the number of blocks that contain p and intersect D in three points and one point, respectively. We then obtain

$$x + y = 11,$$
$$3x + y = 15,$$

with the unique solution $x = 2$, $y = 9$. □

Lemma 12.19 further implies that the blocks that intersect D in three points can be completed in a unique way – any pair of such blocks that is nonintersecting in the D part intersects in the F part. We now know the contents of 15 rows and 21 columns of the incidence matrix.

Lemma 12.20. *A block that intersects D in one point intersects E in exactly four points and F in exactly six points.*

Proof. A block that intersects D in one point intersects two of the first six blocks in that point (Lemma 12.17). Moreover, the intersection of that block with the other four of the first six blocks is in E, so it intersects E in 4 points. Obviously, the block intersects F in $11 - 1 - 4 = 6$ points. □

At this stage, we have exhausted the known combinatorial arguments and start using computing power. The results from here on are due to Denniston [150] and MacWilliams et al. [389].

The search proceeds by completing the remaining 90 blocks. These blocks are considered in sets of six, where all blocks in a set intersect D in point i, $1 \leq i \leq 15$. Starting with the blocks that intersect D in point 1, it turns out [150, 389] that there are 344 choices for such a block if we ignore its four points in E. To find all six blocks in the set, we find all solutions of an instance of

Exact Covers. This is yet another example where an instance of Cliques can also be solved – that framework is used in [389] – but see the comments in Sect. 5.2.

After isomorph rejection, 1021 solutions for the first six blocks is reported in [389]. One may continue in this way, finding one set of blocks at a time. In [150] three sets (18 blocks) are computed directly starting from all nonisomorphic pairs of distinct blocks that intersect D in point 1; thereafter combinatorial arguments (for example, related to symmetry) are applied to reduce the number of solutions to 40. The order in which these sets are considered affects the computing time; in the searches in [150] and [389], the sets are considered in the following orders, listing the point of D that they contain: 1, 10, 15, 5; and 1, 10, 15, 11, 14, respectively.

Even if the points in E are ignored, properties of these can be used to restrict the search space. In this manner, [389] reports that the five aforementioned sets cannot be completed. The approach in [150] ignores the points in E completely and terminates without a completion of the four indicated sets. The discrepancy between the number of sets completed is due to the combinatorial arguments applied in [150] after three completed sets and the order of the sets considered. Since these searches took only hours in the late 1960s and early 1970s, modern computers should be able to carry out the search with little effort.

Theorem 12.21. *The code of a projective plane of order* 10 *has no codewords of weight* 15.

Research Problem 12.22. Repeat the described search. Instead of proceeding sets by sets, try to use Exact Covers for the whole final step (possibly starting after the first set is completed and isomorph rejection has been carried out). The objects to cover are pairs of distinct points in F and pairs of one point in D and one in F.

12.3.3 There are No Codewords of Weight 16

The weight-16 case was settled by results of Carter [96] and Lam et al. [349]. Our treatment follows those sources and [243]. Some partial results were obtained by Prince [490]. By now the reader should have a feeling for the vein of this case as well.

Theorem 12.23. *A codeword* $\mathbf{c} \in C$ *with* $\mathrm{wt}(\mathbf{c}) = 16$ *intersects the words in* W *in* 0, 2, *or* 4 *coordinates.*

Proof. By Theorem 12.5 a codeword $\mathbf{c} \in C$ of weight 16 intersects a word $\mathbf{w} \in W$ in an even number of coordinates. If $\mathbf{c}$ and $\mathbf{w}$ intersect in at least 6 coordinates, then the weight $\mathrm{wt}(\mathbf{c}+\mathbf{w})$ is odd and at most 15, so by previous results we must have $\mathrm{wt}(\mathbf{c}+\mathbf{w}) = 11$, and $\mathbf{c}$ and $\mathbf{w}$ intersect in 8 coordinates. But then $\mathbf{w}$ and $\mathbf{c}+\mathbf{w} \in W$ intersect in 3 coordinates, which is impossible. □

For a codeword **c** of weight 16, the combinatorial structure in these points is a PBD(16, {2, 4}) with at most 111 blocks. Denoting the number of blocks that intersect **c** in exactly i coordinates by b_i, we obtain the following B-equations:

$$\begin{aligned} b_0 + \ b_2 + \ b_4 &= 111, \\ 2b_2 + 4b_4 &= 176, \\ b_2 + 6b_4 &= 120. \end{aligned}$$

This system has the unique solution

$$b_0 = 31, \ b_2 = 72, \ b_4 = 8.$$

Lemma 12.24. *Every point occurs in exactly two of the eight blocks of weight 4 in a* PBD(16, {2, 4})) *with* 111 *blocks.*

Proof. For a given point p, let x and y denote the number of blocks that contain p and have weight 2 and 4, respectively. The point p occurs in 11 blocks and in 15 pairs of distinct points, so $x + y = 11$ and $x + 3y = 15$. The unique solution is $x = 9$, $y = 2$. □

If we take a set of eight blocks of weight 4, there is only one way of completing this set to a PBD(16, {2, 4}) by adding blocks of weight 2. The problem of classifying the particular PBD(16, {2, 4}) is therefore essentially a problem of classifying sets of eight blocks of weight 4 fulfilling Lemma 12.24, that is, equireplicate constant weight error-correcting codes. Carter [96] showed that there are six nonisomorphic such objects, which are presented in Fig. 12.5. These can be obtained by slightly modifying either an algorithm for classifying BIBDs (Sect. 6.1.1; $v = 8$, $k = 2$, $b = 16$, and $r = 4$ with inner products of rows of the incidence matrix at most 1) or an algorithm for classifying constant weight error-correcting codes (Sect. 7.1.3; an equireplicate $(16, 8, 6)_2$ code with constant weight 4).

The numbering of the cases in Fig. 12.5 follows that of Carter [96] (to which [349] refers), but note that a different order is used in [243]. The correspondence between the cases considered is shown in Table 12.1. Carter [96] eliminated Cases II to V and Case VI in part. Lam et al. [349] finally settled the remaining cases.

Table 12.1. Correspondence between cases considered by Carter and Hall

Carter	Hall
I	I
II	II
III	VI
IV	III
V	V
VI	IV

Case I	Case II	Case III	Case IV	Case V	Case VI
11000000	11000000	11000000	11000000	11000000	11000000
10100000	10100000	10100000	10100000	10100000	10100000
10010000	10010000	10010000	10010000	10010000	10010000
10001000	10001000	10001000	10001000	10001000	10001000
01000100	01100000	01100000	01100000	01100000	01100000
01000010	01010000	01010000	01010000	01010000	01010000
01000001	01000100	01000100	01000100	01001000	01000100
00100100	00110000	00101000	00100010	00100100	00100010
00100010	00100010	00100010	00100001	00100010	00100001
00100001	00010001	00010100	00010010	00010010	00010100
00010100	00001100	00010001	00010001	00010001	00010001
00010010	00001010	00001010	00001100	00001100	00001100
00010001	00001001	00001001	00001010	00001001	00001010
00001100	00000110	00000110	00001001	00000110	00001001
00001010	00000101	00000101	00000110	00000101	00000110
00001001	00000011	00000011	00000101	00000011	00000011

Fig. 12.5. Incidence matrix parts corresponding to a weight-16 codeword

We will now take a closer look at Cases I to VI. For reasons of space we omit some of the details (which are essential for the best possible performance of an implementation); the interested reader is referred to [96, 349].

After fixing the 16×8 matrix $\mathbf{M}_{16}$ for one of the Cases I to VI, the incidence matrix to be completed can be divided into parts as shown in Fig. 12.6.

$\mathbf{M}_{16}$	$\mathbf{A}_1$	$\mathbf{0}$
$\mathbf{A}_2$	$\mathbf{A}_3$	$\mathbf{A}_4$
$\mathbf{0}$	$\mathbf{A}_5$	$\mathbf{A}_6$

Fig. 12.6. Structure of incidence matrix with weight-16 codeword

Given $\mathbf{M}_{16}$, the submatrix $\mathbf{A}_1$ is a uniquely determined 16×72 matrix with two 1s in each column. As the earlier calculations show, there are $111-8-72 = 31$ empty columns in the first 16 rows of the incidence matrix. To get a better understanding of the process of adding more rows to the incidence matrix of a putative projective plane, one may view the columns of $\mathbf{M}_{16}$ as vertices of a graph G with its rows (with exactly two 1s) determining the edges of G. This graph is obviously 4-regular. Actually, as the complement of such a graph is 3-regular, it is more convenient to focus on the complement graph instead.

The six complement graphs are depicted in Fig. 12.7; the reader is encouraged to find mappings between the columns in Fig. 12.5 and the vertices of the graphs. From these graphs, it is also easy to determine the automorphisms of the 16×111 submatrix, which can be used to speed up the search.

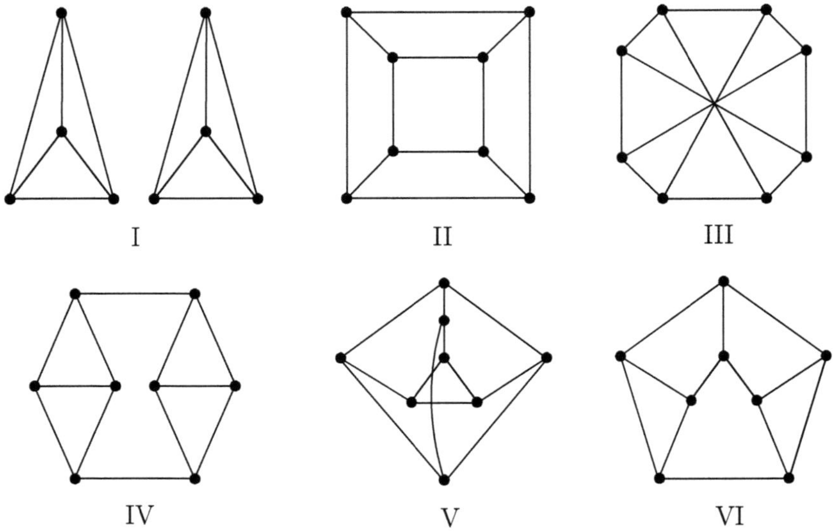

Fig. 12.7. Complement graphs of six incidence structures

In the submatrix $\mathbf{A}_2$, the 1s in a row determine a clique in the graph $\bar{G}$, and the set of all rows in $\mathbf{A}_2$ form a partition of $\bar{G}$ into cliques. The graphs II and III have no cliques of size greater than 2, so $\mathbf{A}_2$ is uniquely determined in those cases. For graphs I, IV, V, one may conclude that a row in $\mathbf{A}_2$ corresponding to a clique of size 3 cannot be completed (in the $\mathbf{A}_3$ part); see [96, Proposition 4.1] for details (this result is due to Berlekamp). Consequently, $\mathbf{A}_2$ is uniquely determined also for the graphs IV and V.

We discuss the search for a completion of the incidence matrix in Case I (one may proceed analogously in the other cases). In this case there are three nonisomorphic completions of $\mathbf{A}_2$, so the case can be divided into three subcases, called the *Two Distinguished Points* case, the *One Distinguished Point* case, and the *No Distinguished Point* case. These cases are shown in Fig. 12.8, in this order. We only list the rows with at least two 1s. It is important to observe that the columns have been ordered to emphasize the correspondence between these and the vertices of graph I in Fig. 12.7; a different order is required when starting from the submatrix of Fig. 12.5 in the setting of Fig. 12.6.

The rows with four 1s (the distinguished points) cannot have 1s in the submatrix $\mathbf{A}_3$, so they are ignored. In all cases, the search proceeds by completing one row at a time, occasionally disregarding the parts $\mathbf{A}_4$ and $\mathbf{A}_6$. The rows are completed in sets, corresponding to the 1s of one of the first eight columns. The automorphism group of the initial 16×111 matrix is utilized in isomorph rejection, for example, after the first set is ready. To find the rows, Cliques or Exact Covers can be used.

11110000
00001111

11110000
00001100
00001010
00001001
00000110
00000101
00000011

11000000
10100000
10010000
01100000
01010000
00110000
00001100
00001010
00001001
00000110
00000101
00000011

Fig. 12.8. The three subcases of Case I

Research Problem 12.25. Nontrivial heuristics are used in [349] in the search for completions. Is it possible to use EXACT COVERS – starting after the first set is completed and isomorph rejection is carried out – in a more straightforward way?

In Case VI there are also three subcases, depending on whether the clique partition of $\bar{G}$ contains 0, 1, or 2 cliques of size 3 (the corresponding points are called P and Q in [243]). In [96] the case with 0 cliques was eliminated, and Case VI was finally settled in [349].

Theorem 12.26. *The code of a projective plane of order* 10 *has no codewords of weight* 16.

Prince [491] recognized that the existence of codewords of weight 16 implies existence of Steiner triple systems of order 19 with three restrictive properties. This could possibly lead to an alternative proof of this case. The following open problem includes just one of these properties; the classified objects can be further tested with respect to the additional two properties; cf. [491].

Research Problem 12.27. Consider an STS(19) and delete one point and the blocks containing that point. Partition the remaining 48 blocks into 8 parallel classes, and arrange the blocks into two 6×4 arrays with one parallel class in each column and with disjoint blocks in each row. Classify all such objects, if any.

12.3.4 There are No Codewords of Weight 19

A proof of nonexistence of projective planes of order 10 is essentially a proof that its code cannot have codewords of weight 19. Whereas the results presented so far can be obtained without too much effort – computational (at least in the 21st century) or combinatorial – the computations related to the weight-19 case are challenging even for modern computers. This is the main reason why an independent verification of this celebrated proof has not yet, at the moment of writing, been carried out.

The result that the code of a projective plane of order 10 cannot have codewords of weight 19 was announced by Lam et al. [351] in the late 1980s, and was obtained using an extensive amount of CPU time on the best available computers (CRAY1A). Obviously, the approach for the search was also polished – with many preliminary experiments – and the published results are indispensable for anyone interested in attacking this problem.

As usual, we commence by studying possible intersection patterns for a codeword of given weight and the blocks of a projective plane.

Theorem 12.28. *A codeword* $\mathbf{c} \in C$ *with* $\mathrm{wt}(\mathbf{c}) = 19$ *intersects the codewords in* W *in* 1, 3, *or* 5 *coordinates.*

Proof. By Theorem 12.5 a codeword $\mathbf{c} \in C$ of weight 19 intersects a word $\mathbf{w} \in W$ in an odd number of coordinates. If $\mathbf{c}$ and $\mathbf{w}$ intersect in at least 7 coordinates, then the weight $\mathrm{wt}(\mathbf{c} + \mathbf{w})$ is even and at most 16, but we know that no such codewords exists. □

For a codeword $\mathbf{c}$ of weight 19, the combinatorial structure in these 19 points is then a $\mathrm{PBD}(19, \{1, 3, 5\})$ with 111 blocks, and the corresponding B-equations are

$$\begin{aligned} b_1 + b_3 + b_5 &= 111,\\ b_1 + 3b_3 + 5b_5 &= 209,\\ 3b_3 + 10b_5 &= 171. \end{aligned}$$

The unique solution of this system is

$$b_1 = 68, \quad b_3 = 37, \quad b_5 = 6.$$

It turns out that there is a very large number of $\mathrm{PBD}(19, \{1, 3, 5\})$ with 111 blocks. In classifying these, one may start from all possible sets of six blocks of weight 5; there are 66 nonisomorphic such sets, which can be obtained as $(19, 6, 8)_2$ codes of constant weight 5. We list these objects in Figs. 12.9 to 12.13; the numbering follows that used by Lam et al. [345, 351]. As many as 21 of these are eliminated in [345] by various arguments, some of which will now be considered.

Lemma 12.29. *If three of the six blocks intersecting a weight-*19 *codeword in five points are pairwise nonintersecting within those* 19 *points, then there is a codeword of weight* 16.

Proof. The three blocks intersect outside the points of the weight-19 codeword. They cannot, however, intersect in one common point, since then the 1s of that row would occur together with at least $3 \times 5 + 8 \times 1 = 23$ (> 19) 1s of the first 19 rows. Adding the three blocks and the weight-19 codeword, we get a codeword of weight 16. □

Case 1	Case 2	Case 3	Case 4	Case 5	Case 6	Case 7
111100	111100	111100	111100	111000	111000	111000
100011	100010	100010	100010	100110	100110	100110
100000	100001	100001	100001	100001	100001	100001
100000	100000	100000	100000	100000	100000	100000
100000	100000	100000	100000	100000	100000	100000
010010	010010	010010	010010	010101	010101	010101
010001	010001	010001	010001	010010	010010	010010
010000	010000	010000	010000	010000	010000	010000
010000	010000	010000	010000	010000	010000	010000
001010	001010	001010	001010	001100	001100	001100
001001	001001	001001	001001	001011	001010	001010
001000	001000	001000	001000	001000	001001	001001
001000	001000	001000	001000	001000	001000	001000
000110	000110	000110	000110	000100	000100	000100
000101	000101	000101	000100	000100	000100	000100
000100	000100	000100	000100	000010	000011	000010
000100	000100	000100	000100	000010	000010	000010
000010	000011	000010	000011	000001	000001	000001
000001	000000	000001	000001	000001	000000	000001

Case 8	Case 9	Case 10	Case 11	Case 12	Case 13	Case 14
111000	111000	111000	111000	111000	111000	111000
100110	100110	100110	100110	100110	100110	100110
100001	100001	100001	100001	100000	100000	100000
100000	100000	100000	100000	100000	100000	100000
100000	100000	100000	100000	100000	100000	100000
010100	010100	010100	010100	010101	010100	010100
010010	010010	010010	010010	010010	010010	010010
010001	010001	010001	010001	010000	010001	010001
010000	010000	010000	010000	010000	010000	010000
001100	001100	001100	001100	001100	001100	001100
001010	001010	001010	001010	001010	001010	001010
001001	001001	001001	001000	001001	001001	001001
001000	001000	001000	001000	001000	001000	001000
000101	000101	000100	000101	000100	000101	000101
000100	000100	000100	000100	000100	000100	000100
000011	000010	000010	000010	000011	000011	000010
000010	000010	000010	000010	000010	000010	000010
000000	000001	000001	000001	000001	000001	000001
000000	000000	000001	000001	000001	000000	000001

Fig. 12.9. Seeds of six blocks of weight 5

111000	111000	111000	111000	111000	111000	111000
100110	100110	100110	100110	100110	100100	100100
100001	100001	100001	100000	100001	100010	100010
100000	100000	100000	100000	100000	100001	100001
100000	100000	100000	100000	100000	100000	100000
010100	010100	010100	010100	010100	010100	010100
010010	010010	010010	010010	010001	010010	010010
010001	010001	010001	010001	010000	010001	010001
010000	010000	010000	010000	010000	010000	010000
001100	001100	001100	001100	001100	001100	001100
001001	001001	001001	001001	001001	001010	001010
001000	001000	001000	001000	001000	001001	001001
001000	001000	001000	001000	001000	001000	001000
000101	000101	000100	000101	000101	000111	000110
000100	000100	000100	000100	000100	000100	000101
000011	000010	000011	000011	000011	000010	000011
000010	000010	000010	000010	000010	000001	000000
000010	000010	000010	000010	000010	000000	000000
000000	000001	000001	000001	000010	000000	000000
Case 15	Case 16	Case 17	Case 18	Case 19	Case 20	Case 21

111000	111000	111000	111000	111000	111000	111000
100100	100100	100100	100100	100100	100100	100100
100010	100010	100010	100010	100010	100010	100010
100001	100001	100001	100001	100001	100001	100001
100000	100000	100000	100000	100000	100000	100000
010100	010100	010100	010100	010100	010100	010100
010010	010010	010010	010010	010010	010010	010010
010001	010001	010001	010001	010001	010001	010000
010000	010000	010000	010000	010000	010000	010000
001100	001100	001100	001100	001100	001100	001100
001010	001010	001010	001010	001010	001010	001010
001001	001001	001000	001000	001000	001000	001000
001000	001000	001000	001000	001000	001000	001000
000110	000110	000111	000110	000110	000110	000111
000101	000100	000100	000101	000101	000100	000100
000010	000010	000010	000011	000010	000010	000010
000001	000001	000001	000001	000001	000001	000001
000000	000001	000001	000000	000001	000001	000001
000000	000000	000000	000000	000000	000001	000001
Case 22	Case 23	Case 24	Case 25	Case 26	Case 27	Case 28

Fig. 12.10. Seeds of six blocks of weight 5 (cont.)

111000	111000	111000	111000	111000	111000	111000
100100	100100	100100	100100	100100	100100	100100
100010	100010	100010	100010	100010	100010	100010
100001	100001	100000	100001	100001	100001	100001
100000	100000	100000	100000	100000	100000	100000
010100	010100	010100	010100	010100	010100	010100
010010	010010	010010	010010	010010	010010	010010
010000	010000	010000	010001	010001	010001	010000
010000	010000	010000	010000	010000	010000	010000
001100	001100	001100	001100	001100	001100	001100
001010	001010	001010	001010	001010	001010	001010
001000	001000	001000	001001	001000	001000	001000
001000	001000	001000	001000	001000	001000	001000
000110	000110	000110	000100	000101	000101	000101
000101	000101	000101	000100	000100	000100	000100
000011	000010	000011	000010	000011	000010	000011
000001	000001	000001	000010	000010	000010	000010
000001	000001	000001	000001	000001	000001	000001
000000	000001	000001	000001	000000	000001	000001
Case 29	Case 30	Case 31	Case 32	Case 33	Case 34	Case 35

111000	111000	111000	111000	111000	111000	111000
100100	100100	100100	100100	100100	100100	100100
100011	100010	100010	100010	100010	100010	100010
100000	100001	100001	100001	100001	100001	100001
100000	100000	100000	100000	100000	100000	100000
010100	010100	010100	010100	010100	010100	010100
010010	010010	010010	010010	010010	010010	010010
010000	010001	010001	010001	010000	010000	010000
010000	010000	010000	010000	010000	010000	010000
001100	001100	001100	001100	001100	001100	001100
001001	001000	001000	001000	001001	001001	001001
001000	001000	001000	001000	001000	001000	001000
001000	001000	001000	001000	001000	001000	001000
000110	000110	000110	000110	000111	000110	000110
000101	000101	000101	000100	000100	000101	000101
000010	000011	000010	000011	000010	000011	000010
000010	000010	000010	000010	000010	000010	000010
000001	000001	000001	000001	000001	000001	000001
000001	000000	000001	000001	000001	000000	000001
Case 36	Case 37	Case 38	Case 39	Case 40	Case 41	Case 42

Fig. 12.11. Seeds of six blocks of weight 5 (cont.)

Case 43	Case 44	Case 45	Case 46	Case 47	Case 48	Case 49
111000	111000	111000	111000	111000	111000	110000
100100	100100	100100	100100	100100	100100	101000
100010	100010	100010	100010	100010	100010	100100
100001	100001	100000	100001	100001	100000	100010
100000	100000	100000	100000	100000	100000	100001
010100	010100	010100	010100	010100	010100	011000
010010	010010	010010	010010	010010	010001	010100
010000	010000	010000	010001	010000	010000	010010
010000	010000	010000	010000	010000	010000	010001
001100	001100	001100	001000	001001	001010	001100
001001	001000	001001	001000	001000	001001	001010
001000	001000	001000	001000	001000	001000	001001
001000	001000	001000	001000	001000	001000	000110
000110	000110	000110	000110	000110	000110	000101
000100	000101	000101	000101	000101	000101	000011
000011	000011	000011	000100	000100	000100	000000
000010	000010	000010	000011	000011	000011	000000
000001	000001	000001	000010	000010	000010	000000
000001	000001	000001	000001	000001	000001	000000

Case 50	Case 51	Case 52	Case 53	Case 54	Case 55	Case 56
110000	110000	110000	110000	110000	110000	110000
101000	101000	101000	101000	101000	101000	101000
100100	100100	100100	100100	100100	100100	100100
100010	100010	100010	100010	100010	100010	100010
100001	100001	100001	100001	100001	100001	100001
011000	011000	011000	011000	011000	011000	011000
010100	010100	010100	010100	010100	010100	010100
010010	010010	010010	010010	010010	010010	010010
010001	010001	010001	010000	010001	010001	010001
001100	001100	001100	001100	001100	001100	001100
001010	001010	001010	001010	001010	001010	001010
001001	001001	001000	001000	001001	001000	001000
000110	000110	000110	000110	000100	000101	000101
000101	000100	000100	000100	000100	000100	000100
000010	000010	000010	000010	000010	000011	000010
000001	000001	000001	000001	000010	000010	000010
000000	000001	000001	000001	000001	000001	000001
000000	000000	000001	000001	000001	000000	000001
000000	000000	000000	000001	000000	000000	000000

Fig. 12.12. Seeds of six blocks of weight 5 (cont.)

Case 57	Case 58	Case 59	Case 60	Case 61	Case 62	Case 63
110000	110000	110000	110000	110000	110000	110000
101000	101000	101000	101000	101000	101000	101000
100100	100100	100100	100100	100100	100100	100100
100010	100010	100010	100010	100010	100010	100010
100001	100001	100001	100000	100001	100001	100000
011000	011000	011000	011000	011000	011000	011000
010100	010100	010100	010100	010100	010100	010100
010010	010010	010010	010010	010010	010010	010010
010001	010000	010000	010000	010001	010000	010000
001100	001100	001100	001100	001100	001100	001100
001010	001010	001010	001010	001000	001001	001001
001000	001000	001000	001000	001000	001000	001000
000100	000101	000101	000101	000100	000100	000101
000100	000100	000100	000100	000100	000100	000100
000010	000011	000010	000011	000011	000011	000011
000010	000010	000010	000010	000010	000010	000010
000001	000001	000001	000001	000010	000010	000010
000001	000001	000001	000001	000001	000001	000001
000001	000000	000001	000001	000001	000001	000001

Case 64	Case 65	Case 66
110000	110000	110000
101000	101000	101000
100100	100100	100100
100010	100010	100010
100001	100000	100000
011000	011000	011000
010100	010100	010100
010010	010001	010001
010000	010000	010000
001001	001010	001010
001000	001001	001001
001000	001000	001000
000101	000110	000110
000100	000101	000101
000100	000100	000100
000011	000011	000010
000010	000010	000010
000010	000001	000001
000001	000000	000001

Fig. 12.13. Seeds of six blocks of weight 5 (cont.)

Cases identified by Lemma 12.29 can now be eliminated because we know that there are no codewords of weight 16.

Lemma 12.30. *If three of the six blocks intersecting a weight-*19 *codeword in five points do not intersect a fourth block within those* 19 *points, and one of the* 19 *points does not occur in any of the six blocks, then there is no completion to an incidence matrix for a projective plane of order* 10.

Proof. If three blocks – say blocks number 2, 3, and 4 – do not intersect the first block in the first 19 points, then we may assume that they intersect the first block in points 20 to 22. (We indeed need three points since we may assume that blocks 2, 3, and 4 intersect pairwise in the first 19 points; otherwise Lemma 12.29 can be applied.) Since the 1s of rows 20 to 22 must occur in the same column as 19 1s of the first 19 rows, we get that the final nine 1s of rows 20 to 22 must occur in columns with one 1 in the first 19 rows.

The point that does not occur in any of the blocks with five 1s in the first 19 rows occurs in $(19-1)/2 = 9$ blocks with three 1s in those rows and in $11-9=2$ blocks with one 1 in those rows. But the two columns of the latter type should (by the first part of the proof) have 1s in rows 20 to 22, so one of them must have at least two 1s in those rows, which is impossible. □

Without further details it is mentioned in [345] that a few more cases can be eliminated by ad hoc arguments.

The remaining sets of six blocks are completed with a set of triples that cover the pairs of distinct points that are not contained in the initial blocks. The obvious way of doing this – cf. Sect. 6.1.3 – is by EXACT COVERS. Actually, for some sets of six blocks there is no completion. In [345] these instances are included in those eliminated by ad hoc arguments. However, as the computation time for processing these is negligible, we prefer not to treat them separately here.

Even after isomorph rejection, we have almost one million 19×111 matrices of the 19 rows corresponding to a codeword of weight 19; these are the starting points for the final effort. The exact number for the various cases is denoted by A2 in [351, Tables 1 and 2]. The subdivision of the incidence matrix in Fig. 12.14 is done in a manner analogous to that in Fig. 12.6. The matrices $\mathbf{M}_{19}$, $\mathbf{A}_1$, and $\mathbf{A}_2$ correspond to the parts of the first 19 rows where the number of 1s in the columns is 5, 3, and 1, respectively.

From each of the partial 19×111 incidence matrices, the search proceeds by completing one block at a time. Since by Lemma 12.29 there are no three blocks that are pairwise nonintersecting in $\mathbf{M}_{19}$, the completions of the first six blocks are unique up to isomorphism. These completions are in part $\mathbf{A}_3$ of Fig. 12.14.

$\mathbf{M}_{19}$	$\mathbf{A}_1$	$\mathbf{A}_2$
$\mathbf{A}_3$	$\mathbf{A}_4$	$\mathbf{A}_5$
$\mathbf{0}$	$\mathbf{A}_6$	

Fig. 12.14. Structure of incidence matrix with weight-19 codeword

For each point of a given block in the first six columns, we now complete those rows of the incidence matrix. This procedure is then continued for another block, and so on. For the first block we have $11-5=6$ rows to complete, but for subsequent choices of blocks this number may be smaller. If a row to be completed intersects x, y, and z of the blocks whose intersection with the first 19 rows has cardinality 5, 3, and 1, respectively, then

$$\begin{aligned} x + y + z &= 11, \\ 5x + 3y + z &= 19, \end{aligned}$$

with three solutions:

$$x=0,\ y=4,\ z=7; \quad x=1,\ y=2,\ z=8; \quad x=2,\ y=0,\ z=9.$$

For the rows to be constructed, this result gives the number of 1s in the parts $\mathbf{A}_4$ and $\mathbf{A}_5$, which is 2 and 8, or 0 and 9. With the latter distribution and when completing the first block, the 1s in $\mathbf{A}_5$ can be placed in a unique way up to isomorphism. We complete such rows first. Also with the former distribution the 1s in $\mathbf{A}_5$ are uniquely determined after the 1s in $\mathbf{A}_4$ have been determined. The 1s in $\mathbf{A}_4$ can be computed as follows.

Consider only the triples in the first 19 rows that do not intersect the first block in those 19 points. Denote the number of such triples by t. Take all nonintersecting pairs of these triples, and use Exact Covers to partition the t triples into $t/2$ such pairs.

From the second block onwards, we precompute the set of all rows compatible with earlier rows before completing the blocks. Then either Cliques or preferably Exact Covers can be used for finding the completions (the rows must intersect in exactly one position; the blocks that do not intersect the block to be completed in the previous points must do it in these points).

The number of partial incidence matrices for each case and each completed block is presented in [351, Tables 1 and 2]. In the computer search of [351], no completion of five blocks was encountered. It should be noted, however, that the order in which the blocks are completed has a big impact on the size of the search space. By [351] it is desirable to choose a block to complete next among such blocks that do not intersect the earlier completed blocks in the first 19 rows (and therefore must intersect them in the rows to be completed).

It is not possible to discuss all techniques that were used by Lam et al. [351] to speed up the search. As an example, consider two blocks that intersect outside the first 19 rows. Then the sum of these blocks and the weight-19 codeword is another codeword of weight 19. We can then check whether the partial incidence matrix in those 19 rows corresponds to a case that has already been considered.

Theorem 12.31. *The code of a projective plane of order* 10 *has no codewords of weight* 19.

The results of this chapter sum up to the following corollary – a worthy conclusion of this book.

Theorem 12.32. *There is no projective plane of order* 10.

References

[1] E. Agrell, A. Vardy, and K. Zeger, Upper bounds for constant-weight codes, *IEEE Trans. Inform. Theory* 46 (2000), 2373–2395.
[2] O. Aichholzer, Extremal properties of 0/1-polytopes of dimension 5, in *Polytopes – Combinatorics and Computation* (G. Kalai and G. M. Ziegler, Eds.), Birkhäuser, Basel, 2000, pp. 111–130.
[3] J. L. Allston, R. W. Buskens, and R. G. Stanton, An examination of the nonisomorphic solutions to a problem in covering designs on fifteen points, *J. Combin. Math. Combin. Comput.* 4 (1988), 189–206.
[4] W. O. Alltop, An infinite class of 4-designs, *J. Combin. Theory* 6 (1969), 320–322.
[5] W. O. Alltop, Extending t-designs, *J. Combin. Theory Ser. A* 18 (1975), 177–186.
[6] L. D. Andersen, Factorizations of graphs, in *The CRC Handbook of Combinatorial Designs* (C. J. Colbourn and J. H. Dinitz, Eds.), CRC Press, Boca Raton, Fla., 1996, pp. 653–667.
[7] I. Anderson, *Combinatorial Designs and Tournaments*, Oxford University Press, Oxford, 1997.
[8] O. Anglada and J.-F. Maurras, Another complete invariant for Steiner triple systems of order 15, *J. Combin. Des.* 13 (2005), 388–391.
[9] K. Appel and W. Haken, Every planar map is four colorable, *Bull. Amer. Math. Soc.* 82 (1976), 711–712.
[10] D. Applegate, E. M. Rains, and N. J. A. Sloane, On asymmetric coverings and covering numbers, *J. Combin. Des.* 11 (2003), 218–228.
[11] E. Arnold, *Äquivalenzklassen linearer Codes*, MSc Thesis, Universität Bayreuth, 1993.
[12] V. Arvind and J. Torán, Isomorphism testing: Perspective and open problems, *Bull. Eur. Assoc. Theor. Comput. Sci. EATCS* 86 (2005), 66–84.
[13] E. F. Assmus, Jr. and J. D. Key, *Designs and Their Codes*, Cambridge University Press, Cambridge, 1992.

[14] E. F. Assmus, Jr. and H. F. Mattson, Jr., Algebraic theory of codes II: Part II – On the possibility of a projective plane of order 10, Air Force Cambridge Research Laboratories Report AFCRL-71-0013, Sylvania Electronic Systems Group, Needham Heights, Mass., 1970.

[15] D. Avis and K. Fukuda, Reverse search for enumeration, *Discrete Appl. Math.* 65 (1996), 21–46.

[16] A. Avizienis, The n-version approach to fault tolerant software, *IEEE Trans. Software Engrg.* 11 (1985), 1491–1501.

[17] L. Babai, Moderately exponential bound for graph isomorphism, in *Fundamentals of Computation Theory* (F. Gécseg, Ed.), Springer-Verlag, Berlin, 1981, pp. 34–50.

[18] L. Babai, Automorphism groups, isomorphism, reconstruction, in *Handbook of Combinatorics* (R. L. Graham, M. Grötschel, and L. Lovász, Eds.), Vol. 2, Elsevier, Amsterdam, 1995, pp. 1447–1540.

[19] L. Babai and E. M. Luks, Canonical labeling of graphs, in *Proc. 15th ACM Symposium on Theory of Computing*, (Boston, Apr. 25–27, 1983), ACM Press, New York, 1983, pp. 171–183.

[20] L. Babai, E. M. Luks, and Á. Seress, Fast management of permutation groups. I., *SIAM J. Comput.* 26 (1997), 1310–1342.

[21] L. Babai and S. Moran, Arthur-Merlin games: A randomized proof system, and a hierarchy of complexity classes, *J. Comput. System Sci.* 36 (1988), 254–276.

[22] L. Babel, I. V. Chuvaeva, M. Klin, and D. V. Pasechnik, Algebraic combinatorics in mathematical chemistry. Methods and algorithms. II. Program implementation of the Weisfeiler-Leman algorithm, Technical Report TUM-M9701, Fakultät für Mathematik, Technische Universität München, 1997.

[23] S. Bacchelli, E. Barcucci, E. Grazzini, and E. Pergola, Exhaustive generation of combinatorial objects by ECO, *Acta Inform.* 40 (2004), 585–602.

[24] T. Baicheva and E. Kolev, Binary codes of length eight, minimum distance three and twenty codewords, in *Proc. 2nd International Workshop on Optimal Codes and Related Topics*, (Sozopol, Bulgaria, June 9–15, 1998), Institute of Mathematics and Informatics, Bulgarian Academy of Sciences, Sofia, 1998, pp. 5–8.

[25] E. Barcucci, A. Del Lungo, E. Pergola, and R. Pinzani, A methodology for plane tree enumeration, *Discrete Math.* 180 (1998), 45–64.

[26] E. Barcucci, A. Del Lungo, E. Pergola, and R. Pinzani, ECO: A methodology for the enumeration of combinatorial objects, *J. Differ. Equations Appl.* 5 (1999), 435–490.

[27] A. Barg, Complexity issues in coding theory, in *Handbook of Coding Theory* (V. S. Pless and W. C. Huffman, Eds.), Vol. 1, Elsevier, Amsterdam, 1998, pp. 649–754.

[28] J. A. Barrau, On the combinatory problem of Steiner, *Proceedings of the Section of Sciences. Koninklijke Akademie van Wetenschappen te Amsterdam* 11 (1908), 352–360.

[29] J. A. Bate, M. Hall, Jr., and G. H. J. van Rees, Structures within $(22, 33, 12, 8, 4)$-designs, *J. Combin. Math. Combin. Comput.* 4 (1988), 115–122.

[30] J. A. Bate and G. H. J. van Rees, Some results on $N(4, 6, 10)$, $N(4, 6, 11)$, and related coverings, *Congr. Numer.* 48 (1985), 25–45.

[31] R. A. Beezer, Counting configurations in designs, *J. Combin. Theory Ser. A* 96 (2001), 341–357.

[32] R. Bertolo, P. R. J. Östergård, and W. D. Weakley, An updated table of binary/ternary mixed covering codes, *J. Combin. Des.* 12 (2004), 157–176.

[33] H. U. Besche, B. Eick, and E. A. O'Brien, A millennium project: Constructing small groups, *Internat. J. Algebra Comput.* 12 (2002), 623–644.

[34] M. R. Best, Binary codes with a minimum distance of four, *IEEE Trans. Inform. Theory* 26 (1980), 738–742.

[35] M. R. Best, A. E. Brouwer, F. J. MacWilliams, A. M. Odlyzko, and N. J. A. Sloane, Bounds for binary codes of length less than 25, *IEEE Trans. Inform. Theory* 24 (1978), 81–93.

[36] T. Beth, D. Jungnickel, and H. Lenz, *Design Theory*, 2nd ed., 2 vols., Cambridge University Press, Cambridge, 1999.

[37] K. Betsumiya, T. A. Gulliver, and M. Harada, Binary optimal linear rate 1/2 codes, in *Applied Algebra, Algebraic Algorithms and Error-Correcting Codes* (M. Fossorier, H. Imai, S. Lin, and A. Poli, Eds.), Springer-Verlag, Berlin, 1999, pp. 462–471.

[38] A. Betten, *Gruppenaktionen auf Verbänden und die Konstruktion kombinatorischer Objekte*, MSc Thesis, Universität Bayreuth, 1995.

[39] A. Betten, R. Laue, S. Molodtsov, and A. Wassermann, Steiner systems with automorphism groups $\mathrm{PSL}(2, 71)$, $\mathrm{PSL}(2, 83)$, and $\mathrm{P\Sigma L}(2, 3^5)$, *J. Geom.* 67 (2000), 35–41.

[40] A. Betten, R. Laue, and A. Wassermann, Simple 7-designs with small parameters, *J. Combin. Des.* 7 (1999), 79–94.

[41] A. Betten, R. Laue, and A. Wassermann, DISCRETA: A tool for constructing t-designs, in *Computer Algebra Handbook: Foundations, Applications, Systems* (J. Grabmeier, E. Kaltofen, and V. Weispfenning, Eds.), Springer-Verlag, Berlin, 2003, pp. 372–375.

[42] A. Betten and A. Wassermann, {0,1}-solutions of integer linear equation systems, in *Parallel Virtual Machine – EuroPVM'96* (A. Bode, J. Dongarra, T. Ludwig, and V. Sunderam, Eds.), Springer-Verlag, Berlin, 1996, pp. 311–314.

[43] A. Beutelspacher, Classical geometries, in *The CRC Handbook of Combinatorial Designs* (C. J. Colbourn and J. H. Dinitz, Eds.), CRC Press, Boca Raton, Fla., 1996, pp. 694–708.

[44] T. Beyer and S. M. Hedetniemi, Constant time generation of rooted trees, *SIAM J. Comput.* 9 (1980), 706–712.

[45] V. N. Bhat, Non-isomorphic solutions of some balanced incomplete block designs. II, *J. Combin. Theory Ser. A* 12 (1972), 217–224.

[46] R. T. Bilous, Enumeration of the binary self-dual codes of length 34, submitted for publication.
[47] R. T. Bilous and G. H. J. van Rees, An enumeration of binary self-dual codes of length 32, *Des. Codes Cryptogr.* 26 (2002), 61–86.
[48] R. T. Bilous and G. H. J. van Rees, Self-dual codes and the $(22, 8, 4)$ balanced incomplete block design, *J. Combin. Des.* 13 (2005), 363–376.
[49] J. R. Bitner and E. M. Reingold, Backtrack programming techniques, *Comm. ACM* 18 (1975), 651–656.
[50] A. Blass and Y. Gurevich, Equivalence relations, invariants, and normal forms, *SIAM J. Comput.* 13 (1984), 682–689.
[51] U. Blass and S. Litsyn, Several new lower bounds on the size of codes with covering radius one, *IEEE Trans. Inform. Theory* 44 (1998), 1998–2002.
[52] A. Blokhuis and C. W. H. Lam, More coverings by rook domains, *J. Combin. Theory Ser. A* 36 (1984), 240–244.
[53] G. T. Bogdanova, A. E. Brouwer, S. N. Kapralov, and P. R. J. Östergård, Error-correcting codes over an alphabet of four elements, *Des. Codes Cryptogr.* 23 (2001), 333–342.
[54] B. Bollobás, *Modern Graph Theory*, Springer-Verlag, New York, 1998.
[55] J. A. Bondy, Basic graph theory: Paths and circuits, in *Handbook of Combinatorics* (R. L. Graham, M. Grötschel, and L. Lovász, Eds.), Vol. 1, Elsevier, Amsterdam, 1995, pp. 3–110.
[56] K. S. Booth and C. J. Colbourn, Problems polynomially equivalent to graph isomorphism, Technical Report CS-77/04, Department of Computer Science, University of Waterloo, 1979.
[57] E. Boros, D. Jungnickel, and S. A. Vanstone, The existence of nontrivial hyperfactorizations of K_{2n}, *Combinatorica* 11 (1991), 9–15.
[58] R. C. Bose, A note on the resolvability of balanced incomplete block designs, *Sankhyā* 6 (1942), 105–110.
[59] I. Bouyukliev, S. Bouyuklieva, T. A. Gulliver, and P. R. J. Östergård, Classification of optimal binary self-orthogonal codes, *J. Combin. Math. Combin. Comput.*, to appear.
[60] I. Bouyukliev and D. B. Jaffe, Optimal binary linear codes of dimension at most seven, *Discrete Math.* 226 (2001), 51–70.
[61] I. Bouyukliev and P. R. J. Östergård, Classification of self-orthogonal codes over $\mathbb{F}_3$ and $\mathbb{F}_4$, *SIAM J. Discrete Math.*, to appear.
[62] I. Bouyukliev and J. Simonis, Some new results for optimal ternary linear codes, *IEEE Trans. Inform. Theory* 48 (2002), 981–985.
[63] S. Bouyuklieva, Some optimal self-orthogonal and self-dual codes, *Discrete Math.* 287 (2004), 1–10.
[64] E. D. Boyer, D. L. Kreher, S. P. Radziszowski, and A. Sidorenko, On $(n, 5, 3)$-Turán systems, *Ars. Combin.* 37 (1994), 13–31.
[65] M. Braun, A. Kohnert, and A. Wassermann, Optimal linear codes from matrix groups, submitted for publication.

[66] D. R. Breach, The 2-$(9,4,3)$ and 3-$(10,5,3)$ designs, *J. Combin. Theory Ser. A* 27 (1979), 50–63.

[67] W. G. Bridges, M. Hall, Jr., and J. L. Hayden, Codes and designs, *J. Combin. Theory Ser. A* 31 (1981), 155–174.

[68] G. Brightwell and P. Winkler, Counting linear extensions, *Order* 8 (1991), 225–242.

[69] G. Brinkmann, Isomorphism rejection in structure generation programs, in *Discrete Mathematical Chemistry* (P. Hansen, P. Fowler, and M. Zheng, Eds.), Amer. Math. Soc., Providence, R.I., 2000, pp. 25–38.

[70] G. Brinkmann and B. D. McKay, Posets on up to 16 points, *Order* 19 (2002), 147–179.

[71] A. E. Brouwer, H. O. Hämäläinen, P. R. J. Östergård, and N. J. A. Sloane, Bounds on mixed binary/ternary codes, *IEEE Trans. Inform. Theory* 44 (1998), 140–160.

[72] A. E. Brouwer, J. B. Shearer, N. J. A. Sloane, and W. D. Smith, A new table of constant weight codes, *IEEE Trans. Inform. Theory* 36 (1990), 1334–1380.

[73] J. W. Brown, Enumeration of Latin squares with application to order 8, *J. Combin. Theory* 5 (1968), 177–184.

[74] R. H. Bruck and H. J. Ryser, The nonexistence of certain finite projective planes, *Canadian Journal of Mathematics* 1 (1949), 88–93.

[75] A. Bruen and J. C. Fisher, Blocking sets, k-arcs and nets of order ten, *Adv. Math.* 10 (1973), 317–320.

[76] G. Brunel, Sur les deux systèmes de triades de trieze éléments, *Journal de Mathématiques Pures et Appliquées. Cinquième Série* 7 (1901), 305–330.

[77] A. Brüngger, A. Marzetta, K. Fukuda, and J. Nievergelt, The parallel search bench ZRAM and its applications, *Ann. Oper. Res.* 90 (1999), 45–63.

[78] F. Buekenhout, Ed., *Handbook of Incidence Geometry*, North-Holland, Amsterdam, 1995.

[79] M. Buratti and F. Zuanni, The 1-rotational Kirkman triple systems of order 33, *J. Statist. Plann. Inference* 86 (2000), 369–377.

[80] W. Burau, Über die zur Kummerkonfiguration analogen Schemata von 16 Punkten und 16 Blöcken und ihre Gruppen, *Abh. Math. Sem. Univ. Hamburg* 26 (1963/1964), 129–144.

[81] S. Butenko, P. Pardalos, I. Sergienko, V. Shylo, and P. Stetsyuk, Estimating the size of correcting codes using extremal graph problems, in *Optimization: Structure and Applications* (E. Hunt and C. E. M. Pearce, Eds.), Kluwer, Dordrecht, the Netherlands, to appear.

[82] G. Butler, *Fundamental Algorithms for Permutation Groups*, Springer-Verlag, Berlin, 1991.

[83] G. Butler and C. W. H. Lam, A general backtrack algorithm for the isomorphism problem of combinatorial objects, *J. Symbolic Comput.* 1 (1985), 363–381.

[84] D. de Caen, D. L. Kreher, S. P. Radziszowski, and W. H. Mills, On the covering of t-sets with $(t+1)$-sets: $C(9,5,4)$ and $C(10,6,5)$, *Discrete Math.* 92 (1991), 65–77.

[85] J.-Y. Cai, M. Fürer, and N. Immerman, An optimal lower bound on the number of variables for graph identification, *Combinatorica* 12 (1992), 389–410.

[86] P. J. Cameron, *Parallelisms of Complete Designs*, Cambridge University Press, Cambridge, 1976.

[87] P. J. Cameron, *Combinatorics: Topics, Techniques, Algorithms*, Cambridge University Press, Cambridge, 1994.

[88] P. J. Cameron, *Permutation Groups*, Cambridge University Press, Cambridge, 1999.

[89] P. J. Cameron and J. H. van Lint, *Designs, Graphs, Codes and Their Links*, Cambridge University Press, Cambridge, 1991.

[90] P. J. Cameron, R. Solomon, and A. Turull, Chains of subgroups in symmetric groups, *J. Algebra* 127 (1989), 340–352.

[91] A. R. Camina, A survey of the automorphism groups of block designs, *J. Combin. Des.* 2 (1994), 79–100.

[92] P. Camion, Linear codes with given automorphism groups, *Discrete Math.* 3 (1972), 33–45.

[93] E. R. Canfield and S. G. Williamson, A loop-free algorithm for generating the linear extensions of a poset, *Order* 12 (1995), 57–75.

[94] W. A. Carnielli, Hyper-rook domain inequalities, *Stud. Appl. Math.* 82 (1990), 59–69.

[95] R. Carraghan and P. M. Pardalos, An exact algorithm for the maximum clique problem, *Oper. Res. Lett.* 9 (1990), 375–382.

[96] J. L. Carter, *On the Existence of a Projective Plane of Order Ten*, PhD Thesis, University of California, Berkeley, 1974.

[97] K. Cattell, F. Ruskey, J. Sawada, M. Serra, and C. R. Miers, Fast algorithms for generating necklaces, unlabeled necklaces, and irreducible polynomials over GF(2), *J. Algorithms* 37 (2000), 267–282.

[98] V. Ćepulić, On symmetric block designs $(40,13,4)$ with automorphisms of order 5, *Discrete Math.* 128 (1994), 45–60.

[99] V. Ćepulić, On symmetric block designs $(45,12,3)$ with automorphisms of order 5, *Ars. Combin.* 37 (1994), 33–48.

[100] V. Ćepulić, On biplanes $(56,11,2)$ with automorphism groups of order four, *Glas. Mat. Ser. III* 31(51) (1996), 201–207.

[101] V. Ćepulić, On symmetric block designs $(40,13,4)$ with automorphisms of order 13, *Glas. Mat. Ser. III* 31(51) (1996), 11–23.

[102] V. Ćepulić, Symmetric block designs $(61,16,4)$ admitting an automorphism of order 15, *Glas. Mat. Ser. III* 35(55) (2000), 233–244.

[103] V. Ćepulić and M. Essert, Biplanes $(56,11,2)$ with automorphism group $Z_2 \times Z_2$ fixing some point, *J. Combin. Theory Ser. A* 48 (1988), 239–246.

[104] V. Ćepulić and M. Essert, Biplanes $(56,11,2)$ with automorphisms of order 4 fixing some point, *Discrete Math.* 71 (1988), 9–17.

[105] V. Ćepulić and M. Essert, Biplanes and their automorphisms, *Studia Sci. Math. Hungar.* 24 (1989), 437–446.
[106] Y. Chen, The Steiner system $S(3,6,26)$, *J. Geom.* 2 (1972), 7–28.
[107] S. Chowla and H. J. Ryser, Combinatorial problems, *Canad. J. Math.* 2 (1950), 93–99.
[108] D. Clarke, S. Devadas, M. van Dijk, B. Gassend, and G. E. Suh, Incremental multiset hash functions and their application to memory integrity checking, Technical Report MIT-LCS-TR-899, Laboratory for Computer Science, Massachusetts Institute of Technology, Cambridge, 2003.
[109] G. Cohen, I. Honkala, S. Litsyn, and A. Lobstein, *Covering Codes*, Elsevier, Amsterdam, 1997.
[110] M. B. Cohen, C. J. Colbourn, L. A. Ives, and A. C. H. Ling, Kirkman triple systems of order 21 with nontrivial automorphism group, *Math. Comp.* 71 (2002), 873–881.
[111] C. J. Colbourn, Embedding partial Steiner triple systems is NP-complete, *J. Combin. Theory Ser. A* 35 (1983), 100–105.
[112] C. J. Colbourn, The complexity of completing partial Latin squares, *Discrete Appl. Math.* 8 (1984), 25–30.
[113] C. J. Colbourn and M. J. Colbourn, Isomorphism problems involving self-complementary graphs and tournaments, *Congr. Numer.* 22 (1978), 153–164.
[114] C. J. Colbourn and M. J. Colbourn, Combinatorial isomorphism problems involving 1-factorizations, *Ars. Combin.* 9 (1980), 191–200.
[115] C. J. Colbourn, M. J. Colbourn, J. J. Harms, and A. Rosa, A complete census of $(10,3,2)$ block designs and of Mendelsohn triple systems of order ten. III. $(10,3,2)$ block designs without repeated blocks, *Congr. Numer.* 37 (1983), 211–234.
[116] C. J. Colbourn and J. H. Dinitz, Eds., *The CRC Handbook of Combinatorial Designs*, CRC Press, Boca Raton, Fla., 1996.
[117] C. J. Colbourn and J. H. Dinitz, Latin squares, in *The CRC Handbook of Combinatorial Designs* (C. J. Colbourn and J. H. Dinitz, Eds.), CRC Press, Boca Raton, Fla., 1996, pp. 97–110.
[118] C. J. Colbourn and J. H. Dinitz, Applications of combinatorial designs to communications, cryptography, and networking, in *Surveys in Combinatorics, 1999* (J. D. Lamb and D. A. Preece, Eds.), Cambridge University Press, Cambridge, 1999, pp. 37–100.
[119] C. J. Colbourn and A. C. H. Ling, Kirkman triple systems of orders 27, 33 and 39, *J. Combin. Math. Combin. Comput.* 43 (2002), 3–8.
[120] C. J. Colbourn, S. S. Magliveras, and R. A. Mathon, Transitive Steiner and Kirkman triple systems of order 27, *Math. Comp.* 58 (1992), 441–449.
[121] C. J. Colbourn, S. S. Magliveras, and D. R. Stinson, Steiner triple systems of order 19 with nontrivial automorphism group, *Math. Comp.* 59 (1992), 283–295.

[122] C. J. Colbourn and P. C. van Oorschot, Applications of combinatorial designs in computer science, *ACM Computing Surv.* 21 (1989), 223–250.

[123] C. J. Colbourn and A. Rosa, *Triple Systems*, Clarendon Press, Oxford, 1999.

[124] M. J. Colbourn, An analysis technique for Steiner triple systems, *Congr. Numer.* 23 (1979), 289–303.

[125] M. J. Colbourn, Algorithmic aspects of combinatorial designs: A survey, *Ann. Discrete Math.* 26 (1985), 67–136.

[126] M. J. Colbourn and C. J. Colbourn, Concerning the complexity of deciding isomorphism of block designs, *Discrete Appl. Math.* 3 (1981), 155–162.

[127] M. J. Colbourn and R. A. Mathon, On cyclic Steiner 2-designs, *Ann. Discrete Math.* 7 (1980), 215–253.

[128] F. N. Cole, Kirkman parades, *Bulletin of the American Mathematical Society* 28 (1922), 435–437.

[129] F. N. Cole, L. D. Cummings, and H. S. White, The complete enumeration of triad systems in 15 elements, *Proceedings of the National Academy of Sciences of the United States of America* 3 (1917), 197–199.

[130] Condor Team, *Condor Version 6.4.7 Manual*, Computer Sciences Department, University of Wisconsin-Madison, 2003.

[131] I. Constantinescu and W. Heise, On the concept of code-isomorphy, *J. Geom.* 57 (1996), 63–69.

[132] J. H. Conway, A. Hulpke, and J. McKay, On transitive permutation groups, *LMS J. Comput. Math.* 1 (1998), 1–8.

[133] J. H. Conway and V. Pless, On the enumeration of self-dual codes, *J. Combin. Theory Ser. A* 28 (1980), 26–53.

[134] J. H. Conway, V. Pless, and N. J. A. Sloane, Self-dual codes over GF(3) and GF(4) of length not exceeding 16, *IEEE Trans. Inform. Theory* 25 (1979), 312–322.

[135] J. H. Conway, V. Pless, and N. J. A. Sloane, The binary self-dual codes of length up to 32: A revised enumeration, *J. Combin. Theory Ser. A* 60 (1992), 183–195.

[136] T. H. Cormen, C. E. Leiserson, R. L. Rivest, and C. Stein, *Introduction to Algorithms*, 2nd ed., MIT Press, Cambridge, 2001.

[137] D. G. Corneil, *Graph Isomorphism*, PhD Thesis, University of Toronto, 1968.

[138] D. G. Corneil and C. C. Gotlieb, An efficient algorithm for graph isomorphism, *J. Assoc. Comput. Mach.* 17 (1970), 51–64.

[139] H. S. M. Coxeter, Self-dual configurations and regular graphs, *Bull. Amer. Math. Soc.* 56 (1950), 413–455.

[140] D. Crnković, Symmetric $(70, 24, 8)$ designs having $\mathrm{Frob}_{21} \times Z_2$ as an automorphism group, *Glas. Mat. Ser. III* 34(54) (1999), 109–121.

[141] R. Davies and G. F. Royle, Graph domination, tabu search and the football pool problem, *Discrete Appl. Math.* 74 (1997), 217–228.

[142] A. Davydov, S. Marcugini, and F. Pambianco, Minimal 1-saturating sets and complete caps in binary projective geometries, in *Proc. 9th International Workshop on Algebraic and Combinatorial Coding Theory (ACCT'2004)*, (Kranevo, Bulgaria, June 19–25, 2004), 2004, pp. 113–119.

[143] V. De Pasquale, Sui sistemi ternari di 13 elementi, *Rendiconti. Reale Istituto Lombardo di Science e Lettere. Serie II.* 32 (1899), 213–221.

[144] A. Del Lungo, E. Duchi, A. Frosini, and S. Rinaldi, On the generation and enumeration of some classes of convex polyominoes, *Electron. J. Combin.* 11 (2004) no. 1, #R60, 46pp.

[145] P. Delsarte, Bounds for unrestricted codes, by linear programming, *Philips Res. Rep.* 27 (1972), 272–289.

[146] P. Delsarte, An algebraic approach to the association schemes of coding theory, *Philips Res. Rep. Suppl.*, no. 10, 1973.

[147] P. Delsarte and J.-M. Goethals, Unrestricted codes with the Golay parameters are unique, *Discrete Math.* 12 (1975), 211–224.

[148] P. Dembowski, Verallgemeinerungen von Transitivitätsklassen endlicher projektiver Ebenen, *Math. Z.* 69 (1958), 59–89.

[149] J. Dénes and A. D. Keedwell, *Latin Squares: New Developments in the Theory and Applications*, North-Holland, Amsterdam, 1991.

[150] R. H. F. Denniston, Non-existence of a certain projective plane, *J. Austral. Math. Soc.* 10 (1969), 214–218.

[151] R. H. F. Denniston, Enumeration of symmetric designs $(25, 9, 3)$, *Discrete Math.* 15 (1982), 111–127.

[152] P. C. Denny, Search and enumeration techniques for incidence structures, Research Report CDMTCS-085, Centre for Discrete Mathematics and Theoretical Computer Science, University of Auckland, 1998.

[153] P. C. Denny and P. B. Gibbons, Case studies and new results in combinatorial enumeration, *J. Combin. Des.* 8 (2000), 239–260.

[154] P. C. Denny and R. Mathon, A census of t-$(t+8, t+2, 4)$ designs, $2 \leq t \leq 4$, *J. Statist. Plann. Inference* 106 (2002), 5–19.

[155] M. Deza, Une propriété extrémale des plans projectifs finis dans une classe de codes équidistants, *Discrete Math.* 6 (1973), 343–352.

[156] M. Deza, Solution d'un problème de Erdős–Lovász, *J. Combin. Theory Ser. B* 16 (1974), 166–167.

[157] L. E. Dickson and F. H. Safford, Solution to problem 8 (group theory), *The American Mathematical Monthly* 13 (1906), 150–151.

[158] I. Diener, On cyclic Steiner systems $S(3, 4, 22)$, *Ann. Discrete Math.* 7 (1980), 301–313.

[159] R. Diestel, *Graph Theory*, 2nd ed., Springer-Verlag, New York, 2000.

[160] Y. Ding, S. Houghten, C. Lam, S. Smith, L. Thiel, and V. D. Tonchev, Quasi-symmetric 2-$(28, 12, 11)$ designs with an automorphism of order 7, *J. Combin. Des.* 6 (1998), 213–223.

[161] J. H. Dinitz, Starters, in *The CRC Handbook of Combinatorial Designs* (C. J. Colbourn and J. H. Dinitz, Eds.), CRC Press, Boca Raton, Fla., 1996, pp. 467–473.

[162] J. H. Dinitz, D. K. Garnick, and B. D. McKay, There are 526,915,620 nonisomorphic one-factorizations of K_{12}, *J. Combin. Des.* 2 (1994), 273–285.

[163] J. D. Dixon and B. Mortimer, *Permutation Groups*, Springer-Verlag, New York, 1996.

[164] S. M. Dodunekov, Minimal block length of a linear q-ary code with specified dimension and code distance (in Russian), *Problemy Peredachi Informatsii* 20 (1984) no. 4, 11–22. English translation in *Problems Inform. Transmission* 20 (1984), 239–249.

[165] S. M. Dodunekov and S. B. Encheva, Uniqueness of some linear subcodes of the binary extended Golay code (in Russian), *Problemy Peredachi Informatsii* 29 (1993) no. 1, 45–51. English translation in *Problems Inform. Transmission* 29 (1993), 38–43.

[166] R. Dougherty and H. Janwa, Covering radius computations for binary cyclic codes, *Math. Comp.* 57 (1991), 415–434.

[167] J. Doyen, A note on reverse Steiner triple systems, *Discrete Math.* 1 (1972), 315–319.

[168] J. Doyen, Designs and automorphism groups, in *Surveys in Combinatorics, 1989* (J. Siemons, Ed.), Cambridge University Press, Cambridge, 1989, pp. 75–83.

[169] J. Doyen and A. Rosa, An extended bibliography and survey of Steiner systems, *Congr. Numer.* 20 (1978), 297–361.

[170] J. Doyen and A. Rosa, An updated bibliography and survey of Steiner systems, *Ann. Discrete Math.* 7 (1980), 317–349.

[171] I. Dumer, D. Micciancio, and M. Sudan, Hardness of approximating the minimum distance of a linear code, *IEEE Trans. Inform. Theory* 49 (2003), 22–37.

[172] T. Easton and R. G. Parker, On completing Latin squares, *Discrete Appl. Math.* 113 (2001), 167–181.

[173] G. Ehrlich, Loopless algorithms for generating permutations, combinations, and other combinatorial configurations, *J. Assoc. Comput. Mach.* 20 (1973), 500–513.

[174] P. van Emde Boas, Machine models and simulations, in *Handbook of Theoretical Computer Science* (J. van Leeuwen, Ed.), Vol. A, Elsevier, Amsterdam, 1990, pp. 1–66.

[175] D. W. Erbach, The code associated with the projective plane of order four, *Arch. Math. (Basel)* 28 (1977), 669–672.

[176] Z. Eslami and G. B. Khosrovshahi, A complete classification of 3-(11, 4, 4) designs with nontrivial automorphism group, *J. Combin. Des.* 8 (2000), 419–425.

[177] Z. Eslami, G. B. Khosrovshahi, and M. M. Noori, Enumeration of t-designs through intersection matrices, *Des. Codes Cryptogr.* 32 (2004), 185–191.

[178] T. Etzion, Optimal constant weight codes over Z_k and generalized designs, *Discrete Math.* 169 (1997), 55–82.

[179] T. Etzion, Nonequivalent q-ary perfect codes, *Discrete Math.* 213 (2000), 283–290.

[180] L. Euler, Recherches sur une nouvelle espèce de quarrés magiques, *Verhandelingen uitgegeven door het zeeuwsch Genootschap der Wetenschappen te Vlissingen* 9 (1782), 85–239. Reprinted in L. Euler, *Opera Omnia*, Ser. 1, Vol. 7, *Commentationes algebraicae ad theoriam combinationum et probabilitatum pertinentes* (L. G. du Pasquier, Ed.), B. G. Teubner, Leipzig, 1923, pp. 291-392.

[181] M. van Eupen and P. Lisoněk, Classification of some optimal ternary linear codes of small length, *Des. Codes Cryptogr.* 10 (1997), 63–84.

[182] S. A. Evdokimov and I. N. Ponomarenko, Recognition and verification of an isomorphism of circulant graphs in polynomial time (in Russian), *Algebra i Analiz* 15 (2003), 1–34. English translation in *St. Petersburg Math. J.* 15 (2004), 813–835.

[183] G. Fang and H. C. A. van Tilborg, Bounds and constructions of asymmetric or unidirectional error-correcting codes, *Appl. Algebra Engrg. Comm. Comput.* 3 (1992), 269–300.

[184] K.-T. Fang and G. Ge, A sensitive algorithm for detecting the inequivalence of Hadamard matrices, *Math. Comp.* 73 (2004), 843–851.

[185] I. A. Faradžev, Constructive enumeration of combinatorial objects, in *Problèmes Combinatoires et Théorie des Graphes*, (Université d'Orsay, July 9–13, 1977), CNRS, Paris, 1978, pp. 131–135.

[186] N. J. Finizio and J. T. Lewis, Enumeration of maximal codes, *Congr. Numer.* 102 (1994), 139–145.

[187] L. M. Finkelstein and W. M. Kantor, Eds., *Groups and Computation*, Amer. Math. Soc., Providence, R.I., 1993.

[188] L. M. Finkelstein and W. M. Kantor, Eds., *Groups and Computation, II*, Amer. Math. Soc., Providence, R.I., 1997.

[189] R. A. Fisher, An examination of the different possible solutions of a problem in incomplete blocks, *Annals of Eugenics* 10 (1940), 52–75.

[190] M. Frances and A. Litman, On covering problems of codes, *Theory Comput. Syst.* 30 (1997), 113–119.

[191] F. Franek, R. Mathon, R. C. Mullin, and A. Rosa, A census of dicyclic $(v, 4, 2)$-designs for $v \leq 22$, *J. Combin. Math. Combin. Comput.* 8 (1990), 89–96.

[192] H. Fredricksen and I. J. Kessler, An algorithm for generating necklaces of beads in two colors, *Discrete Math.* 61 (1986), 181–188.

[193] T. C. Frenz and D. L. Kreher, An algorithm for enumerating distinct cyclic Steiner systems, *J. Combin. Math. Combin. Comput.* 11 (1992), 23–32.

[194] H. Fripertinger, Enumeration of isometry-classes of linear (n, k)-codes over GF(q) in SYMMETRICA, *Bayreuth. Math. Schr.* 49 (1995), 215–223.

[195] H. Fripertinger, Zyklenzeiger linearer Gruppen und Abzählung linearer Codes, *Sém. Lothar. Combin.* 33 (1995), #B33d, 11pp.

[196] H. Fripertinger, Enumeration of linear codes by applying methods from algebraic combinatorics, *Grazer Math. Ber.* 328 (1996), 31–42.

[197] H. Fripertinger and A. Kerber, Isometry classes of indecomposable linear codes, in *Applied Algebra, Algebraic Algorithms and Error-Correcting Codes* (G. Cohen, M. Giusti, and T. Mora, Eds.), Springer-Verlag, Berlin, 1995, pp. 194–204.

[198] Y. Fujii, T. Namikawa, and S. Yamamoto, Classification of two-symbol orthogonal arrays of strength t, $t+3$ constraints and index 4. II, *SUT J. Math.* 25 (1989), 161–177.

[199] K. Fukuda and T. Matsui, Finding all the perfect matchings in bipartite graphs, *Appl. Math. Lett.* 7 (1994), 15–18.

[200] S. Furino, Y. Miao, and J. Yin, *Frames and Resolvable Designs: Uses, Constructions, and Existence*, CRC Press, Boca Raton, Fla., 1996.

[201] M. R. Garey and D. S. Johnson, *Computers and Intractability: A Guide to the Theory of NP-Completeness*, W. H. Freeman, San Francisco, 1979.

[202] E. N. Gelling and R. E. Odeh, On 1-factorizations of the complete graph and the relationship to round-robin schedules, *Congr. Numer.* 9 (1974), 213–221.

[203] I. M. Gessel, Counting Latin rectangles, *Bull. Amer. Math. Soc. (N.S.)* 16 (1987), 79–83.

[204] P. B. Gibbons, *Computing Techniques for the Construction and Analysis of Block Designs*, PhD Thesis, University of Toronto, 1976.

[205] P. B. Gibbons, Computational methods in design theory, in *The CRC Handbook of Combinatorial Designs* (C. J. Colbourn and J. H. Dinitz, Eds.), CRC Press, Boca Raton, Fla., 1996, pp. 718–740.

[206] P. B. Gibbons, R. A. Mathon, and D. G. Corneil, Computing techniques for the construction and analysis of block designs, *Util. Math.* 11 (1977), 161–192.

[207] C. Godsil and G. Royle, *Algebraic Graph Theory*, Springer-Verlag, New York, 2001.

[208] J.-M. Goethals, The extended Nadler code is unique, *IEEE Trans. Inform. Theory* 23 (1977), 132–135.

[209] M. J. E. Golay, Notes on digital coding, *Proceedings of the Institute of Radio Engineers* 37 (1949), 657.

[210] L. A. Goldberg, Efficient algorithms for listing unlabeled graphs, *J. Algorithms* 13 (1992), 128–143.

[211] L. A. Goldberg, *Efficient Algorithms for Listing Combinatorial Structures*, Cambridge University Press, Cambridge, 1993.

[212] L. A. Goldberg, Computation in permutation groups: Counting and randomly sampling orbits, in *Surveys in Combinatorics, 2001* (J. W. P.

Hirschfeld, Ed.), Cambridge University Press, Cambridge, 2001, pp. 109–143.

[213] M. Goldberg, The graph isomorphism problem, in *Handbook of Graph Theory* (J. L. Gross and J. Yellen, Eds.), CRC Press, Boca Raton, Fla., 2004, pp. 68–78.

[214] S. W. Golomb, On the classification of Boolean functions, *IRE Trans. Inform. Theory* 5, Special Supplement (1959), 176–186.

[215] S. W. Golomb and L. D. Baumert, Backtrack programming, *J. Assoc. Comput. Mach.* 12 (1965), 516–524.

[216] D. M. Gordon, O. Patashnik, J. Petro, and H. Taylor, Minimum $(12, 6, 3)$ covers, *Ars. Combin.* 40 (1995), 161–177.

[217] R. L. Graham, M. Grötschel, and L. Lovász, Eds., *Handbook of Combinatorics*, 2 vols., Elsevier, Amsterdam, 1995.

[218] R. L. Graham and J. MacWilliams, On the number of information symbols in difference-set cyclic codes, *Bell System Tech. J.* 45 (1966), 1057–1070.

[219] M. J. Grannell and T. S. Griggs, Configurations in Steiner triple systems, in *Combinatorial Designs and Their Applications* (F. C. Holroyd, K. A. S. Quinn, C. Rowley, and B. S. Webb, Eds.), Chapman & Hall/CRC, Boca Raton, Fla., 1999, pp. 103–126.

[220] M. J. Grannell, T. S. Griggs, and E. Mendelsohn, A small basis for four-line configurations in Steiner triple systems, *J. Combin. Des.* 3 (1995), 51–59.

[221] M. J. Grannell, T. S. Griggs, and J. P. Murphy, Some new perfect Steiner triple systems, *J. Combin. Des.* 7 (1999), 327–330.

[222] J. E. Graver and W. B. Jurkat, The module structure of integral designs, *J. Combin. Theory Ser. A* 15 (1973), 75–90.

[223] R. L. Griess, Jr., *Twelve Sporadic Groups*, Springer-Verlag, Berlin, 1998.

[224] H.-D. O. F. Gronau, Über $(2p-1)$-$(4p, 2p, \lambda)$-Blockpläne, *Rostock. Math. Kolloq.* 11 (1979), 67–74.

[225] H.-D. O. F. Gronau, The 2-$(11, 5, 2)$ and 3-$(12, 6, 2)$ designs, *Rostock. Math. Kolloq.* 15 (1980), 77–80.

[226] H.-D. O. F. Gronau, The 2-$(10, 4, 2)$ designs, *Rostock. Math. Kolloq.* 16 (1981), 5–10.

[227] H.-D. O. F. Gronau, A survey of results on the number of t-(v, k, λ) designs, *Ann. Discrete Math.* 26 (1985), 209–219.

[228] H.-D. O. F. Gronau and J. Prestin, Some results on designs with repeated blocks, *Rostock. Math. Kolloq.* 21 (1982), 15–37.

[229] H. Gropp, The history of Steiner systems $S(2, 3, 13)$, *Mitt. Math. Ges. Hamburg* 12 (1991), 849–861.

[230] R. Grund, A. Kerber, and R. Laue, MOLGEN, ein Computeralgebra-System für die Konstruktion molekularer Graphen, *Match* 27 (1992), 87–131.

[231] T. Grüner, R. Laue, and M. Meringer, Algorithms for group actions applied to graph generation, in *Groups and Computation, II* (L. Finkel-

stein and W. M. Kantor, Eds.), Amer. Math. Soc., Providence, R.I., 1997, pp. 113–122.

[232] T. A. Gulliver and P. R. J. Östergård, Binary optimal linear rate 1/2 codes, *Discrete Math.* 283 (2004), 255–261.

[233] T. A. Gulliver, P. R. J. Östergård, and N. Senkevitch, Optimal linear rate 1/2 codes over $\mathbb{F}_5$ and $\mathbb{F}_7$, *Discrete Math.* 265 (2003), 59–70.

[234] T. A. Gulliver, P. R. J. Östergård, and N. I. Senkevitch, Optimal quaternary linear rate-1/2 codes of length ≤ 18, *IEEE Trans. Inform. Theory* 49 (2003), 1540–1543.

[235] T. A. Gulliver and N. Senkevitch, Optimal ternary linear rate 1/2 codes, *Des. Codes Cryptogr.* 23 (2001), 167–171.

[236] M. Guregová and A. Rosa, Using the computer to investigate cyclic Steiner quadruple systems, *Mat. Časopis Sloven. Akad. Vied* 18 (1968), 229–239.

[237] V. Guruswami, D. Micciancio, and O. Regev, The complexity of the covering radius problem on lattices and codes, *Comput. Complexity*, to appear.

[238] H. Haanpää and P. Kaski, The near resolvable 2-$(13,4,3)$ designs and thirteen-player whist tournaments, *Des. Codes Cryptogr.* 35 (2005), 271–285.

[239] E. Haberberger, A. Betten, and R. Laue, Isomorphism classification of t-designs with group theoretical localisation techniques applied to some Steiner quadruple systems on 20 points, *Congr. Numer.* 142 (2000), 75–96.

[240] L. Habsieger and A. Plagne, New lower bounds for covering codes, *Discrete Math.* 222 (2000), 125–149.

[241] M. Hall, Jr., Hadamard matrices of order 16, Research Summary 36-10, Vol. 1, Jet Propulsion Laboratory, Pasadena, Calif., 1961, pp. 21–26.

[242] M. Hall, Jr., Hadamard matrices of order 20, Technical Report 32-761, Jet Propulsion Laboratory, Pasadena, Calif., 1965.

[243] M. Hall, Jr., Configurations in a plane of order ten, *Ann. Discrete Math.* 6 (1980), 157–174.

[244] M. Hall, Jr., *Combinatorial Theory*, 2nd ed., Wiley, New York, 1986.

[245] M. Hall, Jr. and W. S. Connor, An embedding theorem for balanced incomplete block designs, *Canad. J. Math.* 6 (1953), 35–41.

[246] M. Hall, Jr., R. Roth, G. H. J. van Rees, and S. A. Vanstone, On designs $(22,33,12,8,4)$, *J. Combin. Theory Ser. A* 47 (1988), 157–175.

[247] M. Hall, Jr. and J. D. Swift, Determination of Steiner triple systems of order 15, *Math. Tables Aids Comput.* 9 (1955), 146–152.

[248] M. Hall, Jr., J. D. Swift, and R. J. Walker, Uniqueness of the projective plane with 57 points, *Proc. Amer. Math. Soc.* 4 (1953), 912–916; and 5 (1954), 994–997.

[249] M. Hall, Jr., J. D. Swift, and R. J. Walker, Uniqueness of the projective plane of order eight, *Math. Tables Aids Comput.* 10 (1956), 186–194.

[250] H. Hämäläinen and S. Rankinen, Upper bounds for football pool problems and mixed covering codes, *J. Combin. Theory Ser. A* 56 (1991), 84–95.
[251] R. W. Hamming, Error detecting and error correcting codes, *Bell System Tech. J.* 29 (1950), 147–160.
[252] M. Harada and P. R. J. Östergård, Self-dual and maximal self-orthogonal codes over $\mathbb{F}_7$, *Discrete Math.* 256 (2002), 471–477.
[253] M. Harada and P. R. J. Östergård, On the classification of self-dual codes over $\mathbb{F}_5$, *Graphs Combin.* 19 (2003), 203–214.
[254] F. Harary, *Graph Theory*, Addison-Wesley, Reading, Mass., 1969.
[255] F. Harary, The automorphism group of a hypercube, *J. UCS* 6 (2000), 136–138.
[256] A. S. Hedayat, E. Seiden, and J. Stufken, On the maximal number of factors and the enumeration of 3-symbol orthogonal arrays of strength 3 and index 2, *J. Statist. Plann. Inference* 58 (1997), 43–63.
[257] A. S. Hedayat, N. J. A. Sloane, and J. Stufken, *Orthogonal Arrays: Theory and Applications*, Springer-Verlag, New York, 1999.
[258] Z. Hedrlín and A. Pultr, On full embeddings of categories of algebras, *Illinois J. Math.* 10 (1966), 392–406.
[259] D. Held and M.-O. Pavčević, Symmetric $(79, 27, 9)$-designs admitting a faithful action of a Frobenius group of order 39, *European J. Combin.* 18 (1997), 409–416.
[260] D. Held and M.-O. Pavčević, Some new Hadamard designs with 79 points admitting automorphisms of order 13 and 19, *Discrete Math.* 238 (2001), 61–65.
[261] T. Helleseth, A characterization of codes meeting the Griesmer bound, *Inform. and Control* 50 (1981), 128–159.
[262] L. A. Hemaspaandra and M. Ogihara, *The Complexity Theory Companion*, Springer-Verlag, Berlin, 2001.
[263] P. J. Higgins, *Notes on Categories and Groupoids*, Van Nostrand Reinhold, London, 1971.
[264] R. Hill, *A First Course in Coding Theory*, Oxford University Press, Oxford, 1986.
[265] H. Hitotumatu and K. Noshita, A technique for implementing backtrack algorithms and its application, *Inform. Process. Lett.* 8 (1979), 174–175.
[266] C. M. Hoffmann, *Group-Theoretic Algorithms and Graph Isomorphism*, Springer-Verlag, Berlin, 1982.
[267] I. Holyer, The NP-completeness of edge-coloring, *SIAM J. Comput.* 10 (1981), 718–720.
[268] J. D. Horton, A hyperfactorization of order 8, index 2, *Discrete Math.* 92 (1991), 127–129.
[269] S. Houghten, C. Lam, and L. Thiel, Construction of $(48, 24, 12)$ doubly-even self-dual codes, *Congr. Numer.* 103 (1994), 41–53.

[270] S. K. Houghten, C. W. H. Lam, L. H. Thiel, and J. A. Parker, The extended quadratic residue code is the only $(48, 24, 12)$ self-dual doubly-even code, *IEEE Trans. Inform. Theory* 49 (2003), 53–59.

[271] S. K. Houghten, L. H. Thiel, J. Janssen, and C. W. H. Lam, There is no $(46, 6, 1)$ block design, *J. Combin. Des.* 9 (2001), 60–71.

[272] W. C. Huffman, Codes and groups, in *Handbook of Coding Theory* (V. S. Pless and W. C. Huffman, Eds.), Vol. 2, Elsevier, Amsterdam, 1998, pp. 1345–1440.

[273] W. C. Huffman and V. Pless, *Fundamentals of Error-Correcting Codes*, Cambridge University Press, Cambridge, 2003.

[274] D. R. Hughes and F. C. Piper, *Design Theory*, Cambridge University Press, Cambridge, 1985.

[275] J. F. Humphreys, *A Course in Group Theory*, Oxford University Press, Oxford, 1996.

[276] Q. M. Husain, On the totality of the solutions for the symmetrical incomplete block designs: $\lambda = 2$, $k = 5$ or 6, *Sankhyā* 7 (1945), 204–208.

[277] N. Ito, J. S. Leon, and J. Q. Longyear, Classification of 3-$(24, 12, 5)$ designs and 24-dimensional Hadamard matrices, *J. Combin. Theory Ser. A* 31 (1981), 66–93.

[278] A. V. Ivanov, Constructive enumeration of incidence systems, *Ann. Discrete Math.* 26 (1985), 227–246.

[279] M. T. Jacobson and P. Matthews, Generating uniformly distributed random Latin squares, *J. Combin. Des.* 4 (1996), 405–437.

[280] D. B. Jaffe, Optimal binary linear codes of length ≤ 30, *Discrete Math.* 223 (223), 135–155.

[281] M. Jerrum, A compact representation for permutation groups, *J. Algorithms* 7 (1986), 60–78.

[282] M. R. Jerrum, L. G. Valiant, and V. V. Vazirani, Random generation of combinatorial structures from a uniform distribution, *Theoret. Comput. Sci.* 43 (1986), 169–188.

[283] D. S. Johnson, M. Yannakakis, and C. H. Papadimitriou, On generating all maximal independent sets, *Inform. Process. Lett.* 27 (1988), 119–123.

[284] S. M. Johnson, A new upper bound for error-correcting codes, *IRE Trans. Inform. Theory* 8 (1962), 203–207.

[285] S. M. Johnson, A new lower bound for coverings by rook domains, *Util. Math.* 1 (1972), 121–140.

[286] J. T. Joichi, D. E. White, and S. G. Williamson, Combinatorial Gray codes, *SIAM J. Comput.* 9 (1980), 130–141.

[287] G. Jones, M. Klin, and F. Lazebnik, Automorphic subsets of the n-dimensional cube, *Beiträge Algebra Geom.* 41 (2000), 303–323.

[288] D. Jungnickel and S. A. Vanstone, Hyperfactorizations of graphs and 5-designs, *J. Univ. Kuwait Sci.* 14 (1987), 213–224.

[289] M. Kaib and H. Ritter, Block reduction for arbitrary norms, Technical Report, Mathematische Informatik, Universität Frankfurt am Main, 1994.

[290] M. Kaikkonen, Codes from affine permutation groups, *Des. Codes Cryptogr.* 15 (1998), 183–186.

[291] J. G. Kalbfleisch and R. G. Stanton, On a certain set of linear inequalities, *Canad. Math. Bull.* 11 (1968), 681–690.

[292] J. G. Kalbfleisch and R. G. Stanton, A combinatorial problem in matching, *J. London Math. Soc.* 44 (1969), 60–64; and *J. London Math Soc. (2)* 1 (1969), 398.

[293] J. G. Kalbfleisch, R. G. Stanton, and J. D. Horton, On covering sets and error-correcting codes, *J. Combin. Theory Ser. A* 11 (1971), 233–250.

[294] H. J. L. Kamps and J. H. van Lint, The football pool problem for 5 matches, *J. Combin. Theory* 3 (1967), 315–325.

[295] H. J. L. Kamps and J. H. van Lint, A covering problem, *Colloq. Math. Soc. János Bolyai* 4 (1970), 679–685.

[296] W. M. Kantor and Á. Seress, Eds., *Groups and Computation, III*, Walter de Gruyter, Berlin, 2001.

[297] S. N. Kapralov, Enumeration of the binary linear [24, 7, 10] codes, in *Proc. 5th International Workshop on Algebraic and Combinatorial Coding Theory*, (Sozopol, Bulgaria, June 1–7, 1996), Unicorn, Shumen, Bulgaria, 1996, pp. 151–156.

[298] P. Kaski, Isomorph-free exhaustive generation of combinatorial designs, Research Report A70, Laboratory for Theoretical Computer Science, Helsinki University of Technology, Espoo, 2001.

[299] P. Kaski, Isomorph-free exhaustive generation of designs with prescribed groups of automorphisms, *SIAM J. Discrete Math.*, to appear.

[300] P. Kaski, Nonexistence of perfect Steiner triple systems of orders 19 and 21, *Bayreuth. Math. Schr.*, to appear.

[301] P. Kaski, L. B. Morales, P. R. J. Östergård, D. A. Rosenblueth, and C. Velarde, Classification of resolvable 2-(14, 7, 12) and 3-(14, 7, 5) designs, *J. Combin. Math. Combin. Comput.* 47 (2003), 65–74.

[302] P. Kaski and P. R. J. Östergård, There exists no (15, 5, 4) RBIBD, *J. Combin. Des.* 9 (2001), 357–362.

[303] P. Kaski and P. R. J. Östergård, Miscellaneous classification results for 2-designs, *Discrete Math.* 280 (2004), 65–75.

[304] P. Kaski and P. R. J. Östergård, The Steiner triple systems of order 19, *Math. Comp.* 73 (2004), 2075–2092.

[305] P. Kaski and P. R. J. Östergård, One-factorizations of regular graphs of order 12, *Electron. J. Combin.* 12 (2005) no. 1, #R2, 25pp.

[306] P. Kaski, P. R. J. Östergård, and O. Pottonen, The Steiner quadruple systems of order 16, submitted for publication.

[307] P. Kaski, P. R. J. Östergård, S. Topalova, and R. Zlatarski, Steiner triple systems of order 19 and 21 with subsystems of order 7, *Discrete Math.*, to appear.

[308] A. Kerber, *Applied Finite Group Actions*, 2nd ed., Springer-Verlag, Berlin, 1999.

[309] A. Kerber and R. Laue, Group actions, double cosets, and homomorphisms: Unifying concepts for the constructive theory of discrete structures, *Acta Appl. Math.* 52 (1998), 63–90.

[310] G. Kéri and P. R. J. Östergård, On the covering radius of small codes, *Studia Sci. Math. Hungar.* 40 (2003), 243–256.

[311] G. Kéri and P. R. J. Östergård, Further results on the covering radius of small codes, submitted for publication.

[312] H. Kharaghani and B. Tayfeh-Rezaie, A Hadamard matrix of order 428, *J. Combin. Des.*, to appear.

[313] M. Khatirinejad and P. Lisonek, Classification and constructions of complete caps in binary spaces, *Des. Codes Cryptogr.*, to appear.

[314] G. B. Khosrovshahi, M. Mohammad-Noori, and B. Tayfeh-Rezaie, Classification of 6-$(14, 7, 4)$ designs with nontrivial automorphism groups, *J. Combin. Des.* 10 (2002), 180–194.

[315] H. Kimura, Hadamard matrices of order 28 with automorphism groups of order two, *J. Combin. Theory Ser. A* 43 (1986), 98–102.

[316] H. Kimura, New Hadamard matrix of order 24, *Graphs Combin.* 5 (1989), 235–242.

[317] H. Kimura, Classification of Hadamard matrices of order 28, *Discrete Math.* 133 (1994), 171–180.

[318] H. Kimura, Classification of Hadamard matrices of order 28 with Hall sets, *Discrete Math.* 128 (1994), 257–268.

[319] F. Klein, *Das Erlanger Programm (1872). Vergleichende Betrachtungen über neuere geometrische Forschungen. Einleitung und Anmerkung von H. Wußing*, 3rd ed., Harri Deutsch, Frankfurt am Main, 1997.

[320] A. R. Klivans and D. van Melkebeek, Graph nonisomorphism has subexponential size proofs unless the polynomial-time hierarchy collapses, *SIAM J. Comput.* 31 (2002), 1501–1526.

[321] D. E. Knuth, Estimating the efficiency of backtrack programs, *Math. Comp.* 29 (1975), 121–136.

[322] D. E. Knuth, *The Art of Computer Programming, Volume 3: Sorting and Searching*, 2nd ed., Addison-Wesley, Reading, Mass., 1998.

[323] D. E. Knuth, Dancing links, in *Millennial Perspectives in Computer Science* (J. Davies, B. Roscoe, and J. Woodcock, Eds.), Palgrave, Basingstoke, England, 2000, pp. 187–214.

[324] J. Köbler, U. Schöning, and J. Torán, *The Graph Isomorphism Problem: Its Structural Complexity*, Birkhäuser, Boston, 1993.

[325] W. Kocay, On writing isomorphism programs, in *Computational and Constructive Design Theory* (W. D. Wallis, Ed.), Kluwer, Dordrecht, the Netherlands, 1996, pp. 135–175.

[326] W. Kocay and D. L. Kreher, *Graphs, Algorithms, and Optimization*, Chapman & Hall/CRC, Boca Raton, Fla., 2005.

[327] W. L. Kocay, D. R. Stinson, and S. A. Vanstone, On strong starters in cyclic groups, *Discrete Math.* 56 (1985), 45–60.

[328] G. Kolesova, C. W. H. Lam, and L. Thiel, On the number of 8×8 Latin squares, *J. Combin. Theory Ser. A* 54 (1990), 143–148.

[329] E. Kolev, Lower bounds for mixed covering codes of length 5, *C. R. Acad. Bulgare Sci.* 46 (1993) no. 8, 9–11.

[330] J. F. Korsh and P. S. Lafollette, Loopless generation of linear extensions of a poset, *Order* 19 (2002), 115–126.

[331] B. Korte and J. Vygen, *Combinatorial Optimization. Theory and Algorithms*, 2nd ed., Springer-Verlag, Berlin, 2002.

[332] B. K. Kostova and N. L. Manev, A $[25, 8, 10]$ code does not exist, *C. R. Acad. Bulgare Sci.* 43 (1990) no. 3, 41–44.

[333] E. S. Kramer, D. W. Leavitt, and S. S. Magliveras, Construction procedures for t-designs and the existence of new simple 6-designs, *Ann. Discrete Math.* 26 (1985), 247–273.

[334] E. S. Kramer, S. S. Magliveras, and R. Mathon, The Steiner systems $S(2, 4, 25)$ with nontrivial automorphism group, *Discrete Math.* 77 (1989), 137–157.

[335] E. S. Kramer and D. M. Mesner, t-designs on hypergraphs, *Discrete Math.* 15 (1976), 263–296.

[336] V. Krčadinac, Steiner 2-designs $S(2, 4, 28)$ with nontrivial automorphisms, *Glas. Mat. Ser. III* 37(59) (2002), 259–268.

[337] V. Krčadinac, Steiner 2-designs $S(2, 5, 41)$ with automorphisms of order 3, *J. Combin. Math. Combin. Comput.* 43 (2002), 83–99.

[338] D. L. Kreher, D. de Caen, S. A. Hobart, E. S. Kramer, and S. P. Radziszowski, The parameters 4-$(12, 6, 6)$ and related t-designs, *Australas. J. Combin.* 7 (1993), 3–20.

[339] D. L. Kreher and S. P. Radziszowski, The existence of simple 6-$(14, 7, 4)$ designs, *J. Combin. Theory Ser. A* 43 (1986), 237–243.

[340] D. L. Kreher and S. P. Radziszowski, Finding simple t-designs by using basis reduction, *Congr. Numer.* 55 (1986), 235–244.

[341] D. L. Kreher and S. P. Radziszowski, Simple 5-$(28, 6, \lambda)$ designs from $\mathrm{PSL}_2(27)$, *Ann. Discrete Math.* 37 (1987), 315–318.

[342] D. L. Kreher and D. R. Stinson, *Combinatorial Algorithms: Generation, Enumeration, and Search*, CRC Press, Boca Raton, Fla., 1999.

[343] C. W. H. Lam, How reliable is a computer-based proof? *Math. Intelligencer* 12 (1990) no. 1, 8–12.

[344] C. W. H. Lam, The search for a finite projective plane of order 10, *Amer. Math. Monthly* 98 (1991), 305–318.

[345] C. Lam, S. Crossfield, and L. Thiel, Estimates of a computer search for a projective plane of order 10, *Congr. Numer.* 48 (1985), 253–263.

[346] C. W. H. Lam, G. Kolesova, and L. Thiel, A computer search for finite projective planes of order 9, *Discrete Math.* 92 (1991), 187–195.

[347] C. W. H. Lam and L. Thiel, Backtrack search with isomorph rejection and consistency check, *J. Symbolic Comput.* 7 (1989), 473–485.

[348] C. Lam, L. Thiel, and S. Swiercz, A feasibility study of a search for ovals in a projective plane of order 10, in *Combinatorial Mathematics IX*

(E. J. Billington, S. Oates-Williams, and A. P. Street, Eds.), Springer-Verlag, Berlin, 1982, pp. 349–352.

[349] C. W. H. Lam, L. Thiel, and S. Swiercz, The nonexistence of code words of weight 16 in a projective plane of order 10, *J. Combin. Theory Ser. A* 42 (1986), 207–214.

[350] C. W. H. Lam, L. H. Thiel, and S. Swiercz, A computer search for a projective plane of order 10, in *Algebraic, Extremal and Metric Combinatorics, 1986* (M. M. Deza, P. Frankl, and I. G. Rosenberg, Eds.), Cambridge University Press, Cambridge, 1988, pp. 155–165.

[351] C. W. H. Lam, L. Thiel, and S. Swiercz, The nonexistence of finite projective planes of order 10, *Canad. J. Math.* 41 (1989), 1117–1123.

[352] C. W. H. Lam, L. Thiel, S. Swiercz, and J. McKay, The nonexistence of ovals in a projective plane of order 10, *Discrete Math.* 45 (1983), 319–321.

[353] C. Lam and V. D. Tonchev, Classification of affine resolvable 2-$(27, 9, 4)$ designs, *J. Statist. Plann. Inference* 56 (1996), 187–202; and 86 (2000), 277–278.

[354] E. S. Lander, *Symmetric Designs: An Algebraic Approach*, Cambridge University Press, Cambridge, 1983.

[355] S. Lang, *Algebra*, 3rd ed., Springer-Verlag, New York, 2002.

[356] R. Laue, Construction of combinatorial objects – A tutorial, *Bayreuth. Math. Schr.* 43 (1993), 53–96.

[357] R. Laue, Constructing objects up to isomorphism, simple 9-designs with small parameters, in *Algebraic Combinatorics and Applications* (A. Betten, A. Kohnert, R. Laue, and A. Wassermann, Eds.), Springer-Verlag, Berlin, 2001, pp. 232–260.

[358] R. Laue, Solving isomorphism problems for t-designs, in *Designs 2002: Further Computational and Constructive Design Theory* (W. D. Wallis, Ed.), Kluwer, Boston, 2003, pp. 277–300.

[359] E. L. Lawler, J. K. Lenstra, and A. H. G. Rinnooy Kan, Generating all maximal independent sets: NP-hardness and polynomial-time algorithms, *SIAM J. Comput.* 9 (1980), 558–565.

[360] F. W. Lawvere and S. H. Schanuel, *Conceptual Mathematics: A First Introduction to Categories*, Cambridge University Press, Cambridge, 1997.

[361] J. van Leeuwen, Ed., *Handbook of Theoretical Computer Science*, 2 vols., Elsevier, Amsterdam, 1990.

[362] Z. Leko-Božikov, The classification of Hadamard block designs $H(27)$ on which an elementary abelian Singer group operates, *Punime Mat.* (1986), 43–48.

[363] A. K. Lenstra, H. W. Lenstra, Jr., and L. Lovász, Factoring polynomials with rational coefficients, *Math. Ann.* 261 (1982), 515–534.

[364] J. S. Leon, An algorithm for computing the automorphism group of a Hadamard matrix, *J. Combin. Theory Ser. A* 27 (1979), 289–306.

[365] J. S. Leon, Computing automorphism groups of error-correcting codes, *IEEE Trans. Inform. Theory* 28 (1982), 496–511.

[366] J. S. Leon, Computing automorphism groups of combinatorial objects, in *Computational Group Theory* (M. D. Atkinson, Ed.), Academic Press, London, 1984, pp. 321–335.

[367] J. S. Leon, Permutation group algorithms based on partitions, I: Theory and algorithms, *J. Symbolic Comput.* 12 (1991), 533–583.

[368] J. S. Leon, Partitions, refinements, and permutation group computation, in *Groups and Computation, II* (L. Finkelstein and W. M. Kantor, Eds.), Amer. Math. Soc., Providence, R.I., 1997, pp. 123–158.

[369] J. S. Leon, V. Pless, and N. J. A. Sloane, Self-dual codes over GF(5), *J. Combin. Theory Ser. A* 32 (1982), 178–194.

[370] M. J. Letourneau and S. K. Houghten, Optimal ternary $(10, 7)$ error-correcting codes, *Congr. Numer.* 155 (2002), 71–80.

[371] M. J. Letourneau and S. K. Houghten, Optimal ternary $(11, 7)$ and $(14, 10)$ error-correcting codes, *J. Combin. Math. Combin. Comput.* 51 (2004), 159–164.

[372] R. Lidl and H. Niederreiter, *Finite Fields*, 2nd ed., Cambridge University Press, Cambridge, 1997.

[373] S. Lin and D. J. Costello, Jr., *Error Control Coding: Fundamentals and Applications*, Prentice Hall, Englewood Cliffs, N.J., 1983.

[374] J. H. van Lint, *Introduction to Coding Theory*, 3rd ed., Springer-Verlag, Berlin, 1999.

[375] J. H. van Lint, H. C. A. van Tilborg, and J. R. Wiekema, Block designs with $v = 10$, $k = 5$, $\lambda = 4$, *J. Combin. Theory Ser. A* 23 (1977), 105–115.

[376] J. H. van Lint and R. M. Wilson, *A Course in Combinatorics*, 2nd ed., Cambridge University Press, Cambridge, 2001.

[377] J. H. van Lint, Jr., *Covering Radius Problems*, MSc Thesis, Eindhoven University of Technology, 1988.

[378] S. Litsyn and A. Vardy, The uniqueness of the Best code, *IEEE Trans. Inform. Theory* 40 (1994), 1693–1698.

[379] L. Lovász, *Combinatorial Problems and Exercises*, North-Holland, New York, 1979.

[380] J. X. Lù, An existence theory for resolvable balanced incomplete block designs (in Chinese), *Acta Math. Sinica* 27 (1984), 458–468. English translation in T. C. Y. Lee and S. C. Furino, A translation of J. X. Lu's "An existence theory for resolvable balanced incomplete block designs," *J. Combin. Des.* 3 (1995), 321–340.

[381] A. Lubiw, Some NP-complete problems similar to graph isomorphism, *SIAM J. Comput.* 10 (1981), 11–21.

[382] E. M. Luks, Isomorphism of graphs of bounded valence can be tested in polynomial time, *J. Comput. System Sci.* 25 (1982), 42–65.

[383] E. M. Luks, Permutation groups and polynomial-time computation, in *Groups and Computation* (L. Finkelstein and W. M. Kantor, Eds.), Amer. Math. Soc., Providence, R.I., 1993, pp. 139–175.

[384] M. M-Noori and B. Tayfeh-Rezaie, Backtracking algorithm for finding t-designs, *J. Combin. Des.* 11 (2003), 240–248.

[385] K. L. Ma, *Solving the Dominating Set Problem: A Group Theory Approach*, PhD Thesis, Concordia University, Montreal, 1998.
[386] C. R. MacInnes, Finite planes with less than eight points on a line, *The American Mathematical Monthly* 14 (1907), 171–174.
[387] S. MacLane, *Categories for the Working Mathematician*, 2nd ed., Springer-Verlag, New York, 1998.
[388] F. J. MacWilliams and N. J. A. Sloane, *The Theory of Error-Correcting Codes*, North-Holland, Amsterdam, 1977.
[389] F. J. MacWilliams, N. J. A. Sloane, and J. G. Thompson, On the existence of a projective plane of order 10, *J. Combin. Theory Ser. A* 14 (1973), 66–78.
[390] C. L. Mallows, V. Pless, and N. J. A. Sloane, Self-dual codes over GF(3), *SIAM J. Appl. Math.* 31 (1976), 649–666.
[391] C. L. Mallows and N. J. A. Sloane, Weight enumerators of self-orthogonal codes, *Discrete Math.* 9 (1974), 391–400.
[392] S. A. Malyugin, On a lower bound on the number of perfect binary codes, *Discrete Appl. Math.* 135 (2004), 157–160.
[393] H. B. Mann and H. J. Ryser, Systems of distinct representatives, *Amer. Math. Monthly* 60 (1953), 397–401.
[394] L. Marangunić, The classification of biplanes $(56, 11, 2)$ admitting an automorphism of order 8 which fixes some point, *Rad. Mat.* 2 (1986), 99–111.
[395] F. Margot, Small covering designs by branch-and-cut, *Math. Program.* 94B (2003), 207–220.
[396] V. Martinetti, Un' osservazione relativa alla configurazione di Kummer, *Giornale di Matematiche di Battaglini* 34 (1896), 192–194.
[397] R. Mathon, A note on the graph isomorphism counting problem, *Inform. Process. Lett.* 8 (1979), 131–132.
[398] R. Mathon, The partial geometries pg$(5, 7, 3)$, *Congr. Numer.* 31 (1981), 129–139.
[399] R. Mathon, Symmetric $(31, 10, 3)$ designs with nontrivial automorphism group, *Ars. Combin.* 25 (1988), 171–183.
[400] R. Mathon, Computational methods in design theory, in *Surveys in Combinatorics, 1991* (A. D. Keedwell, Ed.), Cambridge University Press, Cambridge, 1991, pp. 101–117. Reprinted in *Computational and Constructive Design Theory* (W. D. Wallis, Ed.), Kluwer, Dordrecht, the Netherlands, 1996, pp. 29–48.
[401] R. Mathon, Searching for spreads and packings, in *Geometry, Combinatorial Designs and Related Structures* (J. W. P. Hirschfeld, S. S. Magliveras, and M. J. de Resmini, Eds.), Cambridge University Press, Cambridge, 1997, pp. 161–176.
[402] R. Mathon and D. Lomas, A census of 2-$(9, 3, 3)$ designs, *Australas. J. Combin.* 5 (1992), 145–158.
[403] R. Mathon and A. Rosa, A census of Mendelsohn triple systems of order nine, *Ars. Combin.* 4 (1977), 309–315.

[404] R. Mathon and A. Rosa, Some results on the existence and enumeration of BIBD's, Mathematics Report 125-Dec-1985, Department of Mathematics and Statistics, McMaster University, Hamilton, 1985.

[405] R. Mathon and A. Rosa, Tables of parameters of BIBDs with $r \leq 41$ including existence, enumeration, and resolvability results, *Ann. Discrete Math.* 26 (1985), 275–307.

[406] R. Mathon and A. Rosa, Tables of parameters of BIBDs with $r \leq 41$ including existence, enumeration and resolvability results: An update, *Ars. Combin.* 30 (1990), 65–96.

[407] R. Mathon and A. Rosa, 2-(v, k, λ) designs of small order, in *The CRC Handbook of Combinatorial Designs* (C. J. Colbourn and J. H. Dinitz, Eds.), CRC Press, Boca Raton, Fla., 1996, pp. 3–41.

[408] R. A. Mathon, K. T. Phelps, and A. Rosa, Small Steiner triple systems and their properties, *Ars. Combin.* 15 (1983), 3–110; and 16 (1983), 286.

[409] R. J. McEliece, *Finite Fields for Computer Scientists and Engineers*, Kluwer, Boston, 1987.

[410] B. D. McKay, Hadamard equivalence via graph isomorphism, *Discrete Math.* 27 (1979), 213–214.

[411] B. D. McKay, Practical graph isomorphism, *Congr. Numer.* 30 (1981), 45–87.

[412] B. D. McKay, *nauty* user's guide (version 1.5), Technical Report TR-CS-90-02, Computer Science Department, Australian National University, Canberra, 1990.

[413] B. D. McKay, `autoson` – A distributed batch system for UNIX workstation networks (version 1.3), Technical Report TR-CS-96-03, Computer Science Department, Australian National University, Canberra, 1996.

[414] B. D. McKay, Isomorph-free exhaustive generation, *J. Algorithms* 26 (1998), 306–324.

[415] B. D. McKay, A. Meynert, and W. Myrvold, Small Latin squares, quasigroups, and loops, submitted for publication.

[416] B. McKay, W. Myrvold, and J. Nadon, Fast backtracking principles applied to find new cages, in *Proc. 9th ACM-SIAM Symposium on Discrete Algorithms*, (San Francisco, Jan. 25–27, 1998), ACM Press, New York, 1998, pp. 188–191.

[417] B. D. McKay and S. P. Radziszowski, The nonexistence of 4-$(12, 6, 6)$ designs, in *Computational and Constructive Design Theory* (W. D. Wallis, Ed.), Kluwer, Dordrecht, the Netherlands, 1996, pp. 177–188.

[418] B. D. McKay and S. P. Radziszowski, Towards deciding the existence of 2-$(22, 8, 4)$ designs, *J. Combin. Math. Combin. Comput.* 22 (1996), 211–222.

[419] B. D. McKay and S. P. Radziszowski, 2-$(22, 8, 4)$ designs have no blocks of type 3, *J. Combin. Math. Combin. Comput.* 30 (1999), 251–253.

[420] B. D. McKay and E. Rogoyski, Latin squares of order 10, *Electron. J. Combin.* 2 (1995), #N3, 4pp.

[421] B. D. McKay and I. M. Wanless, On the number of Latin squares, *Ann. Comb.* 9 (2005), 335–344.
[422] A. M. McLoughlin, The complexity of computing the covering radius of a linear code, *IEEE Trans. Inform. Theory* 30 (1984), 800–804.
[423] N. S. Mendelsohn and S. H. Y. Hung, On the Steiner systems $S(3,4,14)$ and $S(4,5,15)$, *Util. Math.* 1 (1972), 5–95.
[424] M. Meringer, *Erzeugung regulärer Graphen,* MSc Thesis, Universität Bayreuth, 1996.
[425] M. Meringer, Fast generation of regular graphs and construction of cages, *J. Graph Theory* 30 (1999), 137–146.
[426] G. L. Miller, On the $n^{\log n}$ isomorphism technique (a preliminary report), in *Proc. 10th ACM Symposium on Theory of Computing,* (San Diego, May 1–3, 1978), ACM Press, New York, 1978, pp. 51–58.
[427] G. L. Miller, Graph isomorphism, general remarks, *J. Comput. System Sci.* 18 (1979), 128–142.
[428] W. H. Mills, Review (0509.05031) of [551], *Zbl* 509 (1983), 35.
[429] W. H. Mills and R. C. Mullin, Coverings and packings, in *Contemporary Design Theory: A Collection of Surveys* (J. H. Dinitz and D. R. Stinson, Eds.), Wiley, New York, 1992, pp. 371–399.
[430] T. Miyazaki, The complexity of McKay's canonical labeling algorithm, in *Groups and Computation, II* (L. Finkelstein and W. M. Kantor, Eds.), Amer. Math. Soc., Providence, R.I., 1997, pp. 239–256.
[431] L. B. Morales and C. Velarde, A complete classification of $(12,4,3)$-RBIBDs, *J. Combin. Des.* 9 (2001), 385–400.
[432] L. B. Morales and C. Velarde, Enumeration of resolvable 2-$(10,5,16)$ and 3-$(10,5,6)$ designs, *J. Combin. Des.* 13 (2005), 108–119.
[433] E. J. Morgan, Isomorphism classes of some small block designs, *Ars. Combin.* 4 (1977), 25–35.
[434] E. J. Morgan, Some small quasi-multiple designs, *Ars. Combin.* 3 (1977), 233–250.
[435] P. Mulder, *Kirkman-systemen,* PhD Thesis, Rijksuniversiteit Groningen, 1917.
[436] M. Muzychuk, A solution of the isomorphism problem for circulant graphs, *Proc. London Math. Soc. (3)* 88 (2004), 1–14.
[437] S. Nakano and T. Uno, Constant time generation of trees with specified diameter, in *Graph-Theoretic Concepts in Computer Science* (J. Hromkovic, M. Nagl, and B. Westfechtel, Eds.), Springer-Verlag, Berlin, 2004.
[438] T. Namikawa, Y. Fujii, and S. Yamamoto, Computational study on the classification of two-symbol orthogonal arrays of strength t, $m = t + e$ constraints for $e \leq 3$, *SUT J. Math.* 25 (1989), 179–195.
[439] H. K. Nandi, A further note on non-isomorphic solutions of incomplete block designs, *Sankhyā* 7 (1946), 313–316.
[440] J. R. Nechvatal, Asymptotic enumeration of generalized Latin rectangles, *Util. Math.* 20 (1981), 273–292.

[441] P. M. Neumann, A lemma that is not Burnside's, *Math. Sci.* 4 (1979), 133–141.

[442] J. Nievergelt, N. Deo, and A. Marzetta, Memory-efficient enumeration of constrained spanning trees, *Inform. Process. Lett.* 72 (1999), 47–53.

[443] S. Niskanen and P. R. J. Östergård, Cliquer user's guide, version 1.0, Technical Report T48, Communications Laboratory, Helsinki University of Technology, Espoo, 2003.

[444] A. W. Nordstrom and J. P. Robinson, An optimum nonlinear code, *Inform. and Control* 11 (1967), 613–616.

[445] C. W. Norman, A characterization of the Mathieu group M_{11}, *Math. Z.* 106 (1968), 162–166.

[446] H. W. Norton, The 7×7 squares, *Annals of Eugenics* 9 (1939), 269–307.

[447] K. J. Nurmela, M. K. Kaikkonen, and P. R. J. Östergård, New constant weight codes from linear permutation groups, *IEEE Trans. Inform. Theory* 43 (1997), 1623–1630.

[448] J. Olsson, *Linear Codes with Performance Close to the Singleton Bound*, PhD Thesis, Linköpings universitet, 1999.

[449] P. R. J. Östergård, The football pool problem, *Congr. Numer.* 114 (1996), 33–43.

[450] P. R. J. Östergård, On binary/ternary error-correcting codes with minimum distance 4, in *Applied Algebra, Algebraic Algorithms and Error-Correcting Codes* (M. Fossorier, H. Imai, S. Lin, and A. Poli, Eds.), Springer-Verlag, Berlin, 1999, pp. 472–481.

[451] P. R. J. Östergård, Classification of binary/ternary one-error-correcting codes, *Discrete Math.* 223 (2000), 253–262.

[452] P. R. J. Östergård, Enumeration of 2-$(12, 3, 2)$ designs, *Australas. J. Combin.* 22 (2000), 227–231.

[453] P. R. J. Östergård, A new algorithm for the maximum-weight clique problem, *Nordic J. Comput.* 8 (2001), 424–436.

[454] P. R. J. Östergård, There are 270,474,142 nonisomorphic 2-$(9, 4, 6)$ designs, *J. Combin. Math. Combin. Comput.* 37 (2001), 173–176.

[455] P. R. J. Östergård, Classifying subspaces of Hamming spaces, *Des. Codes Cryptogr.* 27 (2002), 297–305.

[456] P. R. J. Östergård, A fast algorithm for the maximum clique problem, *Discrete Appl. Math.* 120 (2002), 195–205.

[457] P. R. J. Östergård, A 2-$(22, 8, 4)$ design cannot have a 2-$(10, 4, 4)$ subdesign, *Des. Codes Cryptogr.* 27 (2002), 257–260.

[458] P. R. J. Östergård, Constructing combinatorial objects via cliques, in *Surveys in Combinatorics, 2005* (B. S. Webb, Ed.), Cambridge University Press, Cambridge, 2005, pp. 57–82.

[459] P. R. J. Östergård, T. Baicheva, and E. Kolev, Optimal binary one-error-correcting codes of length 10 have 72 codewords, *IEEE Trans. Inform. Theory* 45 (1999), 1229–1231.

[460] P. R. J. Östergård and U. Blass, On the size of optimal binary codes of length 9 and covering radius 1, *IEEE Trans. Inform. Theory* 47 (2001), 2556–2557.

[461] P. R. J. Östergård and H. O. Hämäläinen, A new table of binary/ternary mixed covering codes, *Des. Codes Cryptogr.* 11 (1997), 151–178.

[462] P. R. J. Östergård and M. K. Kaikkonen, New single-error-correcting codes, *IEEE Trans. Inform. Theory* 42 (1996), 1261–1262.

[463] P. R. J. Östergård and P. Kaski, Enumeration of 2-$(9, 3, \lambda)$ designs and their resolutions, *Des. Codes Cryptogr.* 27 (2002), 131–137.

[464] P. R. J. Östergård, J. Quistorff, and A. Wassermann, New results on codes with covering radius 1 and minimum distance 2, *Des. Codes Cryptogr.* 35 (2005), 241–250.

[465] P. R. J. Östergård and M. Svanström, Ternary constant weight codes, *Electron. J. Combin.* 9 (2002) no. 1, #R41, 23pp.

[466] P. R. J. Östergård and A. Wassermann, A new lower bound for the football pool problem for 6 matches, *J. Combin. Theory Ser. A* 99 (2002), 175–179.

[467] P. R. J. Östergård and W. D. Weakley, Constructing covering codes with given automorphisms, *Des. Codes Cryptogr.* 16 (1999), 65–73.

[468] P. R. J. Östergård and W. D. Weakley, Classification of binary covering codes, *J. Combin. Des.* 8 (2000), 391–401.

[469] P. R. J. Östergård and W. D. Weakley, Classifying optimal ternary codes of length 5 and covering radius 1, *Beiträge Algebra Geom.* 43 (2002), 445–449.

[470] J. G. Oxley, *Matroid Theory*, Oxford University Press, Oxford, 1992.

[471] C. H. Papadimitriou, *Computational Complexity*, Addison-Wesley, Reading, Mass., 1994.

[472] M.-O. Pavčević, Symmetric designs of Menon series admitting an action of Frobenius groups, *Glas. Mat. Ser. III* 31(51) (1996), 209–223.

[473] C. A. F. Peters, *Briefwechsel zwischen C. F. Gauss und H. C. Schumacher*, Vol. 4, Holt, Rinehart & Winston, Altona, 1862.

[474] E. Petrank and R. M. Roth, Is code equivalence easy to decide? *IEEE Trans. Inform. Theory* 43 (1997), 1602–1604.

[475] A. J. Petrenjuk, On the constructive enumeration of packings and coverings of index one, *Discrete Math.* 77 (1989), 237–254.

[476] S. Pfaff, Classification of $(78, 22, 6)$ designs having the full automorphism group $E_8 \cdot F_{21}$, *Glas. Mat. Ser. III* 28(48) (1993), 3–9.

[477] K. T. Phelps, On cyclic Steiner systems $S(3, 4, 20)$, *Ann. Discrete Math.* 7 (1980), 277–300.

[478] K. T. Phelps and A. Rosa, Steiner triple systems with rotational automorphisms, *Discrete Math.* 33 (1981), 57–66.

[479] W. A. Pierce, The impossibility of Fano's configuration in a projective plane with eight points per line, *Proc. Amer. Math. Soc.* 4 (1953), 908–912.

[480] C. Pietsch, *Über die Enumeration von Inzidenzstrukturen*, PhD Thesis, Universität Rostock, 1993.

[481] C. Pietsch, On the enumeration of 2-$(7, 3, \lambda)$ block designs, *J. Combin. Math. Combin. Comput.* 16 (1994), 103–114.

[482] V. Pless, On the uniqueness of the Golay codes, *J. Combin. Theory* 5 (1968), 215–228.

[483] V. Pless, A classification of self-orthogonal codes over GF(2), *Discrete Math.* 3 (1972), 209–246.

[484] V. Pless, *Introduction to the Theory of Error-Correcting Codes*, 3rd ed., Wiley, New York, 1998.

[485] V. Pless and N. J. A. Sloane, On the classification and enumeration of self-dual codes, *J. Combin. Theory Ser. A* 18 (1975), 313–335.

[486] V. Pless, N. J. A. Sloane, and H. N. Ward, Ternary codes of minimum weight 6 and the classification of the self-dual codes of length 20, *IEEE Trans. Inform. Theory* 26 (1980), 305–316.

[487] V. S. Pless and W. C. Huffman, Eds., *Handbook of Coding Theory*, 2 vols., Elsevier, Amsterdam, 1998.

[488] V. S. Pless and V. D. Tonchev, Self-dual codes over GF(7), *IEEE Trans. Inform. Theory* 33 (1987), 723–727.

[489] M. Plotkin, Binary codes with specified minimum distance, *IRE Trans. Inform. Theory* 6 (1960), 445–450.

[490] A. R. Prince, On vectors of weight 16 in the code of a projective plane of order 10, *Proc. Roy. Soc. Edinburgh Sect. A* 95 (1983), 137–146.

[491] A. R. Prince, Steiner triple systems of order 19 associated with a certain type of projective plane of order 10, *Period. Math. Hungar.* 17 (1986), 177–184.

[492] A. Proskurowski, F. Ruskey, and M. Smith, Analysis of algorithms for listing equivalence classes of k-ary strings, *SIAM J. Discrete Math.* 11 (1998), 94–109.

[493] G. Pruesse and F. Ruskey, Generating linear extensions fast, *SIAM J. Comput.* 23 (1994), 373–386.

[494] P. W. Purdom, Tree size by partial backtracking, *SIAM J. Comput.* 7 (1978), 481–491.

[495] P. W. Purdom, Jr. and C. A. Brown, *The Analysis of Algorithms*, Holt, Rinehart & Winston, New York, 1985.

[496] D. Raghavarao, *Constructions and Combinatorial Problems in Design of Experiments*, Wiley, New York, 1971.

[497] E. M. Rains and N. J. A. Sloane, Self-dual codes, in *Handbook of Coding Theory* (V. S. Pless and W. C. Huffman, Eds.), Vol. 1, Elsevier, Amsterdam, 1998, pp. 177–294.

[498] L. B. Rall, Tools for mathematical computation, in *Computer Aided Proofs in Analysis* (K. R. Meyer and D. S. Schmidt, Eds.), Springer-Verlag, New York, 1991, pp. 217–228.

[499] D. K. Ray-Chaudhuri and R. M. Wilson, Solution of Kirkman's schoolgirl problem, in *Combinatorics* (T. S. Motzkin, Ed.), Amer. Math. Soc., Providence, R.I., 1971, pp. 187–203.
[500] R. C. Read, Every one a winner; or, How to avoid isomorphism search when cataloguing combinatorial configurations, *Ann. Discrete Math.* 2 (1978), 107–120.
[501] R. C. Read and D. G. Corneil, The graph isomorphism disease, *J. Graph Theory* 1 (1977), 339–363.
[502] G. H. J. van Rees, A note on $C(10,4,2)$ and $C(11,5,3)$, *Congr. Numer.* 99 (1994), 271–275.
[503] G. H. J. van Rees, $(22,33,12,8,4)$-BIBD, an update, in *Computational and Constructive Design Theory* (W. D. Wallis, Ed.), Kluwer, Dordrecht, the Netherlands, 1996, pp. 337–357.
[504] R. S. Rees and W. D. Wallis, Kirkman triple systems and their generalizations: A survey, in *Designs 2002: Further Computational and Constructive Design Theory* (W. D. Wallis, Ed.), Kluwer, Boston, 2003, pp. 317–368.
[505] E. M. Reingold, J. Nievergelt, and N. Deo, *Combinatorial Algorithms: Theory and Practice*, Prentice Hall, Englewood Cliffs, N.J., 1977.
[506] A. Rosa, On reverse Steiner triple systems, *Discrete Math.* 2 (1972), 61–71.
[507] A. Rosa and D. R. Stinson, One-factorizations of regular graphs and Howell designs of small order, *Util. Math.* 29 (1986), 99–124.
[508] P. R. Rosenbaum, Sampling the leaves of a tree with equal probabilities, *J. Amer. Statist. Assoc.* 88 (1993), 1455–1457.
[509] J. J. Rotman, *An Introduction to the Theory of Groups*, 4th ed., Springer-Verlag, New York, 1995.
[510] G. F. Royle, An orderly algorithm and some applications in finite geometry, *Discrete Math.* 185 (1998), 105–115.
[511] F. Ruskey and T. C. Hu, Generating binary trees lexicographically, *SIAM J. Comput.* 6 (1977), 745–758.
[512] F. Ruskey, C. Savage, and T. M. Y. Wang, Generating necklaces, *J. Algorithms* 13 (1992), 414–430.
[513] H. J. Ryser, A note on a combinatorial problem, *Proc. Amer. Math. Soc.* 1, 422–424.
[514] A. Sade, An omission in Norton's list of 7×7 squares, *Ann. Math. Statistics* 22 (1951), 306–307.
[515] C. J. Salwach and J. A. Mezzaroba, The four biplanes with $k = 9$, *J. Combin. Theory Ser. A* 24 (1978), 141–145.
[516] C. Savage, A survey of combinatorial Gray codes, *SIAM Rev.* 39 (1997), 605–629.
[517] J. Sawada, Generating bracelets in constant amortized time, *SIAM J. Comput.* 31 (2001), 259–268.
[518] B. Schmalz, Verwendung von Untergruppenleitern zur Bestimmung von Doppelnebenklassen, *Bayreuth. Math. Schr.* 31 (1990), 109–143.

[519] B. Schmalz, The t-designs with prescribed automorphism group, new simple 6-designs, *Congr. Numer.* 88 (1992), 33–37.

[520] B. Schmalz, t-Designs zu vorgegebener Automorphismengruppe, *Bayreuth. Math. Schr.* 41 (1992), 1–164.

[521] B. Schmalz, The t-designs with prescribed automorphism group, new simple 6-designs, *J. Combin. Des.* 1 (1993), 125–170.

[522] E. Seah and D. R. Stinson, An enumeration of nonisomorphic one-factorizations and Howell designs for the graph K_{10} minus a one-factor, *Ars. Combin.* 21 (1986), 145–161.

[523] E. Seah and D. R. Stinson, Some perfect one-factorizations of K_{14}, *Ann. Discrete Math.* 34 (1987), 419–436.

[524] E. Seah and D. R. Stinson, On the enumeration of one-factorizations of complete graphs containing prescribed automorphism groups, *Math. Comp.* 50 (1988), 607–618.

[525] J. Seberry and M. Yamada, Hadamard matrices, sequences, and block designs, in *Contemporary Design Theory: A Collection of Surveys* (J. H. Dinitz and D. R. Stinson, Eds.), Wiley, New York, 1992, pp. 431–560.

[526] N. V. Semakov and V. A. Zinov'ev, Equidistant q-ary codes with maximal distance and resolvable balanced incomplete block designs (in Russian), *Problemy Peredachi Informatsii* 4 (1968) no. 2, 3–10. English translation in *Problems Inform. Transmission* 4 (1968) no. 2, 1–7.

[527] N. V. Semakov and V. A. Zinov'ev, Balanced codes and tactical configurations (in Russian), *Problemy Peredachi Informatsii* 5 (1969) no. 3, 28–37. English translation in *Problems Inform. Transmission* 5 (1968) no. 3, 22–28.

[528] Á. Seress, *Permutation Group Algorithms*, Cambridge University Press, Cambridge, 2003.

[529] E. C. Sewell, A branch and bound algorithm for the stability number of a sparse graph, *INFORMS J. Comput.* 10 (1998), 438–447.

[530] C. E. Shannon, A mathematical theory of communication, *Bell System Technical Journal* 27 (1948), 379–423, 623–656.

[531] J. Shao and W. Wei, A formula for the number of Latin squares, *Discrete Math.* 110 (1992), 293–296.

[532] S. S. Shrikhande, Affine resolvable balanced incomplete block designs: A survey, *Aequationes Math.* 14 (1976), 251–269.

[533] A. F. Sidorenko, On the Turán numbers $T(n, 5, 4)$ and numbers of monochromatic 4-cliques in 2-colored 3-graphs (in Russian), *Voprosy Kibernet. (Moscow)* 64 (1980), 117–124.

[534] J. Simonis, A description of the $[16, 7, 6]$ codes, in *Applied Algebra, Algebraic Algorithms and Error-Correcting Codes* (S. Sakata, Ed.), Springer-Verlag, Berlin, 1991, pp. 24–35.

[535] J. Simonis, The $[18, 9, 6]$ code is unique, *Discrete Math.* 106/107 (1992), 439–448.

[536] J. Simonis, The $[23, 14, 5]$ Wagner code is unique, *Discrete Math.* 213 (2000), 269–282.

[537] C. C. Sims, Computational methods in the study of permutation groups, in *Computational Problems in Abstract Algebra* (J. Leech, Ed.), Pergamon Press, Oxford, 1970, pp. 169–183.
[538] C. C. Sims, Computation with permutation groups, in *Proc. 2nd ACM Symposium on Symbolic and Algebraic Manipulation*, (Los Angeles, March 23–25, 1971), ACM Press, New York, 1971, pp. 23–28.
[539] D. Slepian, On the number of symmetry types of boolean functions of n variables, *Canad. J. Math.* 5 (1953), 185–193.
[540] D. Slepian, Some further theory of group codes, *Bell System Tech. J.* 39 (1960), 1219–1252.
[541] N. J. A. Sloane and J. G. Thompson, Cyclic self-dual codes, *IEEE Trans. Inform. Theory* 29 (1983), 364–366.
[542] S. L. Snover, *The Uniqueness of the Nordstrom-Robinson and the Golay Binary Codes*, PhD Thesis, Michigan State University, 1973.
[543] F. I. Solov'eva, S. V. Avgustinovich, T. Honold, and W. Heise, On the extendability of code isometries, *J. Geom.* 61 (1998), 3–16.
[544] E. Spence, Symmetric $(31, 10, 3)$-designs with a nontrivial automorphism of odd order, *J. Combin. Math. Combin. Comput.* 10 (1991), 51–64.
[545] E. Spence, A complete classification of symmetric $(31, 10, 3)$ designs, *Des. Codes Cryptogr.* 2 (1992), 127–136.
[546] E. Spence, Symmetric $(41, 16, 6)$-designs with a nontrivial automorphism of odd order, *J. Combin. Des.* 1 (1993), 193–211.
[547] E. Spence, Classification of Hadamard matrices of order 24 and 28, *Discrete Math.* 140 (1995), 185–243.
[548] E. Spence, Construction and classification of combinatorial designs, in *Surveys in Combinatorics, 1995* (P. Rowlinson, Ed.), Cambridge University Press, Cambridge, 1995, pp. 191–213.
[549] E. Spence, The complete classification of Steiner systems $S(2, 4, 25)$, *J. Combin. Des.* 4 (1996), 295–300.
[550] R. G. Stanton, Isomorphism classes of small covering designs with block size five, *Ann. Discrete Math.* 34 (1987), 441–448.
[551] R. G. Stanton, J. L. Allston, W. D. Wallis, and D. D. Cowan, The number of nonisomorphic solutions to a problem in covering designs, *Util. Math.* 21A (1982), 119–136.
[552] R. G. Stanton and J. A. Bate, A computer search for B-coverings, in *Combinatorial Mathematics VII* (R. W. Robinson, G. W. Southern, and W. D. Wallis, Eds.), Springer-Verlag, Berlin, 1980, pp. 37–50.
[553] R. G. Stanton and L. O. James, From covering designs to graphs, *Aequationes Math.* 8 (1972), 76–81.
[554] R. G. Stanton and J. G. Kalbfleisch, Covering problems for dichotomized matchings, *Aequationes Math.* 1 (1968), 94–103.
[555] R. G. Stanton and J. G. Kalbfleisch, Intersection inequalities for the covering problem, *SIAM J. Appl. Math.* 17 (1969), 1311–1316.

[556] R. G. Stanton and R. C. Mullin, Uniqueness theorems in balanced incomplete block designs, *J. Combin. Theory* 7 (1969), 37–48.

[557] R. G. Stanton, R. C. Mullin, and J. A. Bate, Isomorphism classes of a set of prime BIBD parameters, *Ars. Combin.* 2 (1976), 251–264.

[558] R. G. Stanton, M. J. Rogers, R. F. Quinn, and D. D. Cowan, Bipackings of pairs into triples, and isomorphism classes of small bipackings, *J. Austral. Math. Soc. Ser. A* 34 (1983), 214–228.

[559] D. R. Stinson, A short proof of the nonexistence of a pair of orthogonal Latin squares of order six, *J. Combin. Theory Ser. A* 36 (1984), 373–376.

[560] D. R. Stinson, Isomorphism testing of Steiner triple systems: Canonical forms, *Ars. Combin.* 19 (1985), 213–218.

[561] D. R. Stinson, A survey of Kirkman triple systems and related designs, *Discrete Math.* 92 (1991), 371–393.

[562] D. R. Stinson, *Combinatorial Designs: Constructions and Analysis*, Springer-Verlag, New York, 2004.

[563] D. R. Stinson and E. Seah, 284 457 Steiner triple systems of order 19 contain a subsystem of order 9, *Math. Comp.* 46 (1986), 717–729.

[564] A. P. Street and D. J. Street, *Combinatorics of Experimental Design*, Clarendon Press, Oxford, 1987.

[565] J. Surányi, Some combinatorial problems of geometry (in Hungarian), *Mat. Lapok* 22 (1971), 215–230.

[566] M. Svanström, P. R. J. Östergård, and G. T. Bogdanova, Bounds and constructions for ternary constant-composition codes, *IEEE Trans. Inform. Theory* 48 (2002), 101–111.

[567] J. D. Swift, Isomorph rejection in exhaustive search techniques, in *Combinatorial Analysis* (R. Bellman and M. Hall, Jr., Eds.), Amer. Math. Soc., Providence, R.I., 1960, pp. 195–200.

[568] R. Tarjan, Enumeration of the elementary circuits of a directed graph, *SIAM J. Comput.* 2 (1973), 211–216.

[569] G. Tarry, Le problème des 36 officiers, in *Compte Rendu de la 29me Session. Association Française pour l'Avancement des Sciences.*, Vol. 2, (Paris, Aug. 2–9, 1900), Paris, 1901, pp. 170–203.

[570] O. Taussky and J. Todd, Covering theorems for groups, *Annales de la Société Polonaise de Mathématique* 21 (1948), 303–305.

[571] L. Teirlinck, The existence of reverse Steiner triple systems, *Discrete Math.* 6 (1973), 301–302.

[572] L. H. Thiel, C. W. H. Lam, and S. Swiercz, Using a CRAY-1 to perform backtrack search, in *Supercomputing '87: Supercomputer Design, Performance Evaluation and Performance Education (Proc. 2nd International Conference on Supercomputing, San Francisco, 1987)* (L. P. Kartashev and S. I. Kartashev, Eds.), Vol. 3, International Supercomputing Institute, St. Petersburg, Fla., 1987, pp. 92–99.

[573] H. van Tilborg, On the uniqueness resp. nonexistence of certain codes meeting the Griesmer bound, *Inform. and Control* 44 (1980), 16–35.

[574] J. A. Todd, A combinatorial problem, *Journal of Mathematics and Physics* 12 (1933), 321–333.
[575] V. D. Tonchev, Hadamard matrices of order 28 with automorphisms of order 13, *J. Combin. Theory Ser. A* 35 (1983), 43–57.
[576] V. D. Tonchev, Hadamard matrices of order 28 with automorphisms of order 7, *J. Combin. Theory Ser. A* 40 (1985), 62–81.
[577] V. D. Tonchev, Hadamard matrices of order 36 with automorphisms of order 17, *Nagoya Math. J.* 104 (1986), 163–174.
[578] V. D. Tonchev, Symmetric 2-$(31, 10, 3)$ designs with automorphisms of order seven, *Ann. Discrete Math.* 34 (1987), 461–464.
[579] V. D. Tonchev, Transitive Steiner triple systems of order 25, *Discrete Math.* 67 (1987), 211–214.
[580] V. D. Tonchev, Codes, in *The CRC Handbook of Combinatorial Designs* (C. J. Colbourn and J. H. Dinitz, Eds.), CRC Press, Boca Raton, Fla., 1996, pp. 517–543.
[581] V. D. Tonchev and R. V. Raev, Cyclic 2-$(13, 5, 5)$ designs, *C. R. Acad. Bulgare Sci.* 35 (1982), 1205–1208.
[582] S. Topalova, Symmetric 2-$(69, 17, 4)$ designs with automorphisms of order 13, *J. Statist. Plann. Inference* 95 (2001), 335–339.
[583] S. Topalova, Classification of Hadamard matrices of order 44 with automorphisms of order 7, *Discrete Math.* 260 (2003), 275–283.
[584] J. Torán, On the hardness of graph isomorphism, *SIAM J. Comput.* 33 (2004), 1093–1108.
[585] P. Turán, On an extremal problem in graph theory (in Hungarian), *Matematikai és Fizikai Lapok* 48 (1941), 436–452.
[586] T. Tymoczko, The four-color problem and its philosophical significance, *J. Philos.* 76 (1979), 57–83.
[587] T. Tymoczko, Computers, proofs and mathematics: A philosophical investigation of the four-color problem, *Math. Mag.* 53 (1980), 131–138.
[588] T. Ueda, Gray codes for necklaces, *Discrete Math.* 219 (2000), 235–248.
[589] T. Uno, Algorithms for enumerating all perfect, maximum and maximal matchings in bipartite graphs, in *Algorithms and Computation* (H. W. Leong, H. Imai, and S. Jain, Eds.), Springer-Verlag, Berlin, 1997, pp. 92–101.
[590] T. Uno, A fast algorithm for enumerating bipartite perfect matchings, in *Algorithms and Computation* (P. Eades and T. Takaoka, Eds.), Springer-Verlag, Berlin, 2001, pp. 367–379.
[591] R. J. M. Vaessens, E. H. L. Aarts, and J. H. van Lint, Genetic algorithms in coding theory – A table for $A_3(n, d)$, *Discrete Appl. Math.* 45 (1993), 71–87.
[592] L. G. Valiant, The complexity of enumeration and reliability problems, *SIAM J. Comput.* 8 (1979), 410–421.
[593] A. Vardy, Algorithmic complexity in coding theory and the minimum distance problem, in *Proc. 29th ACM Symposium on Theory of Com-*

puting, (El Paso, Tex., May 4–6, 1997), ACM Press, New York, 1997, pp. 92–109.

[594] A. Vardy, The intractability of computing the minimum distance of a code, *IEEE Trans. Inform. Theory* 43 (1997), 1757–1766.

[595] G. Verfaillie, M. Lemaître, and T. Schiex, Russian doll search for solving constraint optimization problems, in *Proc. 13th National Conference on Artificial Intelligence (AAAI-96)*, (Portland, Oreg., Aug. 4–8, 1996), AAAI Press, Menlo Park, Calif., 1996, pp. 181–187.

[596] R. J. Walker, An enumerative technique for a class of combinatorial problems, in *Combinatorial Analysis* (R. Bellman and M. Hall, Jr., Eds.), Amer. Math. Soc., Providence, R.I., 1960, pp. 91–94.

[597] W. D. Wallis, *One-Factorizations*, Kluwer, Dordrecht, the Netherlands, 1997.

[598] A. Wassermann, Finding simple t-designs with enumeration techniques, *J. Combin. Des.* 6 (1998), 79–90.

[599] A. Wassermann, Attacking the market split problem with lattice basis reduction, *J. Comb. Optim.* 6 (2002), 5–16.

[600] G. J. M. van Wee, Improved sphere bounds on the covering radius of codes, *IEEE Trans. Inform. Theory* 34 (1988), 237–245.

[601] B. Weisfeiler, Ed., *On Construction and Identification of Graphs*, Springer-Verlag, Berlin, 1976.

[602] B. Yu. Weisfeiler and A. A. Leman, A reduction of a graph to a canonical form and an algebra arising during this reduction (in Russian), *Nauchn.-Tekhn. Informatsiya Ser. 2* (1968) no. 9, 12–16.

[603] D. Welsh, *Codes and Cryptography*, Oxford University Press, Oxford, 1988.

[604] D. B. West, *Introduction to Graph Theory*, 2nd ed., Prentice Hall, Upper Saddle River, N.J., 2001.

[605] H. Whitney, 2-isomorphic graphs, *American Journal of Mathematics* 55 (1933), 245–254.

[606] S. B. Wicker, *Error Control Systems for Digital Communication and Storage*, Prentice Hall, Upper Saddle River, N.J., 1995.

[607] H. Wielandt, *Finite Permutation Groups*, Academic Press, New York, 1964.

[608] H. S. Wilf, *Combinatorial Algorithms: An Update*, SIAM, Philadelphia, 1989.

[609] R. M. Wilson, The necessary conditions for t-designs are sufficient for something, *Util. Math.* 4 (1973), 207–215.

[610] R. M. Wilson, Nonisomorphic Steiner triple systems, *Math. Z.* 135 (1974), 303–313.

[611] P. M. Winkler, Isometric embeddings in products of complete graphs, *Discrete Appl. Math.* 7 (1984), 221–225.

[612] E. Witt, Über Steinersche Systeme, *Abhandlungen aus dem Mathematischen Seminar der Universität Hamburg* 12 (1938), 265–275.

[613] R. A. Wright, B. Richmond, A. Odlyzko, and B. D. McKay, Constant time generation of free trees, *SIAM J. Comput.* 15 (1986), 540–548.

[614] S. Yamamoto, Y. Fujii, Y. Hyodo, and H. Yumiba, Classification of two-symbol orthogonal arrays of strength 2, size 16, 15 (maximal) constraints and index 4, *SUT J. Math.* 28 (1992), 47–59.

[615] S. Yamamoto, Y. Fujii, Y. Hyodo, and H. Yumiba, Classification of two-symbol orthogonal arrays of strength 2, size 20 and 19 (maximal) constraints, *SUT J. Math.* 28 (1992), 191–209.

[616] S. Yamamoto, Y. Fujii, Y. Hyodo, and H. Yumiba, Connection between the numbers of nonisomorphic solutions of Hadamard matrices, orthogonal arrays and balanced incomplete block designs, *SUT J. Math.* 29 (1993), 143–165.

[617] S. Yamamoto, Y. Fujii, T. Namikawa, and M. Mitsuoka, Three-symbol orthogonal arrays of strength t having $t+2$ constraints, *SUT J. Math.* 27 (1991), 93–111.

[618] S. Yamamoto, Y. Hyodo, M. Mitsuoka, H. Yumiba, and T. Takahashi, Algorithm for the construction and classification of orthogonal arrays and its feasibility, *J. Combin. Inform. System Sci.* 23 (1998), 71–83.

[619] S. Yamamoto, Y. Hyodo, H. Yumiba, and T. Takahashi, Enumeration and classification of two-symbol orthogonal arrays of strength t and $m = t + 4$ constraints, *J. Japan Statist. Soc.* 29 (1999), 135–145.

[620] S. K. Zaremba, Covering problems concerning abelian groups, *J. London Math. Soc.* 27 (1952), 242–246.

[621] Z. Zhang, Linear inequalities for covering codes: Part I – Pair covering inequalities, *IEEE Trans. Inform. Theory* 37 (1991), 573–582.

[622] Z. Zhang and C. Lo, Linear inequalities for covering codes: Part II – Triple covering inequalities, *IEEE Trans. Inform. Theory* 38 (1992), 1648–1662.

[623] G. M. Ziegler, Lectures on 0/1-polytopes, in *Polytopes – Combinatorics and Computation* (G. Kalai and G. M. Ziegler, Eds.), Birkhäuser, Basel, 2000, pp. 1–41.

Problem Index

Index

MIX
Papier aus verantwortungsvollen Quellen
Paper from responsible sources
FSC® C105338

If you have any concerns about our products,
you can contact us on
ProductSafety@springernature.com

In case Publisher is established outside the EU,
the EU authorized representative is:
Springer Nature Customer Service Center GmbH
Europaplatz 3, 69115 Heidelberg, Germany

Printed by Libri Plureos GmbH
in Hamburg, Germany